UNIVERSITY OF NOTTINGHAM
WIT
FROM THE
10 0312581 0
D1179453

THE PHYSICS OF ASTROPHYSICS

Volume II: Gas Dynamics

A SERIES OF BOOKS IN ASTRONOMY

Editors
Donald E. Osterbrock
Joseph S. Miller

Bless, ***Discovering the Cosmos***

Gilmore, King, & van der Kruit, *The Milky Way as a Galaxy*

Miller, *Astrophysics of Active Galaxies and Quasi-Stellar Objects*

Osterbrock, *Astrophysics of Gaseous Nebulae and Active Galactic Nuclei*

Shu, *The Physical Universe*

Shu, *The Physics of Astrophysics, Volumes I and II*

THE PHYSICS OF ASTROPHYSICS

Volume II

GAS DYNAMICS

FRANK H. SHU

PROFESSOR OF ASTRONOMY
UNIVERSITY OF CALIFORNIA, BERKELEY

UNIVERSITY SCIENCE BOOKS
Sausalito, California

University Science Books
55D Gate Five Road
Sausalito, CA 94965
Fax: (415) 332-5393

Copy editor: Aidan Kelly
TEX formatter and illustrator: Ed Sznyter
Text and jacket designer: Robert Ishi
Indexer: BevAnne Ross
Printer and binder: The Maple-Vail Book Manufacturing Group

Library of Congress Catalog Number: 91-65168

ISBN 0-935702-65-2

Printed in the United States of America
10 9 8 7 6 5 4 3

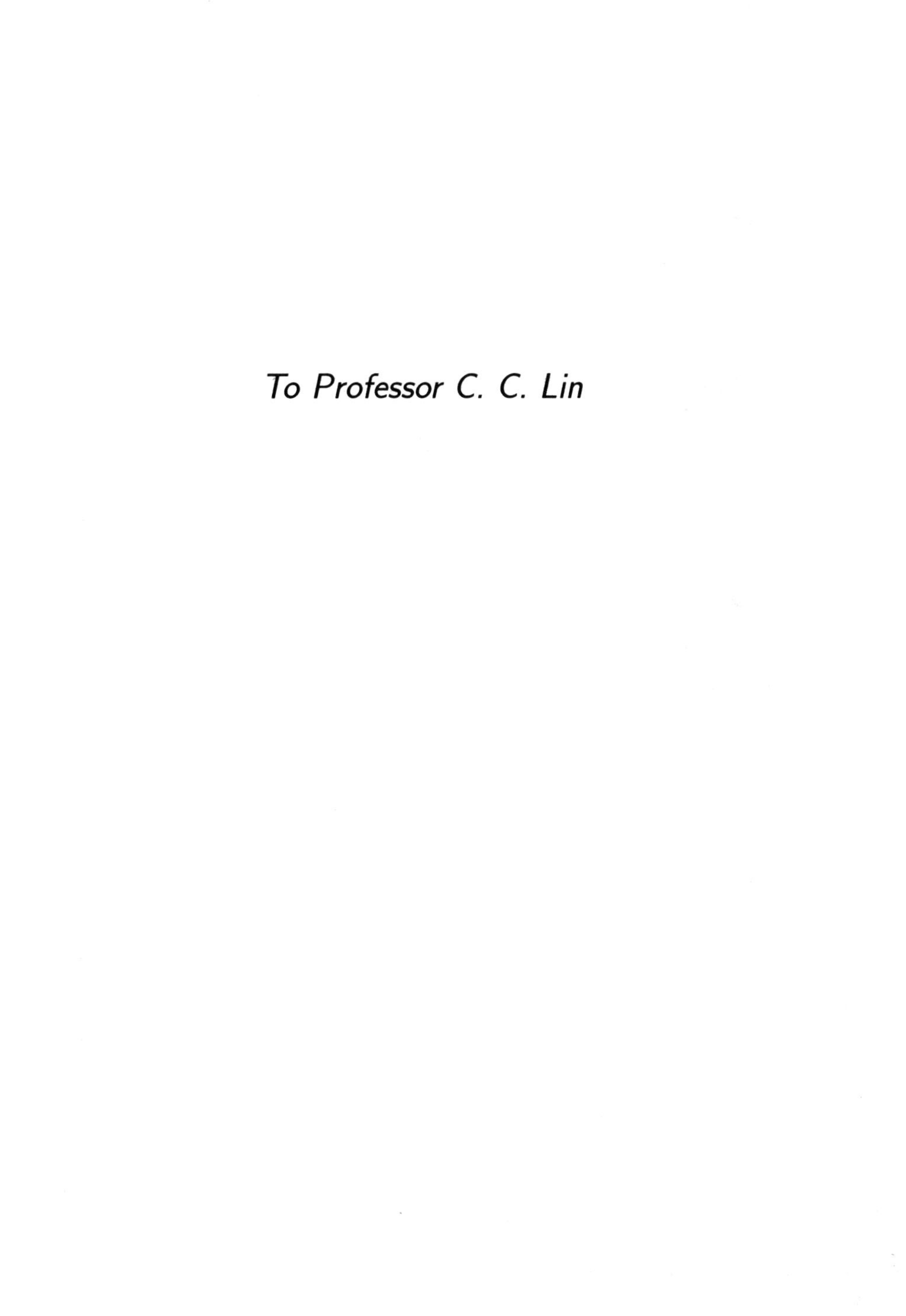

To Professor C. C. Lin

Contents

Preface

Modern astronomers need to know a lot of physics. Unfortunately, the typical curriculum in most graduate astronomy departments leaves little time for the student to learn this material through formal course work. Ideally, while in graduate school, a prospective astronomer with a thorough preparation in undergraduate physics and mathematics would take a two-semester graduate sequence in quantum mechanics (leading up to Dirac's equation), a two-semester sequence in classical electrodynamics (including multipole radiation and the theory of special relativity), a semester of statistical mechanics (including Gibbs's ensembles and some applications in quantum statistical mechanics), a semester or more of gas dynamics and plasma physics (including shockwave theory and magnetohydrodynamics), and a year or more of advanced mathematical methods (including the theory of complex variables, ordinary and partial differential equations, numerical analysis, Fourier analysis, and various asymptotic approximation techniques). Unfortunately, such a program would take the budding professional astronomer through the first two years of graduate school without time for a single course in astronomy!

To cope with this dilemma, teachers of graduate courses in astronomy have traditionally included the relevant physics and mathematics as part of the background lecture material. This compromise, however, is very inefficient (e.g., there is a lot of repetition of subject matter such as radiative transfer or atomic and molecular physics) during an era when a veritable explosion of phenomenological knowledge increasingly demands more lecture time for exposure to the frontiers of the field.

At many universities this challenge has given birth to one or more basic courses in the physics prerequisite to many different subject areas in astronomy. I have taught such courses at the State University of New York at Stony Brook (1968–1973) and the University of California at Berkeley (1973–present). Innovations urged by J. Arons have caused the format at Berkeley to evolve from a single-semester course on "astrophyical processes" to the present format of a year-long sequence—the first semester of which deals with "radiation processes" and the second with "gas dynamics and magnetohydrodynamics."

This two-volume text on *The Physics of Astrophysics* grew from my lecture notes for the reorganized two-semester sequence. It is aimed at first-year graduate students and well-prepared seniors in astronomy and physics.

The first volume deals with *Radiation*, the second with *Gas Dynamics*. The excellent text by G. Rybicki and A. Lightman on *Radiative Processes in Astrophysics* (Wiley) was my model for the first half of the sequence, but my presentation of the basic material pays more attention to low-energy phenomena (e.g., radio astronomy) and statistical astrophysics (e.g., rate equations).

Although these two volumes were written to form a coherent whole, they can be decoupled for use in separate courses. In particular, when I teach the two-course sequence, the first is not a prerequisite for the second. In writing both volumes, I have been guided by the following pedagogical assumptions and philosophy.

A. Although my discussion emphasizes processes rather than objects, so that the topics are arranged in a sequence formed more by the tradition of physics than by that of astronomy, I usually try to motivate the development by using concrete examples from astrophysics. When faced with a choice between abstract principles or practical applications, I always opt for the practical approach.

B. I have tried to make explanations detailed enough to indicate the important points in the reasoning, but I refrain from displaying every step of a derivation, in order not to divert the attention of the reader from more serious matters. The student may wish to read straight through each chapter to get the central thrust of the physical ideas, and then—as an aid to long-term memory and detailed understanding—return with paper and pencil to work out the missing steps in the formal mathematics.

C. For beginning students, who may not know in which subfields they wish to specialize, I believe it better to cover a lot of ground coherently than to delve deeply into any particular subject. Astronomers of the future will need tools that allow them to explore in many different directions.

D. Along the same lines, I prefer to assign long problems that require a sustained attack on a practical astronomical situation than to contrive short problems that require only a few simple steps to demonstrate a limited objective. I hope I have supplied enough hints along the way so that the student does not become frustrated by getting stuck in the midst of an extended calculation.

To use the book effectively, the student should have had some atomic physics and kinetic theory, with exposure to the concepts of the mean free path and the equilibrium thermodynamic distributions of matter and radiation; electrodynamics with the expression of Maxwell's equations in their differential form; special relativity, with some exposure to the notion of Lorentz transformations of electric and magnetic fields at low velocities; and classical mechanics at a level that uses the Lagrangian and Hamiltonian formulations of the subject. Mathematical methods useful to areas other than just fluid mechanics, e.g., the theory of characteristics for the solution of hyperbolic partial differential equations, will receive elaboration as a part of the development of the material in the text. Some attention is also given to numerical methods suitable for use on modern computers (see especially the Problem Sets), but we leave to specialized texts the in-depth presentation of general-purpose methods for the numerical simulation of the equations of fluid mechanics (e.g., finite-difference or smooth-particle-hydrodynamics techniques). For many readers, this book offers a first exposure to the ideas of fluid dynamics; for them, developing insight is more important than getting numbers.

This book adopts boldface symbols, e.g., $\mathbf{A}$, for vectors in ordinary three-space; in contrast to the notation of Volume I, we use double-arrowed variables to represent 3×3 spatial tensors, e.g., $\overleftrightarrow{\pi}$ to represent the viscous stress tensor. We also make use of dyadic notation if a tensor is formed by the direct product of two vectors, e.g., $\mathbf{uu}$, whose Cartesian components are $u_i u_k$. The Appendix contains useful formulae for the expression in orthogonal curvilinear coordinates, including explicit display in cylindrical and spherical coordinates, of the vector and tensor quantities most often found in gas dynamics and magnetohydrodynamics.

Vector calculus in three spatial dimensions constitutes a mathematical tool used freely throughout the book. Apart from the standard theorems of Green, Stokes, etc., the student should be conversant with the identity involving the triple vector product: $\mathbf{A} \times (\mathbf{B} \times \mathbf{C}) = (\mathbf{A} \cdot \mathbf{C})\mathbf{B} - (\mathbf{A} \cdot \mathbf{B})\mathbf{C}$, and the mnemonic that the *minus* sign accompanies the dot product that involves the vector in the *middle*. I also assume that he or she knows the Cartesian-tensor analogs for this expression: $\epsilon_{ikm}\epsilon_{mj\ell} = \delta_{ij}\delta_{k\ell} - \delta_{i\ell}\delta_{kj}$, but infrequent use is made of the latter formula. The reader should also know that the dot and the cross can be interchanged in the triple scalar product: $\mathbf{A} \cdot (\mathbf{B} \times \mathbf{C}) = (\mathbf{A} \times \mathbf{B}) \cdot \mathbf{C}$. Finally, I assume that the student knows how to modify these expressions if one or more of the quantities $\mathbf{A}$, $\mathbf{B}$, $\mathbf{C}$ represents the del operator ∇.

References to research papers occur in the text where they first appear. Books to which the reader may wish to refer often are listed in the Bibliography at the end.

Frank H. Shu
Berkeley, California

THE PHYSICS OF ASTROPHYSICS

Volume II: Gas Dynamics

PART I

SINGLE-FLUID THEORY, INCLUDING RADIATIVE PROCESSES

1

Overview

Reference: Spitzer, *Physics of Fully Ionized Gases*, pp. 131–136.

In this volume, we shall be concerned with the dynamics of gases under astrophysical conditions. In our deliberations, we shall assume that the microscopic properties of individual atoms or molecules are known or can be calculated (see Volume I). Two aspects of the problem of astrophysical gas dynamics distinguishes the subject from similar considerations under terrestrial conditions.

(a) Nontrivial flows on Earth are often generated by the obstructions offered by solid bodies (e.g., a ship plowing through water, or a jet plane through air). Obstructing bodies in otherwise uniform flow (as seen from the body) form a small subset of the problems that concern astrophysicists; of more general interest often are flows generated by gravitational forces, or by interaction with a strong radiation field, or by violent events (e.g., explosions).

(b) Conventional fluid mechanics usually addresses the study of neutral gases or of liquids (water). Astrophysical fluids (any state of matter that *flows*) are frequently at least partially ionized; thus, electromagnetic forces can be more important for the macroscopic dynamics than is typically stressed in conventional textbooks. The subject goes by the name *magnetohydrodynamics* when the methods of continuum mechanics apply, and by *plasma physics* when they do not.

THE APPLICABILITY OF A FLUID APPROACH

To begin our discussion, therefore, we ask the natural question: Under what circumstances can we justify the treatment of a collection of free particles (a gas) as a continuum? The answer to this question lies in comparing the collisional *mean free path* ℓ with the macroscopic length scale L of interest

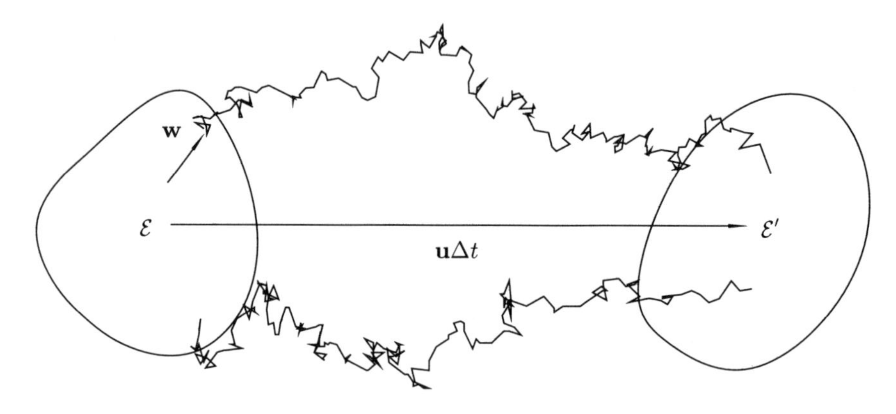

FIGURE 1.1
Any given particle's velocity $\mathbf{v}$ will generally differ by a random contribution $\mathbf{w}$ from the mean value $\mathbf{u}$ of all the particles that constitute a fluid element $\mathcal{E}$. However, if the mean free path ℓ for particle collisions is small in comparison with the dimensions of the fluid element, the particle will be swept along by the motion of the fluid element at the bulk velocity $\mathbf{u}$.

in the problem. If $\ell \ll L$, it makes good physical sense to introduce the concept of a *fluid element* $\mathcal{E}$, with volume V, whose linear size is small in comparison with L but large in comparison with ℓ. The number of particles that make up the material of this fluid element is large, and we may effectively define a mean (or bulk or fluid) velocity $\mathbf{u}$ for the collection, with individual particle motions $\mathbf{v}$ possessing a random component $\mathbf{w}$ above the mean (see Figure 1.1):

$$\mathbf{v} = \mathbf{u} + \mathbf{w}. \tag{1.1}$$

Because $\ell \ll L$, however, the random velocity does not succeed immediately in carrying any particle very far away from its neighbors, because incessant collisions continually redirect $\mathbf{w}$ and cause the particle to perform a random walk about the mean motion $\mathbf{u}$. Let the fluid element $\mathcal{E}'$ be defined in terms of the translation of each point of fluid element $\mathcal{E}$ at the locally defined mean velocity $\mathbf{u}(\mathbf{x}, t)$ over some flow time interval Δt. The utility of a fluid description follows from the fact that not only will the fluid element $\mathcal{E}'$ contain the same number of gas particles as fluid element $\mathcal{E}$, but they will virtually all be the *same* particles. Because of the restraining action of the random scatterings, no particle gets very far from

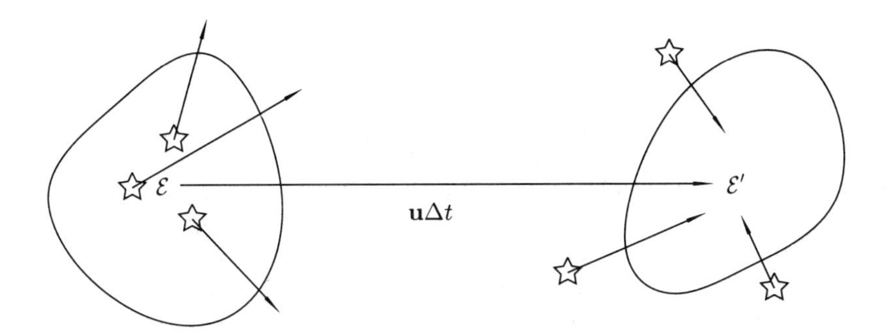

FIGURE 1.2
When the mean free path ℓ exceeds the size of the system, a fluid description fails because particles (stars in a galaxy, for example) generally stream into and out of a fluid element $\mathcal{E}$ as quickly as the element itself moves.

the center of the distribution. A close examination would reveal that a few particles originally within volume V but near to its surface can manage to exchange positions with particles originally outside volume V; however, the effect can be treated as being small, and, as we shall see in Chapter 3, corrections can readily be made via a *diffusion approximation.*

Contrast the preceding situation with the opposite regime, $\ell \gg L$, which holds, for example, for laboratory plasmas and stellar systems (galaxies or star clusters). For such cases, we could still formally introduce the concept of a fluid element whose dimensions are much smaller than L (the size of the system). But the definition would perform little practical service, because the particles that constitute the matter of the "fluid element" would freely stream out of V as a result of their random velocities $\mathbf{w}$ (which are often comparable to $\mathbf{u}$). The free streaming of individual particles out of (and into) the volume V as fast as the element $\mathcal{E}$ itself might move through space can not be even roughly described as a "diffusive" phenomenon (see Figure 1.2). No rigorous perturbational expansion about the mean behavior will be possible under these circumstances, and we must generally attack the problem as one in *kinetic theory* (six-dimensional phase space plus time) rather than as one in *fluid mechanics* (three-dimensional configuration space plus time).

MEAN FREE PATH FOR NEUTRAL ATOMS AND MOLECULES

For a neutral atom or molecule, which interacts with other particles only through short-range forces, the collision mean free path reads

$$\ell = (n\sigma)^{-1}, \tag{1.2}$$

where n is the number density of (typical) collision partners and σ is the associated (elastic) scattering cross section. Typically, $\sigma \sim 10^{-15}\,\text{cm}^2$. Thus, $\ell \sim 10^{-4}$ cm in this room, where $n \sim 10^{19}\,\text{cm}^{-3}$. Interstellar space is much more rarefied; for example, in an atomic hydrogen gas cloud (an H I region in astronomical nomenclature), n might only be $10\,\text{cm}^{-3}$, and ℓ would then equal $\sim 10^{14}$ cm, about the distance from the Sun to Jupiter. This represents an enormous distance by human standards, but, in fact, H I clouds typically span 10^{19} cm or more. Thus, $\ell \ll L$, and a fluid treatment of such a diffuse gas cloud would be valid. A similar justification holds for most of the thermal gas that exists in interstellar space.

PHOTON MEAN FREE PATH VERSUS PARTICLE MEAN FREE PATH

The Thomson formula for electron scattering yields a fiducial cross section for photons to interact with matter:

$$\sigma_\text{T} = \frac{8\pi}{3} r_\text{e}^2, \tag{1.3}$$

where $r_\text{e} \equiv e^2/m_\text{e}c^2$ is the classical radius of the electron. In cgs units, $\sigma_\text{T} = 0.665 \times 10^{-24}\,\text{cm}^2$. Astronomers like to define radiative cross sections per unit mass as the *opacity* κ. In astrophysics, electrons are often stripped off hydrogen atoms. With most of the radiative cross section provided by the free electron and most of the mass provided by the proton, $m_\text{p} = 1.67 \times 10^{-24}$ g, we find that the radiative opacity associated with electron scattering in a completely ionized hydrogen plasma has the value

$$\sigma_\text{T}/m_\text{p} \approx 0.4\,\text{cm}^2\,\text{g}^{-1}. \tag{1.4}$$

This is in fact a typical value for (continuum) gas opacities. If we look at stellar opacity tables (applicable to the temperature and density regimes in the interiors and atmospheres of stars), we will find values of κ ranging, say, from 10^{-3} to $10^5\,\text{cm}^2\,\text{g}^{-1}$. In the deep interiors of high-mass stars, the opacity is almost exclusively given by the Thomson formula. Even at the center of the Sun, electron scattering contributes about half the total radiative opacity.

Equation (1.4) can be usefully contrasted with the analogous expression for the cross section per unit mass for the elastic scattering of hydrogen atoms, which is of order $\sigma/m_\text{H} \sim 10^9\,\text{cm}^2\,\text{g}^{-1}$. A plasma, whose constituent charged particles scatter by long-range Coulomb interactions, would have even larger effective values of σ/m. Thus, except under extreme circumstances (e.g., inside a white dwarf, where free electrons are quantally degenerate and must travel large distances before they can find a vacant state into which to scatter), we may claim as a general rule that photon mean free paths are likely to be much longer than matter mean free paths. Under conditions when the radiation energy density is comparable in value to particle energy densities (true inside stars as well as in many parts of the interstellar medium), the transport of energy by radiation generally dominates the transport by heat conduction (via the random motions of matter). Moreover, because photons fly at the speed of light c, whereas matter generally travels much slower, at speed v, this statement often holds true even when the radiation energy density is not so high. However, the ratio of the energy ϵ carried by a photon to its momentum p is also c, whereas the corresponding ratio of $mv^2/2$ to mv for nonrelativistically moving matter is $v/2$. As a consequence, when $v \ll c$, matter can carry proportionally much more momentum than energy in comparison with photons, and if radiation competes only marginally with matter in carrying energy, then it does not do as well in transporting momentum (i.e., in transmitting force). We may heuristically remember this rule of thumb by the statement that "radiation can heat (or cool), but frequently finds it difficult to push."

MEAN FREE PATHS FOR CHARGED PARTICLES

How do we estimate σ for an ionized gas? As is well-known, Rutherford showed the total cross section for Coulombic interactions to be infinite. Ignoring this difficulty for the time being, we perform a simple order-of-magnitude estimate for the cross section associated with large-angle Coulomb scatterings. To fix ideas, consider an encounter between two electrons (see Figure 1.3).

Clearly, we could be assured of a large angle deflection if two electrons come close enough that their electrostatic energy becomes comparable to their relative kinetic energy. Thus, we can define an effective radius r_{eff} for the sphere of influence involving the collision of two thermal electrons by requiring:

$$\frac{e^2}{r_{\text{eff}}} \sim m_{\text{e}} w^2 \sim kT,$$

where T is the electron temperature and k is Boltzmann's constant. Then

$$\ell \sim (n_{\text{e}} \pi r_{\text{eff}}^2)^{-1} \sim \frac{m_{\text{e}}^2 v_{T\text{e}}^4}{n_{\text{e}} e^4},$$

where $v_{T\text{e}} \equiv (kT/m_{\text{e}})^{1/2}$ is the thermal velocity of the electrons. The associated relaxation time required, for example, for electrons to acquire a Maxwellian distribution if they do not already have one equals

$$t_{\text{c}}(\text{e}-\text{e}) \sim \frac{\ell}{v_{T\text{e}}} \sim \frac{m_{\text{e}}^{1/2}(kT)^{3/2}}{n_{\text{e}} e^4} \sim 0.92 \frac{T^{3/2}}{n_{\text{e}}} \text{ s},$$

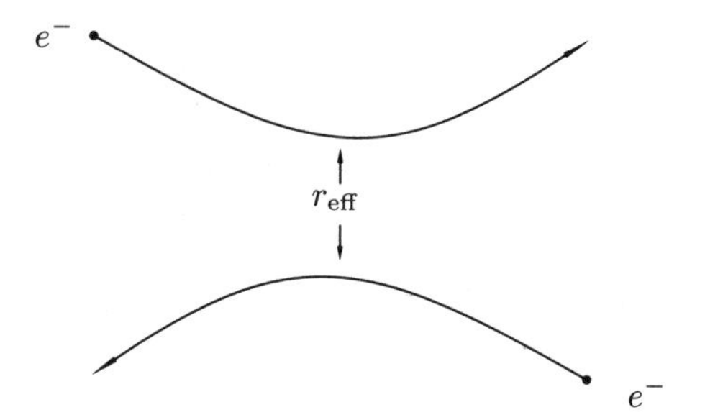

FIGURE 1.3
An effective size $r_{\rm eff}$ characterizes encounters between two charged particles, such as two electrons, that deflect the unperturbed motions by order unity amounts (e.g., turn the trajectories through an angle of the order of one radian). This size $r_{\rm eff}$ makes the potential energy $e^2/r_{\rm eff}$ at closest approach roughly equal to the sum of the relative kinetic energies of the two particles, which is of order kT, where T is the kinetic temperature for a thermal plasma.

when T and $n_{\rm e}$ are expressed in cgs units. A more sophisticated analysis shows that small-angle scatterings (with impact parameters $b \gg r_{\rm eff}$) dominate over large-angle scatterings (with impact parameters $b \sim r_{\rm eff}$) approximately by a factor of the "Coulomb logarithm," $\ln \Lambda$, where Λ roughly equals the ratio of the maximum impact parameter $b_{\rm max}$ to $r_{\rm eff}$. Indeed, Spitzer in the cited reference gives

$$t_{\rm c}({\rm e}-{\rm e}) = \left(\frac{0.290}{\ln \Lambda}\right)\left[\frac{m_{\rm e}^{1/2}(kT)^{3/2}}{n_{\rm e}e^4}\right]. \tag{1.5}$$

The quantity $b_{\rm max}$ must be introduced if we are to cut off the logarithmically divergent contributions to the total Rutherford cross section for Coulomb scattering at large impact parameters. For our problem, we can argue physically for such a cutoff as follows.

THE DEBYE LENGTH

Consider a plasma composed of ions of (a typical) charge $Z_{\rm i}e$ mixed with electrons of charge $-e$. For overall charge neutrality, we require

$$Z_{\rm i}n_{\rm i} = n_{\rm e}. \tag{1.6}$$

(We may perform a sum over i if there is more than one charge species.) We focus our attention on a specific ion, which we place for convenience at the origin of our coordinate system. This ion attracts electrons toward itself and repels other ions. In a steady state, the process will set up an electrostatic potential ϕ that satisfies Coulomb's law:

$$\nabla^2 \phi = -4\pi \rho_e, \tag{1.7}$$

where ρ_e equals the charge-density distribution. This distribution includes the ion at the origin as well as a Boltzmann (spatial) distribution of surrounding electrons and ions:

$$\rho_e = Z_i e \delta(\mathbf{r}) - n_e e \exp(e\phi/kT) + n_i Z_i e \exp(-Z_i e\phi/kT), \tag{1.8}$$

where we have assumed, for simplicity, that electrons and ions have the same temperature T, and where we have let n_e and n_i denote the uniform background mean densities that would result in the absence of the polarization effects being considered here.

We assume the electrostatic energies $-e\phi$ and $Z_i\phi$ to be small in comparison with kT (we consider the conditions necessary for this to be true later), and expand the exponentials for small values of their arguments. The leading terms, 1 times $-n_e e$ and $n_i Z_i e$, respectively, cancel when we make use of equation (1.6). The remaining terms are proportional to ϕ with a coefficient that has the dimensions of inverse length squared, which we write as

$$L_D^{-2} \equiv \frac{4\pi e^2}{kT}(n_e + Z_i^2 n_i). \tag{1.9}$$

The quantity L_D is called the *Debye length.* With the above approximation, we may now write equation (1.7) as

$$\frac{1}{r^2}\frac{d}{dr}\left(r^2 \frac{d\phi}{dr}\right) = -4\pi Z_i e \delta(\mathbf{r}) + L_D^{-2}\phi,$$

where the lack of a preferred direction in the problem lets us look for a spherically symmetric solution. The solution that satisfies $\phi \to Z_i e/r$ as $r \to 0$, and $\phi \to 0$ as $r \to \infty$, reads

$$\phi = \frac{Z_i e}{r}\exp(-r/L_D), \tag{1.10}$$

a formula cited in Chapter 7 of Volume I without proof.

A straightforward physical interpretation holds for equation (1.10). The statistical attraction of electrons and repulsion of ions shield the potential of any ion, reducing the Coulomb value at large r by an exponential factor. We can easily convince ourselves that a similar shielding applies statistically to electrons as well. Notice that ions contribute about as much as the electrons do to the shielding because, although the ions have more inertia (by a factor m_i/m_e), they also move more slowly [by a factor $(m_i/m_e)^{1/2}$] when electrons and ions have the same temperatures; so the mass dependence cancels out in the modification of the spatial distribution that provides the shielding. Indeed, this cancelation implies that ions with effective charges $Z_\mathrm{i} > 1$ make a *larger* relative contribution to the shielding than electrons do because they react more in equilibrium to a given electrostatic potential. In any case, Debye shielding implies that pure Coulomb potentials break down for impact parameters in excess of L_D; consequently, in the Coulomb scattering problem, we should set $b_\mathrm{max} \sim L_\mathrm{D}$.

THE PLASMA PARAMETER

Clearly, the preceding discussion requires that enough particles (e.g., electrons) exist inside a Debye sphere to justify a statistical treatment:

$$n_\mathrm{e} L_\mathrm{D}^3 \gg 1. \tag{1.11}$$

The reciprocal of the dimensionless combination $n_\mathrm{e} L_\mathrm{D}^3$ is called the *plasma parameter*. Since the mean distance between electrons equals $n_\mathrm{e}^{-1/3}$, we see that the mean electrostatic energy between electrons $e^2 n_\mathrm{e}^{1/3}$ has a ratio to kT equal to $\sim (n_\mathrm{e} L_\mathrm{D}^3)^{-2/3}$, which is $\ll 1$ if $n_\mathrm{e} L_\mathrm{D}^3 \gg 1$. Under these circumstances, we expect (a) the expansion of the exponentials carried out in the previous section to have validity, and (b) most encounters between electrons to involve small-angle scatterings. Indeed, the quantity Λ that appears in equation (1.5) can be expressed, for a pure hydrogen plasma, as

$$\Lambda = 24\pi n_\mathrm{e} L_\mathrm{D}^3 = 3L_\mathrm{D}/r_\mathrm{eff}; \tag{1.12}$$

so the Coulomb logarithm, which measures the importance of small-angle scatterings relative to large-angle ones, will be large compared to unity if $n_\mathrm{e} L_\mathrm{D}^3 \gg 1$. We used these facts in Chapter 15 of Volume I to simplify the calculations of thermal *Bremsstrahlung*.

To complete this section, we note that equation (1.5) suggests that the relaxation time required to make the proton distribution function a Maxwellian, if the distribution should be thrown into disequilibrium (say, by a shock wave), should exceed the corresponding time for electron-electron relaxation by a factor equal to the square root of the mass ratio $m_\mathrm{p}/m_\mathrm{e}$, i.e.,

$$t_\mathrm{c}(\mathrm{p}-\mathrm{p}) = (1836)^{1/2} t_\mathrm{c}(\mathrm{e}-\mathrm{e}). \tag{1.13}$$

On the other hand, Spitzer shows that an additional factor of $(m_\mathrm{p}/m_\mathrm{e})^{1/2}$ enters before the electron and proton Maxwellians acquire the same temperature T:

$$t_\mathrm{eq}(\mathrm{e}-\mathrm{p}) = 1836\, t_\mathrm{c}(\mathrm{e}-\mathrm{e}). \tag{1.14}$$

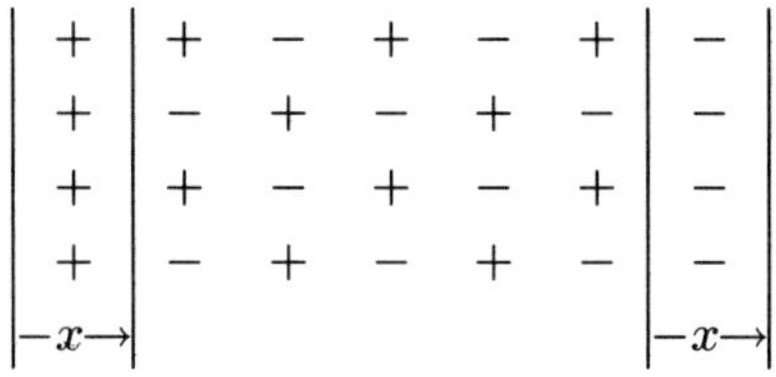

+	+	−	+	−	+	−
+	−	+	−	+	−	−
+	+	−	+	−	+	−
+	−	+	−	+	−	−
$-x\rightarrow$						$-x\rightarrow$

FIGURE 1.4
When electrons of density n_e in an otherwise neutral plasma get displaced by a distance x relative to the positive ions, the charge separation produces an electric field $E = 4\pi n_e e x$ corresponding to that of a capacitor with charge per unit area on the plates equal to $\pm n_e e x$.

In most circumstances, each of these relaxation times is short compared with the evolutionary times of macroscopic astronomical objects. Thus we generally expect that charged particles will locally possess Maxwellian distributions characterized by the same kinetic temperature T.

ELECTRON PLASMA FREQUENCY AND CHARGE SEPARATION

Conceptually, we find it useful to decompose the electron contribution to the inverse square of the Debye length as

$$\omega_{\rm pe}^2/v_{T\rm e}^2,$$

where $v_{T\rm e}$ is the electron thermal velocity and $\omega_{\rm pe}^2$ is the square of the electron *plasma frequency* defined by

$$\omega_{\rm pe}^2 \equiv 4\pi n_e e^2/m_e. \tag{1.15}$$

As we discussed also in Volume I, we envision the simplest context in which the electron plasma frequency arises to be as follows. Imagine displacing the electrons of a plasma relative to the ions by an amount x (see Figure 1.4).

The resulting electric field behaves like that between two capacitor plates with charges per unit area equal to $\pm n_e e x$:

$$E = 4\pi n_e e x.$$

The equation of motion of a displaced electron inside the "capacitor" therefore reads

$$\ddot{x} = -eE/m_e = -\omega_{\rm pe}^2 x,$$

which corresponds to the dynamics of a harmonic oscillator with natural frequency $\omega_{\rm pe}$. Thus, charge separation in a plasma leads to restoring forces

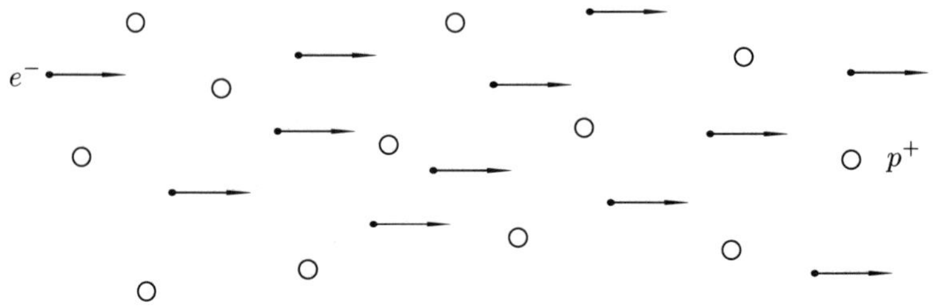

e^- p^+

FIGURE 1.5
An electrically neutral gas may still carry currents if the negatively charged components (usually electrons) drift at a systematic speed with respect to the positively charged components (usually ions). Astrophysically interesting values of magnetic field may be generated in the process even if the drift speeds are very small for all other purposes.

that tend to produce oscillations in the electrons at a natural frequency ω_{pe}. The ions would oscillate with a period slower by the factor $(m_{\text{i}}/Z_{\text{i}}m_{\text{e}})^{1/2}$, and thus can be considered relatively immobile compared to the electrons. For many purposes, therefore, we may calculate the electromagnetic properties of a (partially) ionized gas in terms of a model of mobile electrons moving in a fixed (uniform) background of ions.

Numerically, $\omega_{\text{pe}} = 5.6 \times 10^4 \ n_{\text{e}}^{1/2}$ s if n_{e} is expressed in units of cm^{-3}. Only disturbances with frequencies comparable to or larger than such values will produce appreciable charge separation in a plasma; otherwise fast electrostatic oscillations will be set up that try to cancel the effects of the disturbance. Electromagnetic waves (light) have frequencies that compare with or exceed ω_{pe}, and in Volume I, we discussed the interesting plasma effects that occur in such circumstances. Electromagnetic phenomena on a large scale, however, typically have variations with frequencies $\omega \ll \omega_{\text{pe}}$. Charge separation does not occur for such macroscopic disturbances; this corresponds to the *magnetohydrodynamic regime.*

ELECTRIC CURRENTS AND MAGNETIC FIELDS

In the MHD regime, the condition $\rho_{\text{e}} = 0$ does not imply that we can ignore local material sources for electromagnetic fields. A plasma may be electrically neutral on average, yet still possess electric currents because the electrons drift relative to the positive ions (see Figure 1.5).

Such currents can generate magnetic fields, and, indeed, large-scale magnetic fields appear to be ubiquitous in the cosmos. Particles of charge

q and mass m in the presence of a magnetic field $\mathbf{B}$ do not follow straight-line trajectories, even in the absence of collisions, but gyrate about the field lines with a (nonrelativistic) gyrofrequency given by Larmor's formula,

$$\omega_{\mathrm{L}} \equiv \frac{qB}{mc}. \tag{1.16}$$

For a proton the gyrofrequency $\sim 0.1\ \mathrm{s}^{-1}$ in a magnetic field of strength $B = 10^{-5}$ Gauss (a characteristic value in interstellar clouds). The gyroradius $a = v_{\perp}/\omega_{\mathrm{L}}$ is then negligibly small compared to interesting macroscopic scales for practically all values of $v_{\perp}$. In the presence of magnetic fields, charged particles effectively cling to field lines until they are knocked off by collisions. This tends to bring more coherence to the motion of charged particles than naive estimates might indicate. For example, even when ℓ formally much exceeds L in a dilute plasma, the macroscopic behavior might more resemble that of a fluid than that of a collection of independently moving particles if the plasma is sufficiently magnetized. Random motions in this case, instead of taking the particle a long distance on a straight-line trajectory, become deflected by magnetic fields into tight gyrations that tie the particles together as effectively as string ties together the individual beads of a strand of pearls.

LANDAU DAMPING

When kinetic effects are important for a plasma, surprising physics can often emerge. For example, our simplified derivation for electron plasma oscillations ignored the role of thermal motions. (We effectively assumed the electrons to be *cold* when we ignored the effect of their random motions.) In the presence of thermal motions, we have already witnessed, in a different context, the appearance of a natural length scale, the Debye length,

$$v_{T\mathrm{e}}/\omega_{\mathrm{pe}} \sim L_{\mathrm{D}}.$$

If the wavelength λ characterizing the scale of the charge separation much exceeds L_{D}, neglect of the effects of a finite $v_{T\mathrm{e}}$ should introduce no great error. However, we might expect new phenomena to be introduced when λ becomes comparable to L_{D}. As we shall show in Chapter 29, *Landau damping* of electrostatic plasma oscillations constitutes one such piece of new physics. For now, we merely summarize with the claim that a plasma can behave like a coherent whole on length scales much greater than the Debye length L_{D}, but if we try to set up electrostatic oscillatory disturbances on a shorter or comparable wavelength, the disturbances are quickly damped out, *even without the dissipative help of physical collisions.*

2

Relation of Kinetic Theory to Fluid Mechanics

Reference: Huang, *Statistical Mechanics*, Chapters 3 and 4.

In this chapter, we consider the microscopic basis for a continuum treatment of a collection of particles. The kinetic theory for electrically neutral gases forms the starting point of our discussion. We begin with a derivation of Boltzmann's equation. In what follows, we shall freely interchange between vector notation and Cartesian tensors (with Einstein's summation convention applied to repeated Latin indices). Chapter 3 of Jeffreys and Jeffreys, *Mathematical Physics*, provides a good primer for the technique.

BOLTZMANN'S EQUATION

To describe the motion of gas particles, we may heuristically divide the Hamiltonian H of a particle of mass m with phase-space coordinates $(\mathbf{x}, \mathbf{p})$ at time t into two parts:

$$H = H_{\text{sm}} + H_{\text{irr}}. \tag{2.1}$$

In H_{sm}, we include the effects of smoothly varying external forces, e.g., macroscopic gravitational fields whose properties we have, in principle, complete knowledge about and whose influence we can try to calculate by the conventional techniques of classical mechanics. In H_{irr}, we put the effects of molecular collisions mediated by irregular and unpredictable microscopic forces that we attempt to treat only by statistical means.

We define the single-particle distribution function $f(\mathbf{x}, \mathbf{p}, t)$ so that

$$f(\mathbf{x}, \mathbf{p}, t)\, d^3x\, d^3p$$

equals the number of particles within a configuration-space and velocity-space volume $d^3x\, d^3p$ centered about $(\mathbf{x}, \mathbf{p})$ at time t. Consider now the time rate of change of the number of particles within any small fixed volume V_6

of phase space with surface S_5. The flux of particles associated with the smooth part of the flow across a face of S_5 oriented in the x_i direction equals $f\dot{x}_i = f\partial H_{\rm sm}/\partial p_i$; the flux across a face oriented in the p_i direction, $f\dot{p}_i = -f\partial H_{\rm sm}/\partial x_i$. Violent encounters, however, can lead to sudden appearances and disappearances of particles from the interior of the momentum-space part of the volume V_6 without the particles having to cross the surface S_5 smoothly. In other words, on the mean-free-flight time scale $t_{\rm c}$ *between collisions* in a dilute gas, we assume that the *duration of a collision* $t_{\rm d}$ appears negligible because of the short-range nature of the intermolecular forces between neutral atoms and molecules. When $t_{\rm d} \ll t_{\rm c}$, we take the presence of molecular collisions into account as volume *source* or *sink* terms in the conservation equation for particles:

$$\int_{V_6}\left[\frac{\partial f}{\partial t}+\frac{\partial}{\partial \mathbf{x}}\cdot\left(f\frac{\partial H_{\rm sm}}{\partial \mathbf{p}}\right)+\frac{\partial}{\partial \mathbf{p}}\cdot\left(-f\frac{\partial H_{\rm sm}}{\partial \mathbf{x}}\right)\right]d^3x\,d^3p = \text{sources}-\text{sinks}. \tag{2.2}$$

In equation (2.2), we have used the divergence theorem to convert surface integrals over S_5 into volume integrals over V_6.

Consider, now, the removal of particles of momentum $\mathbf{p}$ by *elastic collisions* with particles of momentum $\mathbf{p}_2$, with the post-encounter pair possessing momenta $\mathbf{p}'$ and $\mathbf{p}_2'$:

$$(\mathbf{p},\mathbf{p}_2) \to (\mathbf{p}',\mathbf{p}_2').$$

Given $\mathbf{p}$ and $\mathbf{p}_2$, the conservation of momentum and energy (we specialize to the nonrelativistic regime),

$$\mathbf{p}'+\mathbf{p}_2' = \mathbf{p}+\mathbf{p}_2, \tag{2.3}$$

$$\frac{1}{2m}[|\mathbf{p}'|^2+|\mathbf{p}_2'|] = \frac{1}{2m}[|\mathbf{p}|^2+|\mathbf{p}_2|^2], \tag{2.4}$$

serve to specify four of the six unknowns in $(\mathbf{p}',\mathbf{p}_2')$. (The short-range character of the interaction forces allows us to assume that the collision occurs essentially at one value of $\mathbf{x}$; so we need not account for changes of the smooth part of the potential energy associated with the external field.) The other two unknowns in $(\mathbf{p}',\mathbf{p}_2')$ are fixed (for featureless molecules) once we specify the impact parameter b and the azimuthal orientation ϕ of the collision (see Figure 2.1).

For an elastic collision, the magnitude of the relative velocity constitutes a collisional invariant:

$$|\mathbf{v}'-\mathbf{v}_2'| = |\mathbf{v}-\mathbf{v}_2|. \tag{2.5}$$

Thus we may specify the remaining two pieces of information concerning the collision in terms of the *change* in the orientation of the relative velocity, i.e., in terms of two angles θ and ϕ. In any case, if $b\,db\,d\phi = \sigma(\mathbf{p},\mathbf{p}_2|\mathbf{p}',\mathbf{p}_2')\,d\Omega$

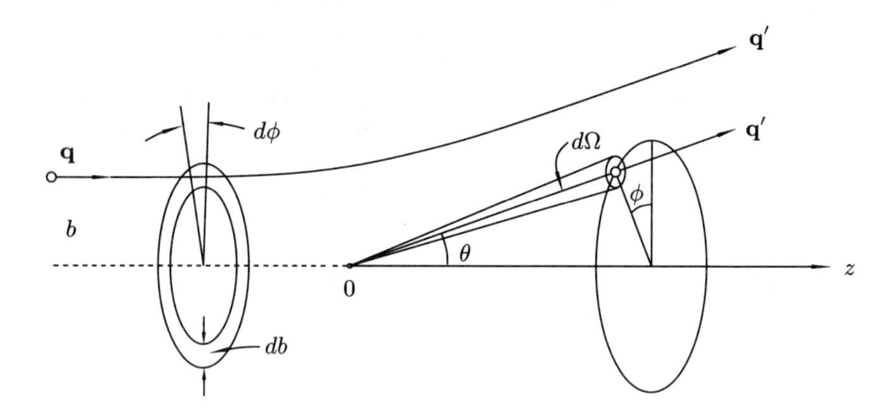

FIGURE 2.1
The geometry of a collision between two molecules of identical mass m and pre-encounter momenta $\mathbf{p} = m\mathbf{v}$ and $\mathbf{p}_2 = m\mathbf{v}_2$, when we reduce the problem to a (classical) description in terms of the relative motion before and after the collision: $\mathbf{q} = \mathbf{v} - \mathbf{v}_2$ and $\mathbf{q}' = \mathbf{v}' - \mathbf{v}_2'$. In an elastic encounter, the collision occurs in a single plane $\phi =$ constant, turning $\mathbf{q}$ through an angle θ into $\mathbf{q}'$ without changing the magnitude of the relative velocity, $|\mathbf{q}'| = |\mathbf{q}|$. For given intermolecular forces, the angle θ will depend on the impact parameter b. The differential cross section for the encounter $\sigma(\mathbf{p}, \mathbf{p}_2|\mathbf{p}', \mathbf{p}_2')$ is defined so that $\sigma\, d\Omega = b\, db d\phi$, where $d\Omega \equiv \sin\theta\, d\theta d\phi$. A quantum-mechanical calculation yields the probability distributions, or $\sigma(\mathbf{p}, \mathbf{p}_2|\mathbf{p}', \mathbf{p}_2') = \sigma(\mathbf{p}', \mathbf{p}_2', \mathbf{p}, \mathbf{p}_2)$, without relying on the notion of well-defined trajectories. Similarly, generalization to incorporate relativistic effects presents no difficulties of principle.

represents the differential collisional cross-section associated with scattering to within solid angle $d\Omega$ of (θ, ϕ), we may now write the collisional sink term as

$$\text{sink} = \int |\mathbf{v} - \mathbf{v}_2| f(\mathbf{p}_2)\, d^3p_2 \sigma(\mathbf{p}, \mathbf{p}_2|\mathbf{p}', \mathbf{p}_2')\, d\Omega f(\mathbf{p})\, d^3x d^3p. \tag{2.6}$$

In the above, we have suppressed the dependence of f on $\mathbf{x}$ and t for simplicity of notation. The integrations over d^3p_2 and $d\Omega$ take place over all possible values, whereas the integration over d^3xd^3p occurs over the small phase-space volume V_6.

By a completely analogous line of reasoning, we may write the source of particles scattered into V_6 centered on $(\mathbf{x}, \mathbf{p})$ due to all encounters of the type

$$(\mathbf{p}', \mathbf{p}_2') \rightarrow (\mathbf{p}, \mathbf{p}_2)$$

as

$$\text{source} = \int |\mathbf{v}' - \mathbf{v}_2'| f(\mathbf{p}_2')\, d^3p_2' \sigma(\mathbf{p}', \mathbf{p}_2' | \mathbf{p}, \mathbf{p}_2)\, d\Omega f(\mathbf{p}')\, d^3x d^3p'. \tag{2.7}$$

Time-reversibility of the process requires equality of the forward-going and backward-going cross sections:

$$\sigma(\mathbf{p}', \mathbf{p}_2' | \mathbf{p}, \mathbf{p}_2) = \sigma(\mathbf{p}, \mathbf{p}_2 | \mathbf{p}', \mathbf{p}_2') \equiv \sigma(\Omega), \tag{2.8}$$

where $\sigma(\Omega)$ represents a shorthand notation for the angle-dependent elastic cross section. Moreover, it is possible to show that the Jacobian of the transformation from primed momenta to unprimed momenta equals unity:

$$d^3p' d^3p_2' = d^3p d^3p_2. \tag{2.9}$$

If we substitute equations (2.5) and (2.9) into equation (2.7), we may now write

$$\text{source} - \text{sink} = \int_{V_6} \left(\frac{\delta f}{\delta t}\right)_{\text{c}} d^3x d^3p, \tag{2.10}$$

where we have defined

$$\left(\frac{\delta f}{\delta t}\right)_{\text{c}} \equiv \int |\mathbf{v} - \mathbf{v}_2| \sigma(\Omega) [f(\mathbf{p}_2') f(\mathbf{p}') - f(\mathbf{p}_2) f(\mathbf{p})]\, d\Omega d^3p_2, \tag{2.11}$$

with the integrals with respect to $d\Omega$ and d^3p_2 taken over all solid angles (for the collision outcome) and momenta (for the collisional partner 2).

In equations (2.2) and (2.10), the phase-space volume V_6 is completely arbitrary; consequently, we may set the integrands of the two sides equal to each other, thereby obtaining *Boltzmann's equation*:

$$\frac{\partial f}{\partial t} + \frac{\partial H_{\text{sm}}}{\partial \mathbf{p}} \cdot \frac{\partial f}{\partial \mathbf{x}} - \frac{\partial H_{\text{sm}}}{\partial \mathbf{x}} \cdot \frac{\partial f}{\partial \mathbf{p}} = \left(\frac{\delta f}{\delta t}\right)_{\text{c}}, \tag{2.12}$$

where the right-hand side is given by equation (2.11). Our derivation of Boltzmann's equation here has followed the standard (heuristic) route; a more rigorous derivation from first principles is possible by the so-called BBGKY hierarchy (see Uhlenbeck and Ford, *Statistical Mechanics*, for an exposition). We forego such a discussion here, apart from mentioning that such a treatment essentially uses the ratio of the time during a collision and the time between collisions, $t_{\text{d}}/t_{\text{c}}$, as a small expansion parameter to truncate the hierarchy of equations that result from integrating over successive coordinates of the N-particle distribution function. The latter describes the state of the system in $6N$-dimensional phase space. (When dealing with *long-range* forces, such as electrostatic interactions in a plasma, we need a different expansion procedure—see Montgomery and Tidman, *Plasma Kinetic Theory*, for a derivation of the so-called *Lenard-Balescu equation*.)

COLLISIONAL INVARIANTS

Boltzmann's equation *predicts* the irreversible increase of the entropy of a thermally isolated gas:

$$S \equiv -k \int f \ln f \, d^3x d^3p, \tag{2.13}$$

where k is Boltzmann's constant and where the integration is taken over all of the six-dimensional phase space available to a gas particle (see Problem Set 1). Apart from this result, it is difficult to extract very general statements about the time evolution of f in six-dimensional phase space by a direct attack on equation (2.12). A more fruitful approach considers a *moment equation* approach.

We may motivate such an approach as follows. The kinetic description contains more information than we often really need. We do not usually care to know the details of the distribution in momentum space (or, as we shall see more accurately in Chapter 3, its small departures from a Maxwellian). As a more modest goal, we may be content with knowledge of the behavior, in space $\mathbf{x}$ and time t, of the first few velocity moments of the distribution function. Of particular interest are the moments associated with the mass, momentum, and kinetic energy about the mean motion:

$$\begin{pmatrix} \rho \\ \rho\mathbf{u} \\ \rho\mathcal{E} \end{pmatrix} \equiv \int \begin{pmatrix} m \\ m\mathbf{v} \\ m|\mathbf{v}-\mathbf{u}|^2/2 \end{pmatrix} f(\mathbf{x},\mathbf{v},t)\, d^3v. \tag{2.14}$$

In equations (2.14) we have modified conventions slightly so as to let f denote the distribution function in *velocity* space rather than in momentum space. Moreover, we suppose that the function f satisfies Boltzmann's equation in the form:

$$\frac{\partial f}{\partial t} + \mathbf{v}\cdot\frac{\partial f}{\partial \mathbf{x}} - \frac{\partial \mathcal{V}}{\partial \mathbf{x}}\cdot\frac{\partial f}{\partial \mathbf{v}} = \left(\frac{\delta f}{\delta t}\right)_{\rm c}, \tag{2.15}$$

where $(\delta f/\delta t)_{\rm c}$ is given by

$$\left(\frac{\delta f}{\delta t}\right)_{\rm c} = \int |\mathbf{v}-\mathbf{v}_2|\sigma(\Omega)[f(\mathbf{v}')f(\mathbf{v}_2') - f(\mathbf{v})f(\mathbf{v}_2)]\, d\Omega\, d^3v_2. \tag{2.16}$$

We have also specialized the consideration of smooth accelerations to those derivable from the negative gradient of a gravitational potential $-\nabla\mathcal{V}$. In $\mathcal{V}$, we allow the possibility of self-gravitation; thus the potential $\mathcal{V}$ satisfies Poisson's equation:

$$\nabla^2\mathcal{V} = 4\pi G(\rho + \rho_{\rm ext}), \tag{2.17}$$

with G being the universal gravitational constant, and with $\rho_{\rm ext}$ being the mass density associated with a distribution of matter (if any) external to our system. In equations (2.14), ρ, $\mathbf{u}$, and $\mathcal{E}$ represent the mass density, the mean (or bulk or fluid) velocity, and specific internal energy (internal energy per unit mass) associated with the gas at $(\mathbf{x}, t)$. In terms of the number density n that we get by integrating f over all $\mathbf{v}$, $\rho = mn$.

We wish to derive equations that govern the spacetime evolution of ρ, $\mathbf{u}$, and $\mathcal{E}$. To do this, consider multiplying equation (2.15) by $\chi(\mathbf{v})$, where $\chi(\mathbf{v})$ represents any constant or power of $\mathbf{v}$, e.g., $\chi(\mathbf{v}) = m$, $m\mathbf{v}$, $m\mathbf{v}\mathbf{v}$, etc. Integrate the result over *all* $\mathbf{v}$ to obtain

$$\int \left(\chi \frac{\partial f}{\partial t} + \chi v_k \frac{\partial f}{\partial x_k} - \chi \frac{\partial \mathcal{V}}{\partial x_k} \frac{\partial f}{\partial v_k} \right) d^3v = \int \chi \left(\frac{\delta f}{\delta t} \right)_{\mathrm{c}} d^3v. \tag{2.18}$$

If we substitute equation (2.16) into the right-hand side of equation (2.18), we get an integrand in which $\mathbf{v}$ and $\mathbf{v}_2$ are dummy variables, with the two entering symmetrically everywhere except in $\chi(\mathbf{v})$. The integral therefore will have the same value if we make the substitution,

$$\chi(\mathbf{v}) \to \frac{1}{2}[\chi(\mathbf{v}) + \chi(\mathbf{v}_2)]. \tag{2.19}$$

Equations (2.5), (2.8), and (2.9) allow us to exchange the roles of the integrations over $d^3v' d^3v_2'$ and $d^3v d^3v_2$. If we subsequently rename primed variables as unprimed variables, and vice versa, we would recover the original expression, except that the term $[f(\mathbf{v}')f(\mathbf{v}_2') - f(\mathbf{v})f(\mathbf{v}_2)]$ changes sign and the right-hand side of equation (2.19) acquires primes on its arguments. In other words, the integral on the right-hand side of equation (2.18) can be evaluated with the substitution,

$$\chi(\mathbf{v}) \to \frac{1}{4}[\chi(\mathbf{v}) + \chi(\mathbf{v}_2) - \chi(\mathbf{v}') - \chi(\mathbf{v}_2')]. \tag{2.20}$$

If χ is a conserved quantity in a collision, the right-hand side of equation (2.20) equals zero. For elastic collisions involving short-range forces in the nonrelativistic regime, there exist exactly five such independent χ's in general; they are the mass, momentum, and (kinetic) energy of a particle:

$$\chi = m; \qquad \chi = mv_i; \qquad \text{and } \chi = \frac{m}{2}|\mathbf{v}|^2. \tag{2.21}$$

If we restrict our attention to such values of χ [or to linear combinations of them, such as $m|\mathbf{v} - \mathbf{u}|^2/2 = m|\mathbf{v}|^2/2 - \mathbf{u} \cdot m\mathbf{v} + m|\mathbf{u}|^2/2$, where $\mathbf{u}(\mathbf{x}, t)$ may be regarded as a constant in any integration over $\mathbf{v}$], we have

$$\int \chi \left(\frac{\delta f}{\delta t} \right)_{\mathrm{c}} d^3v = 0.$$

The above result expresses mathematically the simple notion that collisions can not contribute to the time rate of change of any quantity whose total is conserved in the collisional process.

MOMENT EQUATIONS FOR CONSERVED QUANTITIES

Define the average of any quantity Q by the symbol

$$\langle Q \rangle \equiv n^{-1} \int Q f \, d^3v,$$

where $n \equiv \int f \, d^3v$. Since derivatives with respect to t and $\mathbf{x}$ commute with operations in $\mathbf{v}$, equation (2.18) now yields, if χ represents one of the conserved quantities m, mv_i, or $m|\mathbf{v}|^2/2$:

$$\frac{\partial}{\partial t}(n\langle \chi \rangle) + \frac{\partial}{\partial x_k}(n\langle v_k \chi \rangle) + n\frac{\partial \mathcal{V}}{\partial x_k}\left\langle \frac{\partial \chi}{\partial v_k} \right\rangle = 0. \tag{2.22}$$

To derive the last term of the above equation, we have used the divergence theorem to convert a volume integral (in $\mathbf{v}$) to a surface integral, and assumed that the distribution function f vanishes faster than any power of $\mathbf{v}$ as $\mathbf{v} \to \infty$ on the latter surface.

MASS CONSERVATION

Substitution of $\chi = m$ into equation (2.22) yields the equation of mass conservation,

$$\frac{\partial \rho}{\partial t} + \frac{\partial}{\partial x_k}(\rho u_k) = 0, \tag{2.23}$$

also known as the *continuity equation.* This equation implies more than simple mass conservation, for it states that changes in the local matter content due to fluid flow occur in a *continuous* fashion. When mass disappears from any volume element, it does so by flowing in a well-defined manner across the surface of the volume; it does not do so by suddenly reappearing in some completely disconnected region.

Written in vector notation, the equation of continuity reads

$$\frac{\partial \rho}{\partial t} + \nabla \cdot (\rho \mathbf{u}) = 0, \tag{2.24}$$

which has the usual expression that we associate with a conservation relation (see Chapter 1 of Volume I). Another form in which the continuity equation often appears can be derived by differentiating out the second term, dividing by ρ, and moving the term not involving derivatives of ρ to the right-hand side:

$$\rho^{-1}\frac{D\rho}{Dt} = -\nabla \cdot \mathbf{u},$$

where D/Dt is the time derivative that follows the motion of a *fluid element* of the substance (not equivalent to the motion of a gas particle):

$$\frac{D}{Dt} \equiv \frac{\partial}{\partial t} + \mathbf{u} \cdot \boldsymbol{\nabla}, \tag{2.25}$$

and is known as the *substantial* (or *Lagrangian*) derivative. Since the reciprocal of ρ equals the volume per unit mass, we may also write the desired relation as

$$\rho \frac{D}{Dt}(\rho^{-1}) = \boldsymbol{\nabla} \cdot \mathbf{u},$$

which demonstrates that the divergence of the velocity field yields the fractional time rate of change of the specific volume ρ^{-1}.

MOMENTUM CONSERVATION

Substitution of $\chi = mv_i$ into equation (2.22) yields the equation of momentum conservation:

$$\frac{\partial}{\partial t}(\rho u_i) + \frac{\partial}{\partial x_k}(\rho \langle v_i v_k \rangle) + \rho \frac{\partial \mathcal{V}}{\partial x_i} = 0.$$

We decompose the particle velocity v_i as the sum $u_i + w_i$, and write

$$\langle v_i v_k \rangle = u_i u_k + \langle w_i w_k \rangle,$$

where we have made use of the identity $\langle w_i \rangle = 0$ that follows trivially from the definition of u_i as the average of v_i. Moreover, we follow the usual procedure of separating out the trace of the symmetric dyadic $w_i w_k$; i.e., we write

$$\rho \langle w_i w_k \rangle = P \delta_{ik} - \pi_{ik},$$

where P is the "gas pressure,"

$$P \equiv \frac{1}{3} \rho \langle |\mathbf{w}|^2 \rangle, \tag{2.26}$$

and π_{ik} is the "viscous stress tensor,"

$$\pi_{ik} \equiv \rho \langle \frac{1}{3} |\mathbf{w}|^2 \delta_{ik} - w_i w_k \rangle. \tag{2.27}$$

The *momentum equation*, in its conservation form, now reads

$$\frac{\partial}{\partial t}(\rho u_i) + \frac{\partial}{\partial x_k}(\rho u_i u_k + P \delta_{ik} - \pi_{ik}) = -\rho \frac{\partial \mathcal{V}}{\partial x_i}. \tag{2.28}$$

Notice that the i-th component of the momentum density associated with fluid equals ρu_i. The flux of the i-th component of momentum in the k-th direction consists of the sum of a mean part, $\rho u_i u_k$, and a random part, with the latter composed of an isotropic component $P\delta_{ik}$ and a nonisotropic (traceless) component $-\pi_{ik}$.

By making use of the equation of continuity, we may also manipulate the momentum equation so that it becomes the *force equation*:

$$\rho \frac{D\mathbf{u}}{Dt} = -\rho \nabla \mathcal{V} - \nabla P + \nabla \cdot \overleftrightarrow{\pi} . \tag{2.29}$$

In this form, the equation most closely resembles Newton's second law of dynamics, $m\mathbf{a} = \mathbf{f}$.

ENERGY CONSERVATION

Substitution of $\chi = m|\mathbf{v}|^2/2 = m|\mathbf{u}|^2/2 + m\mathbf{w}\cdot\mathbf{u} + m|\mathbf{w}|^2/2$ into equation (2.22) yields the equation of energy conservation:

$$\frac{\partial}{\partial t}\left[\frac{\rho}{2}(|\mathbf{u}|^2 + \langle|\mathbf{w}|^2\rangle)\right] + \frac{\partial}{\partial x_k}\left[\frac{\rho}{2}\langle(u_k + w_k)(u_i + w_i)^2\rangle\right] + \rho\frac{\partial \mathcal{V}}{\partial x_k}u_k = 0.$$

Expanding the term inside the spatial divergence, we get

$$\langle(u_k + w_k)(u_i + w_i)^2\rangle = |\mathbf{u}|^2 u_k + 2u_i\langle w_i w_k\rangle + u_k\langle|\mathbf{w}|^2\rangle + \langle w_k|\mathbf{w}|^2\rangle.$$

In accordance with equation (2.14), we define the *specific internal energy* $\mathcal{E}$ through

$$\rho\mathcal{E} \equiv \rho\langle\frac{1}{2}|\mathbf{w}|^2\rangle = \frac{3}{2}P, \tag{2.30}$$

where the second relation follows from equation (2.26). We also define the *conduction heat flux* by

$$F_k \equiv \rho\langle w_k \frac{1}{2}|\mathbf{w}|^2\rangle. \tag{2.31}$$

With these definitions, we may write the *total energy equation* in its conservation form as

$$\frac{\partial}{\partial t}\left(\frac{\rho}{2}|\mathbf{u}|^2 + \rho\mathcal{E}\right) + \frac{\partial}{\partial x_k}\left[\frac{\rho}{2}|\mathbf{u}|^2 u_k + u_i(P\delta_{ik} - \pi_{ik}) + \rho\mathcal{E}u_k + F_k\right] = -\rho u_k\frac{\partial \mathcal{V}}{\partial x_k}. \tag{2.32}$$

In equation (2.32) we have used the manipulations of the previous section to rewrite $\rho\langle w_i w_k\rangle$ as $P\delta_{ik} - \pi_{ik}$.

Equation (2.32) states that the total fluid energy density is the sum of a part due to bulk motion $\rho|\mathbf{u}|^2$ and a part due to random motions $\rho\mathcal{E}$. The flux of fluid energy in the k-th direction consists of the translation of the bulk kinetic energy at the k-th component of the mean velocity, $(\rho|\mathbf{u}|^2/2)u_k$, plus the enthalpy (sum of internal energy and pressure) flux $(\rho\mathcal{E} + P)u_k$, plus a viscous contribution, $-u_i\pi_{ik}$, plus the conductive flux, F_k.

We find it convenient for some purposes to express energy conservation in a form that involves only the internal energy plus the PdV work. To do this, first multiply equation (2.28) by u_i and use equation (2.23) to derive the *work equation*:

$$\frac{\partial}{\partial t}\left(\frac{\rho}{2}|\mathbf{u}|^2\right) + \frac{\partial}{\partial x_k}\left(\frac{\rho}{2}|\mathbf{u}|^2 u_k\right) = -\rho u_i \frac{\partial \mathcal{V}}{\partial x_i} - u_i \frac{\partial P}{\partial x_i} + u_i \frac{\partial \pi_{ik}}{\partial x_k}. \tag{2.33}$$

The subtraction of equation (2.33) from equation (2.32) results in the *internal energy equation*:

$$\frac{\partial}{\partial t}(\rho\mathcal{E}) + \frac{\partial}{\partial x_k}(\rho\mathcal{E}u_k) = -P\frac{\partial u_k}{\partial x_k} - \frac{\partial F_k}{\partial x_k} + \Psi, \tag{2.34}$$

where Ψ is the *rate of viscous dissipation*:

$$\Psi \equiv \pi_{ik}\frac{\partial u_i}{\partial x_k}. \tag{2.35}$$

If we use the equation of continuity, we may also write the internal energy equation in the form of the first law of thermodynamics:

$$\rho\frac{D\mathcal{E}}{Dt} = -P\boldsymbol{\nabla}\cdot\mathbf{u} - \boldsymbol{\nabla}\cdot\mathbf{F}_{\text{cond}} + \Psi. \tag{2.36}$$

We recognize $-P\boldsymbol{\nabla}\cdot\mathbf{u} = -P[\rho D(\rho^{-1})/Dt]$ as the rate of doing PdV work, and $-\boldsymbol{\nabla}\cdot\mathbf{F}_{\text{cond}}+\Psi$ as the time rate of adding heat (through heat conduction and the viscous conversion of ordered energy in differential fluid motions to disordered energy in random particle motions).

THE NEED FOR A CLOSURE RELATION

If we include one each of the equivalent forms of the equations for mass conservation (scalar relation), momentum conservation (vector relation), and energy conservation (scalar relation), we have five linearly independent equations. On the other hand, the same equations have thirteen variables, ρ (1), u_i (3), $P = 2\rho\mathcal{E}/3$ (1), π_{ik} (5 for a symmetric traceless tensor), and F_i (3). Hence we have thirteen variables, but only five equations, a common theme of the moment method before we use any physical arguments to derive *closure relations* (see the comments of Chapter 2 in Volume I). Until we introduce a way to obtain a closed set of moment equations, everything that we have done so far concerning the velocity moments, although mathematically exact, has no real physical content. This includes our introduction of suggestive names for various quantities, such as viscous stress tensor, conductive heat flux, etc.

When can we find useful closure conditions? If we follow the example of radiative transfer (Volume I), we might guess: at the extremes. And indeed, at the extreme when the mean free path for collisions is much smaller than macroscopic length scale, $\ell \ll L$ (the material analog of the optically thick regime), we expect the concept of *local thermodynamic equilibrium* (LTE) for the translational degrees of freedom to hold, so that, as we shall see in Chapter 3, the *Chapman-Enskog* procedure allows us to derive useful closure relations. To zeroth order in a systematic expansion for small ℓ/L, we shall find that the eight needed relations take the form $\pi_{ik} \approx 0$ and $F_i \approx 0$. The neglect of diffusive effects in this manner leads to a complete set of fluid relations called the *Euler equations.* To next order, the diffusive terms π_{ik} and F_i are not zero, and we get the so-called *Navier-Stokes equations.*

What about the other extreme, $\ell \gg L$? For radiative transfer, the analogous situation corresponds to optically thin conditions. Under these circumstances, photons travel basically in straight lines unimpeded by the presence of matter, and simplifications become possible for the problem of radiative transfer. With material bodies, however, when the mean free path becomes much larger than the macroscopic length scale (say, the size of the system), we can not assume that particles will travel in straight lines, because they will usually be subject to large-scale macroscopic forces that bend the trajectories into curved orbits. Consequently, the dynamics of material bodies remains quite complicated in the regime $\ell \gg L$ (the state of affairs, e.g., in stellar systems), although the kinetic treatment required will generally be simpler than that which applies in the really awkward regime $\ell \sim L$ (the state of affairs, e.g., in planetary rings).

3

Transport Coefficients for Diffusive Effects

Reference: Lifshitz and Pitaevskii, *Physical Kinetics*, pp. 1–34.

In this chapter we wish to derive closure relations for the higher-order velocity moments of Boltzmann's equation. Our formal procedure corresponds to successive approximations for the solution of equation (2.15) in the limit of small

$$\varepsilon \equiv \ell/L, \tag{3.1}$$

where ℓ is the collision mean free path and L is the macroscopic length of the problem. In the first nontrivial order of approximation, we shall find that the closure relations for the viscous stress tensor π_{ik} and heat conduction flux F_i take the forms of Hooke's law (stress $\propto$ strain) and Fourier's law (conduction flux $\propto$ negative gradient of temperature), with the formalism yielding a prescription for the calculation of the proportionality constants (the so-called "transport coefficients"). A simple physical argument then demonstrates that these results arise basically from diffusive effects.

CHAPMAN-ENSKOG PROCEDURE

A straightforward inspection of equation (2.16) produces the order of magnitude estimate,

$$\text{source or sink part of } \left(\frac{\delta f}{\delta t}\right)_{\mathrm{c}} \sim \nu_{\mathrm{c}} f,$$

where ν_{c} is the collision frequency,

$$\nu_c = n\langle \sigma w \rangle. \tag{3.2}$$

For the time being, we do not attempt to define the meaning of the angular brackets above too precisely. On the other hand, the individual terms on the left-hand side of equation (2.15) have intrinsic order of magnitude

$$uf/L,$$

where u and L are a typical flow velocity and length. To the extent that u and w have comparable orders of magnitude, we see that the ratio of terms on the left-hand and right-hand sides of the Boltzmann equation has a value $\sim \varepsilon$. To remind ourselves of this fact, we rewrite equation (2.15) in the suggestive operator form:

$$\mathcal{L}f = \epsilon^{-1}\mathcal{C}(f|f), \tag{3.3}$$

where $\mathcal{L}$ is the left-hand differential operator:

$$\mathcal{L} \equiv \frac{\partial}{\partial t} + \mathbf{v}\cdot\frac{\partial}{\partial \mathbf{x}} - \frac{\partial \mathcal{V}}{\partial \mathbf{x}}\cdot\frac{\partial}{\partial \mathbf{v}}, \tag{3.4}$$

and $\mathcal{C}$ is the collision integral operator:

$$\mathcal{C}(f|g) \equiv \int |\mathbf{v}-\mathbf{v}_2|\sigma(\Omega)[f(\mathbf{v}')g(\mathbf{v}_2') - f(\mathbf{v})g(\mathbf{v}_2)]\, d\Omega d^3v_2. \tag{3.5}$$

In equation (3.3) ϵ would equal ε if we were to nondimensionalize all variables and to perform a series expansion of f in ascending powers of ε. In actual practice, we shall not bother to nondimensionalize, but we expand formally in ascending powers of ϵ anyway. If we set $\epsilon = 1$ at the end, we will have performed the correct ordering as if we had introduced dimensionless variables.

Following the above prescription, we attempt a solution in the form of a series expansion:

$$f = f_0 + \epsilon f_1 + \epsilon^2 f_2 + \cdots \tag{3.6}$$

We require that f at all orders of approximation yields the correct moments (2.14), i.e.,

$$\int \begin{pmatrix} m \\ m\mathbf{v} \\ m|\mathbf{v}-\mathbf{u}|^2/2 \end{pmatrix} f_0\, d^3v = \begin{pmatrix} \rho \\ \rho\mathbf{u} \\ 3nkT/2 \end{pmatrix}, \tag{3.7}$$

$$\int \begin{pmatrix} m \\ m\mathbf{v} \\ m|\mathbf{v}-\mathbf{u}|^2/2 \end{pmatrix} f_N\, d^3v = 0 \qquad \text{for } N = 1,\, 2,\, \ldots \tag{3.8}$$

In equation (3.7) we have defined the kinetic temperature T so that the internal energy per unit volume $\rho\mathcal{E} = 3nkT/2$.

LOWEST-ORDER SOLUTION: EULER EQUATIONS

If we substitute equation (3.6) into equation (3.3), we get, to lowest asymptotic order,

$$\mathcal{C}(f_0|f_0) = 0. \tag{3.9}$$

For equation (3.9) to hold at all $(\mathbf{x}, \mathbf{v}, t)$, we require the integrand to be zero for all possible collision pairs, i.e.,

$$f_0(\mathbf{v}')f_0(\mathbf{v}_2') = f_0(\mathbf{v})f_0(\mathbf{v}_2),$$

for all collisional transformations of $(\mathbf{v}, \mathbf{v}_2) \rightleftharpoons (\mathbf{v}', \mathbf{v}_2')$. If we take logarithms, we require

$$\ln f_0(\mathbf{v}') + \ln f_0(\mathbf{v}_2') = \ln f_0(\mathbf{v}) + \ln f_0(\mathbf{v}_2),$$

which demonstrates that $\ln f_0$ must be an additive linear function of the collision invariants:

$$\ln f_0 = am + \mathbf{b} \cdot m\mathbf{v} + c\frac{m}{2}|\mathbf{v}|^2.$$

In the above we allow a, $\mathbf{b}$, and c to be arbitrary functions of $(\mathbf{x}, t)$. These must be chosen so that we satisfy equation (3.7). A little algebra then shows that f_0 is given by the local Maxwellian:

$$f_0 = n(m/2\pi kT)^{3/2} \exp(-m|\mathbf{v} - \mathbf{u}|^2/2kT). \tag{3.10}$$

To zeroth order, because f_0 is an isotropic function of $\mathbf{w} \equiv \mathbf{v} - \mathbf{u}$, π_{ik} and F_i, as defined by equations (2.27) and (2.31) vanish:

$$\pi_{ik}^{(0)} = 0, \qquad F_i^{(0)} = 0. \tag{3.11}$$

With these closure conditions, the set of fluid equations reads

$$\frac{\partial \rho}{\partial t} + \boldsymbol{\nabla} \cdot (\rho \mathbf{u}) = 0, \tag{3.12}$$

$$\frac{\partial \mathbf{u}}{\partial t} + \boldsymbol{\nabla}\left(\frac{1}{2}|\mathbf{u}|^2\right) + (\boldsymbol{\nabla} \times \mathbf{u}) \times \mathbf{u} = -\boldsymbol{\nabla}\mathcal{V} - \frac{1}{\rho}\boldsymbol{\nabla}P, \tag{3.13}$$

$$\rho\left(\frac{\partial \mathcal{E}}{\partial t} + \mathbf{u} \cdot \boldsymbol{\nabla}\mathcal{E}\right) = -P\boldsymbol{\nabla} \cdot \mathbf{u}. \tag{3.14}$$

To manipulate equation (3.13) into the displayed vector-invariant form, we have made use of the identity

$$(\mathbf{u} \cdot \boldsymbol{\nabla})\mathbf{u} = \boldsymbol{\nabla}\left(\frac{1}{2}|\mathbf{u}|^2\right) + (\boldsymbol{\nabla} \times \mathbf{u}) \times \mathbf{u}.$$

Equations (3.12)–(3.14) are known as the *Euler equations.* They form a closed set when we add the *constitutive relations* implied by equation (3.10):

$$\rho\mathcal{E} = \frac{3}{2}P = \frac{3}{2}nkT. \tag{3.15}$$

Up to an additive constant, the *specific entropy* s of a classical perfect gas then equals (see Problem Set 1):

$$s \equiv c_v \ln(P\rho^{-\gamma}), \tag{3.16}$$

where the specific heat at constant volume $c_v = 3k/2m$ and $\gamma = 5/3$ for our present model of a molecules with no internal structure (a monatomic gas). Making use of equation (3.12) and the definition, equation (2.25), for the substantial derivative, we may manipulate equation (3.14) to read

$$\rho T \frac{Ds}{Dt} = 0; \tag{3.17}$$

i.e., in the absence of viscous and conductive effects, a neutral gas with no internal degrees of freedom (and therefore no ability to interact with radiation) can undergo only adiabatic variations.

NEXT-ORDER APPROXIMATION: NAVIER-STOKES EQUATIONS

To next order, the substitution of equation (3.6) into equation (3.3) gives

$$\mathcal{C}(f_1|f_0) + \mathcal{C}(f_0|f_1) = \mathcal{L}f_0. \tag{3.18}$$

The solution for f_1 from the above linear integral equation—as well as higher-order approximations in the general *Chapman-Enskog* procedure—is discussed in the classic treatise by Chapman and Cowling, *The Mathematical Theory of Nonuniform Gases.* We won't go into details here, but merely quote the results for π_{ik} and F_i in the linear order of approximation,

$$\pi_{ik} = \mu D_{ik}, \qquad F_i = -\mathcal{K}\frac{\partial T}{\partial x_i}, \tag{3.19}$$

where D_{ik} is the deformation rate tensor (traceless symmetric rate of strain),

$$D_{ik} \equiv \frac{\partial u_i}{\partial x_k} + \frac{\partial u_k}{\partial x_i} - \frac{2}{3}(\nabla \cdot \mathbf{u})\delta_{ik}. \tag{3.20}$$

In equation (3.19), μ equals the coefficient of shear viscosity, and $\mathcal{K}$ equals the coefficient of thermal conductivity. For a neutral monatomic gas, these two transport coefficients have a ratio given by *Eucken's constant:*

$$\frac{\mathcal{K}}{\mu} = \frac{5}{2}c_v, \tag{3.21}$$

where $c_v \equiv 3k/2m$. Apart from the above successful prediction (in comparison with experimental measurements), the Chapman-Enskog procedure also yields the value of μ as

$$\mu = \frac{5}{8}\frac{(\pi mkT)^{1/2}}{\sigma_{\text{transport}}}. \tag{3.22}$$

The transport cross section is defined by the integral

$$\sigma_{\text{transport}} \equiv \int_0^\infty dg\, g^7 e^{-g^2} \int_0^{\pi/2} 2\pi\sigma(g,\Theta)\sin^2\Theta\, d\Theta, \tag{3.23}$$

where Θ is the polar scattering angle in the center-of-mass frame, and $g \equiv (m/2kT)^{1/2}|\mathbf{w}-\mathbf{w}_2|$ is the dimensionless relative velocity. In equation (3.23) we have explicitly allowed for the (realistic) possibility that the elastic scattering cross section might depend on the magnitude of the relative velocity (center of mass energy).

The inclusion of the correction f_1 does not modify the relations (3.15) or (3.16). The Navier-Stokes equations then are the same as the Euler equations, except that on the right-hand side of equation (3.13), we must add the viscous force per unit mass,

$$\rho^{-1}\nabla\cdot \overleftrightarrow{\pi},$$

and to the right-hand side of equation (3.17) we must add minus the divergence of the conductive flux and the rate of viscous dissipation,

$$-\nabla\cdot\mathbf{F}_{\text{cond}} + \Psi,$$

where $\pi_{ik} = \mu D_{ik}$ allows us to write

$$\Psi = \frac{1}{2}\mu\|\overleftrightarrow{D}\|^2, \tag{3.24}$$

with the square of the scalar norm of the deformation tensor equaling

$$\|\overleftrightarrow{D}\|^2 \equiv D_{ik}D_{ik}. \tag{3.25}$$

To derive equation (3.24), we use the fact that π_{ik} equals a symmetric tensor to infer

$$\Psi \equiv \pi_{ik}\frac{\partial u_i}{\partial x_k} = \pi_{ik}\frac{1}{2}\left(\frac{\partial u_i}{\partial x_k} + \frac{\partial u_k}{\partial x_i}\right) = \frac{1}{2}\pi_{ik}D_{ik},$$

where we may take the last step because $\pi_{ik} = \mu D_{ik}$, with D_{ik} defined by equation (3.20) being traceless.

Since D_{ik} is a symmetric tensor, we can always rotate the coordinate axes so that D_{ik} is diagonal (see Chapter 3 of Jeffreys and Jeffreys, *Mathematical Physics*, for a simple discussion). In this so-called *principal axes frame*,

$$\|\overleftrightarrow{D}\|^2 = D_{11}^2 + D_{22}^2 + D_{33}^2$$

is clearly positive definite. However, the contraction of any tensor with itself is a scalar independent of the orientation of the coordinate axes used to

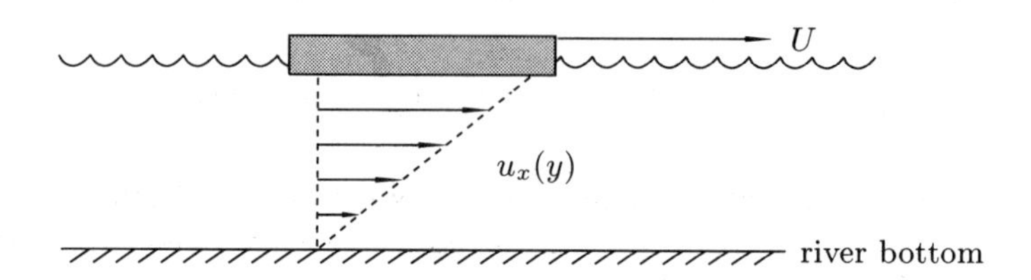

FIGURE 3.1
If we drag a large barge in the horizontal direction x at speed U relative to a stationary river, we will set up a velocity profile $u_x(y)$ in the river underneath the barge that depends on depth y, with $u_x(y)$ increasing from 0 at the river bottom to the full value U just beneath the barge. The shear of any layer of fluid with respect to adjacent layers will transmit a frictional force, layer by layer, that ultimately tries to slow down the barge with respect to the nonflowing river bottom. Newton and Hooke both surmised (correctly) that the frictional force per unit area across the interface of two adjacent layers should be proportional to the rate of strain, $\partial u_x/\partial y$, with the constant of proportionality now called the coefficient of shear viscosity μ.

define its components; thus the expression in equation (3.25) must be positive definite in all frames, and the presence of nonzero viscous dissipation can only increase the entropy of the system. Except for heat flow through the boundaries of the system, the same deduction can be made concerning the action of heat conduction (see Problem Set 1) in the *heat equation*:

$$\rho T \frac{Ds}{Dt} = -\nabla \cdot \mathbf{F}_{\rm cond} + \Psi. \tag{3.26}$$

This conclusion represents the macroscopic manifestation of Boltzmann's "H theorem" (the irreversible increase of the system entropy due to elastic molecular encounters).

SIMPLE ORDER-OF-MAGNITUDE ESTIMATE

Except for numerical coefficients of order unity, equation (3.22) gives the coefficient of shear viscosity as

$$\mu \sim m v_T/\sigma, \tag{3.27}$$

where v_T is the thermal speed $(kT/m)^{1/2}$ and σ is the typical collision cross section. Notice in particular that μ is independent of the gas density n; i.e., the frictional force exerted by a dilute gas in a container does not depend on how many particles the container holds! This conclusion so surprised Maxwell when he first derived it that he immediately concocted a pendulum experiment to verify the result. A simple physical argument suffices to reproduce the reasoning (see Figures 3.1 and 3.2).

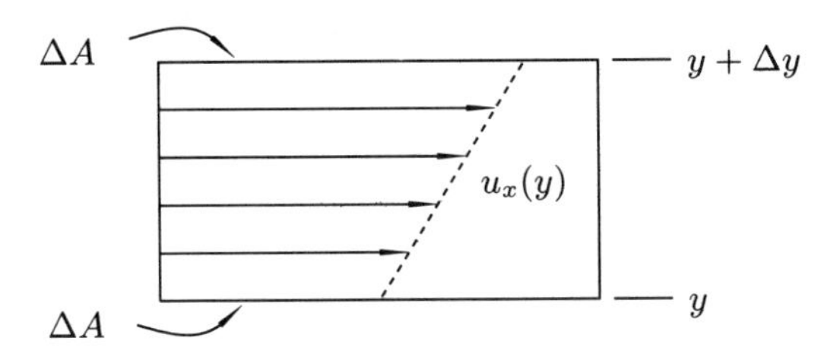

FIGURE 3.2
The idealized situation envisaged by Maxwell to derive the expression $\mu \sim mv_T/\sigma$ for the coefficient of shear viscosity in a dilute gas.

Consider the case of a plane-parallel shear flow, as might be set up approximately, for example, by dragging a barge up the Thames river. Both Newton and Hooke recognized in an implicit fashion that such a situation should give rise to a frictional stress that would be proportional to the shear rate $\partial u_x/\partial y$ experienced by a fluid element. (For this reason, a viscous force law of the form $\pi_{xy} = \mu\partial u_x/\partial y$ is often referred to as "Newtonian.") It took Maxwell, however, to realize how to calculate the proportionality "constant" μ for a dilute gas. Stripped to its basic essence, his argument proceeds as follows (see Figure 3.2).

Define f_x to equal the viscous force per unit volume acting on a fixed region bounded by two faces of area ΔA located at y and $y + \Delta y$ (see Figure 3.3).

Equal but opposite fluxes of gas molecules $\sim nv_T/2$ cross the upper face because of random thermal motions in the y direction. After traveling a mean distance ℓ, they collide with other molecules and share whatever properties they carried characteristic of the region from where they came. The difference in x-momentum per particle between the molecules coming from above and those leaving from below equals $\sim 2m\ell\partial u_x/\partial y$ if there exists a mean shear flow. Thus the exchange of particles from inside and outside the fluid element across the upper face contributes a positive time rate of change of the x-component of the momentum equal to $[(nv_T\Delta A)m\ell\,\partial u_x/\partial y]_{y+\Delta y}$, where the subscript denotes evaluation at $y+\Delta y$. Similarly, the exchange of particles across the bottom face contributes a negative time rate of change of the x-momentum equal to $-[(nv_T\Delta A)m\ell\,\partial u_x/\partial y]_y$. The sum of the expressions at y and $y+\Delta y$ equals the contribution of particle diffusion across both surfaces to the x frictional force exerted on the fluid element by its surroundings:

$$f_x\Delta A\Delta y \sim \left[\frac{\partial}{\partial y}\left(mnv_T\ell\frac{\partial u_x}{\partial y}\right)\right]\Delta A\Delta y,$$

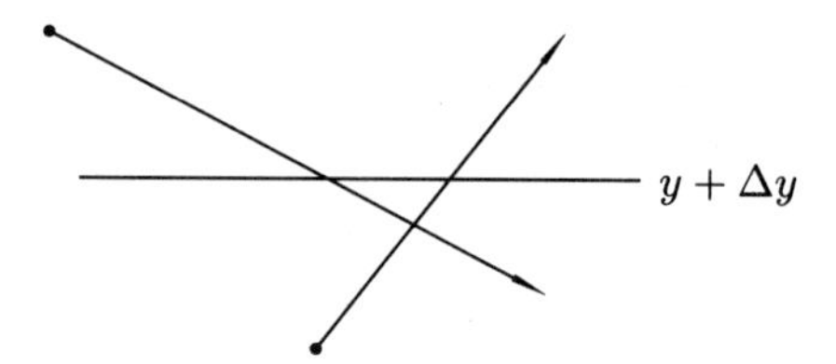

FIGURE 3.3
Equal but opposite fluxes of gas molecules statistically cross any interface located at a vertical position $y + \Delta y$ fixed locally with respect to the bulk motion of a fluid. Although this random interchange yields no net modification in the mass of the fluid on either side, a net transfer of transverse momentum can take place if the mean velocity u_x possesses a gradient in the y direction, because the molecules then crossing from the top possess on average a different x-momentum than the molecules crossing from the bottom. Upon collision with ambient particles, such molecules deposit their excess (or deficit) x-component of momentum into the local gas, effectively transmitting a drag force between the layers of fluid on either side of the interface.

where f_x is the frictional force per unit volume. Dividing by the volume $\Delta A \Delta y$, we get

$$f_x = \frac{\partial \pi_{xy}}{\partial y} \qquad \text{where} \qquad \pi_{xy} = \mu \frac{\partial u_x}{\partial y},$$

with μ given by equation (3.27). (Q.E.D.)

With the above derivation we can easily understand why μ does not depend on n. Higher gas densities yield a higher flux of gas particles, $\sim n v_T$ to carry excesses or deficits of momenta, Δp, but the distance $\ell \sim (n\sigma)^{-1}$ that they can carry $\Delta p \sim \ell m \partial u_x / \partial y$ before suffering a collision diminishes as n increases. For a given shear rate, the magnitude of the discrepancy carried per particle compared to the local mean varies as n^{-1}. The dependence on n therefore cancels out in the product, $\mu \sim n v_T \, \Delta p$.

ALTERNATIVE DERIVATION VIA VELOCITY CORRELATIONS

According to equation (2.27), we may identify

$$\pi_{xy} = -\rho \langle w_x w_y \rangle \sim \rho v_T \ell \frac{\partial u_x}{\partial y},$$

since w_x and w_y are correlated in a shear flow on the length scale ℓ. With $\pi_{xy} = \mu \partial u_x / \partial y$, this again produces the estimate (3.27).

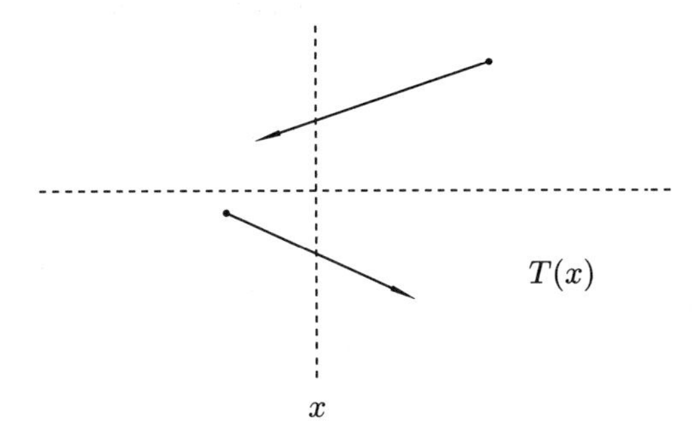

FIGURE 3.4
Equal but opposite fluxes of gas molecules statistically cross any interface located at a position x fixed locally with respect to the bulk motion of the fluid. A net transfer of energy can take place if the mean temperature T possesses a gradient in the x direction, because the molecules then crossing from the left possess on average a different thermal energy than the molecules crossing from the right. Upon collisions with ambient particles, such molecules deposit their excess (or deficit) thermal energy into the local gas, giving rise to a conductive flux between the layers of fluid on either side of the interface.

The same technique allows a quick estimate for the coefficient of thermal conductivity. Consider a gas with a temperature gradient in the x direction (see Figure 3.4). Kinetic theory yields the identification

$$F_x = \rho\langle \frac{1}{2} w_x |\mathbf{w}|^2 \rangle \sim -n v_T \ell \, \frac{3}{2} k \frac{\partial T}{\partial x}.$$

With $F_x = -\mathcal{K}\partial T/\partial x$, we obtain

$$\mathcal{K} \sim n v_T \ell \frac{3}{2} k = v_T 3k/2\sigma,$$

which implies

$$\mathcal{K} \sim c_v \mu,$$

in agreement with equation (3.21) apart from a factor of 5/2.

KROOK'S EQUATION

Although precise derivations of transport coefficients require the use of Boltzmann's equation, with its complex collision integral, Max Krook and his colleagues discovered that calculations of acceptable accuracy often result by making the following simple replacement:

$$\left(\frac{\delta f}{\delta t}\right)_{\mathrm{c}} \to -\nu_{\mathrm{c}}(f - f_0), \tag{3.28}$$

where ν_{c} is some appropriately defined collision frequency and f_0 is the local Maxwellian given by equation (3.10). The distribution function f then satisfies the Krook equation:

$$\frac{\partial f}{\partial t} + \mathbf{v} \cdot \frac{\partial f}{\partial \mathbf{x}} - \frac{\partial \mathcal{V}}{\partial \mathbf{x}} \cdot \frac{\partial f}{\partial \mathbf{v}} = -\nu_{\mathrm{c}}(f - f_0), \tag{3.29}$$

with the first few velocity moments of f required exactly to yield the quantities n, $\mathbf{u}$, and kT that appear in the definition of f_0.

We may heuristically justify equation (3.29) as follows. Collisions occurring at the frequency ν_{c} tend to remove particles (per unit phase-space volume) at the rate $-\nu_{\mathrm{c}} f$ and replace them with a thermalized distribution at the rate $+\nu_{\mathrm{c}} f_0$. The process in this scheme looks analogous to the source and sink terms of radiative transfer, wherein the absorption of photons occurs at a rate $-c\rho\kappa_\nu I_\nu$, and the (LTE) emission of photons occurs at a rate $+c\rho\kappa_\nu B_\nu(T)$, with $B_\nu(T)$ being the Planck function (see Chapter 2 of Volume I). For collisional scattering treated as the "absorption" and "emission" of particles, the Krook procedure replaces $c\rho\kappa_\nu$ by ν_{c} [see equation (3.2)] and $B_\nu(T)$ by f_0.

The mathematical advantage of equation (3.29) over equation (2.15) is that the former is a partial differential equation instead of a integro-partial-differential equation. Moreover, if we can regard ν_{c} as given (instead of being itself dependent on f), equation (3.29) is formally linear in f (to the extent that n, $\mathbf{u}$, and T can be regarded also as known). These properties allow much easier computations for transport coefficients, not only in the hydrodynamical limit $\ell \ll L$, but also in the awkward regime $\ell \sim L$. (For further details, see the astrophysical literature and Problem Set 1.)

GENERAL UNIMPORTANCE OF PARTICLE DIFFUSIVE EFFECTS

From our discussion at the beginning, we expect that the terms associated with π_{ik} and F_i will be unimportant in most astronomical settings. To see this explicitly, consider, for example, the ratio of the inertial term to the viscous term in the momentum equation (no summation convention below):

$$\frac{\partial(\rho u_i u_k)/\partial x_k}{\partial \pi_{ik}/\partial x_k} \sim \frac{\rho U^2/L}{\mu U/L^2} = \frac{UL}{\nu} \equiv \mathrm{Re},$$

where U is a typical flow speed, Re is the *Reynolds number*, and $\nu \equiv \mu/\rho$ is called the *kinematic viscosity*. The kinematic viscosity has the unit of velocity times length; indeed,

$$\nu \sim m v_T/\sigma\rho = v_T/n\sigma = v_T\ell.$$

The last formula has the usual form for a diffusion coefficient: random velocity of diffusing particle times its mean free path. We shall see in Chapters 6 and 21 that ν is associated with the diffusion of *vorticity*, i.e., with the diffusion of the curl of the velocity field $\nabla \times \mathbf{u}$.

In any case, we see that

$$\mathrm{Re} \sim \frac{UL}{v_T\ell} \sim \varepsilon^{-1} \gg 1 \qquad \text{when} \qquad U \sim v_T.$$

In other words, for sonic or supersonic flows (a frequent occurrence in astrophysics), the Reynolds number must be large if a fluid description holds in the first place for a dilute neutral gas. When Re $\gg 1$, viscous forces are much less important than inertial effects.

Consider next the ratio of the viscous-dissipation and heat-conduction terms in equation (3.26):

$$\frac{\Psi}{\nabla \cdot \mathbf{F}_{\text{cond}}} \sim \frac{\mu U^2/L^2}{\mathcal{K}T/L^2} \sim \frac{U^2}{c_v T} \sim \frac{U^2}{v_T^2}.$$

Since $U/v_T \sim M$, where M is the *Mach number* (the ratio of the fluid speed to the sound speed), we expect viscous dissipation to dominate over thermal conduction at high Mach numbers, and vice versa at low Mach numbers. For $M \sim 1$, we can not consistently ignore one without ignoring the other.

Consider now the advection of heat (due to the flux of enthalpy) compared to the conduction of heat:

$$\frac{\partial[(\rho\mathcal{E} + P)u_k]/\partial x_k}{\nabla \cdot \mathbf{F}_{\text{cond}}} \sim \frac{\rho c_P T U/L}{\mathcal{K}T/L^2} = \frac{\rho U L}{\mathcal{K}/c_P} = \frac{UL}{\chi},$$

where $c_P = 5k/2m$ is the specific heat at constant pressure, and $\chi \equiv \mathcal{K}/\rho c_P$ is the *thermal diffusivity*. The dimensionless quantity UL/χ, called the *Peclet number* Pe, can be written as the product

$$\mathrm{Pe} = \mathrm{Pr}\ \mathrm{Re},$$

where $\mathrm{Pr} = \nu/\chi$ is the dimensionless ratio of two diffusivities and is called the *Prandtl number*. For a neutral monatomic gas, $\mathrm{Pr} = \mu c_P/\mathcal{K} = 2/3$ when we make use of equation (3.21), together with $c_P/c_v = \gamma = 5/3$. Thus $\mathrm{Pe} = (2/3)\,\mathrm{Re} \gg 1$, and the advection of heat (e.g., by thermal convection), when it occurs, generally dominates over heat conduction.

Many such dimensionless numbers arise in fluid mechanics when one considers various competing processes. The above definitions give only some of the simplest examples. In astrophysics, we usually conclude that diffusive effects can not compete with flow effects. The difference from terrestrial examples arises because astrophysical systems are comparatively large. In this case, diffusion with a diffusivity $\mathcal{D}$ occupies a natural time scale $L^2/\mathcal{D}$, which usually takes much longer than the corresponding flow time L/U. Exceptions to this rule of thumb can occur, but normally only when the relevant transport coefficient has an anomalously large value compared with the standard estimate from kinetic theory. The empirical evidence for the presence of such "anomalous transport coefficients" (e.g., in the theories of accretion disks or magnetic reconnection) is often quite indirect; this subject lies at one of the forefronts of astrophysical research.

HIERARCHY OF TIME SCALES

To finish, we summarize our deliberations in terms of a hierarchy of time scales for a typical system composed of a neutral monatomic gas. On a time of the order of the duration of a collision, t_d, we can telescope an N-particle description down to a probabilistic description in terms of a single-particle distribution function f. This represents the *kinetic regime*, and the Boltzmann equation constitutes the archetype for the governing equation. On a time of the order of the time between collisions, $t_\text{c} \gg t_\text{d}$, the distribution relaxes in velocity space to a local Maxwellian, but inhomogeneities in space do not yet have enough time to iron themselves out. Unbalanced macroscopic forces take a flow time, $t_\text{flow} \sim L/U \gg t_\text{c}$, to come into mechanical equlibria. This represents the *hydrodynamic regime*, and the Euler equations constitute the archetype for the governing set. On a diffusion time scale $t_\text{diff} \sim L^2/\chi \gg t_\text{flow}$, processes like heat conduction act to establish conditions of complete thermodynamic equilibrium (e.g., uniformity of temperature). Because the last *diffusive relaxation time* can be very long, other, more effective processes may enter to interfere with the completion of the final phase—the heat death of the system. Examination of the action of some of these processes is one goal of this volume.

4

Fluids as Continua

References: Jeffreys and Jeffreys, *Mathematical Physics*, Chapters 2 and 3; Landau and Lifshitz, *Fluid Mechanics*, pp. 1–14, 47–54.

In Chapters 2 and 3, we derived the basic equations of fluid mechanics under the restrictive conditions of a rarefied monatomic gas with no internal degrees of freedom. The advantage conferred by an idealized model that allows us to proceed from first principles is offset by our inability to consider, for example, interactions with radiation, or the flows of imperfect gases and liquids. In this chapter we rectify these shortcomings by adopting the viewpoint of fluids as continua rather than as collections of particles. We begin with a discussion of the *kinematics* of fluid motion.

KINEMATICS OF FLUID MOTION

To gain a deeper insight into the relative motions possible for neighboring pieces of a fluid, consider a point P at $\mathbf{x}$ and another point Q at $\mathbf{x}+\mathbf{R}$, where $\mathbf{R}$ is small in comparison with the length scale L over which the flow varies by an amount of order unity (see Figure 4.1).

Consider the relative motion of any neighboring point Q with respect to the center of mass P of a fluid element. We wish to prove that the most general kinematics for the fluid element correspond to the superposition of four kinds of motion: (1) a uniform translation at the center-of-mass velocity, (2) a uniform rotation about the center of mass, (3) a uniform dilation or contraction, and (4) a distortion of the fluid element without a change of volume.

To express the above idea mathematically, expand the instantaneous velocity difference between Q and P as

$$\delta u_i \equiv u_i(\mathbf{x}+\mathbf{R},t) - u_i(\mathbf{x},t) = \frac{\partial u_i}{\partial x_k} R_k. \tag{4.1}$$

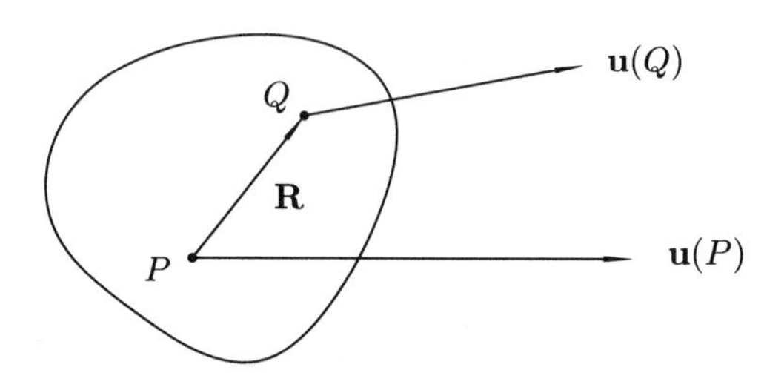

FIGURE 4.1
The fluid velocity $\mathbf{u}(Q)$ at a point Q displaced by a small amount $\mathbf{R}$ from a point P will differ from $\mathbf{u}(P)$ by a small amount $\delta\mathbf{u}$ that we may express in terms of a Taylor series expansion.

We find it advantageous to write the rate-of-strain tensor, $\partial u_i/x_k$ in terms of its symmetric and antisymmetric parts:

$$\frac{\partial u_i}{\partial x_k} = \frac{1}{2}\left(\frac{\partial u_i}{\partial x_k} + \frac{\partial u_k}{\partial x_i}\right) + \frac{1}{2}\left(\frac{\partial u_i}{\partial x_k} - \frac{\partial u_k}{\partial x_i}\right). \tag{4.2}$$

With the antisymmetric part we can associate a (pseudo)vector, the vorticity $\boldsymbol{\omega} \equiv \boldsymbol{\nabla} \times \mathbf{u}$; in Cartesian tensor notation,

$$\omega_m = \epsilon_{mps}\frac{\partial u_s}{\partial x_p}, \qquad i.e., \qquad \frac{\partial u_i}{\partial x_k} - \frac{\partial u_k}{\partial x_i} = \epsilon_{kim}\omega_m, \tag{4.3}$$

where we have used the well-known identity for the Levi-Cevita tensor, $\epsilon_{kim}\epsilon_{mps} = \delta_{kp}\delta_{is} - \delta_{ks}\delta_{ip}$. The contribution of the antisymmetric part to the differential velocity reads

$$\frac{1}{2}\left(\frac{\partial u_i}{\partial x_k} - \frac{\partial u_k}{\partial x_i}\right)R_k = \frac{1}{2}\epsilon_{kim}\omega_m R_k = \epsilon_{imk}\Omega_m R_k, \tag{4.4}$$

where $\boldsymbol{\Omega} \equiv \boldsymbol{\omega}/2$. Since the last expression in equation (4.4) equals the i-th component of $\boldsymbol{\Omega} \times \mathbf{R}$, we identify one-half of the vorticity, $(\boldsymbol{\nabla} \times \mathbf{u})/2$, as the (instantaneous) angular velocity of a fluid element with respect to its center of mass.

With the symmetric part of the rate of strain tensor,

$$E_{ik} = \frac{1}{2}\left(\frac{\partial u_i}{\partial x_k} + \frac{\partial u_k}{\partial x_i}\right), \tag{4.5}$$

we may associate a quadratic form,

$$\Phi \equiv \frac{1}{2}E_{km}R_k R_m. \tag{4.6}$$

The quantity Φ is a scalar, in the sense that it remains invariant under rotations of the coordinate axes; moreover, its gradient with respect to $\mathbf{R}$ yields the contribution of the symmetric part of the rate-of-strain tensor to the relative velocity:

$$\frac{\partial \Phi}{\partial R_i} = \frac{1}{2}(E_{im}R_m + E_{ki}R_k) = \frac{1}{2}\left(\frac{\partial u_i}{\partial x_k} + \frac{\partial u_k}{\partial x_i}\right) R_k. \tag{4.7}$$

For this reason, we may call Φ the *velocity potential* that generates the *irrotational* part of the velocity field. (In other words, if the velocity field has no vorticity, $\nabla \times \mathbf{u} = 0$, then there exists a scalar function Φ such that $\mathbf{u} = \nabla\Phi$.)

A theorem in linear algebra states that we may always rotate the coordinate axes so as to make a symmetric tensor diagonal. In the principal axes frame of E_{ik}, which we denote by a prime, Φ takes the form of a sum of squares:

$$\Phi = \frac{1}{2}E'_{j\ell}R'_jR'_\ell = \frac{1}{2}\left(E'_{11}R'^2_1 + E'_{22}R'^2_2 + E'_{33}R'^2_3\right). \tag{4.8}$$

(We do not need to put a prime on Φ, because it is a scalar.) In this frame all the strains are *extensional*:

$$E'_{11} = \frac{\partial u'_1}{\partial x'_1}, \qquad E'_{22} = \frac{\partial u'_2}{\partial x'_2}, \qquad E'_{33} = \frac{\partial u'_3}{\partial x'_3}. \tag{4.9}$$

In other words, $\partial u'_1/\partial x'_1$ represents the rate of elongation of the fluid element in the $1'$ direction, etc. If $E'_{11} = E'_{22} = E'_{33}$, we have a uniform dilation (if positive) or compression (if negative). Hence we may associate the trace of E'_{ik} with the time rate of change of volume of the fluid element due to uniform expansion or contraction. The trace of a tensor is a scalar and does not depend on the coordinate representation; in particular,

$$E_{ii} = \nabla \cdot \mathbf{u},$$

which we found in Chapter 2 to give the time rate of change of a fluid's specific volume.

The above discussion motivates us to write E_{ik} as the sum of a (traceless) deformation rate tensor D_{ik} and a part associated with $\nabla \cdot \mathbf{u}$,

$$E_{ik} = \frac{1}{2}D_{ik} + \frac{1}{3}(\nabla \cdot \mathbf{u})\delta_{ik}, \tag{4.10}$$

where D_{ik} is defined by equation (3.20):

$$D_{ik} \equiv \frac{\partial u_i}{\partial x_k} + \frac{\partial u_k}{\partial x_i} - \frac{2}{3}(\nabla \cdot \mathbf{u})\delta_{ik}.$$

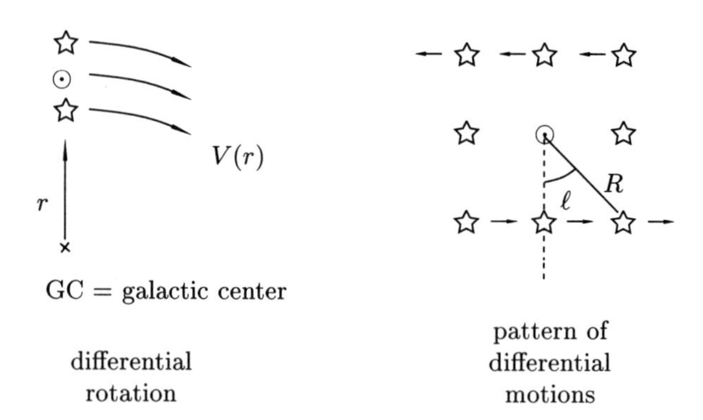

FIGURE 4.2
In an idealized model, differential rotation occurs in the Galaxy with a rotational velocity $V(\varpi)$ that depends only on the distance ϖ from the rotation axis. In this case, an observer at $\odot$ moving with the local standard of rest sees a planar pattern of differential motions, in both the radial and tangential velocities, that exhibits a double sinusoidal dependence on galactic longitude ℓ and that increases at fixed ℓ with distance R from $\odot$.

Collecting all terms, we can rewrite equation (4.1) as

$$\mathbf{u}(Q) = \mathbf{u}(P) + \boldsymbol{\Omega} \times \mathbf{R} + K\mathbf{R} + \boldsymbol{\nabla}_R F, \tag{4.11}$$

where $\boldsymbol{\Omega} \equiv \boldsymbol{\nabla} \times \mathbf{u}/2$, $K \equiv (\boldsymbol{\nabla} \cdot \mathbf{u})/3$, and $F \equiv D_{ik}R_iR_k/4$. Equation (4.11) is known as *Stokes' theorem*: The most general differential motion of a fluid element corresponds to a uniform translation (first term) plus a uniform rotation (second term) plus a uniform expansion or contraction (third term) plus a distortion without a change in volume (fourth term).

APPLICATION TO DIFFERENTIAL MOTIONS IN THE GALAXY

Jan Oort gave the most famous application of Stokes' theorem to astronomy when he analyzed the pattern of mean velocities [as measured by optical Doppler shifts and proper motions (displacements perpendicular to the line of sight)] of stars within a kpc or two of the Sun's position in our Galaxy (see Figure 4.2). Oort found that the radial and tangential velocity (in the plane of the Galactic disk) varies with radial distance R from the solar position and Galactic longitude ℓ as

$$\delta u_R = AR\sin(2\ell), \qquad \delta u_\ell = AR\cos 2\ell + BR.$$

It is often said that the observed sinusoidal-variation dependence of δu_R and δu_ℓ on *twice* the galactic longitude ℓ proves that the Galaxy rotates differentially about a center (located in the direction of $\ell = 0$). In terms of an axisymmetric model in which the rotation velocity at a distance ϖ from the axis of symmetry passing through the Galactic center equals $V(\varpi)$, Oort's constants A and B are given by

$$A = -\left[\frac{\varpi}{2}\frac{d}{d\varpi}(\varpi^{-1}V)\right]_\odot , \tag{4.12}$$

$$B = -\left[\frac{1}{2\varpi}\frac{d}{d\varpi}(\varpi V)\right]_\odot , \tag{4.13}$$

where the subscript $\odot$ denotes that the derivatives are to be evaluated at the solar circle. Conventional values of A and B are $A = 15\,\mathrm{km\,s^{-1}\,kpc^{-1}}$ and $B = -10\,\mathrm{km\,s^{-1}\,kpc^{-1}}$.

In fact, as first shown by E.A. Milne, *any* general field of mean (fluid) motion will give rise to a double sinusoidal dependence for δu_R and δu_ℓ. The term A arises from the distortional tensor D_{ik}; the term B, from the vorticity $\nabla \times \mathbf{u}$. What distinguishes differential rotation as the correct conclusion from Oort's analysis are (a) that the zero point for ℓ lies in the *same* direction as that which Shapley found for the center of the spatial distribution of globular clusters (the putative center of the Galaxy), and (b) that a contribution of the form KR to δu_R, corresponding to a locally uniform expansion or contraction (like Hubble flow), is apparently small. The modern view suggests that the Galaxy possesses spiral structure. Unless we have a special location with respect to the spiral arms, local expansional or compressional motions in the Galactic spiral density wave will produce a nonvanishing K term (for the directions parallel to the Galactic plane).

LINEAR MOMENTUM OF A FLUID ELEMENT

The linear momentum $\mathbf{p}$ of a fluid element equals the fluid velocity $\mathbf{u}(Q)$ integrated over the mass of the element:

$$\mathbf{p} = \int \mathbf{u}(Q)\,dm. \tag{4.14}$$

If we substitute equation (4.11) into equation (4.14), we get

$$\mathbf{p} = \mathbf{u}(P)\int dm + \frac{1}{2}(\nabla\times\mathbf{u})_P \times \int \mathbf{R}\,dm + \frac{1}{3}(\nabla\cdot\mathbf{u})_P \int \mathbf{R}\,dm + \int \nabla_R F\,dm.$$

If P equals the center of mass of the fluid element, then the second and third terms on the right-hand side vanish, since $\int \mathbf{R}\,dm = 0$. Moreover,

$$(\nabla_R F)_i = \frac{1}{2}D_{ik}R_k,$$

where D_{ik} is evaluated at P and does not enter in the integration over dm. As a consequence,

$$\int \nabla_R F = 0 \qquad \text{because} \qquad \int R_k \, dm = 0.$$

Hence, for a fluid element, the linear momentum equals the mass times the center-of-mass velocity,

$$\mathbf{p} = m\mathbf{u}(P),$$

just as it does for a collection of particles.

ANGULAR MOMENTUM OF A FLUID ELEMENT

The instantaneous angular momentum $\mathbf{J}$ of a fluid element with respect to a point P at its center of mass equals

$$\mathbf{J} \equiv \int [\mathbf{R} \times \mathbf{u}(Q)] \, dm. \tag{4.15}$$

Consider the principal axes frame of E_{ik} (or D_{ik}):

$$J_1' = \int [R_2' u_3'(Q) - R_3' u_2'(Q)] \, dm,$$

where

$$u_3'(Q) = u_3'(P) + (\Omega_1' R_2' - \Omega_2' R_1') + E_{33}' R_3',$$
$$u_2'(Q) = u_2'(P) + (\Omega_3' R_1' - \Omega_1' R_3') + E_{22}' R_2',$$

with $\boldsymbol{\Omega} \equiv \nabla \times \mathbf{u}/2$ and E_{ik} evaluated at P. Using $\int \mathbf{R}' \, dm = 0$, we obtain, after some algebra,

$$J_1' = I_{11}'\Omega_1' + I_{12}'\Omega_2' + I_{13}'\Omega_3' + I_{23}'(E_{22}' - E_{33}'),$$

where $I_{j\ell}'$ is the moment of inertia tensor:

$$I_{j\ell}' \equiv \int (|\mathbf{R}'|^2 \delta_{j\ell} - R_j' R_\ell') \, dm.$$

Notice that $I_{j\ell}'$ is not diagonal in the primed frame unless the principal axes of I_{ik} happen to coincide with those of E_{ik}.

The difference $E_{22}' - E_{33}'$ may also be written as $(D_{22}' - D_{33}')/2$, since the isotropic part of $E_{j\ell}'$ does not enter into the difference. In a similar fashion we may write

$$J_2' = I_{2\ell}'\Omega_\ell' + \frac{1}{2} I_{31}'(D_{33}' - D_{11}'),$$

$$J_3' = I_{3\ell}'\Omega_\ell' + \frac{1}{2} I_{12}'(D_{11}' - D_{22}'),$$

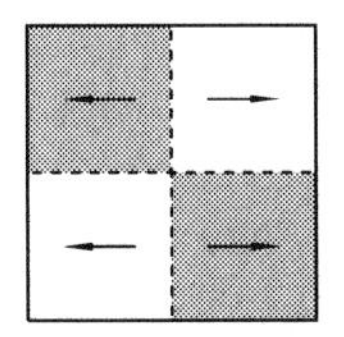

FIGURE 4.3
A fluid element with only extensional strain possesses no vorticity, yet it may still have spin angular momentum if the principal axes of the moment-of-inertia tensor do not align with the principal axes of the rate-of-strain tensor, e.g., if the shaded quadrants have more mass than the unshaded quadrants.

with a summation over the repeated ℓ's. For a solid body we would have gotten $J_j' = I_{j\ell}'\Omega_\ell'$. An extra contribution from the extensional strain arises if the principal axes of the moment-of-inertia tensor do not coincide with those of D_{ik} (or E_{ik}). Notice, in particular, that a fluid element can have angular momentum with respect to its center of mass without possessing spinning motion, i.e., even if $\boldsymbol{\Omega} = \nabla \times \mathbf{u}/2 = 0$. A simple example of this situation is depicted schematically in Figure 4.3 for a fluid element possessing only extensional strain.

The vorticity equals zero by construction, but the element will contain angular momentum if its mass distribution does not have the same symmetry as the strain pattern, e.g., if the shaded quadrants have more mass than the unshaded quadrants.

APPLICATION TO GALAXY FORMATION

A popular idea holds that gravitational torques associated with density inhomogeneities in the environment (see Figure 4.4) cause rotation in a protogalaxy which would otherwise condense from a purely expanding gaseous background (the Hubble flow of the universe).

In fact, we shall show in Chapter 6 that forces which arise from the gradient of a scalar potential can not impart vorticity (spin) to any element of fluid. Since the present-day rotation of our Galaxy (like that of other spiral galaxies) possesses nonzero vorticity, the process must involve more than just gravitational interactions. Gravitational torques *can* give angular momentum to an initially nonrotating cloud, but $\mathbf{J}$ must primarily take the form of distortional contributions like that described at the end of the previous section. The generation of vorticity in such a situation must rely on hydrodynamical processes, such as passage through curved shock fronts

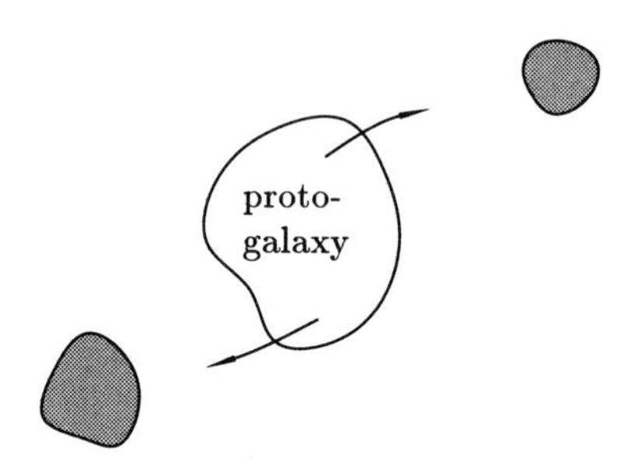

FIGURE 4.4
Gravitational accelerations from neighboring density inhomogenieties, being derivable from the gradient of a scalar potential, cannot generate vorticity in the gas of a protogalaxy, but they can exert torques and impart spin angular momentum of the type illustrated in Figure 4.3. The production of the vorticity seen in actual spiral galaxies must arise from later hydrodynamical action (e.g., the appearance of curved shocks).

during the collapse process. We shall return to this topic (see the discussion of Crocco's theorem in Chapter 6).

EULERIAN VERSUS LAGRANGIAN DESCRIPTION

The preceding discussion finishes our treatment of fluid kinematics; we wish now to rederive the basic equations of fluid dynamics from the point of view of continuum mechanics. Our basic approach follows an *Eulerian description*, in which we characterize the flow by the density ρ, velocity $\mathbf{u}$, pressure P, temperature T, etc. of the fluid as functions of time t when seen by observer at different fixed locations $\mathbf{x}$. This methodology can be contrasted with a *Lagrangian description*, in which we attempt to calculate the dynamics by following the motion of individual fluid elements. Thus a Lagrangian description follows the tradition of Newtonian mechanics, whereas, historically, the Eulerian description provided the model for a *field theory* that Maxwell pioneered in his development of classical electrodynamics.

Consider an arbitrary volume V bounded by a surface A, both of which are *fixed* in space. We express the change of any quantity in terms of volumetric contributions (e.g., external sources minus sinks) plus surface effects (e.g. flow of quantity past the surface A). Schematically, then, we have (see Figure 4.5)

$$\text{time-rate of change} = \text{volumetric contributions} + \text{surface effects}$$

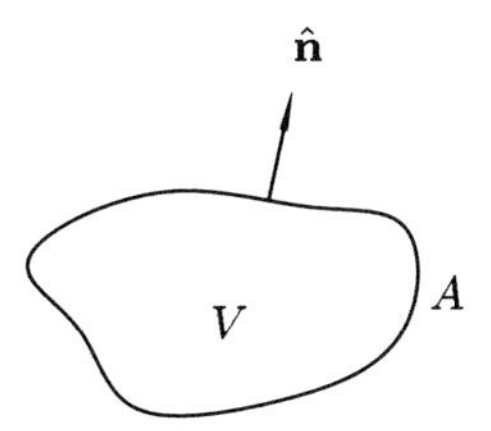

FIGURE 4.5
In an Eulerian description we consider the budget of fluid quantities as matter inside an arbitrary volume V undergoes explicit changes in time, and as matter inside and outside V flows across the fixed surface A with local unit normal $\hat{\mathbf{n}}$.

MASS CONSERVATION

In the absence of volumetric sinks and sources of matter, the time rate of change of the mass in volume V must equal minus the rate of mass flux $\rho\mathbf{u}$ past the element of area $\hat{\mathbf{n}}\,dA$ integrated over the entire area A:

$$\frac{d}{dt}\int_V \rho\,dV = -\oint_A \rho\mathbf{u}\cdot\hat{\mathbf{n}}\,dA = -\int_V \boldsymbol{\nabla}\cdot(\rho\mathbf{u})\,dV,$$

where we have made use of the divergence theorem to obtain the last expression. If we differentiate inside the integral on the left-hand side and transpose everything to one side, we may write

$$\int_V \left[\frac{\partial\rho}{\partial t} + \boldsymbol{\nabla}\cdot(\rho\mathbf{u})\right] dV = 0.$$

Since V is completely arbitrary, we require the integrand to vanish, thereby obtaining the equation of continuity:

$$\frac{\partial\rho}{\partial t} + \boldsymbol{\nabla}\cdot(\rho\mathbf{u}) = 0. \tag{4.16}$$

MOMENTUM CONSERVATION

The time rate of change of the fluid momentum contained in volume V equals minus the surface integral of the momentum flux due to fluid flow across A, plus the influence of internal stresses (forces per unit area) acting on the fluid across the surface A, plus the contribution of volumetric forces (such as gravity) acting on every point of V:

$$\frac{d}{dt}\int_V \rho u_i\,dV = -\oint_A (\rho u_i)u_k n_k\,dA + \oint_A P_{ik}n_k\,dA + \int_V \rho g_i\,dV. \tag{4.17}$$

In equation (4.17), P_{ik} is the force per unit area exerted by the outside on the inside in the i-th direction across a face whose normal is oriented in the k-th direction. For a dilute gas $P_{ik} = -\rho\langle w_i w_k\rangle$ in the kinetic-theory notation of Chapter 2. We have used g_i to denote the i-th component of the gravitational acceleration:

$$g_i = -\frac{\partial \mathcal{V}}{\partial x_i}. \tag{4.18}$$

If we apply the divergence theorem to equation (4.17), we obtain the momentum equation:

$$\frac{\partial}{\partial t}(\rho u_i) + \frac{\partial}{\partial x_k}(\rho u_i u_k) - \frac{\partial P_{ik}}{\partial x_k} = \rho g_i. \tag{4.19}$$

ENERGY CONSERVATION

The time rate of change of the total fluid energy (kinetic energy of fluid motion plus internal energy) equals minus the surface integral of the energy flux (kinetic plus internal), plus the surface integral of the rate of doing work by the internal stresses, plus the volume integral of the rate of doing work by the local body forces (e.g., gravitation), minus the heat loss by conduction across the surface A, plus volumetric gains minus volumetric losses of energy due to local sources and sinks (e.g., radiation):

$$\frac{d}{dt}\int_V \left(\frac{1}{2}\rho|\mathbf{u}|^2 + \rho\mathcal{E}\right) dV = -\oint_S \left[\left(\frac{1}{2}\rho|\mathbf{u}|^2 + \rho\mathcal{E}\right)\mathbf{u}\right]\cdot\hat{\mathbf{n}}\, dA$$

$$+\oint_A u_i P_{ik} n_k\, dA + \int_V \mathbf{u}\cdot\mathbf{g}\rho\, dV - \oint_A \mathbf{F}_{\text{cond}}\cdot\hat{\mathbf{n}}\, dA + \int_V (\Gamma - \Lambda)\, dV. \tag{4.20}$$

If we apply the divergence theorem, we obtain the total energy equation:

$$\frac{\partial}{\partial t}\left[\rho\left(\frac{1}{2}|\mathbf{u}|^2 + \mathcal{E}\right)\right] + \frac{\partial}{\partial x_k}\left[\rho\left(\frac{1}{2}|\mathbf{u}|^2 + \mathcal{E}\right)u_k - u_i P_{ik} + F_k\right] = \rho\mathbf{g}\cdot\mathbf{u} + \Gamma - \Lambda. \tag{4.21}$$

GRAVITATIONAL STRESSES

The fluid equations derived above have the standard expression,

$$\frac{\partial}{\partial t}(\text{density of quantity}) + \boldsymbol{\nabla}\cdot(\textbf{flux} \text{ of quantity}) = \text{sources minus sinks.}$$

What we choose to call "sources and sinks," however, depends on what we wish to consider as part of the system and what as part of the external world. Thus, if we choose to regard gravitation as an external force, it appears on the right-hand side of equation (4.19) as a volumetric source for changing the momentum content of the fluid. On the other hand, if $\mathbf{g}$ arises as the *self-gravity* of the system, then we can manipulate the equations to make its effects appear as the divergence of an internal surface stress. To do this, begin with Poisson's equation (2.17):

$$\nabla \cdot \mathbf{g} = -4\pi G\rho.$$

With the above we may write the term ρg_i that appears on the right-hand side of equation (4.19) as

$$\rho g_i = -\frac{1}{4\pi G}\frac{\partial g_k}{\partial x_k} g_i = -\frac{1}{4\pi G}\left[\frac{\partial}{\partial x_k}(g_k g_i) - g_k \frac{\partial g_i}{\partial x_k}\right].$$

But equation (4.18) implies that

$$\frac{\partial g_i}{\partial x_k} = -\frac{\partial^2 \mathcal{V}}{\partial x_k x_i} = \frac{\partial g_k}{\partial x_i},$$

which allows us to write

$$g_k \frac{\partial g_i}{\partial x_k} = g_k \frac{\partial g_k}{\partial x_i} = \frac{1}{2}\frac{\partial}{\partial x_k}(g_m g_m \delta_{ik}).$$

Collecting expressions, we find that the self-gravitational force per unit volume can be expressed as the divergence of a stress tensor,

$$\rho g_i = \frac{\partial G_{ik}}{\partial x_k},$$

where G_{ik} is the gravitational analog of Maxwell's stress tensor in electrostatics, with $-4\pi G$ replacing 4π (in cgs units) and the gravitational field g_i replacing the electric field E_i,

$$G_{ik} = -\frac{1}{4\pi G}\left(g_i g_k - \frac{1}{2}|\mathbf{g}|^2 \delta_{ik}\right). \tag{4.22}$$

If we move the term $\partial G_{ik}/\partial x_k$ to the left-hand side of equation (4.19), we get the momentum equation in its pure conservation form, i.e., without any "sources or sinks" on the right-hand side. (Q.E.D.)

Notice also that if we use the potential $\mathcal{V}$ to characterize the (Newtonian) gravitational field, we cannot localize its contribution to the total *energy* budget. The first term on the right-hand side of equation (4.21) then becomes

$$-\rho\mathbf{u} \cdot \nabla\mathcal{V} = -\nabla \cdot (\rho\mathbf{u}\mathcal{V}) - \frac{\partial}{\partial t}(\rho\mathcal{V}) + \rho\frac{\partial \mathcal{V}}{\partial t}, \tag{4.23}$$

if we make use of the equation of continuity. The first and second terms on the right-hand side of (4.23) can be absorbed into the second and first terms of the left-hand side of equation (4.21), respectively, but the third term in equation (4.23) can be expressed neither as the explicit time derivative of the density of a quantity nor as the divergence of the flux of a quantity. Only when we integrate over all space does the expression give the time derivative of the total gravitational potential energy,

$$\mathcal{W} \equiv \frac{1}{2} \int \rho \mathcal{V} \, d^3x,$$

with the factor 1/2 necessary to avoid counting pairs of fluid elements twice for a self-gravitating configuration. Indeed, because

$$\mathcal{V}(\mathbf{x}, t) = -G \int \frac{\rho(\mathbf{x}', t) \, d^3x'}{|\mathbf{x} - \mathbf{x}'|}$$

for a self-gravitating density distribution $\rho(\mathbf{x}, t)$, we have

$$\int \rho \frac{\partial \mathcal{V}}{\partial t} \, d^3x = \int \frac{\partial \rho}{\partial t} \mathcal{V} \, d^3x = \frac{1}{2} \int \frac{\partial}{\partial t} (\rho \mathcal{V}) \, d^3x,$$

and it is only the combination of the second and third terms on the right-hand side of equation (4.23) that gives $-d\mathcal{W}/dt$ with the correct sign for total energy conservation. In other words, the long-range nature of Newtonian gravity makes it somewhat of a sham to pretend that the effects of gravitation can be localized.

RADIATIVE FORCE AND WORK

In some circumstances, we can not ignore $\mathbf{f}_{\text{rad}}$, the force per unit volume exerted by radiation on matter,

$$\mathbf{f}_{\text{rad}} = \frac{\rho}{c} \int_0^\infty \kappa_\nu \mathbf{F}_\nu \, d\nu, \tag{4.24}$$

where κ_ν is the radiative opacity (absorption plus scattering) and $\mathbf{F}_\nu$ is the monochromatic radiative energy flux (see Chapter 2 of Volume I). In this case, we need to add

$$\mathbf{f}_{\text{rad}}$$

to the right-hand side of equation (4.19), and we also need to include the rate of doing work on the matter by this force,

$$\mathbf{u} \cdot \mathbf{f}_{\text{rad}},$$

on the right-hand side of equation (4.21). If we do make these modifications, we need to carefully distinguish between the contributions to the radiative sources or sinks of heat that we include in $\Gamma - \Lambda$, and the radiative rate of work $\mathbf{u} \cdot \mathbf{f}_{\text{rad}}$ that arises because of the systematic transfer of momentum from photons to matter.

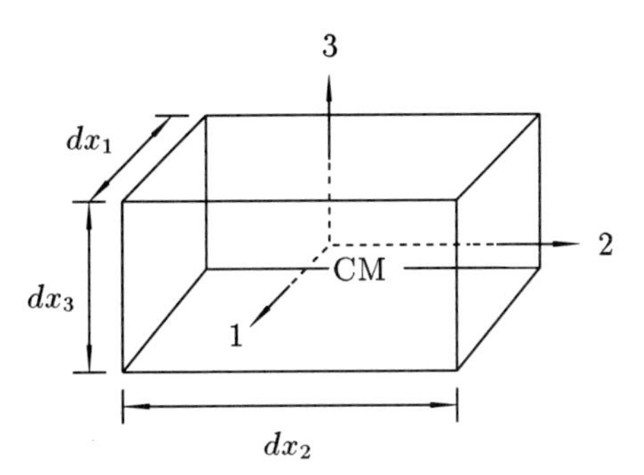

FIGURE 4.6
A paralleliped fluid element of dimensions dx_1, dx_2, and dx_3 with a center of mass at CM. The front and back faces with area $dx_2\, dx_3$ have unit normals that point in the ± 1 directions, etc.

SYMMETRY OF STRESS TENSOR

Equations (4.16), (4.19), (4.21) do not form a closed set until we specify constitutive relations that allow us to calculate P_{ik}, $\mathbf{F}_{\text{cond}}$, etc. We start with P_{ik}, which we assert without proof transforms as a rank-2 tensor under rotations of the coordinate axes (i.e., transforms as the direct product of two vectors; see any textbook on fluid mechanics). We now wish to show that P_{ik} must be a symmetric tensor, independent of any of the details of the properties of the fluid (e.g., whether it is a gas or a liquid). To see this, consider a paralleliped fluid element (Figure 4.6).

Since the volume torques and inertial terms depend on dx to one higher power than the surface torques, the latter must sum to zero; otherwise the surface torques would impart infinite angular momentum to the fluid element in the limit $dx \to 0$. (The surface torques do not cancel to next order, but their differences together with the volume torques combine to give the fluid element a finite time rate change of angular momentum.) Consider now the contribution from the 1-forces to the 3-torques. The average lever arm equals $dx_1/2$, whereas, if we ignore the difference in the point of application, the 2-force acting on the $+1$ face equals $P_{21}dx_2dx_3$ and the 2-force acting on the -1 face equals $-P_{21}dx_2dx_3$. Consequently, the 3-torque equals

$$\frac{1}{2}dx_1(P_{21}dx_2dx_3) - \frac{1}{2}dx_1(-P_{21}dx_2dx_3) = P_{21}dx_1dx_2dx_3.$$

In a similar manner, the contribution from the 2-forces to the 3-torques reads

$$-\frac{1}{2}(P_{12}dx_3dx_1) + \frac{1}{2}dx_2(-P_{12}dx_3dx_1) = -P_{12}dx_1dx_2dx_3.$$

The total surface 3-torques equal

$$(P_{21} - P_{12})dx_1dx_2dx_3,$$

which vanish (to the order of approximation that we are considering) if

$$P_{21} = P_{12}.$$

In a similar manner, we may deduce for each i and k:

$$P_{ki} = P_{ik}. \tag{4.25}$$

PRESSURE AND VISCOUS STRESS TENSOR

The second step in specifying the symmetric tensor P_{ik} is to separate out the scalar pressure P:

$$P_{ik} = -P\delta_{ik} + \pi_{ik}. \tag{4.26}$$

The terms on the right-hand side of equation (4.26) contain a sign difference. In the usual definitions, this difference arises because the pressure P represents the normal force per unit area exerted by a fluid element on its surroundings, whereas the viscous stress tensor π_{ik} represents the force per unit area exerted by the surroundings on the fluid element (in the direction i on the face whose normal lies along k).

The real trick comes in identifying P, not as $-1/3$ times the trace of P_{ik}, but as the *thermodynamic pressure*, related to the internal energy per unit mass of the fluid, $\mathcal{E}$, and the specific entropy, s, by the fundamental law of thermodynamics:

$$d\mathcal{E} = T\,ds - P\,d(\rho^{-1}). \tag{4.27}$$

It is implicit to our fluid formulation, then, that the concept of *local thermodynamic equilibrium* (LTE) can be applied to at least the translational degrees of freedom. (See Chapter 9 of Volume I for a more detailed exposition of this point.) If the assumption of LTE for the translational degrees of freedom does not hold, we would generally need to abandon a fluid approach in favor of a kinetic description (see, e.g., Chapters 28–30 on the methods of plasma physics).

Even after adopting a fluid approach, we still need to specify $\pi_{ik} = \pi_{ki}$. In the general case, and unlike the situation in Chapter 3, π_{ii} need not equal

0, because the negative trace of $P_{ik}/3$ need not equal the thermodynamic pressure. Nevertheless, we may confine our attention to *Newtonian* fluids where the viscous stress π_{ik} is linearly proportional to the rate of strain $\partial u_i/\partial x_k$ ("Hooke's law"). For an isotropic fluid, the most general form for such a relationship reads

$$\pi_{ik} = \mu D_{ik} + \beta(\nabla \cdot \mathbf{u})\delta_{ik}, \tag{4.28}$$

where D_{ik} is the deformation-rate tensor given by equation (3.20), and μ and β are respectively called the *shear* and *bulk* coefficients of viscosity. On a microscopic level, the latter arises because a nonideal gas has internal degrees of freedom that can suffer excitation or damping through changes in the volume of the fluid. The resultant coupling to the translational degrees of freedom can give rise to an effective frictional force that opposes such changes of volume.

CONDUCTIVE FLUX AND HEAT EQUATION

For the conductive flux, we invoke Fourier's law,

$$\mathbf{F}_{\text{cond}} = -\mathcal{K}\nabla T, \tag{4.29}$$

where $\mathcal{K}$ is the coefficient of thermal conductivity. Such a form arises as a general result from the considerations of *irreversible thermodynamics*, as long as the departures from LTE are small (see Callen, *Thermodynamics*). For a very hot and diffuse gas, the mean free path for thermal conduction can become longer than the macroscopic length scale of interest. In this case, equation (4.29) may lose validity, and the conductive flux may *saturate* at a value comparable to the enthalpy per unit volume, $\rho h \equiv P + \rho\mathcal{E}$, times the thermal velocity, $v_T = (kT/m)^{1/2}$, of the most mobile component (usually free electrons). The evaporation of clouds by the (possibly saturated) conduction of heat from a hot intercloud medium constitutes a problem of astrophysical interest to theories of the interstellar medium and cooling flows in galaxy clusters. (See the astronomical literature for more detailed discussions of this topic.)

If we apply equation (4.27), we may subtract the work equation from the total energy equation in the manner of Chapter 2 and obtain the heat equation for the fluid,

$$\rho T \frac{Ds}{Dt} = -\nabla \cdot \mathbf{F}_{\text{cond}} + \Psi + \Gamma - \Lambda, \tag{4.30}$$

where Ψ equals the (positive definite) rate of viscous dissipation:

$$\Psi = \pi_{ik}\frac{\partial u_i}{\partial x_k} = \frac{\mu}{2}||D_{ik}||^2 + \beta(\nabla \cdot \mathbf{u})^2. \tag{4.31}$$

Notice that the rate of doing work by the radiative force, $\mathbf{u} \cdot \mathbf{f}_{\text{rad}}$, has canceled out in the heat equation (4.30).

RADIATIVE EFFECTS

Including radiative effects in a relativistically correct treatment is very complicated. (For a detailed exposition, consult Mihalas and Mihalas, *Foundations of Radiation Hydrodynamics.*) A heuristic nonrelativistic approach suffices for many purposes. In the presence of fluid motions for which $|\mathbf{u}| \ll c$, the frequency-integrated zeroth angular moment for the equation of radiative transfer reads

$$\frac{\partial E_{\rm rad}}{\partial t} + \nabla \cdot [\mathbf{F}_{\rm rad} + (E_{\rm rad} + P_{\rm rad})\mathbf{u}] = -\mathbf{f}_{\rm rad} \cdot \mathbf{u} - \Gamma + \Lambda, \qquad (4.32)$$

where we have applied the principle that what acts as a source of heat for the matter is a sink for the radiation field, and vice versa. In equation (4.32), $E_{\rm rad}$ and $P_{\rm rad}$ are the energy density and pressure of the radiation field, and we have distinguished between the radiative enthalpy flux $(E_{\rm rad} + P_{\rm rad})\mathbf{u}$ carried by the matter motion and the energy flux $\mathbf{F}_{\rm rad}$ due to the diffusion of photons relative to the matter (i.e., the radiative energy flux in the local frame of rest of the fluid):

$$\mathbf{F}_{\rm rad} = \int_0^\infty \mathbf{F}_\nu \, d\nu. \qquad (4.33)$$

To the same order of approximation, the frequency-integrated first angular moment for the equation of radiative transfer reads

$$-\nabla P_{\rm rad} = \mathbf{f}_{\rm rad}, \qquad (4.34)$$

where $\mathbf{f}_{\rm rad}$ is given by equation (4.24). (In general, we cannot reduce the radiative pressure *tensor* to a scalar pressure except when the medium is *optically thick*; however, for most hydrodynamic applications, a full treatment of radiative transfer effects is often impractical.) In any case, if we use equation (4.34), we may rewrite equation (4.32) as

$$\frac{\partial E_{\rm rad}}{\partial t} + \nabla \cdot [\mathbf{F}_{\rm rad} + E_{\rm rad}\mathbf{u}] = -P_{\rm rad}\nabla \cdot \mathbf{u} - \Gamma + \Lambda. \qquad (4.35)$$

Adding the internal energy equation of the matter to equation (4.35) and using the equation of continuity, we get the total internal energy equation for matter and radiation:

$$\rho \frac{D}{Dt}\left(\frac{E_{\rm rad}}{\rho} + \mathcal{E}\right) = -(P + P_{\rm rad})\nabla \cdot \mathbf{u} - \nabla \cdot (\mathbf{F}_{\rm cond} + \mathbf{F}_{\rm rad}) + \Psi + \rho\epsilon. \qquad (4.36)$$

To the right-hand side of the above we have added a possible volumetric source of nuclear energy, $\rho\epsilon$, which contributes to the combined increase of internal energy in both radiation and matter.

OPTICALLY THICK CONDITIONS

Under optically thick conditions, we have good thermal coupling to the matter and the following closure relations for the radiation field,

$$P_{\text{rad}} = \frac{1}{3}E_{\text{rad}} = \frac{1}{3}aT^4, \tag{4.37}$$

where a is the radiation constant and T is the matter temperature. In this situation, the *radiation conduction approximation* gives for the diffusive flux (see Chapter 2 of Volume I)

$$\mathbf{F}_{\text{rad}} = -\frac{c}{3\rho\kappa_R}\boldsymbol{\nabla}(aT^4), \tag{4.38}$$

where κ_R is the *Rosseland mean opacity.* Moreover, we may introduce an entropy per unit mass associated with the radiation field,

$$s_{\text{rad}} \equiv \frac{4aT^3}{3\rho}, \tag{4.39}$$

which satisfies the thermodynamic relation,

$$d(E_{\text{rad}}/\rho) = T ds_{\text{rad}} - P_{\text{rad}} d(\rho^{-1}).$$

The substitution of the above relations into the total internal energy equation (4.36) yields the *total heat equation,*

$$\rho T \frac{Ds_{\text{tot}}}{Dt} = -\boldsymbol{\nabla} \cdot \mathbf{F}_{\text{tot}} + \Psi + \rho\epsilon, \tag{4.40}$$

where $s_{\text{tot}} = s + s_{\text{rad}}$ and $\mathbf{F}_{\text{tot}}$ is the total diffusive flux,

$$\mathbf{F}_{\text{tot}} = \mathbf{F}_{\text{cond}} + \mathbf{F}_{\text{rad}}.$$

Equation (4.40) has useful applications for theories of stellar structure as well as the early universe.

SUMMARY

The preceding discussion completes our formal presentation of the basic equations that govern the behavior of single-component fluids, including the effects of radiation fields. Beginning with Chapter 5 we shall consider some astrophysical applications.

5

Equilibria of Self-Gravitating Spherical Masses

Reference: Shapiro and Teukolsky, *Black Holes, White Dwarfs, and Neutron Stars.*

With this chapter we begin our consideration of applications of the single-component fluid equations. We start with the simplest case: systems in which motion is absent altogether, or at least has no dynamic effects. States of *hydrostatic equilibria* in the presence of self-gravitation include the classic objects of astronomy, planets and stars, and thus form a venerable departure point for our study of astrophysical fluid mechanics.

ASSUMPTION OF SPHERICAL SYMMETRY

Our analysis begins with a justification of the neglect of the dynamical effects of fluid motion in the theory of the interiors of planets and stars. We assume spherical symmetry, so that any fluid velocity present has the component decomposition in spherical polar coordinates:

$$\mathbf{u}(\mathbf{x}, t) = [u(r, t), 0, 0]. \tag{5.1}$$

In making this assumption, we give up any hope of treating convection, which necessarily involves nonspherical *overturning* motions, on an *a priori* basis. Such motions, which constitute an important heat-transport mechanism in many models are generally treated semi-empirically via *mixing length* theory in current discussions. We shall defer exposition of this method for handling the convective transport of heat to Chapter 10; in what follows, we allow formally for the presence of a nonzero convective flux F_{conv} in addition to radiative diffusion F_{rad} and heat conduction F_{cond} (most significantly, by degenerate electrons).

If we assume spherical symmetry and ignore the effects of viscosity, the basic equations read

$$\frac{\partial \rho}{\partial t} + \frac{1}{r^2}\frac{\partial}{\partial r}(r^2 \rho u) = 0, \tag{5.2}$$

$$\frac{Du}{Dt} = -\frac{1}{\rho}\frac{\partial P_{\text{tot}}}{\partial r} + g, \tag{5.3}$$

$$T\frac{Ds_{\text{tot}}}{Dt} = -\frac{1}{\rho r^2}\frac{\partial}{\partial r}[r^2(F_{\text{rad}} + F_{\text{cond}} + F_{\text{conv}})] + \epsilon, \tag{5.4}$$

$$F_{\text{rad}} = -\frac{c}{3\kappa_{\text{R}}\rho}\frac{\partial}{\partial r}(aT^4), \tag{5.5}$$

$$F_{\text{cond}} = -\mathcal{K}\frac{\partial T}{\partial r}, \tag{5.6}$$

$$\frac{1}{r^2}\frac{\partial}{\partial r}(r^2 g) = -4\pi G\rho. \tag{5.7}$$

In the above equations, $D/Dt = \partial/\partial t + u\partial/\partial r$, and P_{tot} and s_{tot} refer to the sum of radiation pressures and specific entropies (for which we yet need constitutive relations, giving their dependences on density, temperature, and chemical composition). The sum of F_{cond} and F_{rad} is often written as a single term equal to radiative diffusion with an effective opacity κ_{eff} defined by the occurrence of radiation and matter conduction *in parallel*:

$$\frac{1}{\kappa_{\text{eff}}} = \frac{1}{\kappa_{\text{R}}} + \left(\frac{3\rho}{4caT^3}\right)\mathcal{K}, \tag{5.8}$$

where κ_{R} is the Rosseland-mean radiative opacity and $\mathcal{K}$ is the conductivity due to matter (e.g., degenerate electrons). The form of equation (5.8) looks a little odd only because radiative opacity is defined as a *resistance* to flow, whereas thermal conductivity is defined as a *transmittance*.

TRANSFORMATION TO A LAGRANGIAN DESCRIPTION

The basic equations (5.2)–(5.7) are written above in an Eulerian formulation; for reasons to become clear below, evolutionary calculations are best made in a Lagrangian formulation. To transform to a Lagrangian description, we first introduce the concept of the mass $M(r,t)$ contained inside a sphere of radius r at time t:

$$M(r,t) \equiv \int_0^r \rho(r',t)\, 4\pi r'^2 dr'. \tag{5.9}$$

For what follows, we find it advantageous to write the above in the differential form,

$$\frac{\partial M}{\partial r} = 4\pi r^2 \rho, \qquad \text{with} \qquad M = 0 \qquad \text{at} \qquad r = 0. \tag{5.10}$$

Poisson's equation for the gravitational field g can now be integrated:

$$r^2 g = -GM \qquad \Rightarrow \qquad g = -\frac{GM}{r^2}, \tag{5.11}$$

which represents Newton's theorem that the gravitational field of a spherical mass distribution can be calculated by considering only the mass interior to r, and as if that interior mass were all concentrated at a single point at the origin. In terms of $M(r,t)$, the equation of continuity (5.2) can be written as

$$\frac{1}{4\pi r^2}\frac{\partial}{\partial r}\left(\frac{\partial M}{\partial t} + u\frac{\partial M}{\partial r}\right) = 0.$$

We integrate the above over r and obtain the conclusion that DM/Dt can, at best, equal a universal arbitrary function of t at all r. If we do not allow matter to appear or disappear at the origin, we require

$$\frac{DM}{Dt} = 0; \tag{5.12}$$

i.e., the mass interior to us does not change if we follow the motion of a spherical shell (e.g., as it slowly expands or contracts on an evolutionary time scale).

The defining property of a Lagrangian description, equation (5.12), suggests that it would be convenient for us to transform variables from (r,t) to (M,t). In this transformation,

$$\frac{D}{Dt} \equiv \frac{\partial}{\partial t} + u\frac{\partial}{\partial r} \to \left(\frac{\partial}{\partial t}\right)_M.$$

Thus,

$$\frac{Dr}{Dt} \equiv \left(\frac{\partial r}{\partial t}\right)_r + u\left(\frac{\partial r}{\partial r}\right)_t = 0 + u = \left(\frac{\partial r}{\partial t}\right)_M.$$

Furthermore,

$$\left(\frac{\partial}{\partial r}\right)_t = \left(\frac{\partial M}{\partial r}\right)_t\left(\frac{\partial}{\partial M}\right)_t = 4\pi r^2\rho\left(\frac{\partial}{\partial M}\right)_t.$$

With the above formulae, equations (5.2)–(5.6) become

$$\left(\frac{\partial r}{\partial M}\right)_t = \frac{1}{4\pi r^2\rho}, \tag{5.13}$$

$$\left(\frac{\partial^2 r}{\partial t^2}\right)_M = -4\pi r^2\left(\frac{\partial P_{\text{tot}}}{\partial M}\right)_t - \frac{GM}{r^2}, \tag{5.14}$$

$$T\left(\frac{\partial s_{\text{tot}}}{\partial t}\right)_M = -\left(\frac{\partial L}{\partial M}\right)_t + \epsilon, \tag{5.15}$$

$$L \equiv 4\pi r^2(F_{\rm rad} + F_{\rm cond} + F_{\rm conv}) = -\frac{64\pi^2 r^4 caT^3}{3\kappa_{\rm eff}}\left(\frac{\partial T}{\partial M}\right)_t + 4\pi r^2 F_{\rm conv}. \tag{5.16}$$

For use in equation (5.16), we need a (mixing-length) prescription for computing $F_{\rm conv}$ (see Chapter 10) given the other variables of the problem (in particular, the gradient of the specific entropy). Finally, to close the set of equations we require constitutive relations that give the pressure $P_{\rm tot}$, specific entropy $s_{\rm tot}$, and nuclear energy generation rate per unit mass ϵ, in terms of ρ, T, and the chemical composition of the material medium. We also need evolutionary equations for the latter, which require rate equations for various processes that can change the relative abundances of different species (ionization, nuclear transformation, etc.).

BOUNDARY CONDITIONS

The presence of four derivatives in M demands the imposition of four spatial boundary conditions. Two apply at the center of the star (or planet):

$$r = 0 \qquad \text{and} \qquad L = 0 \qquad \text{at} \qquad M = 0; \tag{5.17}$$

and two at the surface:

$$P_{\rm tot} = \frac{1}{3}aT^4 \qquad \text{and} \qquad L = 4\pi r^2 \sigma T^4 \qquad \text{at} \qquad M = M_*, \tag{5.18}$$

where M_* is the total mass of the object.

The constraint $P_{\rm tot} = P_{\rm rad}$ guarantees that the matter density ρ has dropped to zero at $M = M_*$, a practical definition of what we mean by the "surface." The constraint $L = 4\pi r^2 \sigma T^4$ at $M = M_*$ reflects the preconceived notion that eventually all the energy emergent from the interior must be radiated in photons from the surface effectively as if the latter were a blackbody at the surface temperature T. A more sophisticated analysis—see Chapter 4 of Volume I—demonstrates that the effective temperature $T_{\rm eff}$ of the equivalent blackbody is actually reached at a Rosseland-mean optical depth equal to 2/3 beneath the surface of the star; thus, more accurate surface boundary conditions are often imposed by attaching interior integrations to atmospheric models. This refinement becomes especially important to carry out if the subphotospheric layers prove to be convective. Experience shows that failure to apply precise outer boundary conditions introduces little error for the interior structure of stars with radiative outer envelopes. In such cases, the two surface boundary conditions given in equation (5.18) can even be replaced by the simpler relations, $P_{\rm tot} = 0$ and $T = 0$ at $M = M_*$, referred to in the literature as the *radiative zero boundary conditions.* The physical basis for this replacement lies with the fact that realistic surface pressures and temperatures (what matters actually is

T^4) are so much lower than interior values that we might as well set them equal to zero. The simplification fails for convective envelopes, because their structure depends sensitively on the value of the specific entropy reached in the deeper layers, and misleading results for the latter would be obtained if we assign the surface value too cavalierly. (See specialized textbooks for further discussion of this point.)

THE ASSUMPTION OF QUASIHYDROSTATIC EQUILIBRIUM

So far, except for the neglect of viscous effects (needed here, if at all, only for the mediation of shocks), our discussion has made few assumptions. In particular, equation (5.14) allows the possibility of dynamical (radial) motions. If we were concerned with rapid phases of stellar evolution (e.g., core collapse in a supernova) or with stellar pulsations (e.g., in a Cepheid variable), we would need to retain the acceleration term $(\partial^2 r/\partial t^2)_M$. For all other phases of stellar evolution, we may ignore this term in comparison with either of the two terms on the right-hand side of equation (5.14). In particular, the ratio of the gravitational term to the inertial term has the typical order of magnitude

$$(G\bar{\rho})t_{\text{evol}}^2, \tag{5.19}$$

where $\bar{\rho} \sim M_*/R^3$ is the mean density of the configuration and t_{evol} is its characteristic evolutionary time scale. We recognize the combination $(G\bar{\rho})^{-1/2}$ as the free-fall timescale t_{ff} if the self-gravity of the object were unopposed by any other forces; therefore, the expression (5.19) corresponds to $(t_{\text{evol}}/t_{\text{ff}})^2$, which much exceeds unity for most phases of stellar evolution. For example, the evolutionary time scale of the Sun equals $\sim 10^{10}$ yr, whereas $t_{\text{ff}} \sim 1$ hr. In such cases, we may very safely ignore the inertial term $(\partial^2 r/\partial t^2)_M$, implying for equation (5.14) the condition of hydrostatic equilibrium (in Lagrangian coordinates):

$$\left(\frac{\partial P_{\text{tot}}}{\partial M}\right)_M = -\frac{GM}{4\pi r^4}. \tag{5.20}$$

We emphasize that the condition of hydrostatic equilibrium does *not* imply that the star has no changes of radius. Temporal changes [on thermal (Kelvin-Helmholtz) or nuclear time scales] can still occur through the term $(\partial s/\partial t)_M$ in equation (5.15)—or through the time derivatives that enter in the rate equations governing compositional changes, and these changes will generally lead (implicitly) to variations of the Lagrangian radius r at each M as the star evolves. *Quasihydrostatic equilibrium* (the approximation that drops the time derivatives in the equation of motion, but retains them everywhere else), then, refers only to the assumption that such slow changes have negligible dynamical effects, not that they are absent altogether.

THE MECHANICAL EQUILIBRIA OF COLD SELF-GRAVITATING SPHERES

For an interesting subclass of astronomical objects, the interior temperatures are so low that we may effectively take T to be zero for any practical considerations of the mechanical equilibrium. Examples of such objects include planets, white dwarfs, and neutron stars. The temperatures inside these objects do not actually equal absolute zero, but the problem of their thermal structure—as governed by equations (5.15) and (5.16), generalized to include the possibility of convection—can be decoupled from the problem of their mechanical structure—as governed by equations (5.13) and (5.20). The decoupling occurs when we ignore T for the latter problem because P can then depend only on one remaining thermodynamic variable, which we take to be ρ:

$$P = P(\rho). \tag{5.21}$$

Fluids that satisfy equation (5.21) are said to have a *barotropic* equation of state.

An important example of such a barotropic relationship involves a plasma whose internal pressure is dominated by electron degeneracy effects. In Problem Set 2 you are asked to show that the relationship between the pressure and the number density in a completely degenerate electron gas has the parametric form:

$$n_{\rm e} = (2s+1)\frac{4\pi}{3}\left(\frac{m_{\rm e}c}{h}\right)^3 x^3, \qquad P_{\rm e} = \frac{1}{6}(2s+1)\pi\left(\frac{m_{\rm e}c}{h}\right)^3 m_{\rm e}c^2 f(x), \tag{5.22}$$

where $f(x)$ is the function,

$$f(x) \equiv x(x^2+1)^{1/2}(2x^2-3) + 3\,\text{arcsinh}\, x. \tag{5.23}$$

In the above equations, $s = 1/2$ represents the quantum number associated with electron spin, and $h/m_{\rm e}c$, where h is Planck's constant, represents the characteristic spatial separation (as dictated by the uncertainty principle) between two electrons characterized by moderately relativistic degeneracy.

For $x \ll 1$, $f(x) \approx 8x^5/5$, implying, when we eliminate the parameter x, the well-known relationship for a nonrelativistic completely degenerate electron gas,

$$P_{\rm e} = \left(\frac{8\pi h^2}{15m_{\rm e}}\right)\left(\frac{3n_{\rm e}}{8\pi}\right)^{5/3} \propto \rho^{5/3}, \tag{5.24}$$

if we assume a completely ionized medium with a definite chemical composition. For $x \gg 1$, $f(x) \approx 2x^4$, implying, when we eliminate the parameter x, the equally well-known relationship for an ultrarelativistic completely degenerate electron gas,

$$P_{\rm e} = \left(\frac{2\pi hc}{3}\right)\left(\frac{3n_{\rm e}}{8\pi}\right)^{4/3} \propto \rho^{4/3}. \tag{5.25}$$

The slower increase of pressure P_e with increasing density ρ reflects the fact that the electron speeds have saturated at c, while their individual momenta p continue to increase on average, because of the larger Δp that must be maintained when the mean interelectron spacing decreases.

In Problem Set 2, you are asked to integrate the set of structure equations (since t does not enter explicitly below, we may use total derivatives with respect to the variation with M):

$$\frac{dr}{dM} = \frac{1}{4\pi r^2 \rho}, \tag{5.26}$$

$$\frac{dP}{dM} = -\frac{GM}{4\pi r^4}, \tag{5.27}$$

subject to the two-point boundary conditions:

$$r = 0 \quad \text{at} \quad M = 0, \qquad \text{and} \qquad P = 0 \quad \text{at} \quad M = M_*,$$

when $P = P(\rho)$ has the form appropriate for the (electron degenerate) matter inside a white dwarf. Although we do not need to worry about the problem here, special numerical methods have been developed to attack the difficulties introduced by the boundary conditions being posed at opposite ends of the integration range. The first—the *shooting method*—is more direct and more accurate, but also more difficult to converge to the correct solution. The second—the *Henyey technique*—incorporates the two-point nature of the boundary conditions explicitly at the outset in its formulation, but the method requires a reasonable guess as the initial iterate for successive relaxations to the correct answer. Thus, the two methods are quite complementary (see Chapter 13 for a brief discussion), and both are utilized in full studies of stellar evolution theory.

MASS-RADIUS RELATIONSHIP

To see the gist of the physical results without the benefit of the full integrations, let us examine the scaling relationships implied by equations (5.26) and (5.27). We want to see how the radius R of the star (where $M = M_*$) scales with M_*. Equation (5.26) shows that $\rho \propto M/R^3$; hence $P \propto M^{5/3}/R^5$ if $P \propto \rho^{5/3}$. In the nonrelativistic regime, then, equation (5.27) implies

$$M^{2/3}/R^5 \propto GM/R^4 \qquad \Rightarrow \qquad R \propto M^{-1/3}.$$

R. H. Fowler gave the first demonstration that nonrelativistically degenerate white dwarfs have radii that decrease as $M_*^{-1/3}$ as the mass of the star M_* increases. The increasing self-gravitation of the star compresses

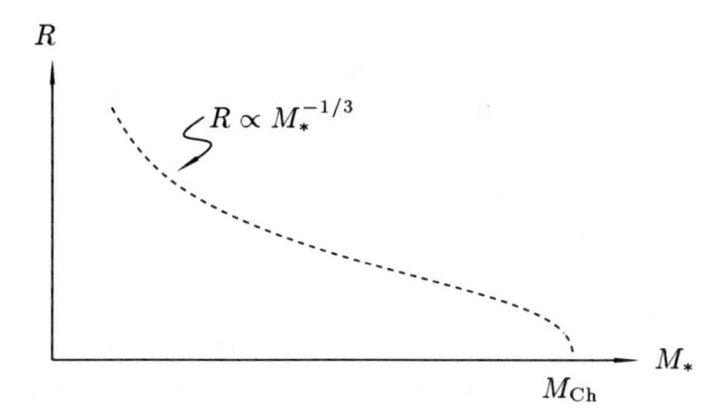

FIGURE 5.1
The mass-radius relationship for cold self-gravitating objects supported by electron-degeneracy pressure.

the matter to higher internal densities until the rising electron degeneracy pressure can find a balance at a decreased size for the star.

Chandrasekhar found that the balance would acquire a different character when the electrons in the white dwarf compress to densities high enough to acquire ultrarelativistic degeneracy. In this case, $P \propto \rho^{4/3} \propto M^{4/3}/R^4$, and equation (5.27) scales as

$$M^{1/3}/R^4 \propto GM/R^4 \qquad \Rightarrow \qquad M \propto \text{constant}.$$

The fact that R cancels out in the scaling relationship has the following important physical interpretation. When the electrons inside a white dwarf become ultrarelativistic, the balance between degeneracy pressure and self-gravity no longer has any sensitivity to radius R; more precisely, the two forces scale as the same power of R, so that imbalance at one radius means imbalance at all other radii. Hence, if M_* represents a mass in which the available pressure in degeneracy barely equals the required pressure for hydrostatic equilibrium, an increase in the mass of the white dwarf will result in an imbalance that the star attempts to cure by lowering its radius. Decreasing R for the ultrarelativistic regime, however, increases the self-gravitation in the same proportion as the degeneracy pressure; thus the star remains unbalanced and must contract further. The critical value of M_* for which the equilibrium value of R shrinks to zero is called the *Chandrasekhar limit*, $M_{\rm Ch}$; you will find its numerical value in Problem Set 2.

GIANT PLANETS AND NEUTRON STARS

To summarize our deliberations, we plot a schematic curve of R versus M_* for the Fowler-Chandrasekhar theory of white dwarfs (see Figure 5.1). For a mass $M_* \sim 0.5\ M_\odot$, a white dwarf has a radius $\sim 10^9$ cm.

With this plot in mind, we ask the natural question: Would the radius of a cold self-gravitating spherical mass really (a) increase without limit as $M_*^{-1/3}$ as $M_* \to 0$? (b) decrease to zero as $M_* \to M_{\rm Ch}$?

With regard to the first issue, experience in the solar system informs us that bodies of relatively small mass usually halve their volumes when their masses are halved. On the other hand, low-mass white dwarfs *double* their volumes when their masses are halved. At what point does the cross-over in behavior occur? To answer this question, perform the following thought experiment. Imagine gradually lowering the mass of a white dwarf. As the white dwarf decompresses, the average electron degeneracy energy (the so-called "Fermi energy") gets reduced, dropping eventually to tens of electron volts. At $\sim$ 10 eV, electrons tend to attach themselves to atoms rather than to remain free. The "degeneracy pressure," however, does not abruptly vanish because the electronic shells of neutral atoms still can not be pushed on top of one another. This happens for a mass $M_* \sim 10^{-3}\ M_\odot$, at which point the white-dwarf mass-radius relation predicts a radius $R \sim 10^{-1}\ R_\odot$, implying a mean density of $\sim 1\,{\rm g\,cm^{-3}}$. Liquids and solids composed of neutral atoms and molecules have this sort of density; indeed, astronomers believe Jupiter to have essentially a liquid hydrogen interior, and it has a mass $\sim 10^{-3}\ M_\odot$ and a radius $\sim 10^{-1}\ R_\odot$. In other words, Jupiter has close to the *maximum size* possible for any cold object held together completely by its self-gravity. Objects less massive than Jupiter have smaller sizes, eventually satisfying the mass-radius relationship $R \propto M^{1/3}$ that we associate with asteroids and spherical cows.

At the high-mass end, the average Fermi energy becomes very large as the material of a white dwarf suffers indefinite compression near the Chandrasekhar limit. When electrons acquire Fermi energies of about 50 MeV per particle, they tend to combine with the protons of the white-dwarf matter to form neutrons. White dwarfs more massive than $M_{\rm Ch}$ must become, therefore, neutron stars. The P-ρ relationship for neutron-star matter is more complex (and less certain) than that for an ideal Fermi-Dirac gas. Moreover, the Newtonian equation for force balance,

$$\frac{dP}{dr} = -\frac{GM\rho}{r^2},$$

where we have reverted to an Eulerian description, must be replaced by the Oppenheimer-Volkoff equation,

$$\frac{dP}{dr} = -\frac{G(M + 4\pi r^3 P/c^2)(\rho + P/c^2)}{r^2(1 - 2GM/c^2 r)}, \tag{5.28}$$

because the gravitational fields are so strong as to require a general relativistic treatment. In the above, we have used the Schwarzschild metric, in which $4\pi r^2$ equals the surface area of the sphere that encloses the *gravitational mass* $M(r)$,

$$M(r) = \int_0^r \rho\, 4\pi r^2 dr. \tag{5.29}$$

The mass $M(r)$ does *not* equal the enclosed baryonic mass,

$$M_{\rm B}(r) = m_{\rm B} \int_0^r n_{\rm B} \frac{4\pi r^2 dr}{(1 - 2GM/c^2 r)^{1/2}}, \tag{5.30}$$

which we get by integrating the total number of baryons (each with mass $m_{\rm B}$) contained within the enclosed volume. The difference between $M(r)$ and $M_{\rm B}(r)$ represents the *mass deficit* associated with the relativistically correct gravitational binding. For a Chandrasekhar-mass neutron star, the mass deficit amounts typically to about 10 percent of the baryonic mass; i.e., the binding energy equals about 10 percent of the rest energy of the completely disassembled star, but the precise value depends on the adopted P-ρ relationship. (For models, refer to G. Baym and C. Pethick, 1979, *Ann. Rev. Astr. Ap.*, **17**, 415.)

6

Inviscid Barotropic Flow

Reference: Landau and Lifshitz, *Fluid Mechanics*, pp. 14–24, 310–317.

In this chapter we investigate the flow of fluids in which we ignore the effects of viscosity. In addition to the *inviscid* assumption, we also suppose that the energetics allows us to replace the heat equation with a barotropic equation of state:

$$P = P(\rho). \tag{6.1}$$

Such a replacement considerably simplifies many dynamical discussions, and its formal justification can arise in many ways. Chapter 5 introduced one example, in which the matter of the fluid is quantally degenerate. Another is when all heat-transport mechanisms can be ignored, so that the heat equation becomes a constraint of adiabatic flow:

$$\frac{Ds}{Dt} = 0.$$

Moreover, if all streamlines originate from a region of uniform fluid properties, then the conservation of s along each streamline would imply the constancy of s on all streamlines at the single value of $s = s_0$ that applies to the initial uniform region. With one thermodynamic variable known, the pressure P of a single-component fluid can depend on only one other variable, which we may take to be the density ρ.

A third example involves the opposite regime, when the (radiative) heating and cooling of the fluid occurs so efficiently that, to lowest order, the heat equation corresponds to the near balance of the two volumetric rates, Γ and Λ:

$$\Gamma - \Lambda \equiv -\rho\mathcal{L} = 0. \tag{6.2}$$

For gas of a given chemical composition under optically thin conditions, the net heat-loss function per unit mass, $\mathcal{L}$, can often be expressed as a function of only density ρ and temperature T (see Problem Set 3). The constraint of thermal equilibrium,

$$\mathcal{L}(\rho, T) = 0, \tag{6.3}$$

then associates a unique T for a given ρ, which implies, through the equation of state for an ideal gas, $P = P(\rho)$.

In summary, enough situations permit the hypothesis, equation (6.1), to motivate an examination of the general properties of fluid flow under this simplification. We begin with the derivation of two general theorems associated with the force equation: *Kelvin's circulation theorem* and *Bernoulli's theorem.*

KELVIN'S CIRCULATION THEOREM

When we discard the viscous stress tensor $\overleftrightarrow{\pi}$ and ignore radiative forces, we may write the force equation as

$$\frac{\partial \mathbf{u}}{\partial t} + \nabla\left(\frac{1}{2}|\mathbf{u}|^2\right) + (\nabla \times \mathbf{u}) \times \mathbf{u} = \mathbf{g} - \frac{1}{\rho}\nabla P.$$

If we take the curl of the above equation and write $\boldsymbol{\omega}$ for $\nabla \times \mathbf{u}$, we obtain

$$\frac{\partial \boldsymbol{\omega}}{\partial t} + \nabla \times (\boldsymbol{\omega} \times \mathbf{u}) = \nabla \times \mathbf{g} + \frac{1}{\rho^2}\nabla\rho \times \nabla P, \tag{6.4}$$

where we have used the vector identity that the curl of the gradient of any function equals zero. A classical gravitational field $\mathbf{g} = -\nabla\mathcal{V}$ satisifes this property, $\nabla \times \mathbf{g} = 0$; so gravitational fields can not contribute to the generation or destruction of vorticity in equation (6.4).

If $P = P(\rho)$, ∇P can be written as $P'(\rho)\nabla\rho$, and the second term on the right-hand side of equation (6.4) also equals zero. In this case, we have

$$\frac{\partial \boldsymbol{\omega}}{\partial t} + \nabla \times (\boldsymbol{\omega} \times \mathbf{u}) = 0. \tag{6.5}$$

We may associate each value of $\boldsymbol{\omega}$ with a certain number of vortex lines per unit area, just as in magnetostatics we may associate with each value of $\mathbf{B}$ a certain number of magnetic field lines per unit area. With such a picture, we may give the following geometric interpretation of equation (6.5), the physical essence of *Kelvin's circulation theorem*: *The number of vortex lines that thread any element of area, that moves with the fluid, remains unchanged in time for inviscid barotropic flow* (Figure 6.1).

Proof: Define the circulation Γ around a circuit C by the line integral,

$$\Gamma \equiv \oint_C \mathbf{u} \cdot d\boldsymbol{\ell}. \tag{6.6}$$

Notice that Γ has a nonvanishing value if the circulatory component of the fluid velocity does not average to zero in the closed circuit. Transforming

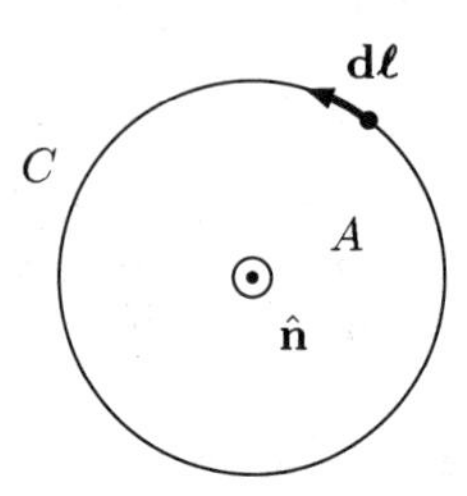

FIGURE 6.1
The geometry referred to in Kelvin's circulation theorem: a circuit C bounds an area A that moves with the fluid motion. The direction of the line element $\mathbf{d}\boldsymbol{\ell}$ on the circuit C is given in terms of the unit normal $\hat{\mathbf{n}}$ to the area A by the usual right-hand rule.

the line integral to a surface integral over the enclosed area A by Stokes' theorem, we obtain

$$\Gamma = \int_A \boldsymbol{\omega} \cdot \hat{\mathbf{n}}\, dA, \tag{6.7}$$

where $\boldsymbol{\omega} \equiv \boldsymbol{\nabla} \times \mathbf{u}$. Equation (6.7) states that the circulation Γ of the circuit C can be calculated as the number of vortex lines that thread the enclosed area A.

What is the time rate of change of Γ if every point on C moves at the local fluid velocity $\mathbf{u}$? To answer this question, take the time derivative of equation (6.7). The time derivative of the right-hand side has two contributions, one which comes from the explicit time rate of change of $\boldsymbol{\omega}$ when A is held fixed, the other which comes from the time rate of change of the area A when $\boldsymbol{\omega}$ is held fixed. Schematically, then, we have

$$\frac{d\Gamma}{dt} = \int_A \frac{\partial \boldsymbol{\omega}}{\partial t} \cdot \hat{\mathbf{n}}\, dA + \int \boldsymbol{\omega} \cdot (\text{time rate of change of } \mathbf{area}).$$

The latter effect can be expressed mathematically with the help of the Figure 6.2.

Thus we may write

$$\frac{d\Gamma}{dt} = \int_A \frac{\partial \boldsymbol{\omega}}{\partial t} \cdot \hat{\mathbf{n}}\, dA + \oint_C \boldsymbol{\omega} \cdot (\mathbf{u} \times \mathbf{d}\boldsymbol{\ell}).$$

We may interchange the cross and the dot in the triple scalar product to write

$$\boldsymbol{\omega} \cdot (\mathbf{u} \times \mathbf{d}\boldsymbol{\ell}) = (\boldsymbol{\omega} \times \mathbf{u}) \cdot \mathbf{d}\boldsymbol{\ell}.$$

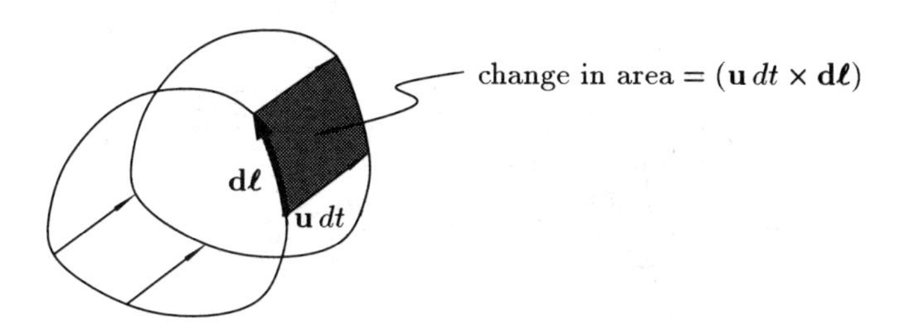

FIGURE 6.2
The time-rate change of an area A moving locally with the fluid velocity $\mathbf{u}$ comes from the parts swept up or left behind by the motion of the boundary C. Each line element $\mathbf{d\ell}$ on C contributes $\mathbf{u} \times \mathbf{d\ell}$ to the time rate of change of the area, an expression that contains both the magnitude of the effect as well as the direction of the local normal to the additional area element.

If we use Stokes's theorem to convert the resulting line integral to a surface integral, we obtain

$$\frac{d\Gamma}{dt} = \int_A \left[\frac{\partial \boldsymbol{\omega}}{\partial t} + \boldsymbol{\nabla} \times (\boldsymbol{\omega} \times \mathbf{u}) \right] \cdot \hat{\mathbf{n}}\, dA. \tag{6.8}$$

The integrand of the right-hand side equals zero by equation (6.5); thus we have the geometric interpretation of Kelvin's circulation theorem,

$$\frac{d\Gamma}{dt} = 0.$$

(Q.E.D.)

APPLICATION TO INCOMPRESSIBLE FLOWS

Notice that Kelvin's circulation theorem follows from equation (6.4), independent of the barotropic assumption (6.1), if the fluid is incompressible, so that $\boldsymbol{\nabla}\rho = 0$. For an incompressible fluid, i.e., a liquid, the variation of pressure P in the force equation equals whatever it needs to so that $\mathbf{u}$ can also satisfy the equation of continuity:

$$\boldsymbol{\nabla} \cdot \mathbf{u} = 0. \tag{6.9}$$

Speaking strictly, we apply the name *hydrodynamics* to the study of the flows of liquids like water. Many problems in hydrodynamics involve the motion of a solid body (e.g., a ship) through water that is stationary at

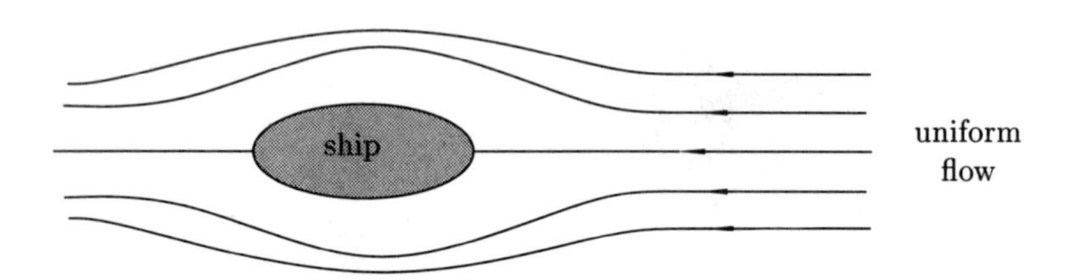

FIGURE 6.3
Streamlines for inviscid incompressible flow around an elliptical cylinder. Changing the sign of the velocity U at infinity would alter the direction of the flow but not the shapes of the streamlines. This fact underlies *d'Alembert's paradox*, namely, that an unbounded nonviscous fluid offers no resistive drag to a body moving uniformly through it.

infinity. From the point of view of an observer fixed on the ship, the water flowing past the ship originates from a steady region of uniform conditions (Figure 6.3).

Uniform flow has no vorticity, and Kelvin's circulation theorem then guarantees (as long as we can ignore the action of viscosity) that none will be subsequently generated in the flow around the ship. If $\nabla \times \mathbf{u} = 0$ everywhere, the flow field $\mathbf{u}$ can be derived from the gradient of a scalar potential Φ:

$$\mathbf{u} = \nabla\Phi. \tag{6.10}$$

If we substitute this relation into the equation of continuity (6.9), we obtain the condition for *potential flow*:

$$\nabla^2\Phi = 0. \tag{6.11}$$

Thus the solution of many problems in hydrodynamics boils down to a solution of Laplace's equation (6.11). The problem is *well-posed* (see Chapter 13) if we specify either the value of Φ on the bounding surfaces of the fluid (Dirichlet boundary conditions) or the value of its normal derivative (Neumann boundary conditions). In this example, we require that water does not penetrate the ship (no normal velocity):

$$\hat{\mathbf{n}} \cdot \nabla\Phi = 0 \qquad \text{on } S, \tag{6.12}$$

where S is the surface of the ship. We also require that the flow become uniform at infinity:

$$\Phi = -Ux \qquad \text{as} \qquad r \to \infty, \tag{6.13}$$

where $\nabla\Phi = -U\hat{\mathbf{x}}$ is the fluid velocity of the water at infinity with respect to the ship (i.e., the negative of the ship's velocity with respect to an observer at rest).

LAMINAR AND TURBULENT BOUNDARY LAYERS

Notice that in the context of the inviscid approximation, we can not impose a no-slip boundary condition,

$$\hat{\mathbf{t}} \cdot \nabla\Phi = 0 \qquad \text{on } S, \tag{6.14}$$

where $\hat{\mathbf{t}}$ is a unit vector tangent to the surface S of the ship. In other words, a well-posed formulation of Laplace's equation does not allow us to specify both the normal and the tangential derivatives of the function Φ on any part of a surface (see, e.g., Chapter 1 of Jackson, *Classical Electrodynamics*).

As we shall discuss in Chapter 13, our governing equation does not possess enough spatial derivatives to impose condition (6.14) on top of condition (6.12). Only if we include in the force equation the viscous term, $\nabla\cdot \overleftrightarrow{\pi}$, where $\pi_{ik} = \mu D_{ik}$ (which introduces *third-order* spatial derivatives of Φ), can we justify, *mathematically and physically*, the additional imposition of the no-slip boundary condition (6.14). The character of the fluid flow near the surface of the ship must be modified tremendously by the imposition of a condition, equation (6.14), which makes the tangential velocity go to zero instead of allowing it to have any finite value consistent with the formulation of the problem of potential flow given in the previous section. What relevance, then, does the inviscid formulation have for the realistic problem where water does have a nonvanishing coefficient of viscosity?

The answer turns out that, except for a thin *boundary layer*, the flow is very nearly given by the inviscid theory, i.e., by potential flow. In the boundary layer, the inviscid approximation breaks down completely, and Kelvin's circulation theorem no longer holds. Generation of vorticity can occur in the boundary layer, and eddies get swept back in the flow to become part of the *wake* behind a ship (see Figure 6.4).

In the boundary layer, the viscous acceleration, $\partial(\nu D_{ik})/\partial x_k$, where $\nu \equiv \mu/\rho$ is the kinematic viscosity, must acquire an order of magnitude (in a direction tangential to the ship's surface) comparable to that of the advective term $u_k\partial u_i/\partial x_k$. Since the tangential velocity has a typical order of magnitude U, a detailed analysis shows that the thickness of the (laminar) boundary layer relative to the typical dimension L of the ship depends on an inverse fractional power of the Reynold's number,

$$\text{Re} \equiv UL/\nu.$$

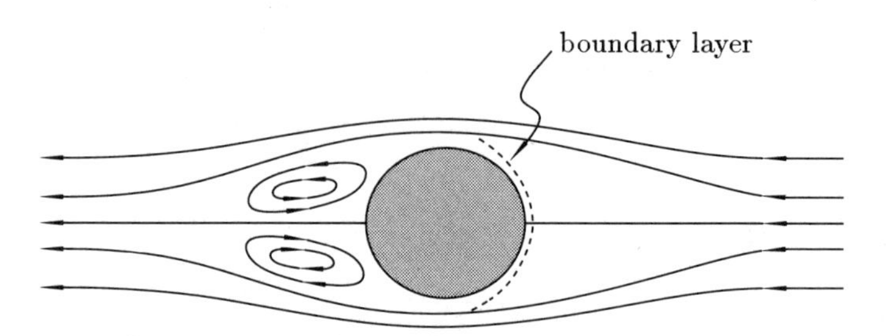

FIGURE 6.4
A real fluid such as water possesses viscosity. Experiments show that a viscous fluid in contact with a solid body remains at rest with respect to it. To reconcile a no-slip boundary condition with the fact that the state well away from the body is well-approximated by potential flow, Prandtl invented the boundary-layer hypothesis: The formation of a thin layer of fluid where the tangential velocity climbs rapidly from zero to the full velocity of the main stream passing the body. The boundary layer typically thickens downstream. The boundary layer's forward inertia may carry the fluid beyond where the flow can monotonically maintain the necessary stagnation point at the rear of the body. Past a certain point, then, the flow may actually reverse its motion and form eddies. (See Milne-Thompson, *Theoretical Hydrodynamics.*)

(In actuality, the boundary layer does not have a constant fractional size, but thickens with increasing distance from the bow of the ship.) The powerful (asymptotic) technique known as *singular perturbation theory* was developed by applied mathematicians to handle the attendant mathematical problem. (For an exposition, see Bender and Orszag, *Advanced Mathematical Methods for Scientists and Engineers.*)

For Re $\gg 1$, then, we expect considerable shear to develop in the boundary layer. The shear rate increases with increasing distance from the ship's bow, and if the ship is large enough (i.e., if the Reynold's number is large enough), a point may come where the boundary layer becomes unstable to the development of *turbulence.* A *line of separation* may then develop that separates the part of the flow that is smooth (*laminar*) from the part that is irregular (*turbulent*) (see Figure 6.5).

Discovering the conditions for the onset of turbulence is a goal of the theory of hydrodynamic stability that we shall investigate in Chapters 8 and 9. For now, we content ourselves with the following moral, which applies in the regime Re $\gg 1$: In a large region of space (the region of potential flow), we may treat viscosity as a small effect. In a small region of space (the boundary layer), viscosity becomes an order-unity effect. Thus, except

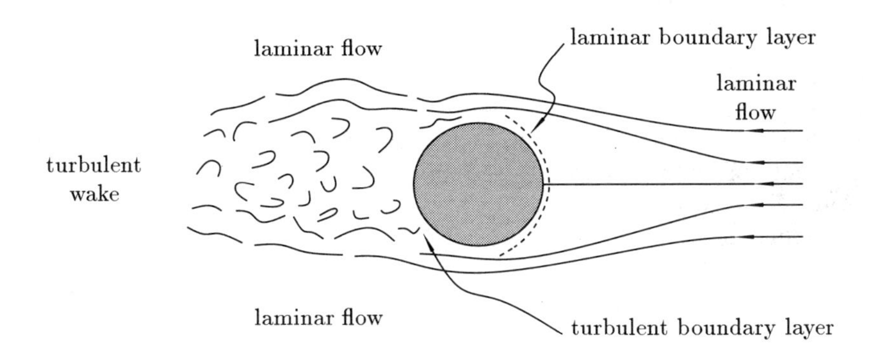

FIGURE 6.5
As the flow velocity U of a viscous fluid past a solid body increases, the line separating a laminar boundary layer from a turbulent boundary layer moves forward. Behind the line of separation lies a turbulent wake in which energy is continually washed downstream with the eddies. This drain of energy results in a net frictional drag on the body that resolves d'Alembert's paradox.

for special regions (boundary layers, shocks, etc.), we may usually ignore the effects of viscosity. The effects of disturbances introduced in the small regions may propagate downstream to affect appreciable quantities of fluid, but viscosity usually plays only a relatively minor role in the affected regions (e.g., the dissipation of turbulence). In particular, the fluctuations present in a turbulent wake can not penetrate far into the region of laminar flow, because the latter is governed by Laplace's equation (6.11), which guarantees that disturbances with a fluctuating (sinusoidal) dependence in one direction (along the length of the turbulent wake) have a damped (exponential) dependence in the other direction (perpendicular to the length of the wake) outside of the wake itself. Eddies (where $\nabla \times \mathbf{u} \neq 0$) dominate a turbulent wake in incompressible flow, but the effects of such eddies subside quickly when we enter the region of laminar flow.

BERNOULLI'S THEOREM

Closely related to Kelvin's circulation theorem we find Bernoulli's theorem for steady barotropic flow. In this circumstance the inviscid force equation reads

$$\nabla\left(\frac{1}{2}|\mathbf{u}|^2\right) + (\nabla \times \mathbf{u}) \times \mathbf{u} = -\nabla\mathcal{V} - \nabla h, \qquad (6.15)$$

where h equals the indefinite integral,

$$h \equiv \int \frac{dP}{\rho}, \tag{6.16}$$

which is well-defined if $P = P(\rho)$. The usefulness of the definition comes from our ability to express the pressure force per unit mass $-\nabla P/\rho$ as the term $-\nabla h$ in equation (6.15).

Define now the Bernoulli function,

$$\mathcal{B} \equiv \frac{1}{2}|\mathbf{u}|^2 + \mathcal{V} + h, \tag{6.17}$$

and write equation (6.15) as

$$\nabla\mathcal{B} + (\nabla \times \mathbf{u}) \times \mathbf{u} = 0. \tag{6.18}$$

If we dot equation (6.18) with $\mathbf{u}$, we get Bernoulli's theorem,

$$\mathbf{u} \cdot \nabla\mathcal{B} = 0, \qquad i.e., \qquad \mathcal{B} = \text{constant on a streamline.} \tag{6.19}$$

Up to an integration constant, the quantity h defined by equation (6.16) equals the specific *enthalpy* if the barotropic relation between P and ρ results because all variations are taken to be adiabatic, i.e., if the fundamental thermodynamic relation for the specific enthalpy h of a chemically homogeneous fluid,

$$dh = T\,ds + \rho^{-1}\,dP,$$

simplifies to $dh = \rho^{-1}\,dP$ because $ds = 0$. For more general barotropic dependences, $P = P(\rho)$, the formal integral function (6.16) need have no relationship to the thermodynamically defined enthalpy h_{thermo}. For example, when we later discuss shock waves in this book, we shall find that the combination $|\mathbf{u}|^2/2 + h_{\text{thermo}} + \mathcal{V}$ is conserved across a (nonmagnetic) viscous shock, but because the specific entropy s increases on passage through the disturbance, the Bernoulli constant $\mathcal{B}$ will suffer a jump across the front, even though h as defined by equation (6.16) may equal h_{thermo} both in front and in back of the shock. This behavior leads to *Crocco's theorem.*

CROCCO'S THEOREM

If all the streamlines originate from a region of uniform flow, $\mathcal{B}$ would equal the same constant everywhere. If $\nabla\mathcal{B} = 0$, equation (6.18) implies that $\omega = \nabla \times \mathbf{u}$ must also generally vanish everywhere. We see, therefore, that the existence of vorticity in the flow in these circumstances connects with the issue of the variation of $\mathcal{B}$ from streamline to streamline. This observation suggests an efficient way to generate vorticity in the absence

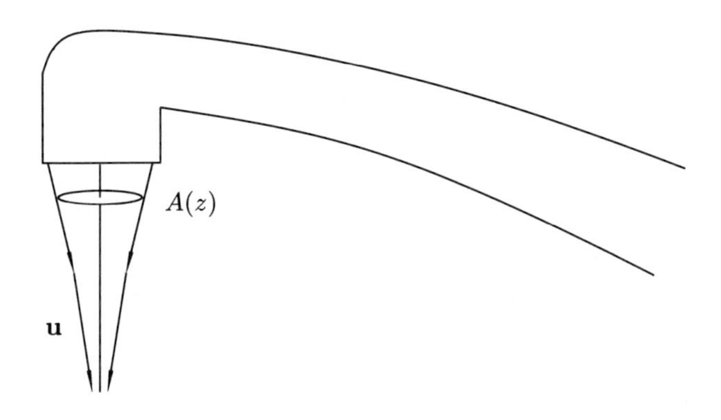

FIGURE 6.6
Flow from a faucet.

of solid bodies and their associated boundary layers. As mentioned in the previous section, the Bernoulli function $\mathcal{B}$ suffers a jump across shocks. If the shock jump has uneven strength across a front, as is usual for a *curved* shock, then even if $\mathcal{B}$ does not vary from streamline to streamline in the preshock gas, it will vary in the postshock region because different jumps in $\mathcal{B}$ occur after passage through different parts of the shock front. The generation of vorticity by this process is known as Crocco's theorem.

A SIMPLE APPLICATION OF BERNOULLI'S THEOREM

Consider what happens when you turn on the kitchen faucet in a trickle (see Figure 6.6). For the quasi-one-dimensional flow of an incompressible fluid, Bernoulli's theorem becomes

$$\frac{1}{2}u^2 + \mathcal{V} + \frac{P}{\rho} = \text{constant}. \tag{6.20}$$

The pressure inside the water stream equilibrates with the atmospheric pressure, which we may take to be a constant over the small variation of height involved in the experiment. Thus, P/ρ equals a constant along the length of the stream of water, and we infer that the fluid speeds up as it drops in the gravitational potential $\mathcal{V}$ of the Earth. If we denote the cross-sectional area of the stream at height z as $A(z)$, the equation of continuity can be written as the constraint of constant mass flux,

$$\rho u A = \text{constant}, \tag{6.21}$$

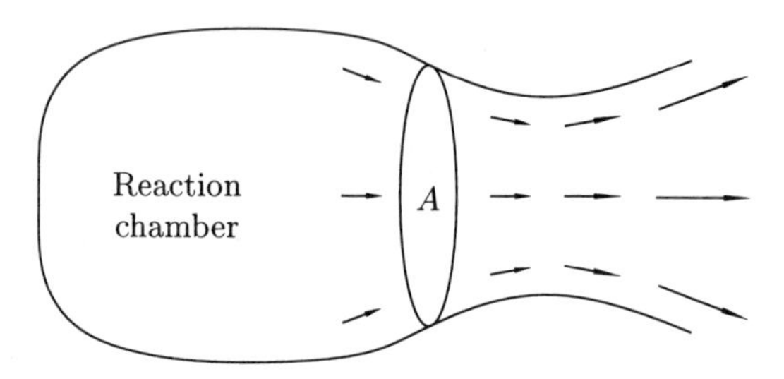

FIGURE 6.7
The de Laval nozzle.

which implies, for constant ρ, that $A(z)$ decreases as $u(z)$ increases.

When the cross-sectional area $A(z)$ diminishes enough, the effects of surface tension can no longer be ignored. The relative slip of water through air also tends to generate oscillatory disturbances on the water-air interface (by the so-called *Kelvin-Helmholtz instability* that we shall discuss in Chapter 8). The combined actions of surface tension and Kelvin-Helmholtz instabilities tend to break up the smooth stream of water into individual falling droplets once the cross-sectional area $A(z)$ diminishes beyond a certain critical value. (It is said that Richard Feynman deduced these consequences of Bernoulli's theorem for the kitchen-sink experiment while still a young boy.)

THE DE LAVAL NOZZLE

To highlight the qualitative differences introduced by the effects of compressibility, let us consider the problem of the *de Laval nozzle.* The problem arises in the operation of a jet engine, in which stationary hot gas generated in a reaction chamber is allowed to flow through a (horizontal) tube of varying cross-sectional area (Figure 6.7).

If we again make the approximation of steady, quasi-one-dimensional, barotropic flow, we may write Bernoulli's theorem and the equation of continuity as

$$\frac{1}{2}u^2 + \int \frac{dP}{\rho} = \text{constant}, \tag{6.22}$$

$$\rho u A = \text{constant}, \tag{6.23}$$

where we have ignored possible variations of the gravitational potential for the fast flow of jet gases over the limited spatial extent relevant for terrestrial applications.

The variation of the area A with x will introduce spatial variations for each of the other quantities. To consider the rate of such variations, take the differential of equation (6.22) to obtain

$$u\,du + \frac{1}{\rho}\frac{dP}{d\rho}d\rho = 0 \qquad \Rightarrow \qquad u\,du + \frac{a^2}{\rho}d\rho = 0, \tag{6.24}$$

where we have defined the square of the speed of sound associated with our barotropic relation as

$$a^2 \equiv \frac{dP}{d\rho}. \tag{6.25}$$

Equation (6.24) implies that the fractional change of density ρ is related to the fractional change of the fluid velocity u by the equation,

$$\frac{d\rho}{\rho} = -M^2\frac{du}{u}, \tag{6.26}$$

where $M \equiv u/a$ is the Mach number. The above equation states that the square of the Mach number provides a measure of the importance of compressibility. In particular, the flow of air at subsonic speeds past terrestrial obstacles can often be approximated as occurring incompressibly, because the fractional change of ρ is negligible in comparison with the fractional change of u if $M \ll 1$. In contrast, supersonic flight past obstacles necessarily involves substantial compressions and expansions.

To relate the change of u to the change of A in our original nozzle problem, take the logarithmic derivative of equation (6.23) to obtain

$$\frac{d\rho}{\rho} + \frac{du}{u} + \frac{dA}{A} = 0. \tag{6.27}$$

If we eliminate $d\rho/\rho$ by using equation (6.26), we get

$$(1 - M^2)\frac{du}{u} = -\frac{dA}{A}. \tag{6.28}$$

Equation (6.28) has the following implications:

(a) At subsonic speeds, i.e., $1 - M^2 > 0$, an increase in velocity, $du > 0$, requires a decrease in area, $dA < 0$. This corresponds to normal experience, e.g., our faucet experiment of the previous section or the speeding up of a river as the channel narrows. (The sound speed associated with an incompressible fluid is formally infinite, so the flow of a liquid always occurs at subsonic speeds in the context of the approximation $\rho =$ constant.)

(b) At supersonic speeds, i.e., $1 - M^2 < 0$, an increase in velocity, $du > 0$, requires an *increase* in the area of the nozzle, $dA > 0$! This counter-intuitive result has a simple explanation: for $M > 1$, the density

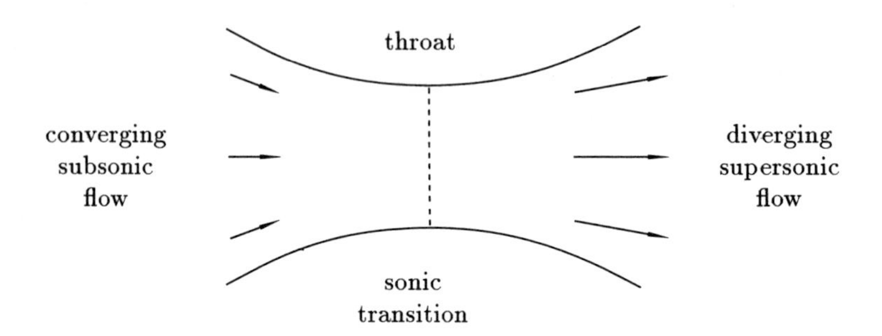

FIGURE 6.8
To produce a transition from subsonic to supersonic flow in a nozzle, we need a converging-diverging geometry.

decreases faster than the area increases [in accordance to the requirements of equations (6.26) and (6.28)], so the velocity must increase to maintain a constant flux of mass.

(c) A sonic transition, i.e., $1 - M^2 = 0$, can be made *smoothly*, i.e., with du/dx finite, only at the throat of the nozzle where $dA = 0$ (see Figure 6.8). To obtain supersonic exhaust, therefore, we must accelerate the reaction gases through a converging-diverging nozzle, a fundamental feature behind the design of jet engines and rockets.

(d) The converse to (c) does not necessarily hold: M does not necessarily equal unity at the throat of a nozzle, where $dA = 0$. If $M \neq 1$, equation (6.28) implies that the fluid velocity reaches a local extremum when the area does, i.e., $du = 0$ where $dA = 0$. Whether the extremum corresponds to a local maximum or minimum depends on whether we have subsonic or supersonic flow, and whether the nozzle has a converging-diverging or a diverging-converging shape (Figure 6.9).

(e) Whether supersonic exhaust is actually achieved in nozzle flow also depends on boundary conditions, in particular, on the pressure of the ambient medium in comparison with the pressure of the reaction chamber. If a sonic transition does occur, the flow behavior depends sensitively on the nozzle conditions, since the coefficient $1 - M^2$ in equation (6.28) becomes arbitrarily small near the transition region. This sensitivity has the following important physical consequence. In the absence of frictional forces (i.e., viscosity), the flow should be time-reversible. In other words, we should be able to reverse—*in principle*—the operation of a de Laval nozzle, and obtain a smooth

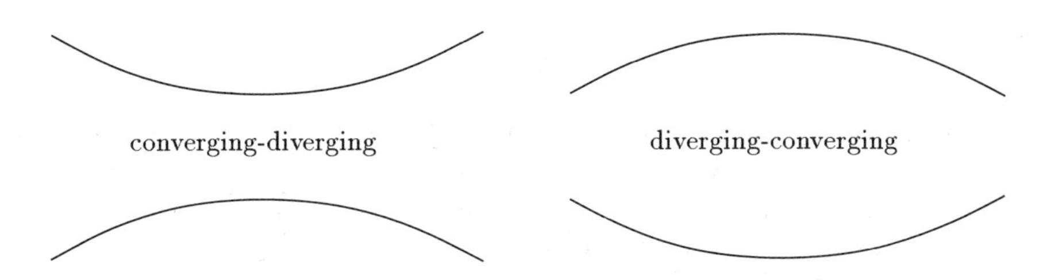

FIGURE 6.9
If a sonic transition does not occur in the nozzle flow, the fluid speed u reaches an extremum at the position where the cross-sectional area A has an extremum. Whether this extremum in u corresponds to a maximum or a minimum depends on whether the flow occurs subsonically or supersonically, and on whether the nozzle has a converging-diverging or a diverging-converging geometry.

deceleration of supersonic flow to subsonic flow. In fact, such an operation cannot be carried out in practice. The deceleration of supersonic flow to subsonic flow almost always produces shocks, in which the transition occurs not smoothly but in one sharp jump. The sensitivity of transsonic flow explains how a normally small perturbation—here viscosity—can acquire an order-unity effect in a limited region of space. The shock transition layer occupies, as we shall see in Chapter 15, a thickness $\sim$ a few mean free paths for elastic gas collisions.

(f) When external body forces are present, we do not need to have a throat to achieve the smooth transition of subsonic flow to supersonic flow. The external forces can provide the requisite acceleration. To see the modification of the nozzle constraint, we consider the astrophysically important example of gravitational forces.

THE BONDI PROBLEM

The Bondi problem involves the spherical accretion of gas by a gravitating point mass M. The self-gravity of the gas is ignored in the usual formulations of this problem. In spherical symmetry, the steady flow of a barotropic gas has the governing set of equations:

$$4\pi r^2 \rho u = \text{constant} \equiv -\dot{M}, \tag{6.29}$$

$$\frac{u^2}{2} + h - \frac{GM}{r} = 0, \tag{6.30}$$

where

$$h \equiv \int_{\rho_\infty}^{\rho} \frac{dP}{\rho}, \tag{6.31}$$

with ρ_∞ being the value of ρ at infinity. We have defined the lower limit in equation (6.31) so that the constant on the right-hand side of equation (6.30) has the value 0 (because we require $u = 0$ and $\rho = \rho_\infty$ at $r = \infty$).

For isothermal flow, $P = c_\infty^2 \rho$, with a constant speed of sound $c_\infty = (kT/m)^{1/2}$, the h function has the form,

$$h = c_\infty^2 \ln(\rho/\rho_\infty); \tag{6.32}$$

whereas, for polytropic flow, $P \propto \rho^\gamma$,

$$h = \frac{\gamma}{\gamma - 1} c_\infty^2 \left[\left(\frac{\rho}{\rho_\infty} \right)^{\gamma-1} - 1 \right], \tag{6.33}$$

where $c_\infty^2 \equiv P_\infty/\rho_\infty$ is again the square of the thermal speed at infinity. In the latter case, notice that the local acoustic speed a is given through

$$a^2 \equiv \frac{dP}{d\rho} = \gamma c_\infty^2 \left(\frac{\rho}{\rho_\infty} \right)^{\gamma-1}. \tag{6.34}$$

A characteristic length scale of the problem is the Bondi radius,

$$r_\mathrm{B} \equiv GM/c_\infty^2, \tag{6.35}$$

which equals the distance from the star where the gravitational potential energy of a gas particle equals 2/3 of the thermal energy. We may introduce dimensionless variables by defining

$$x \equiv r/r_\mathrm{B}, \qquad v \equiv |u|/c_\infty, \qquad \alpha \equiv \rho/\rho_\infty. \tag{6.36}$$

We also define a dimensionless accretion rate λ by measuring $\dot{M}$ in units of a fiducial mass flux $\rho_\infty c_\infty$ past an area $4\pi r_\mathrm{B}^2$:

$$\lambda \equiv \frac{\dot{M}}{4\pi\rho_\infty (GM)^2/c_\infty^3}. \tag{6.37}$$

In terms of these nondimensional quantities, we may write equations (6.29) and (6.30) as

$$x^2 \alpha v = \lambda; \tag{6.38}$$

$$\frac{v^2}{2} + H(\alpha) - \frac{1}{x} = 0, \tag{6.39}$$

where $H(\alpha)$ is given in isothermal flow by

$$H(\alpha) = \ln\alpha \qquad \text{for} \qquad \gamma = 1, \tag{6.40}$$

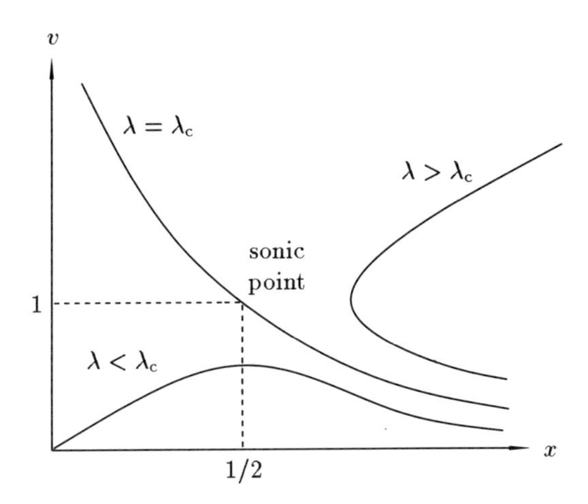

FIGURE 6.10
The schematic character of the flow solutions to the Bondi accretion problem. In this figure, v represents the inflow speed measured in units of the sound speed; x, the radial distance from the central gravitating mass in units of the Bondi radius; λ, a dimensionless accretion rate. The solution topology changes character when λ crosses a critical value λ_c.

and in polytropic flow by

$$H(\alpha) = \frac{\gamma}{\gamma - 1}(\alpha^{\gamma-1} - 1) \qquad \text{for} \qquad \gamma \neq 1. \tag{6.41}$$

In Problem Set 2 you are asked to calculate the value of the dimensionless mass accretion rate, $\lambda = \lambda_c$, that corresponds to a physically acceptable solution containing a transsonic transition. For example, the critical accretion rate in isothermal flow is given by

$$\lambda_c = \frac{1}{4}e^{3/2} = 1.12\ldots$$

More general values of λ lead to solutions that look schematically like Figure 6.10.

The solutions with $\lambda > \lambda_c$ contain sonic transitions, but are unphysical because they are double-valued in v for each value of x. Solutions with $\lambda < \lambda_c$ remain subsonic everywhere. Which solution would nature choose?

In Bondi's original treatment (H. Bondi, 1952, *M.N.R.A.S.*, **112**, 195), he answered this question on the basis of philosophical rhetoric, namely, if all physically acceptable solutions have $\lambda \leq \lambda_c$, why shouldn't nature choose the case with the maximum possible value for the mass accretion rate? This answer, $\lambda = \lambda_c$, turns out to be correct, but a better line of reasoning arises as follows. Since the equations plus boundary conditions at infinity do not provide an obvious distinction between the allowable solutions, consider the behavior of the solutions near the origin. For solutions in which $v \to 0$ as $x \to 0$, equations (6.39) and (6.40) imply

$$\ln \alpha - \frac{1}{x} \to 0, \qquad i.e., \qquad \alpha \to e^{1/x} \qquad \text{as} \qquad x \to 0. \tag{6.42}$$

Equation (6.42) has a simple physical interpretation: When $v \to 0$, the situation approaches quasihydrostatic equilibrium. An isothermal gas in hydrostatic equilibrium in a gravitational potential $\mathcal{V}$ satisfies the *barometric formula*,

$$\rho \propto \exp(-\mathcal{V}/c_\infty^2),$$

whose nondimensional form equals equation (6.42) when we identify $\mathcal{V} = -GM/r$.

For $M = 1\ M_\odot$, $T_\infty = 100\,\text{K}$ (a typical interstellar temperature), $r_B = GMm/kT_\infty = 1.6 \times 10^{16}\,\text{cm}$ for $m = 1.67 \times 10^{-24}\,\text{g}$ (a hydrogen atom). Thus $x = r/r_B \approx 0.5 \times 10^{-5}$ for $r = 7 \times 10^{10}\,\text{cm} = 1R_\odot$. Hence in a typical astrophysical situation for spherical accretion by a star, subsonic flow everywhere requires an increase of the density ρ just outside the stellar surface above the interstellar value ρ_∞ by a factor of

$$\alpha = e^{1/x} \sim \exp(2 \times 10^5) \sim 10^{87,000}.$$

This is an enormous factor—indeed, an impossibly large factor, given that there are only $\sim 10^{80}$ H atoms in the entire observable universe. Since the interstellar medium contains on average ~ 1 H atom per cm^3, to provide enough back pressure to limit the inflow to subsonic speeds everywhere would require many more hydrogen atoms for each cubic centimeter near the star than exists in the entire observable universe. Clearly, under anything that approximates isothermal conditions, spherical mass accretion by a star from the interstellar medium must induce a transition from subsonic flow to supersonic flow.

For the latter case, the dimensional mass-accretion rate equals the Bondi rate

$$\dot{M} = \lambda_c 4\pi\rho_\infty \frac{(GM)^2}{c_\infty^3}. \tag{6.43}$$

Hoyle and Lyttleton have argued that if the star has a velocity V with respect to the interstellar medium, the accretion formula must be modified to read

$$\dot{M} = \lambda_* 4\pi\rho_\infty \frac{(GM)^2}{(V^2 + c_\infty^2)^{3/2}}, \tag{6.44}$$

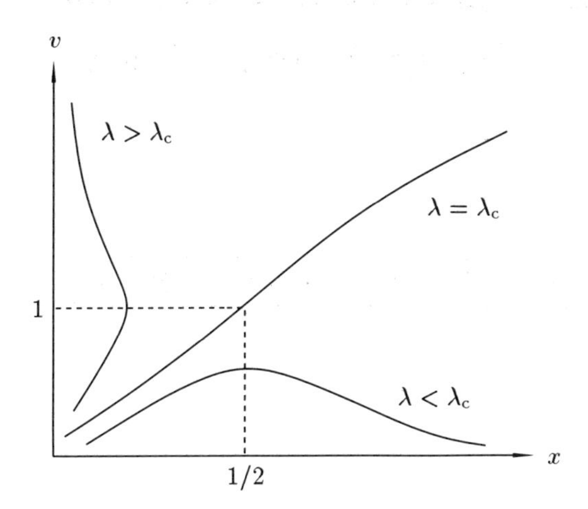

FIGURE 6.11
The schematic character of the flow solutions to the Parker solar-wind problem. In this figure, v represents the outflow speed measured in units of the sound speed; x, the radial distance from the central gravitating mass in units of the Bondi radius; λ, a dimensionless mass-loss rate. The solution topology changes character when λ crosses a critical value λ_c.

where λ_* equals a dimensionless coefficient of order unity. Problem Set 2 asks you to construct an order-of-magnitude estimate to justify this formula, as well as to provide reasonable numerical estimates for the importance of changes of stellar masses by interstellar accretion.

PARKER'S SOLAR-WIND SOLUTION

A problem that has a close mathematical relationship to Bondi's solution for spherical accretion is Parker's solution for a thermally-driven solar wind (see E. N. Parker, *Interplanetary Dynamical Processes*). However, we leave the detailed discussion of the solar-wind case to Problem Set 2 (see also Figure 6.11).

7

Viscous Accretion Disks

Reference: J. E. Pringle, 1981, *Ann. Rev. Astr. Ap.*, **19**, 137.

Astronomers believe that a slow accumulation of matter by a central object via the flow of gas through a thin centrifugally supported disk occurs in many astronomical objects: active galactic nuclei, x-ray binaries, nebular material around young stars, etc. Excess emission of radiation of a spectral character quantitatively and qualitatively different from what would normally be expected for the central object alone (black hole, neutron star, pre-main sequence star, etc.) constitutes the main empirical evidence for energy release by *disk accretion*. Occasionally, one also sees double-horned line profiles that can most simply be interpreted in terms of the presence of a disk whose matter rotates at Keplerian speeds about a gravitating central mass.

Apart from some general considerations, however, astronomers understand the basic cause for the accretion flow rather poorly. For an isolated self-gravitating system with given mass and angular momentum, we can easily show that the configuration of least energy corresponds to one where all the mass resides at the center (with the smallest radii) and a single atom or molecule orbiting at a very large distance carries all the angular momentum. Thus the dissipation of mechanical (and gravitational) energy favors mechanisms that produce an inward transport of matter and an outward transport of angular momentum. In the absence of macroscopic instabilities (which we shall discuss in Chapter 8), however, matter in a rotating disk tends eventually to settle into nonintersecting circular orbits, which minimize the amount of mechanical dissipation that can take place. Why should any outward transport of angular momentum (necessary for an accompanying inward transport of mass) take place in an *axisymmetric* disk?

In the early 1970s, a number of astrophysicists (notably Shakura and Sunyaev, Rees, Lynden-Bell and Pringle) resurrected earlier ideas (by

Maxwell, Jeffreys, and Lust) that disk accretion could result from the internal friction present in the shearing layers of a differentially rotating disk. The structure of *viscous accretion disks* and their radiative properties quickly became a subject of intense theoretical investigation. In this chapter, we consider the basic elements of the theory. We take a physical approach in deriving the governing equations, but the same set could be obtained more laboriously and formally by appropriately manipulating the Navier-Stokes equations for this geometry (see Appendix).

BASIC EQUATIONS

To simplify the important part of the dynamics, let us assume that the disk is spatially thin. We begin by introducing cylindrical coordinates (ϖ, φ, z) and the surface mass density σ obtained by an integration of the volume density ρ over the vertical dimension of the disk:

$$\sigma(\varpi, t) \equiv \int_{-\infty}^{+\infty} \rho(\varpi, z, t)\, dz. \tag{7.1}$$

If the disk has limited vertical extent, the fluid velocity $\mathbf{u}$ varies little with z, and we may integrate the fluid equations over z with $\mathbf{u}$ evaluated in the midplane $z = 0$. To lowest order, the radial force equation then yields centrifugal balance,

$$\varpi\Omega^2(\varpi) = \frac{GM}{\varpi^2}, \tag{7.2}$$

where $\Omega(\varpi)$ is the local angular velocity at radius ϖ in the disk, and M is the mass of the central object. In order for us to ignore the contribution of the pressure gradient in the radial force balance, we have supposed that the orbit speed $\varpi\Omega$ much exceeds the thermal speed $a \equiv (kT/m)^{1/2}$ in the midplane of the disk:

$$\varpi\Omega \gg a. \tag{7.3}$$

Since the disk has a characteristic vertical height $H = a/\Omega$, we see that the approximation represented by equation (7.3) holds to the extent that the disk is indeed spatially thin, $H \ll \varpi$ (Figure 7.1).

In equation (7.2), we have also assumed that the mass of the disk is much less than that of the central object,

$$2\pi \int_{\varpi_{\rm i}}^{\varpi_{\rm o}} \sigma\, \varpi d\varpi \ll M, \tag{7.4}$$

where $\varpi_{\rm i}$ and $\varpi_{\rm o}$ are the radii of the inner and outer edges of the disk, respectively. When equation (7.4) holds, equation (7.2) implies that the circular frequency Ω satisfies Kepler's third law:

$$\Omega \propto \varpi^{-3/2}.$$

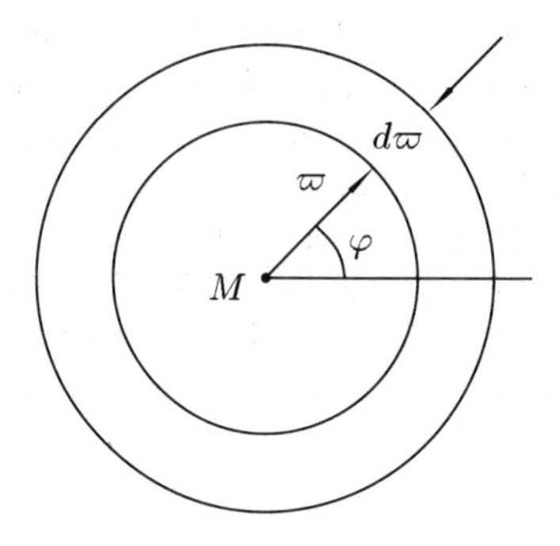

FIGURE 7.1
The geometry of an accretion disk whose gravitation is dominated by a central mass M.

Such a Keplerian disk possesses shear of an amount equal to

$$\varpi \frac{d\Omega}{d\varpi} = -\frac{3}{2}\Omega, \tag{7.5}$$

and it is tempting to suppose that a shear stress $\pi_{\varpi\varphi}$ can be associated with this rate of strain,

$$\pi_{\varpi\varphi} = \mu\varpi \frac{d\Omega}{d\varpi}, \tag{7.6}$$

where μ is the coefficient of viscosity. The definition, equation (7.6), which agrees with the usual fluid-mechanics formulation, differs by a sign from the convention used by most accretion-disk workers. In any case, notice the physical result that a uniformly rotating disk, $d\Omega/d\varpi = 0$, generates no viscous shear stress, $\pi_{\varpi\varphi}$. For $\mu > 0$ in a disk that rotates differentially in a normal way, notice also that the friction of the slower moving material on the outside exerts a surface force in the negative φ direction on neighboring material to the inside. This leads to a viscous torque that promotes an outward transport of angular momentum. Indeed, if we denote the angular momentum per unit mass of the fluid as j,

$$j = \varpi^2\Omega, \tag{7.7}$$

we can show that the φ-component of the force equation can be manipulated to read

$$\rho\left(\frac{\partial j}{\partial t} + u_\varpi \frac{\partial j}{\partial \varpi} + u_z \frac{\partial j}{\partial z}\right) = \frac{1}{\varpi}\frac{\partial}{\partial \varpi}(\varpi^2 \pi_{\varpi\varphi}). \tag{7.8}$$

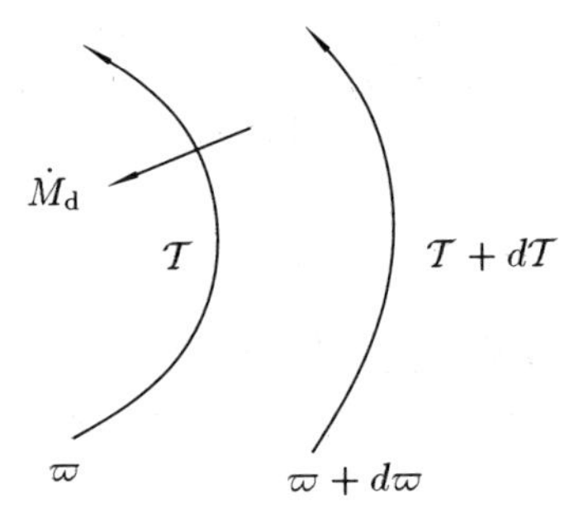

FIGURE 7.2
The difference in viscous torques $\mathcal{T}$ and $\mathcal{T} + d\mathcal{T}$ exerted across circles of radii ϖ and $\varpi + d\varpi$ transports angular momentum radially outward through the disk, and allows mass to flow inward at an accretion rate $\dot{M}_d$.

Since equations (7.2) and (7.7) imply that j has negligible variations with t and z, we may integrate equation (7.8) over z and obtain the angular momentum transport equation,

$$-\dot{M}_d \frac{dj}{d\varpi} = \frac{\partial \mathcal{T}}{\partial \varpi}, \tag{7.9}$$

where we have denoted the mass-accretion rate across a circle of circumference $2\pi\varpi$ in the disk as

$$\dot{M}_d \equiv -2\pi\varpi u_\varpi \sigma, \tag{7.10}$$

and $\mathcal{T}$ as the viscous torque across the same circle:

$$\mathcal{T} \equiv 2\pi\varpi \int_{-\infty}^{+\infty} \varpi \pi_{\varpi\varphi} \, dz, \tag{7.11}$$

with ϖ being the lever arm of the tangential force per unit area $\pi_{\varpi\varphi}$. Equation (7.9) states that the difference between the viscous torque exerted on the outer and inner edges of an annulus of area $2\pi\varpi\, d\varpi$ causes an (inward) flow of mass $\dot{M}_d$ in the disk when the difference in specific angular momentum between the two edges equals dj (see Figure 7.2).

We define the kinematic viscosity in the usual manner,

$$\nu \equiv \mu/\rho, \tag{7.12}$$

so that equation (7.6) becomes

$$\pi_{\varpi\varphi} = \rho\nu\varpi \frac{d\Omega}{d\varpi}.$$

With the (*ad hoc*) assumption that ν varies only slowly with z so that we may approximate it by its value in the midplane, we may perform the z integration in equation (7.11) to obtain

$$\mathcal{T} = 2\pi\varpi\sigma\nu\varpi^2 \frac{d\Omega}{d\varpi}. \tag{7.13}$$

We still need to obtain the variation of σ. The latter satisfies the axisymmetric equation of continuity integrated over z:

$$\frac{\partial\sigma}{\partial t} + \frac{1}{\varpi}\frac{\partial}{\partial\varpi}(\varpi u_\varpi \sigma) = 0,$$

which, with the help of equation (7.10), we may write as

$$\frac{\partial\sigma}{\partial t} - \frac{1}{2\pi\varpi}\frac{\partial \dot{M}_\mathrm{d}}{\partial\varpi} = 0. \tag{7.14}$$

If ν can be regarded as known, equations (7.9), (7.13), and (7.14) constitute a closed set to solve for the ϖ and t dependences of the disk accretion rate, $\dot{M}_\mathrm{d}$, the viscous torque, $\mathcal{T}$, and the disk surface density, σ.

NEED FOR A SOURCE OF ANOMALOUS VISCOSITY

If we eliminate $\dot{M}_\mathrm{d}$ and $\mathcal{T}$ by using equations (7.9) and (7.13), we obtain for equation (7.14)

$$\frac{\partial\sigma}{\partial t} + \frac{1}{\varpi}\frac{\partial}{\partial\varpi}\left[\left(\frac{dj}{d\varpi}\right)^{-1}\frac{\partial}{\partial\varpi}\left(\varpi^3\frac{d\Omega}{d\varpi}\nu\sigma\right)\right] = 0. \tag{7.15}$$

Since $j = \varpi^2\Omega$, equation (7.15) demonstrates that the spreading of σ (inward in the inner parts for mass accretion and outward in the outer parts to conserve total angular momentum) represents a *diffusion* process; i.e., σ satisfies a real PDE which has one derivative in time and two derivatives in space, with a diffusion coefficient given in order of magnitude by the kinematic viscosity ν. The time scale for viscous accretion to reach quasi-steady state in the disk locally must therefore take the order-of-magnitude form

$$t_\mathrm{acc} = \varpi^2/\nu. \tag{7.16}$$

(For concrete examples using a transformation to j as the "spatial" variable in place of ϖ, see D. Lynden-Bell and J. E. Pringle, 1974, *M.N.R.A.S.*, **168**, 603.)

If the only source of viscosity were molecular, then ν would $\sim \ell v_T$, where ℓ is the particle mean free path and v_T is the thermal velocity.

Values appropriate for a nebular disk around a newly born star might be $\varpi \sim 10^{14}\,\mathrm{cm}$, $\ell \sim 10\,\mathrm{cm}$, and $v_T \sim 10^5\,\mathrm{cm\,s^{-1}}$. The viscous accretion time scale would then be $t_{\mathrm{acc}} \sim 10^{22}\,\mathrm{s} \sim 3 \times 10^{14}\,\mathrm{yr}$, longer by 7 to 8 orders of magnitude than the age conventionally ascribed to such disks. Clearly, if viscous accretion explains such objects, there must exist an anomalous source of viscosity. The same conclusion holds for all the other astronomical objects for which the action of viscous accretion disks has been invoked.

What constitutes this source of anomalous viscosity? Many processes have been proposed; some are very specific and unavoidable in relevant circumstances, such as convective turbulence (see, e.g., D. N.C. Lin and J. Papaloizou, 1980, *M.N.R.A.S.*, **191**, 37). Others represent little more than wishful thinking (but may operate in a more general context), such as vague appeals to magnetic stresses. An early assertion—that accretion disks might prove unstable to the generation of fluid turbulence—has turned out to depend on a vain hope. Beginning with the work of Papaloizou and Pringle, interesting fluid instabilities have been discovered for fat disks (or "toroids") and narrow rings, but such instabilties generally depend on the reflection of sound waves off boundaries. Acoustic waves have to propagate too many wavelengths (each $\sim H$) to reach such boundaries in vertically thin but radially extended disks (where $H \ll \varpi$). Small amounts of wave dissipation (e.g., by refraction to high altitudes) would then stabilize the relevant disturbances, rendering the mechanism inoperative (but see Chapter 11 for the case when the self-gravitation of the disk proves important).

A more promising mechanism arises if the disk contains poloidal magnetic fields (those that contain components of either B_ϖ or B_z). Chandrasekhar (see pp. 384–389 of his book *Hydrodynamic and Hydromagnetic Stability*) shows that a differentially rotating, and electrical conducting, cylinder of incompressible fluid, threaded by a sufficiently weak magnetic field parallel to its rotation axis, is unstable with respect to undulating axisymmetric disturbances that alternately displace fluid inward and outward along the length of the (infinite) cylinder if the angular velocity $\Omega(\varpi)$ decreases in magnitude for increasing distance ϖ from the rotation axis. The potential importance of this result (when suitably generalized for less idealized conditions) for the viscosity mechanism of (magnetized) accretion disks went unnoticed until Balbus and Hawley called attention to it in 1991. We shall defer discussion of this instability until Chapter 8.

Plausible sources for the desired anomalous viscosity have been identified, therefore, in some well-understood circumstances when an inviscid disk would be unstable. Even in these cases, however, the procedure for translating an indication of instability to a practical estimate for the effective turbulent viscosity remains uncertain and controversial. People impatient to get on with the job of calculating models for comparison with astronomical observations often bypass this difficulty by writing the effective viscous shear-stress in terms of a dimensionless "alpha parameter,"

$$\pi_{\varpi\varphi} \equiv -\alpha P, \tag{7.17}$$

where P is the pressure. (Occasionally, specialists on this subject debate whether P should equal the gas pressure alone or the gas plus radiation pressure; however, such debates seem pointless as long as we have no physical source for the anomalous viscosity.) In any case, the rationale for the α model resides in the following idea. If a powerful source of fluid turbulence should exist for accretion disks, the turbulent shear stress associated with the fluctuating velocity field $\delta\mathbf{u}$, should have the form

$$\pi_{\varpi\varphi} = -\rho\langle\delta u_\varpi \delta u_\varphi\rangle, \tag{7.18}$$

where the angular brackets refer to some appropriate average over an ensemble of turbulent fluctuations. If we assume that a nonvanishing correlation exists, we may estimate $\langle\delta u_\varpi \delta u_\varphi\rangle$ as a fraction α of v_T^2, where $\alpha < 1$ if the fluctuating velocity components are subsonic in magnitude. Similarly, if the anomalous viscosity has an origin in a fluctuating magnetic field $\delta\mathbf{B}$, we have an associated Maxwell stress,

$$\pi_{\varpi\varphi} = \frac{1}{4\pi}\langle\delta B_\varpi \delta B_\varphi\rangle. \tag{7.19}$$

Again, if we approximate $\langle\delta B_\varpi \delta B_\varphi\rangle/4\pi$ as a fraction $-\alpha$ of the thermal pressure ρv_T^2, we would recover equation (7.17).

The α prescription has the virtue that it satisfies dimensional considerations. One can plausibly argue that α is bounded above by the value unity. Unfortunately, attempts by different groups to compute α from more basic principles produce a relatively large range of plausible values (e.g., $\alpha = 10^{-2}$ to 10^{-4} for disk convective turbulence). Moreover, α will generally vary over the disk in a manner that need not scale directly with P. In what follows, we shall adhere to the conventional hypothesis that α, or equivalently ν, is large enough to make the model astronomically meaningful.

STEADY-STATE VISCOUS ACCRETION AND THE BOUNDARY LAYER

In a steady state, $\partial\sigma/\partial t = 0$, and equation (7.14) implies that the disk accretion rate $\dot{M}_\mathrm{d}$ equals a constant. We may then integrate equation (7.9) to find that the required torque $\mathcal{T}$ must be given by

$$\mathcal{T} = -\dot{M}_\mathrm{d}(j - j_0), \tag{7.20}$$

where j_0 is an integration constant. Lynden-Bell and Pringle (1974) assert that the integration constant j_0 is determined by the condition that the

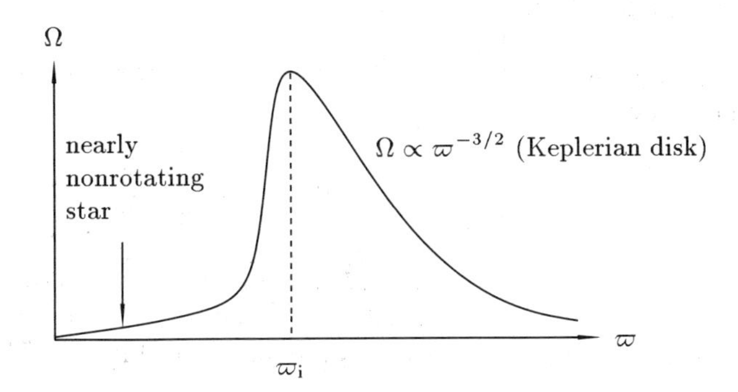

FIGURE 7.3
The angular velocity Ω of the part of the disk that adjoins a slowly spinning star must increase inwards at the Keplerian value $\Omega \propto \varpi^{-3/2}$ for the disk proper, reach a maximum value $\Omega(\varpi_i)$, and decrease thereafter to the small value appropriate for a nearly nonrotating star.

available viscous torque $\mathcal{T} \propto d\Omega/d\varpi$ [see equation (7.13)] vanishes at an inner boundary $\varpi = \varpi_i$, where $d\Omega/d\varpi = 0$ because Ω turns over (in the boundary layer) from its full Keplerian value to a much smaller one (appropriate, e.g., for the surface of a nonrotating star; see Figure 7.3). In other words,

$$j_0 = \varpi_i^2 \Omega(\varpi_i). \tag{7.21}$$

Unfortunately, the usual visualization of this process contains an inconsistency in the treatment. Equation (7.21) substituted into equation (7.20) implies that the effect of the inner boundary extends an order unity distance from ϖ_i since $j = \varpi^2\Omega(\varpi) \propto \varpi^{1/2}$ and j_0 have comparable order of magnitudes for $\varpi \sim \varpi_i$. The whole point of a boundary layer, however, revolves around the notion that it represents a small region of space which mediates a transition between flows where viscosity can be neglected to where it can not. In the theory of viscous accretion disks, however, we postulate the effects of viscosity to be important everywhere, except possibly inside the central object. If a boundary layer arises anywhere, it occurs where the central object (e.g., a star) attaches onto the disk, not (stretching semantics) where the disk attaches to the central object. Even as I was writing this book, several different groups had begun to undertake full-scale hydrodynamical simulations in attempts to find solutions that naturally join the disk accretion onto a more complex pattern of flow inside a slowly rotating star.

BUDGET FOR ENERGY RELEASE

The vertically integrated rate of viscous dissipation (stress times rate of strain) equals

$$\int_{-\infty}^{+\infty} \Psi \, dz = \int_{-\infty}^{+\infty} \pi_{\varpi\varphi} \varpi \frac{d\Omega}{d\varpi} \, dz = \frac{1}{2\pi\varpi} \mathcal{T} \frac{d\Omega}{d\varpi}, \tag{7.22}$$

if we make use of equations (7.6) and (7.13). For thin disks, terms in the heat equation containing z derivatives dominate terms containing ϖ derivatives; hence, the heat equation has the approximate balance

$$\frac{\partial}{\partial z}(F_{\text{rad}} + F_{\text{cond}} + F_{\text{conv}}) = \Psi. \tag{7.23}$$

If we integrate equation (7.23) over all z, and if we assume that the (initially optically thick) diffusive heat flux eventually turns purely radiative, with a rate $\sigma_{\text{SB}} T_{\text{eff}}^4$ (per unit area) for each of the upper and lower faces, equation (7.23) becomes

$$2\sigma_{\text{SB}} T_{\text{eff}}^4 = \frac{\mathcal{T}}{2\pi\varpi} \frac{d\Omega}{d\varpi}, \tag{7.24}$$

with σ_{SB} being the Stefan-Boltzmann constant, and T_{eff} being the local effective temperature of the disk.

For a steady-state accretion disk, far from the inner boundary, equation (7.20) gives

$$\mathcal{T} \approx -\dot{M}_{\text{d}} j \propto \varpi^{1/2},$$

for Keplerian rotation with $j = \varpi^2 \Omega \propto \varpi^{1/2}$. In the disk proper, $\varpi \gg \varpi_{\text{i}}$; and $T_{\text{eff}} \propto \varpi^{-3/4}$ for steady viscous accretion. The basic reason for this result is simple. According to equations (7.20) and (7.22), the rate of viscous dissipation in an annulus, where $\varpi \gg \varpi_{\text{i}}$, equals

$$(2\pi\varpi \, d\varpi) \int_{-\infty}^{+\infty} \Psi \, dz \approx -\dot{M}_{\text{d}} \varpi^2 \Omega \frac{d\Omega}{d\varpi} d\varpi = \frac{3}{2} \frac{GM\dot{M}_{\text{d}}}{\varpi^2} d\varpi$$

for a Keplerian disk. Thus the dissipation rate equals 3/2 times the local release of gravitational energy if we were to merely lower matter at a rate $\dot{M}_{\text{d}}$ across a gravitational potential drop, $d(-GM/\varpi)$. A quantity equal to 1/2 times $GM\dot{M}_{\text{d}} \, d\varpi/\varpi^2$ is readily understandable as the actual usable part of the gravitational energy release, because the other 1/2 goes into speeding up the orbital motion. The remaining part, 1 times $GM\dot{M}_{\text{d}} \, d\varpi/\varpi^2$, results because of the difference in the rate of doing work (angular velocity times torque) on the two edges of the annulus by the viscous forces:

$$d(\Omega\mathcal{T}) \approx -\dot{M}_{\text{d}} \, d(\varpi^2\Omega^2) = \frac{GM\dot{M}_{\text{d}}}{\varpi^2} d\varpi.$$

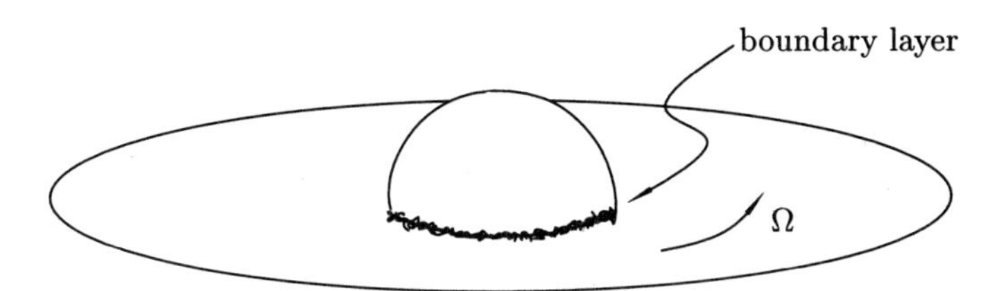

FIGURE 7.4
Conventional thinking places, between the disk proper and a slowly spinning star, a narrow boundary layer where the tangential velocity decreases abruptly from full Keplerian values nearly to zero. The attendant dissipation of orbital kinetic energy in a narrow annular ring heats the gas to high temperatures; so high-frequency radiation is emitted by the boundary layer.

Thus the viscous torque behaves like Robin Hood, stealing from the energy rich (the inner parts of the disk) to give to the energy poor (the outer parts of the disk), so that the local release is three times the value that one might have naively expected, $d(GM\dot{M}_{\rm d}/2\varpi)$. Apart from this factor of three, the energy release locally in each annulus of constant fractional thickness (i.e., an annulus where $d\varpi \propto \varpi$) is proportional to $1/\varpi$; and the energy radiated per unit area $\sigma_{\rm SB}T_{\rm eff}^4$ is proportional to $1/\varpi^3$, yielding the result $T_{\rm eff} \propto \varpi^{-3/4}$. (Q.E.D.)

Near $\varpi_{\rm i}$, where $\mathcal{T} \propto (j - j_0)$, the local energy dissipation falls below the value appropriate for the disk proper. In fact, the total energy dissipated from ∞ to $\varpi_{\rm i}$, without accounting for the boundary layer, equals the integral

$$\int_{\varpi_{\rm i}}^{\infty} 2\pi\varpi\, d\varpi \int_{-\infty}^{+\infty} \Psi\, dz = -\dot{M}_{\rm d} \int_{\varpi_{\rm i}}^{\infty} [\varpi^2\Omega(\varpi) - \varpi_{\rm i}^2\Omega(\varpi_{\rm i})] \frac{d\Omega}{d\varpi}\, d\varpi$$

For a Keplerian disk, this integral equals the total rate of orbital energy release:

$$\frac{GM\dot{M}_{\rm d}}{2\varpi_{\rm i}}.$$

The other half of the gravitational potential energy is stored as Keplerian rotation, and this half can all be released in a boundary layer, if the central object has a surface (in the sense of the outer limit of an inner solution) that remains at rest with respect to an inertial frame (Figure 7.4).

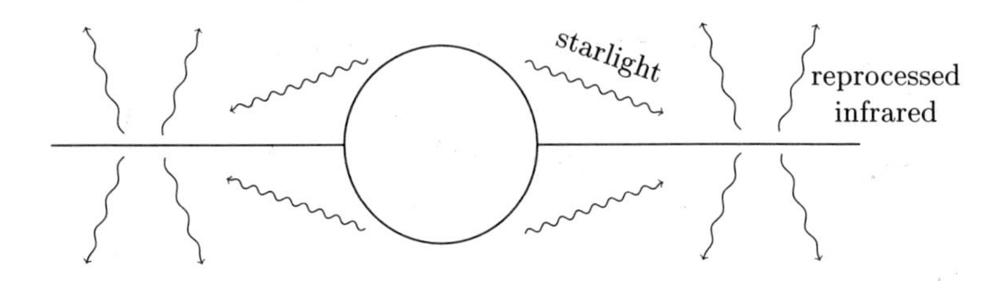

FIGURE 7.5
The reprocessing of starlight shining on a flattened circumstellar disk will lead to a radial distribution of temperature that has the same power-law dependence, $\propto \varpi^{-3/4}$ for $\varpi \gg R_*$, as the steady-state model of viscous accretion.

SPECTRAL ENERGY DISTRIBUTION

A disk with a power-law temperature profile that extends over a large dynamic range in radii tends to radiate a power-law spectral-energy distribution. To see this, notice that each (optically thick) annulus of area $2\pi\varpi\, d\varpi$ contributes mostly photons at the peak of the local blackbody function having frequency $\nu \propto T_{\text{eff}} \propto \varpi^{-3/4}$. Thus the spectral-energy distribution of the emergent radiation has the form

$$L_\nu\, d\nu \propto T_{\text{eff}}^4\, \varpi\, d\varpi \propto \varpi^{-2}\, d\varpi \propto \nu^{1/3}\, d\nu,$$

since $\varpi \propto \nu^{-4/3}$. Consequently,

$$\nu L_\nu \propto \nu^{4/3} \tag{7.25}$$

for the optically thick emission that arises from a classical viscous accretion disk (one that has a Keplerian rotation curve).

At one time, it was thought that a continuum energy distribution of the form given by equation (7.25) would provide the distinguishing spectral feature of a viscous accretion disk. Then a number of researchers independently pointed out that simple reprocessing of light from a central luminous ball illuminating a spatially flat disk would also produce a power-law energy distribution having the dependence $\nu L_\nu \propto \nu^{4/3}$ (Figure 7.5).

The reason is that the amount of light that falls per unit area on a spatially flat disk from a sphere of radius R drops off as the product of the usual dilution factor $1/\varpi^2$ and a projection factor which becomes R/ϖ for $\varpi \gg R$. Thus, $T_{\text{eff}} \propto \varpi^{-3/4}$, and simple reprocessing of the light from an extended central object can mask the effects of true disk accretion.

8

Fluid Instabilities

Reference: Chandrasekhar, *Hydrodynamic and Hydromagnetic Stability.*

So far we have considered only fluids that evolve smoothly in space and time, i.e., *laminar* flows. We wish now to examine the stability of such configurations, existing initially either in hydrostatic equilibrium or in steady flow. Broadly speaking, we foresee a program which concerns itself with the following progressively harder questions:

(a) the criterion for instability of an infinitesimal (inviscid) perturbation,

(b) the growth rate and behavior of the mode of instability (i.e., eigenvalues and eigenfunctions in a normal mode analysis),

(c) the possible stabilizing influence of diffusive effects such as viscosity or heat conduction (or radiative losses, etc.),

(d) the eventual nonlinear resolution of the instability and its back reaction on the basic state.

In this chapter we concern ourselves with physical derivations of goal (a) in the above list for a variety of astrophysically interesting instabilities. Instability criteria can most safely be deduced by a full formal analysis of the governing fluid equations; the cited reference at the beginning gives many excellent examples of this approach. As an introduction to the subject, however, we prefer a more heuristic treatment in order to develop better intuition for the relevant physics.

CONVECTIVE INSTABILITY

We begin with a derivation of *Schwarzschild's criterion* for the convective instability of a homogeneous fluid. In astrophysics, we often need to consider fluids heated from below which develop a vertical temperature stratification (with respect to gravity) that has cold gases overlying hot (see

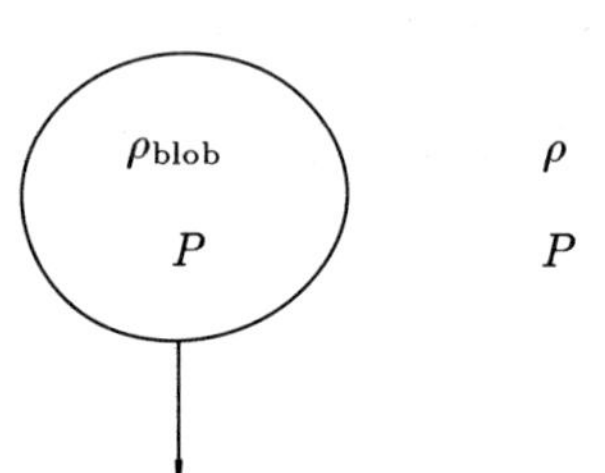

FIGURE 8.1
A heuristic analysis of convective instability considers the displacement of a small blob of gas in pressure equilibrium with its surroundings, vertically stratified as a consequence of hydrostatic equilibrium in a gravitational field. The upward (downward) displacement, assumed to occur adiabatically, grows as a result of the buoyancy excess (deficit) if the density of the blob ρ_{blob} decreases (increases) in response to the changing pressure P faster than the ambient density ρ decreases (increases). On the other hand, if ρ_{blob}—at constant specific entropy s—changes less quickly than the change of ρ—because of, for example, the requirements of stellar structure—the medium is convectively stable, and the displaced blob will be sent back to its original position.

Figure 8.1). (The Earth's atmosphere forms such an example, since the Sun's rays heat the ground, which reradiate infrared radiation that warms the overlying air.) Since warm fluid is buoyant relative to cool, when does such an adverse temperature gradient become unstable with respect to the tendency to develop overturning motions (thermal convection)?

Imagine isolating a blob of gas immersed in the ambient medium. We assume the blob to be initially in thermal and mechanical equilibrium with its surroundings, and we wish to analyze whether a small displacement of the blob would lead to forces that restore the blob to its original position (stability), or that accelerate the blob further away from equilibrium (instability). If the blob is small, the sound-travel time across it to establish pressure equilibrium (same P) with the surroundings will be shorter than the corresponding time required for heat to diffuse across it. In other words, if the blob were to move by a slight amount downward (to make dP positive), say, it would maintain pressure equilibrium by compressing *adiabatically* (constant s) in the new ambient medium of higher P. Thus the change in the internal density of the blob is related to the change of its (and the external medium's) pressure via

$$(d\rho)_{\text{blob}} = \left(\frac{\partial \rho}{\partial P}\right)_s dP. \tag{8.1}$$

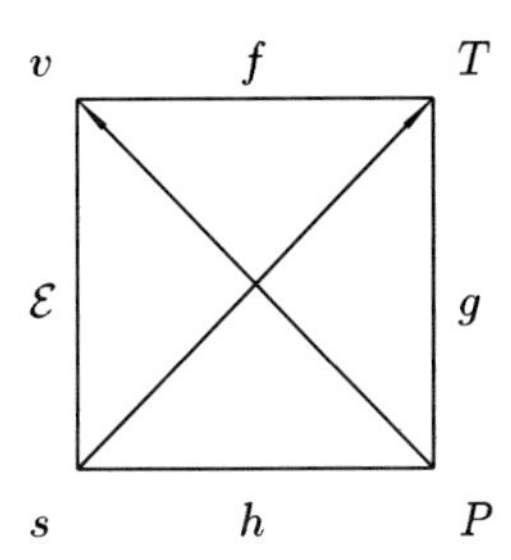

FIGURE 8.2
The thermodynamic square.

The corresponding change in the density of the ambient medium equals

$$(d\rho)_{\text{ambient}} = \left(\frac{\partial \rho}{\partial P}\right)_s dP + \left(\frac{\partial \rho}{\partial s}\right)_P ds, \tag{8.2}$$

where dP and ds are the differences of the pressure and specific entropy of the medium at the new position compared to the old that arise because of the constraints of mechanical and thermal balance (see, e.g., Chapter 5). The blob will continue to move downward if it suffers a buoyancy deficit relative to its new surroundings, i.e., if its density increase exceeds that of the ambient medium: $(d\rho)_{\text{blob}} > (d\rho)_{\text{ambient}}$. Our assumption of a small blob allows us to take the two dP's in equations (8.1) and (8.2) to be equal. Thus *convective instability* will arise if the ambient medium satisfies

$$\left(\frac{\partial \rho}{\partial s}\right)_P ds < 0 \qquad \text{for a downward displacement.} \tag{8.3}$$

The quantity $(\partial\rho/\partial s)_P$ does not lend itself to immediate thermodynamic interpretation; we may transform it to a more familiar term by invoking one of Maxwell's relations,

$$\left(\frac{\partial \rho^{-1}}{\partial s}\right)_P = \left(\frac{\partial T}{\partial P}\right)_s. \tag{8.4}$$

A handy way to remember this and other thermodynamic relations is to draw the square depicted in Figure 8.2 with the accompanying mnemonic for the letters: "Harry Schwartz eagerly votes for the greatest person," with h = specific enthalpy, s = specific entropy, $\mathcal{E}$ = specific internal energy, $v = \rho^{-1}$ = specific volume, f = specific free energy, T = temperature, g = specific Gibbs energy, and P = pressure.

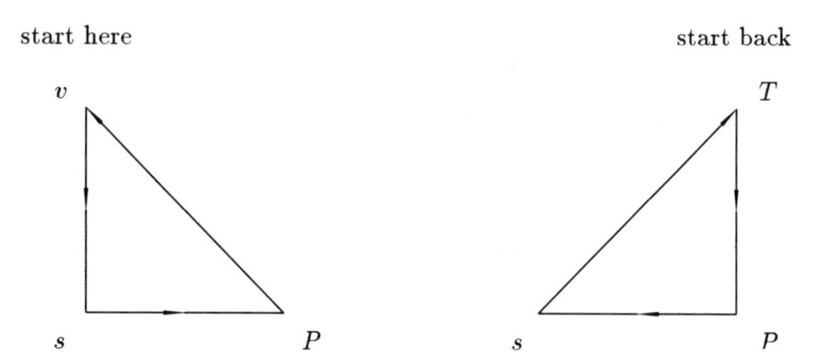

FIGURE 8.3
The four corners of the thermodynamic square, taken three at a time, that yield the Maxwell relation, $(\partial v/\partial s)_P = (\partial T/\partial P)_s$. The two terms have the same sign because the diagonal arrows that connect the quantities being held constant to the variables in the numerators of the partial derivatives go in the same sense.

The fundamental dependence of any energy on its natural variables follows by looking at the appropriate side of the square and its adjacent corners—for example, $\mathcal{E}$ as a side, with s and T as its corners, implies $d\mathcal{E}$ is given in terms of ds and dv. To find the coefficients, follow the diagonals to the opposite corners and use the sense of the arrows to determine the sign; thus,

$$d\mathcal{E} = T\,ds - P\,dv.$$

Similarly, we have $dh = T\,ds + v\,dP$, etc. The Maxwell relations (expressing the equivalence of taking, in either order, the mixed second derivative of an energy with respect to any two variables) follow from examining the corners and arrows of the square (see Figure 8.3); thus equation (8.4)—associated with the mixed second derivative $\partial^2 h/\partial P \partial s$—corresponds to the Maxwell relation, $(\partial v/\partial s)_P = (\partial T/\partial P)_s$, that we have displayed as equation (8.4).

Another Maxwell relation, for example, reads $(\partial s/\partial P)_T = -(\partial v/\partial T)_P$, involving the corners and arrows shown in Figure 8.4.

In any case, all thermodynamically stable substances have temperatures that increase upon adiabatic compression, $(\partial T/\partial P)_s > 0$; so equation (8.4) implies that $(\partial \rho/\partial s)_P < 0$, and equation (8.3) becomes *Schwarzschild's criterion for convective instability*,

$$ds > 0 \qquad \text{in the direction of gravity.} \tag{8.5}$$

To remember this result, associate specific entropy with buoyancy—high-entropy material has high buoyancy. If the entropy of a star or a planetary

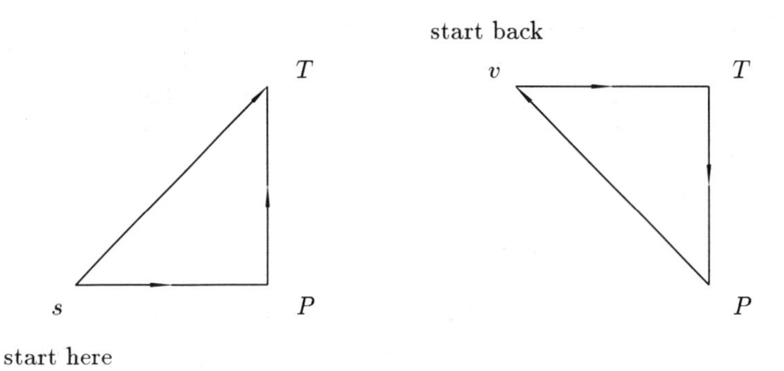

FIGURE 8.4
The four corners of the thermodynamic square, taken three at a time, that yield the Maxwell relation, $(\partial s/\partial P)_T = -(\partial v/\partial T)_P$. The two terms have opposite signs because the diagonal arrows that connect the quantities being held constant to the variables in the numerators of the partial derivatives go in opposite senses.

atmosphere increases inward, then more-buoyant material underlies less-buoyant material, and the medium has a tendency to overturn. The overturning mixes the entropy—high entropy (i.e., hotter than average material) rising and low entropy (i.e., colder than average material) sinking—so there results a net (convective) transport of heat outward. The nonlinear resolution of this mixing must be to produce a more nearly uniform value of s; i.e., if heat transport by radiation plus conduction alone requires a *superadiabatic temperature gradient* (i.e., a gradient that corresponds to s increasing inward), then convection would set in that tends to make the distribution of s more nearly uniform. This line of argument suggests that the convective heat flux $\mathbf{F}_{\text{conv}}$ should be proportional to $-\nabla s$, and the dimensional consideration of Chapter 10 on the "mixing length theory" yields a proportionality constant that can be regarded as large in stellar interiors (and some planetary atmospheres) because the competing effects—heat conduction and radiative diffusion—are so ineffective. For $\mathbf{F}_{\text{conv}}$ to carry away the part of the heat generated in the interior that can not be transported by radiation and conduction, then, requires only a very small superadiabatic gradient, i.e., a small value for $-\nabla s$. In regions where convection is efficient, therefore, it constitutes a good zeroth approximation to take

$$s = \text{constant}. \tag{8.6}$$

For example, in the troposphere of the Earth [beneath the clouds, where the atmosphere is convective (with warm air currents carrying the moisture upward that condenses into clouds) and where airplane rides are bumpy], meteorologists often take the temperature to drop vertically at an *adiabatic lapse rate* that corresponds to the approximation in equation (8.6).

ROTATIONAL INSTABILITY

Next, we wish to derive *Rayleigh's criterion* for the instability of a rotating fluid. Consider an axisymmetric fluid distribution (cylinder or flattened disk) which rotates about its axis under a combination of gravitational attraction and pressure gradient:

$$-\varpi\Omega^2(\varpi) = g - \frac{1}{\rho}\frac{dP}{d\varpi}. \tag{8.7}$$

Extreme cases correspond to interstellar gas in the Galaxy, where the gravitational term $g = -d\mathcal{V}/d\varpi$ totally dominates over the term $-\rho^{-1}dP/d\varpi$; and *Couette flow* between two rotating cylinders in the laboratory, where the only term of importance on the right-hand side in the fluid interior is $-\rho^{-1}dP/d\varpi$.

We now consider the question of whether the rotational equilibrium represented by equation (8.7) is stable. To avoid complications associated with (radial) buoyancy effects, we confine the discussion to an incompressible fluid, ρ = constant. Furthermore, let us restrict our attention to axisymmetric disturbances. Imagine now interchanging the material in two rings of radii ϖ_1 and ϖ_2, with $\varpi_2 > \varpi_1$ for definiteness. Since the system preserves axial symmetry, the specific angular momentum j of a ring, which equals $\varpi^2\Omega(\varpi)$ initially, will be conserved. The centrifugal force on ring 1, now at ϖ_2, equals j_1^2/ϖ_2^3, and the net acceleration acting on this ring has the expression

$$\frac{j_1^2}{\varpi_2^3} + g_2 - \frac{1}{\rho}\frac{dP}{d\varpi}(\varpi_2).$$

From the equilibrium condition (8.7), we have

$$g_2 - \frac{1}{\rho}\frac{dP}{d\varpi}(\varpi_2) = \frac{j_2^2}{\varpi_2^3}.$$

Consequently, the net acceleration acting on ring 1 at its new position equals

$$\frac{j_1}{\varpi_2^3} - \frac{{j_2}^2}{\varpi_2^3} = \frac{1}{\varpi_2^3}(j_1^2 - j_2^2),$$

which points in the outward direction, continuing to push the ring in the direction of its initial displacement, if $j_2^2 < j_1^2$, i.e., if the specific angular momentum decreases outward. Thus *Rayleigh's criterion* reads

$$\frac{d}{d\varpi}\left[(\varpi^2\Omega)^2\right] < 0 \qquad \text{for rotational instability.} \tag{8.8}$$

The square of the epicyclic frequency, defined by

$$\kappa^2 \equiv \frac{1}{\varpi^3}\frac{d}{d\varpi}[(\varpi^2\Omega)^2], \tag{8.9}$$

is sometimes referred to as the *Rayleigh discriminant.* The epicyclic frequency has the following relation to particle orbits (see, e.g., Binney and Tremaine, *Galactic Dynamics*). If a particle traveling in a circular trajectory at angular speed $\Omega(\varpi)$ is kicked (in the plane of the orbit) slightly away from this trajectory, it will gyrate (in an elliptical epicycle) about the radius of its original circle with a frequency equal to κ, assuming κ to be real. When κ^2 is positive, then, the displacements oscillate about a fixed mean position, and the circular orbit is stable to small perturbations. If the frequency κ is imaginary (κ^2 negative), then the resulting orbit will begin to depart exponentially (with a growth rate equal to $|\kappa|$) from the original trajectory, and the circular orbit is unstable. Although we have carried out our derivation for systems satisfying Newtonian mechanics, there exists, in fact, a straightforward relativistic generalization of this result. In particular, circular orbits around a black hole become unstable outside the event horizon because the angular momentum required for smaller circular orbits increases, rather than decreases, as the particle approaches the event horizon (see Misner, Thorne, and Wheeler, *Gravitation*).

G. I. Taylor's theoretical and experimental studies of Couette flow between two concentric cylinders set rotating at two different angular speeds, Ω_i and Ω_o (see Figure 8.5), show that if Rayleigh's criterion for instability is satisfied, the presence of viscosity can keep the flow stable for a little while longer. The critical Reynolds number is typically on the order of 10^2 to 10^3 depending on the parameters of the flow.

Theoretical analysis of the problem produces the stability diagram shown in Figure 8.6 (consult Lin, *The Theory of Hydrodynamical Stability*).

If the parameters of the flow fall well within the region of instability, so many modes of instability set in experimentally that the flow becomes fully turbulent. Consider now flows at low Reynolds numbers (points A, B, and C in Figure 8.6). Imagine keeping the slope $\Omega_\mathrm{i}\varpi_\mathrm{i}^2/\Omega_\mathrm{o}\varpi_\mathrm{o}^2$ constant, but increasing the Reynolds number, $\Omega\varpi^2/\nu$ for either the inner or the outer cylinder, by decreasing ν or increasing Ω_i and Ω_o simultaneously. The points A, B, and C will move in the process along straight lines which extend from the origin through A, B, and C to A', B', and C'. Clearly,

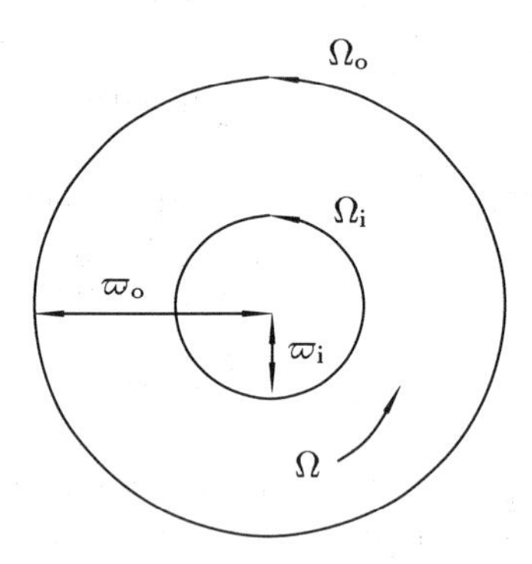

FIGURE 8.5
A face-on view of the situation for Couette flow between an inner cylinder of radius ϖ_i rotating at angular velocity Ω_i and an outer cylinder of radius ϖ_o rotating at angular velocity Ω_o.

points A and B, which satisfy Rayleigh's *inviscid* criterion for instability but are stabilized initially by the effects of viscosity, will eventually move into the unstable regime and give rise to turbulent flow. On the other hand, point C, which is originally stable by Rayleigh's criterion, *continues to remain stable at arbitrarily large Reynolds numbers.* (After all, Rayleigh's criterion was derived for inviscid flow, i.e., for *infinite* Reynolds numbers.)

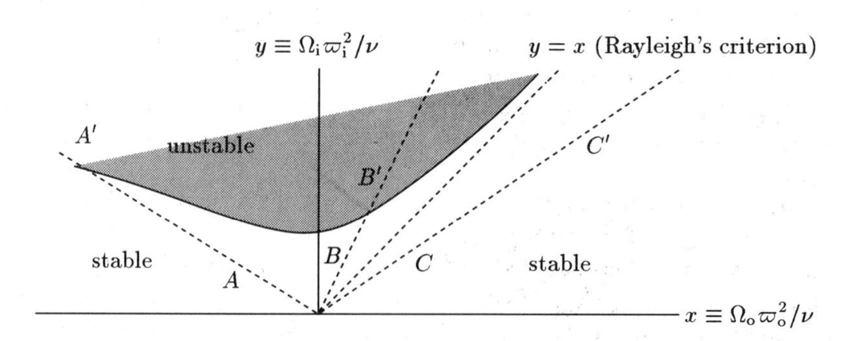

FIGURE 8.6
The stability diagram for the cylindrical Couette flow depicted in Figure 8.5.

The oft-repeated statement that differentially rotating flows become unstable at large Reynolds numbers of order 10^2 or 10^3 therefore refers to cases A and B, not to C, i.e., to flows which are *unstable* according to an inviscid criterion, but which can be stabilized at low Reynolds numbers by the effects of molecular viscosity. Differentially rotating flows that are *stable* to begin with, according to the inviscid criterion, become *even more stable* by the inclusion of the effects of viscosity; i.e., they are stable at *all* Reynolds numbers. In other words, the Reynolds number of the circulation has no relevance to the question of the development of turbulence (that could act as a source of anomalous viscosity) in a stably stratified rotational flow for which the specific angular momentum of the fluid increases outward, *as it does in all models of accretion disks.*

The above result would appear very ominous for the hypothesis of a turbulent origin for the anomalous viscosity of accretion disks if Chandrasekhar had not shown that the inclusion of even a slight magnetic field changes matters completely. (Indeed, the instability cited below disappears if the magnetic field is too strong.) The presence of a weak B_z oriented parallel to the rotation axis of an electrically conducting cylinder of fluid turns out to modify the instability criterion from equation (8.8) to

$$\frac{d}{d\varpi}\left(\Omega^2\right) < 0 \qquad \text{for rotational instability.} \tag{8.10}$$

The importance of this modification comes with the realization that while no theoretical model of an accretion disk has a distribution of specific angular momentum that decreases outward, they (as well as every observationally resolved astronomical disk) do all have angular velocities that decrease outward. We defer further discussion of this interesting issue until Chapter 22, after we have derived the fundamental equations of magnetohydrodynamics and recovered the additional material waves allowed by the presence of embedded magnetic fields.

RAYLEIGH-TAYLOR AND KELVIN-HELMHOLTZ INSTABILITIES

A configuration of two distinct fluids (e.g., air and water) in pressure equilibrium across a common interface can spontaneously develop two common types of instabilities. *Rayleigh-Taylor* instability occurs when a heavy fluid rests on top of a light fluid in an effective gravitational field $\mathbf{g}$. Such an adverse arrangement would seem unlikely to exist as an initial state if $\mathbf{g}$ represents a "real" gravitational field, but the situation can arise naturally if $\mathbf{g}$ were actually associated with acceleration or deceleration of an entire region. For example, Figure 8.7 shows a hydrodynamical simulation of the Rayleigh-Taylor instability resulting from the effective reversal of gravity when a supernova shock wave runs through layers of a star with a radial

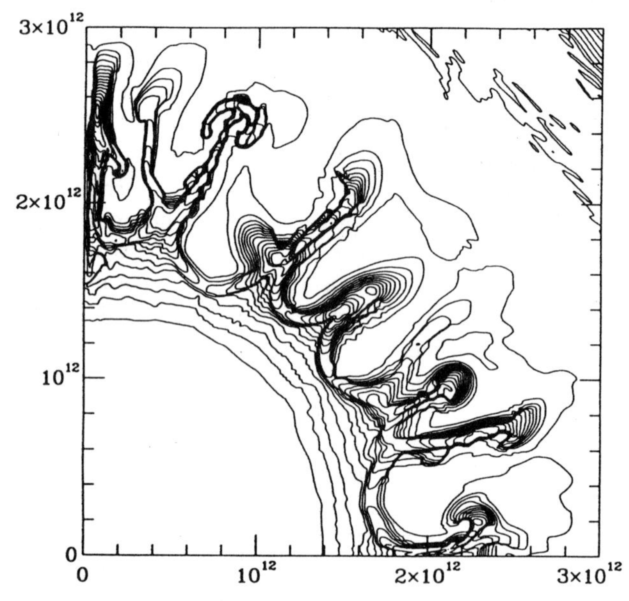

FIGURE 8.7
Density contours in the hydrodynamical simulation of the passage of a supernova shock wave running through a model for SN 1987A with a pre-explosion radius 3×10^{12} cm. The calculation assumes rotational symmetry about the vertical axis, and the shock front has propagated off the grid at a time 9814 s after the initial explosion. Rayleigh-Taylor instability (followed by Kelvin-Helmholtz instability) has led to the penetration of heavy elements into the helium mantle of the star. [From D. Arnett, B. Fryxell, and E. Muller 1989, *Ap. J. (Letters)*, **341**, L63.]

stratification of heavy elements. Energetically, the heavy fluid prefers to be on the "bottom" (outward from the center), and the system wants to overturn. Since the light and heavy fluids can not interpenetrate freely, the overturning gets accomplished via "fingers" that drip past one another. The resulting mixing of heavy elements throughout the envelope of the supernova remnant can have observational consequences. Indeed, the simulation reported in Figure 8.7 was motivated by x-ray and gamma-ray observations of Supernova 1987A which indicated that radioactive cobalt is more thoroughly distributed among the explosive debris than was predicted by model calculations of thin-shell nucleosynthesis in the presupernova star.

Kelvin-Helmholtz instability occurs when two fluids in contact try to slip past one another via a tangential discontinuity. A jet of gas that tries to propagate through an ambient medium (usually taken to be stationary)

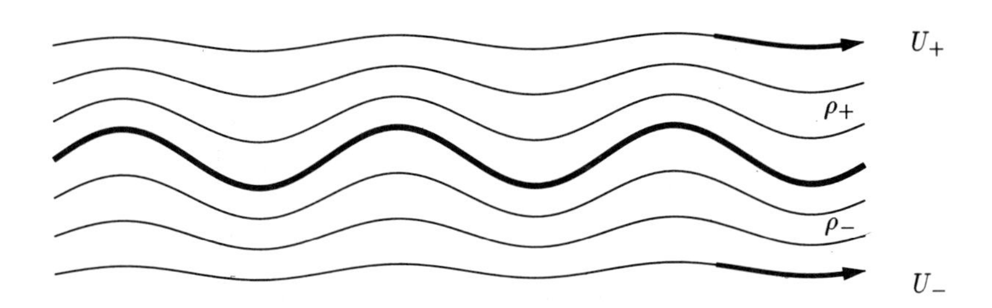

FIGURE 8.8
Consider the problem of one incompressible fluid of density ρ_+ flowing horizontally at speed U_+ over another of density ρ_- at a different speed U_-. The system lies in a vertical gravitational field g, and the interface has an associated surface tension T. Rayleigh-Taylor instability is absent if the lighter fluid overlies the denser, i.e., if $\rho_- > \rho_+$; however, Kelvin-Helmholtz instability will arise when the square of the slip speed satisfies

$$(U_+ - U_-)^2 \geq \frac{2(\rho_+ + \rho_-)}{\rho_+ \rho_-} [Tg(\rho_- - \rho_+)]^{1/2}.$$

For air over water, the critical speed, whose square is given by the right-hand side, equals $650\,\mathrm{cm\,s^{-1}}$ (12.5 knots), the observed value for when the wind will begin to ruffle a quiet sea and seagulls begin to soar.

provides a well-known astrophysical example, although additional complications associated with compressibility and supersonic flow (e.g., internal shock waves) complicates this particular example in comparison with the usual cases cited (e.g., the rippling of the surface of a lake by a wind that blows over it). The centrifugal force accompanying a curvature of either sign of the interface provides the mechanism for increasing any initial undulations (see Figure 8.8).

The two instabilities often appear together in the same problem. For example, Kelvin-Helmholtz instability can distort the opposed fingers of heavy and light fluid that try to slip past one another from the nonlinear development of a Rayleigh-Taylor instability. When we generalize the considerations from sudden jumps of fluid properties across sharp interfaces to more gentle gradients in extended regions, Rayleigh-Taylor instabilities fall in the same class of buoyancy-driven disturbances as thermal convection. Kelvin-Helmholtz instabilities then belong to the same category as the shear-flow disturbances to be discussed in Chapter 9. Chandrasekhar in his book *Hydrodynamic and Hydromagnetic Stability* works out many detailed examples of these two types of instabilities, and we shall not repeat those calculations here. Instead, we content ourselves by deriving via physical argument the *Richardson criterion* for the stability of an inviscid region containing a favorable buoyancy gradient but an adverse velocity gradient (see also Chandrasekhar, pp. 487–492).

Consider a gas that has a stable entropy gradient, $ds/dz < 0$, in a gravitational field g that points in the $-z$ direction. Suppose further that the gas flows in the x-direction with variable velocity $U(z)$, such that the shear dU/dz would give rise to Kelvin-Helmholtz instability, were it not for the stabilizing effect of the entropy gradient ds/dz. The Richardson criterion results from asking: How large can dU/dz get before ds/dz no longer provides a sufficient stabilizing influence?

Our derivation of the Richardson criterion proceeds via energy considerations. Consider first the buoyant work performed on a small blob that moves a distance dz. We suppose that the size of the blob times $|dU/dz|$ is much less than the adiabatic sound speed a_s, so that the blob maintains pressure equilibrium with the ambient medium when displaced. Equations (8.1) and (8.2) then imply that the buoyant force per unit volume at the end of the displacement has the expression

$$g\left[(d\rho)_{\text{blob}} - (d\rho)_{\text{ambient}}\right] = -g\left(\frac{\partial\rho}{\partial s}\right)_P ds,$$

where

$$\left(\frac{\partial\rho}{\partial s}\right)_s = \frac{(\partial\rho/\partial T)_P}{(\partial s/\partial T)_P} = \frac{T}{c_P}\left(\frac{\partial\rho}{\partial T}\right)_P < 0.$$

The average force per unit volume felt by the blob during the displacement equals half of the above value. Thus the work required per unit mass to interchange *two* similar blobs, initially at z and $z + dz$, equals

$$(dW)_{\text{buoy}} = -\frac{g}{\rho}\left(\frac{\partial\rho}{\partial s}\right)_P ds\, dz. \tag{8.11}$$

With $(\partial\rho/\partial s)_P < 0$, $(dW)_{\text{buoy}}$ is positive for convectively stable regions where $ds\, dz > 0$. In other words, it takes work to overturn gas in a region stable to convection.

Consider next the release of shear energy in the interchange of the same two blobs. Let the two blobs initially have speeds U and $U + dU$, and suppose that their mixing yields an average speed $U + (dU/2)$. We then easily compute the difference in kinetic energies per unit mass between the final and initial states as

$$(dK)_{\text{mix}} = \frac{1}{2}\left[U^2 + (U + dU)^2\right] - \frac{1}{2}\left[2\left(U + \frac{dU}{2}\right)^2\right];$$

i.e.,

$$(dK)_{\text{mix}} = \frac{1}{4}(dU)^2, \tag{8.12}$$

a positive-definite expression independent of the sign of dU. In other words, any departure from uniform flow contains extra kinetic energy in the form of shear, which can be tapped, in principle, to drive dynamic mixing (Kelvin-Helmholtz instability). However, the work (8.11) needed to overcome the buoyancy excess will outweigh the extra kinetic energy (8.12) if $(dW)_{\text{buoy}} > (dK)_{\text{mix}}$; i.e., we have stability if

$$-\frac{g}{\rho}\left(\frac{\partial\rho}{\partial s}\right)_P dsdz > \frac{1}{4}(dU)^2.$$

Dividing by $(dU)^2 > 0$, we obtain the Richardson criterion,

$$\text{Ri} \equiv \frac{-g(\partial\rho/\partial s)_P(ds/dz)}{\rho(dU/dz)^2} > \frac{1}{4} \quad \text{for stability.} \tag{8.13}$$

The thermodynamics of compression modifies the naive expression that we might have written down, on dimensional grounds, for the Richardson number,

$$(\text{Ri})_{\text{naive}} = \frac{-g(d\rho/dz)}{\rho(dU/dz)^2},$$

but the competition between shear (destabilizing in the denominator) and buoyancy (stabilizing in the numerator) should be clear nevertheless.

APPLICATION TO THE RADIATIVE CORE OF THE SUN

What happens when application of two or more general criteria yields conflicting answers? For example, will instability arise in the differentially rotating, unmagnetized, radiative interior of a star if it satisfies Rayleigh's criterion for instability, equation (8.8), but the Richardson number Ri for the flow remains larger than 1/4? At first sight, one might tempted to argue that energy is as important a consideration as angular momentum; therefore, if any one constraint suffices to prevent the onset of the dangerous mode of motion, we have stability. In other words, if no net excess energy exists to drive the instability, disturbances cannot grow. This argument overlooks the difficulty that the presence of an adverse angular momentum gradient may modify the energetics of disturbances calculated assuming the absence of rotation, i.e., assuming rectilinear flow. Moreover, displacements in certain directions (e.g., perpendicular to $\mathbf{g}$) and special geometries (e.g., spherical rather than rectilinear) not considered in the simpler analyses may make a crucial difference to the conclusions. Thus, general criteria form useful guides, but when uncertainty develops, nothing can replace a full stability analysis that includes all the relevant factors.

Precisely this sort of ambiguity of interpretation, plus the additional complication of the issue of *Ekman pumping* (secondary circulation induced by boundary-layer effects), occurred during the 1960s over debates whether the Sun possesses a core that rotates much faster than the visible surface. The stakes were high, namely, a contribution to the gravitational quadrupole moment of the Sun that would mar the precise agreement between general relativity and the observed precession of Mercury's perihelion. Helioseismology has settled the argument by demonstrating empirically that the radiative core rotates at a small angular speed equal to the that of the mid-latitudes of the outer convective envelope. The overall result complies with general considerations of stability, but the detailed form of the deduced law of differential rotation (isorotation on conical surfaces in the convection zone and rigid rotation of the radiative core) includes some surprising departures from theoretical expectations [see, e.g., T.L. Duvall, J.W. Harvey, and M.A. Pomerantz, 1986, *Nature*, **321**, 500; W.A. Dziembowski, P.R. Goode, and K.G. Libbrecht, 1988, *Ap. J. (Letters)*, **337**, L53]. Magnetic fields may have an important role in this problem.

THERMAL INSTABILITY

In this section we wish to derive *Field's criterion* for thermal instability (G. B. Field, 1965, *Ap. J.*, **142**, 153). Applications potentially encompass all optically thin gases in which thermal balance between heating and cooling leads to an equilibrium relation between density ρ and temperature T of the form [cf. equation (6.3)]

$$\mathcal{L}(\rho, T) = 0. \tag{8.14}$$

In equation (8.14), $\mathcal{L}$ represents the net loss of heat per unit mass of the gas. The net loss function $\mathcal{L} = 0$ in the $\log(T)$-$\log(\rho)$ plane often has the contour shown in Figure 8.9.

The region above the curve corresponds to $\mathcal{L} > 0$, because cooling exceeds heating if the temperature exceeds the equilibrium value for a given density. Conversely, the region below the curve corresponds to $\mathcal{L} < 0$, because the heating there exceeds the cooling.

The basic shape depicted in the figure arises for good atomic-physics reasons, namely the existence of a quantum universe in which level structures in atoms and molecules come in discrete units (see Problem Set 3). At temperatures of 10–100 K, rotational levels of molecules and fine-structure levels of atoms begin to be excited by collisional impacts. The subsequent radiative de-excitation represents a net source of cooling for the gas. This cooling rate depends sensitively on the exact temperature T because the fraction of particles with particle energy ϵ exceeding the threshold energy

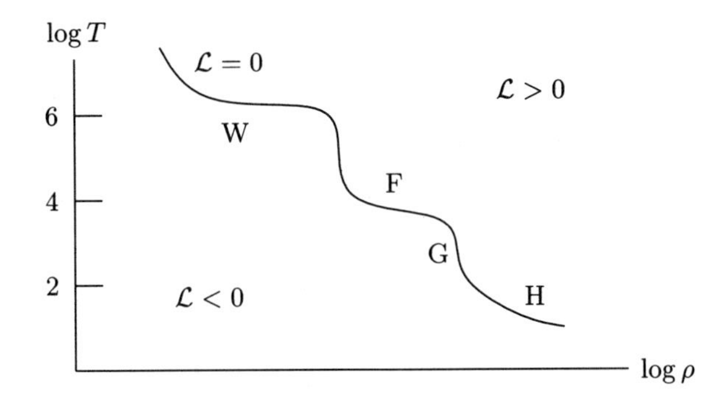

FIGURE 8.9
A schematic contour of thermal balance, $\mathcal{L}(\rho, T) = 0$, separates the log density-log temperature plane into two parts. At temperatures T lower than the value that can be maintained in thermal equilibrium at given density ρ, the gas suffers a net heating, $\mathcal{L} < 0$. At temperatures T greater than the equilibrium value at given ρ, the gas suffers a net radiative loss, $\mathcal{L} > 0$. Plateaus (H, F, W) of temperature at $T \sim 10^2$ K, 10^4 K, and 10^6 K, where heating balances cooling over a relatively wide range of density, arise for the atomic physics reasons explained in the text.

drops off in a Maxwell-Boltzmann distribution as $e^{-\epsilon/kT}$. Thus, for this cooling to balance whatever source of heating exists, T need only change a little (in this range) when ρ changes a lot. As T begins to approach and exceed 10^3 K, however, the Boltzmann factor $e^{-\epsilon/kT}$ approaches unity for values of ϵ relevant to excited rotational or fine-structure levels. At this point, almost all the particles in the thermal distribution can participate anyway in inelastic collisions; so we need large changes in T (to increase the availability pool slightly) to compensate for small changes in ρ if cooling is to balance heating. When T approaches $\sim 10^4$ K, enough particles in the distribution have sufficient energies to collisionally excite the lower-lying electronic states of the common elements, including hydrogen, and small increases in T are again sufficient to balance large decreases in ρ. The same pattern repeats itself as T approaches and exceeds 10^5 K, with temperature sensitivity restored at $\sim 10^6$ K, when the inner shells of elements like oxygen and iron begin to come into play in the radiative cooling processes. Such line cooling again becomes inefficient much above $10^7 K$, when the dominant form of radiative loss usually involves free-free emission.

Imagine now that we have a static, uniform, homogeneous gas in thermal equilibrium at some ρ and T. Consider a small blob embedded in this medium, and perturb it away from the equilibrium curve along a locus

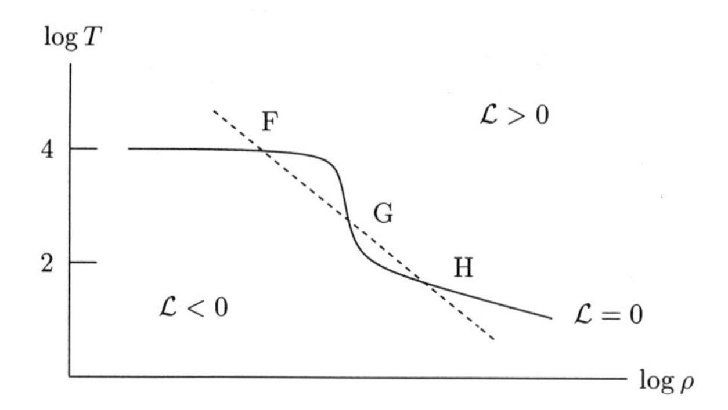

FIGURE 8.10
Thermal equilibrium (solid curve) in the part of the log density-log temperature plane relevant to the two-phase model of the interstellar medium. The dashed line shows a fiducial locus for an isobar, $P \propto \rho T =$ constant, with a choice of the pressure level that yields three possible phases in pressure equilibrium with one another. For the reasons explained in the text (see also Problem Set 3), phases F and H are thermally stable, but phase G is thermally unstable.

$P \propto \rho T =$ constant. [We ignore possible changes of the ionization fraction (and molecular weight) which would complicate the form of the instability criterion we are about to derive.] The assumption of constant P again invokes the idea that acoustic equilibrium with the surroundings can be established for a small blob on a time scale short in comparison with the thermal time (the situation most susceptible to a thermal condensation or rarefaction, because the blob does not have to contend with pressure discrepancies). The locus $P \propto \rho T =$ constant lies along a $-45°$ line in a $\log \rho$-$\log T$ plot (see Figure 8.10).

Consider, first, the situation when the medium exists at point F, and we displace the blob slightly to lower temperatures and higher densities along the locus $P =$ constant. According to the diagram, the blob enters a region where $\mathcal{L} < 0$, i.e., where the heating exceeds the cooling. Thus the blob must heat up again and re-expand back toward point F. Similarly, we easily conclude that the blob will return to F if we initially displace it to higher temperatures and lower densities where $\mathcal{L} > 0$, i.e., where the cooling exceeds heating. Gas placed anywhere in the plateau region F ($T \sim 10^4$ K) is therefore thermally stable. We can similarly conclude the same for the plateau regions H (with $T \sim$ 10-100 K) and W (for Weymann, who first investigated the thermal stability of coronal gas in the Sun with $T \sim 10^6$ K).

Consider, however, if the medium existed in the cliff region G (with $T \sim 10^3\,\mathrm{K}$). If we take a piece of such a medium and displace it toward lower temperatures and higher densities, it will now enter a region where $\mathcal{L} > 0$, i.e., where cooling exceeds heating. Thus, maintaining the same pressure as its surroundings, such a blob would get cooler and cooler, denser and denser, until it makes a phase transition to become H-type gas (interstellar clouds at temperatures of 10–100 K). Conversely, if we were to displace a blob of G-type gas to higher temperatures and lower densities, it will go to where $\mathcal{L} < 0$, i.e., where heating exceeds cooling. Now, maintaining pressure equilibrium with its surroundings, it will expand to become hotter and hotter, more and more rarefied, until it makes a transition to become F-type gas (warm intercloud medium). In other words, gas artificially placed in the *thermally unstable* phase G, would spontaneously separate into two stable phases, warm rarefied gas F and cold dense gas H, in pressure equilibrium with one another at a common P (if we ignore the effects of self-gravity, magnetic fields, etc.). This idea forms the basis of the *two-phase model* of the interstellar medium [see G. B. Field, D. W. Goldsmith, and H. J. Habing, 1969, *Ap. J. (Letters)*, **155**, L149], which continues to illustrate an elegant physical principle at work in diffuse gases, despite its supplantation in modern work on the interstellar medium by more complex scenarios.

With a little thought, we may translate our geometric construction to the mathematical statement known as *Field's criterion*:

$$\left(\frac{\partial \mathcal{L}}{\partial T}\right)_P < 0 \qquad \text{for thermal instability.} \tag{8.15}$$

If gas satisfies equation (8.15), then the typical growth rate equals $-c_P^{-1}(\partial\mathcal{L}/\partial T)_P$, which yields an estimate for the cooling time of the medium (see Problem Set 3).

According to our general comments concerning the nature of line cooling functions, thermally stable, optically thin gas can exist, not only in the temperature range 10–100 K and 10^4 K, but also at $\sim 10^6$ K. The *corona* of the Sun consists of gas (heated probably by magnetic events) at temperatures comparable to the last value. Indeed, the upper atmosphere of the Sun has a temperature structure that looks schematically as drawn in Figure 8.11.

The *chromosphere* of the Sun corresponds to a thin transition layer that transforms (partially optically thick) gas at a temperature $\sim 6 \times 10^3$ K, characteristic of the *photosphere* of the Sun, to the temperature $\sim 2 \times 10^6$ K characteristic of its (optically thin) corona. Clearly, then, the chromosphere must contain gas at intermediate temperatures, say, at 10^5 K. But according to our analysis, gas in this temperature regime should be thermally unstable. How then can the chromosphere exist as a long-lasting entity?

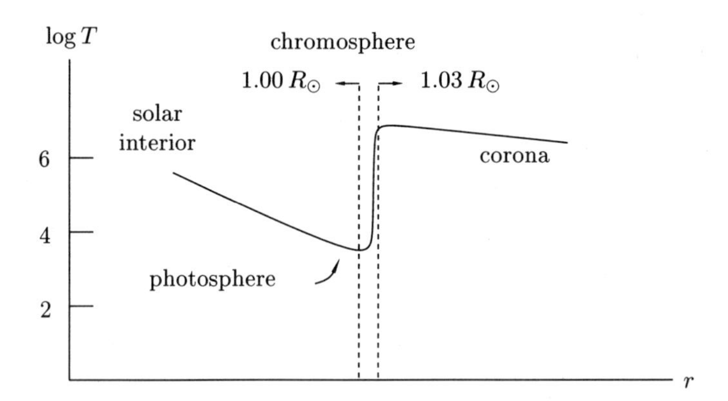

FIGURE 8.11
The solar chromosphere exists at temperatures, from a few thousand to a few hundred thousand K, that would lead us to expect naively that it should be thermally unstable. However, heat conduction from the hot solar corona to the cool solar photosphere strongly affects the structure of the spatially narrow and optically thin chromosphere. This conductive flux probably provides the necessary stabilizing influence for the transition zone from photosphere to corona that the chromosphere represents.

The answer is that in spatially thin layers we cannot ignore the stabilizing effects of diffusive phenomena. At short wavelength scales, thermal runaways can be prevented by heat conduction (see Problem Set 3). The chromosphere exists (in part) because it has a smaller radial extent than the minimum length scale over which thermal instability can overcome heat conduction.

GRAVITATIONAL INSTABILITY

The problem of the gravitational condensation of objects from a uniform background traces its origins to Newton's early speculations; however, Jeans is usually credited with the first quantitative calculation. Within the context of Newtonian physics, however, the problem cannot be posed consistently if Poisson's equation must be satisfied by the equilibrium state:

$$\nabla \cdot \mathbf{g}_0 = -4\pi G\rho_0. \tag{8.16}$$

For a homogeneous and isotropic Newtonian universe, $\mathbf{g}_0 = 0$, since symmetry considerations yield no direction to which the gravitational field could plausibly point. Thus the left-hand side of equation (8.16) must equal zero, which leaves us with an inconsistency if we want the right-hand side to give a nonzero mass density ρ_0 for the medium from which astronomical objects can condense. In his treatment Jeans swept this issue under the rug by ignoring the zeroth-order equations and jumping to the first-order perturbation equations; this course of action is popularly referred to as "the Jeans swindle."

Relativistic cosmology circumvents this problem by letting the universe expand, which modifies the problem of the *Jeans instability* considerably (giving, e.g., algebraic growth with time rather than exponential growth for unstable fragments). A closer analog with Jeans's original formulation, within the context of Newtonian physics, posits that the otherwise static universe rotates uniformly as a whole at an angular velocity $\Omega =$ constant. We may think of this model in terms of a series of self-gravitating uniformly rotating cylinders in which we let the outer radius $R \to \infty$. At each point interior to the cylinder, a centrifugal balance exists between the rotation rate and the self-gravity (given by Gauss's law)

$$-\varpi\Omega^2 = g_0 = -2\pi G\rho_0\varpi;$$

i.e., we choose Ω so that

$$\Omega = (2\pi G\rho_0)^{1/2}. \tag{8.17}$$

Problem Set 3 asks you to consider the problem of acoustic wave propagation, modified by the presence of self-gravitation, in such a configuration. In particular, you are asked to derive the following dispersion relation for waves propagating purely in the z-direction:

$$\omega^2 = k^2a^2 - 4\pi G\rho_0, \tag{8.18}$$

which relates the wave frequency ω and the wave number k of the disturbance. If G were zero, equation (8.18) would yield the usual dispersion relation for (small amplitude) acoustic waves: $\omega^2 = k^2a^2$, where a is the sound speed, and k^2a^2 is the acoustic contribution to the square of the restoring frequency because the excess pressure of compressed regions tends to re-expand the dense portions of the fluctuations into the rarefied areas. The term $-4\pi G\rho_0$ represents, then, the lowering of the natural restoring frequency by the gravitational self-attraction of compressed regions. Relative to pressure effects, self-gravity has its largest effect for small k (large wavelengths) because, after all, gravity is a long-range force. For wavenumbers smaller than a critical value,

$$k_{\rm J} \equiv (4\pi G\rho_0/a^2)^{1/2}, \tag{8.19}$$

the disturbances are unstable to gravitational collapse. The statement that the increased self-gravitation of a dense region for itself overcomes acoustic effects if the wavelength $\lambda = 2\pi/k$ between successive condensations exceeds the Jeans wavelength, $\lambda_J \equiv 2\pi/k_J = (\pi a^2/G\rho_0)^{1/2}$, is known as the *Jeans criterion for gravitational stability.* In one form or another, it constitutes perhaps the most frequently cited result of instability theory in all of astronomy.

9

Viscous Shear Flows and Turbulence

References: Lin, *The Theory of Hydrodynamic Stability.* Landau and Lifshitz, *Fluid Mechanics,* Chapter 3.

Since the beginning of the twentieth century, the stability of viscous shear flows has been studied as a prelude to understanding turbulence. Heisenberg attacked the problem as the topic of his dissertation and contributed greatly to early progress in the field. Even a genius of his order, however, found the subject to be extremely difficult. Indeed, he is reputed to have said that he hoped before he died someone would explain quantum mechanics to him, but after he died, he hoped God would explain turbulence to him. We shall not attempt to survey the vast literature on fluid turbulence here. In this chapter, we confine our discussion to the most basic features of the process; we make no attempt at a fully quantitative development, since none exists for most problems involving turbulent flow.

POISEUILLE FLOW

The simplest example of laminar shear flow occurs when a pressure head drives water through a channel of infinite depth in z (*plane Poiseuille flow*; see Figure 9.1).

Let the pressure head per unit length $= \partial P/\partial x =$ constant $\equiv -\alpha$. In steady state, the x-equation of motion reads

$$\rho u \frac{\partial u}{\partial x} = -\frac{\partial P}{\partial x} + \frac{\partial}{\partial y}\left(\mu \frac{\partial u}{\partial y}\right). \tag{9.1}$$

With $\partial P/\partial x$ and μ equal to constants, we anticipate a solution for u that has no dependence on x, i.e., in which the flow remains invariant to translations down the length of the channel, $u = u(y)$. We apply the no-slip boundary conditions on the walls of the channel,

$$u(\pm h) = 0, \tag{9.2}$$

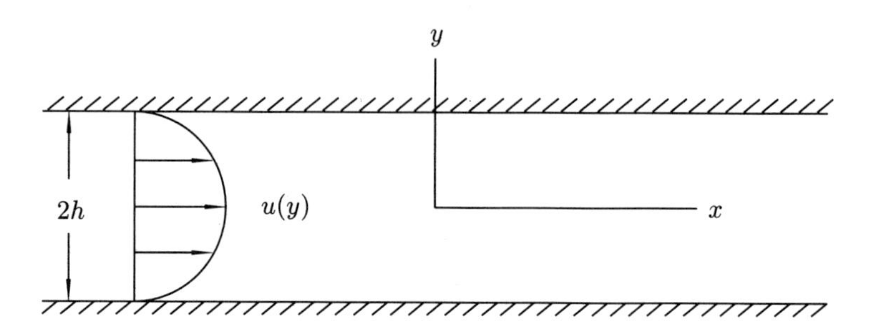

FIGURE 9.1
The flow of water of viscosity μ through a channel of width $2h$, driven by a pressure head per unit length α, has a parabolic velocity profile, $u(y)$, which increases quadratically with distance from zero at the channel walls, $y = \pm h$, to a maximum value $\alpha h^2/2\mu$ in the middle of the channel, $y = 0$.

and we obtain as the solution to equation (9.1) the following parabolic velocity profile:

$$u(y) = \frac{\alpha}{2\mu}(h^2 - y^2). \tag{9.3}$$

We define the average velocity U by

$$U \equiv \frac{1}{2h}\int_{-h}^{+h} u(y)\,dy = \frac{\alpha h^2}{3\mu}, \tag{9.4}$$

which equals 2/3 of the velocity at the center of the channel. Let us also define a Reynolds number for the flow by

$$\text{Re} \equiv \frac{UL}{\nu}, \tag{9.5}$$

where $L \equiv 2h$ is the width of the channel and $\nu \equiv \mu/\rho$ is the kinematic viscosity.

TECHNIQUE FOR STABILITY ANALYSIS

We wish to examine whether the steady viscous shear flow described by equation (9.3) is stable. The lack of x and t dependences in the basic state implies that we may assume small disturbances proportional to

$$Y(y)e^{i(\omega t - kx)}, \tag{9.6}$$

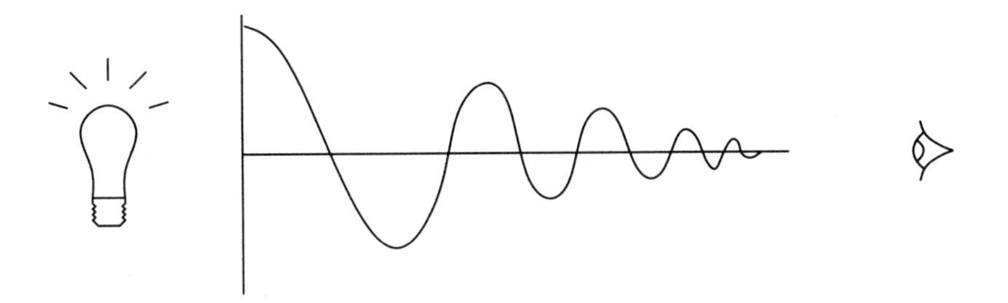

FIGURE 9.2
The propagation of light in a vacuum can show temporal growth at the expense of spatial decay (complex wave frequency and wavenumber, ω and k), if there exists a light source with exponentially growing intensity at the boundary of the propagation region. The temporal growth in this case does not correspond to any true instability of the vacuum to the growth of electromagnetic waves.

where the unknown function $Y(y)$ remains to be determined. For conventional analyses of stability, we wish to examine whether solutions exist in which ω is complex (with a positive growth rate). The advection of disturbances downstream in x, however, compromises such a procedure in that ω complex would automatically make k also complex. Stability analyses with open boundaries that assume complex possibilities for both wave frequencies and wave numbers are extremely vulnerable to misinterpretation. As an example, we note that Maxwell's equations in a vacuum allow free waves that satisfy the dispersion relation:

$$\omega^2 = c^2 k^2. \tag{9.7}$$

If we allow the possibility of complex k, we would find that ω as given by equation (9.7) would also be complex. Does this mean that light propagating in a vacuum can spontaneously amplify in time? No! If we were to examine the boundaries of the system in this thought experiment, we would find a source of light which sends out radiation of progressively larger amplitudes in time. The light which has gotten farther from this source has a smaller amplitude than light close to the source because the former was emitted at an earlier epoch when the light source was weaker (see Figure 9.2).

Clearly, then, the growth in time (complex ω) in the above figure comes at the expense of a decay in space (complex k), and truly free electromagnetic waves propagating in a vacuum suffer no instabilities. In other words, if we assume either k or ω to be real, the other is also automatically real.

Confusion of the above type usually cannot arise for *global normal-mode problems* in which we include homogeneous boundary conditions (no sources or sinks at the boundaries) as part of the modal formulation. In such cases, our boundary conditions guarantee that no hidden wave emitters, which can lead to spurious disturbance growth, lie lurking at the edges of our computational space. If there is flow of matter into or out of the system as in the Poiseuille problem, however, we must adopt a different strategy for a stability analysis. The technique used by workers in the field postulates a *constant-amplitude* disturbance-emitter $\propto e^{i\omega t}$, with ω *real*, to be located at $x = 0$. We then ask, does this disturbance grow in *space* as it is advected downstream by the flow? In other words, if we assume disturbance proportionalities of the form given by equation (9.6), with ω taken to be real, but in which k can be complex, then we say the flow is unstable if we can find solutions with positive values for the imaginary part of k. In this interpretation, decaying solutions (with increasing x), corresponding to negative values for $\text{Im}(k)$, are stable disturbances that have been damped by the action of viscosity.

RESULTS

The formal analysis for the problem posed above is quite involved (refer to Lin's book for details). We content ourselves here with a discussion of the physical results, which we summarize in terms of the following stability diagram (Figure 9.3). We define a parameter space comprised of the dimensionless (real) disturbance frequency $\omega L/U$ and the Reynolds number Re of the basic flow. In the plane defined by $\omega L/U$ as the vertical axis and Re as the horizontal axis, we plot the locus of the parameters that yield marginally stable disturbances, k purely real, i.e., neither spatial growth nor decay. The part of parameter space outside this locus yields stable disturbances, $\text{Im}(k) < 0$; the part inside, unstable disturbances, $\text{Im}(k) > 0$.

At low Reynolds numbers, $\text{Re} < \text{Re}_{\text{cr}} = 7700$, all small-amplitude disturbances damp by the action of viscosity. For $\text{Re} > \text{Re}_{\text{cr}}$, there exists a range of frequencies $\omega L/U$ which yield spatially growing disturbances. High-frequency disturbances yield short spatial-length scales, which are easily damped by viscosity and therefore remain stable. The dangerous disturbances for exciting instability in the laminar flow are those with low temporal and spatial frequencies. Such disturbances occurring in a flow of moderate Reynolds number do not have enough viscous dissipation to prevent overall growth. Notice, however, that at very large Reynolds numbers, we again get stability. Viscosity in plane Poiseuille flow must play, therefore, a dual *stabilizing and destabilizing* role. This dual role does not come as a total surprise, because it is the release of the excess energy contained in the shear that drives the instability. On the other hand, the shear present

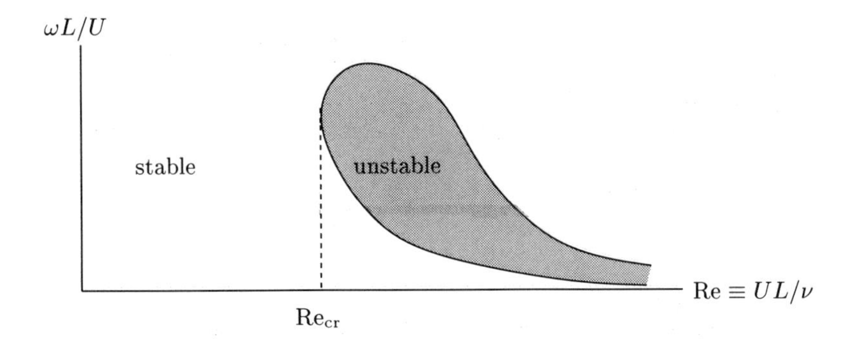

FIGURE 9.3
The stability diagram for viscous channel flow (plane Poiseuille flow). The effect of viscosity is represented by the Reynolds number, $\mathrm{Re} \equiv UL/\nu$, where U is mean flow velocity and $L \equiv 2h$ is the channel width. A small oscillatory disturbance $\propto e^{i\omega t}$ is introduced at the channel inlet. For a given dimensionless frequency $\omega L/U$ below some maximum value, viscosity serves as both a destabilizing and a stabilizing influence. Increasing the Reynolds number along a horizontal cut in the diagram first makes a stable flow become unstable (spatial growth downstream), but the unstable region later turns stable again at very large values of Re. The dual role of viscosity may not come as a complete surprise when we recall that the equilibrium velocity $U = \alpha h^2/3\rho\nu$, and therefore the shear rate of the basic flow is itself a function of the kinematic viscosity ν.

in the basic flow of this problem originates from the viscosity in the first place. Thus, increasing viscosity augments its role both in the perturbation and in the basic flow.

People often forget this dual facet of the problem when they cite Poiseuille flow as an example that shear flows become unstable at large-enough Reynolds numbers. We have no reason to believe that instability develops for differential motions driven by other causes, particularly if the shear arises because of external body forces such as gravitation. The Reynolds number usually tells us nothing about the actual cause of instability in such cases.

ONSET OF TURBULENCE

The stability of viscous parallel layers has a close relation to the problem of the stability of boundary layers (see Chapter 6). A detailed analysis shows that the excess energy contained in the differential motions of the shearing layers can be tapped mechanistically to drive fluctuations only when the

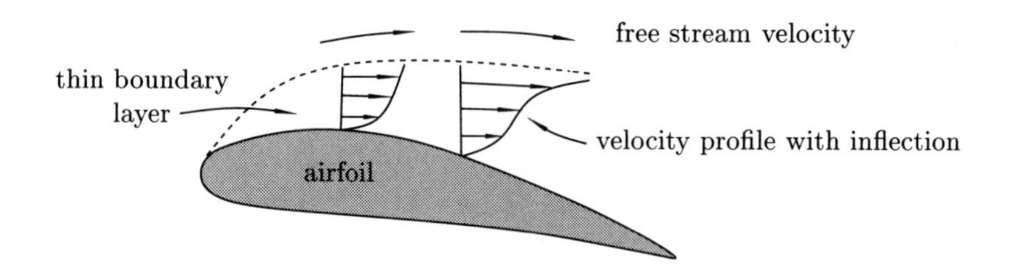

FIGURE 9.4
A laminar boundary layer, such as that which develops in the front part of an airfoil (not drawn to scale), becomes unstable to the onset of turbulence on a line of separation where the velocity profile first achieves a point of inflection.

velocity profile in the laminar boundary layer develops an inflection point (consult Lin's book and Figure 9.4).

Experimentally, such a critical condition always seems to appear for viscous shear flows of sufficiently large Reynolds number around obstructing solid bodies, at which point a boundary layer becomes turbulent. The locus of all such points on the body describes a line, the line of separation (Figure 9.5).

Because turbulent boundary layers plus their associated wakes exert a much bigger drag on moving bodies than their laminar counterparts, a primary design consideration ("streamlining") for ships and airplanes

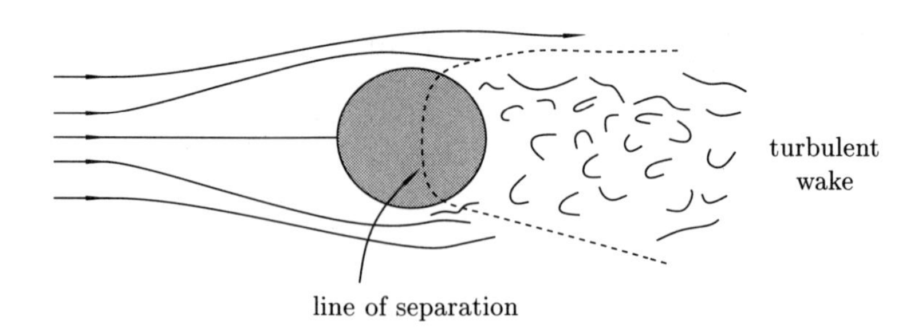

FIGURE 9.5
The drag on a solid body moving through a fluid depends on the size of the turbulent wake. This size can be reduced by streamlining the body shape so as to push the line of separation as far back as possible.

involves pushing the line of separation toward the rear of the craft as far as possible.

To summarize, the appearance of instability in viscous shear flows usually results in the nonlinear outcome: the onset of turbulence. When the Reynolds number of the flow is very large, so many channels for breakdown of the smooth regular flow manifest themselves that the concept of individual modes of instability is not very useful. At one time, people thought in terms of the onset of *chaos* as a process of adding one mode of instability on top of another (see, e.g., Landau and Lifshitz, *Fluid Mechanics*); however, more recent research (on systems of small dimensionality) has focused on the mechanism of *period doubling.* It remains unclear whether the mechanism of period doubling as the route to chaotic behavior continues to hold for systems with effectively an infinite number of degrees of freedom, such as a fluid. The transition to chaos constitutes an active area of current investigation in the field of nonlinear dynamics. An alternative strategy plunges headlong into the problem of *fully developed turbulence.* Although a formidable body of literature exists on this topic, no general theory has emerged except for the idealized case of *homogeneneous isotropic turbulence* in an *incompressible* fluid. In the rest of this chapter, we shall review qualitatively the basic features of what is known about this phenomenon.

ROLE OF EDDIES IN INCOMPRESSIBLE TURBULENCE

Conventional thinking about the nature of fluid turbulence tends to assume that the fluctuations occur in an incompressible medium. Recall from our discussion in Chapter 6 that density variations remain negligible even in air if the turbulent flow field remains subsonic. For an incompressible fluid, the equation of continuity constrains the velocity field to be divergence-free:

$$\nabla \cdot \mathbf{u} = 0. \tag{9.8}$$

This has the following important implication. If we were to Fourier decompose the velocity field,

$$\mathbf{u}(\mathbf{x}, t) = \int \mathbf{v}(\mathbf{k}, \omega) e^{i\mathbf{k}\cdot\mathbf{x}-\omega t}\, d^3k\, d\omega, \tag{9.9}$$

equation (9.8) requires that the individual Fourier components satisfy the constraint that the motions are transverse to the wavenumber $\mathbf{k}$:

$$\mathbf{k} \cdot \mathbf{v} = 0. \tag{9.10}$$

No longitudinal disturbances (e.g., sound waves) occur in an incompressible fluid. For this reason, fluid dynamicists refer to the turbulent elements as *eddies*.

HOMOGENEOUS ISOTROPIC TURBULENCE

In a typical situation involving fluid turbulence, the mean flow contains shear, and energy can be extracted from the differential motions to feed fluctuations ("eddies"). The scale L of the largest eddies is usually comparable to the size of the unstable region according to an instability analysis of the laminar counterpart of the flow. For very large Reynolds numbers, the dynamics of the largest eddies cannot depend to any appreciable extent on the magnitude of the viscosity coefficient. If the turbulence in the comoving frame of the mean flow can be regarded as homogeneous and isotropic, the amplitude U of the fluid velocity fluctuations is the only other dimensional quantity relevant to describing the dynamical characteristic of the turbulent flow field (per unit mass of the fluid) at the largest scale L. The time scale over which the instability mechanism feeds energy into eddies of the largest scale L equals, in order of magnitude, L/U. Per unit mass, then, the rate at which energy is fed into the largest eddies equals

$$\epsilon = \frac{U^2}{L/U} = \frac{U^3}{L},$$

to numerical factors of order unity. We henceforth write such order of magnitude estimates as

$$\epsilon \sim U^3/L. \tag{9.11}$$

In a steady state, the energy fed into the largest eddies can neither accumulate nor dissipate viscously. The only other route is for the energy to be progressively transferred via nonlinear interactions (through the advective term in the equation of motion) to eddies of smaller and smaller scale. Consider eddies of not too small a linear scale λ, which have associated velocities v_λ. The energy that cascades through them per unit mass must equal ϵ, and the relation between ϵ, λ, and v_λ again can not depend on any other quantities, so that dimensional analysis again dictates

$$\epsilon \sim v_\lambda^3/\lambda. \tag{9.12}$$

If we now compare equations (9.11) and (9.12), we obtain *Kolmogorov's law*:

$$v_\lambda \sim U\left(\frac{\lambda}{L}\right)^{1/3}. \tag{9.13}$$

The eddy-cascade process—which forms the dominant feature of incompressible, homogeneous, isotropic turbulence—leads to a velocity spectrum as a function of eddy size λ that depends on the 1/3 power of λ. This law—Kolmogorov's law—demonstrates that the largest eddies have the most velocity (turbulent energy), whereas the smallest eddies carry most of the vorticity, $\sim v_\lambda/\lambda$.

Where does this cascade process end? Clearly, it must do so when eddies have so small a scale λ_0 that the viscous dissipation rate per unit mass, Ψ/ρ, becomes comparable to the energy cascaded downward into this spectral region. The viscous dissipation rate per unit mass at the eddy scale λ_0 equals

$$\Psi/\rho \sim \rho^{-1}\mu\left(\frac{\partial u_i}{\partial x_k}\right)_0^2 \sim \nu\left(\frac{v_{\lambda_0}}{\lambda_0}\right)^2.$$

For Ψ/ρ to equal ϵ, we require

$$\nu\left(\frac{v_{\lambda_0}}{\lambda_0}\right)^2 \sim \frac{v_{\lambda_0}^3}{\lambda_0} \qquad \Rightarrow \qquad v_{\lambda_0}\lambda_0 \sim \nu, \tag{9.14}$$

a result which we might also have guessed by dimensional analysis. Since $\nu \sim v_T\ell$, notice that for very subsonic turbulence, $v_{\lambda_0} \ll v_T$, and therefore the smallest eddy still has a size $\lambda_0 \gg \ell$, the mean free path for elastic collisions.

If we set $\lambda = \lambda_0$ and apply equation (9.13) to equation (9.14), we obtain

$$\lambda_0^{4/3} \sim \frac{\nu L^{1/3}}{U} = \left(\frac{\nu}{UL}\right)L^{4/3} \qquad \Rightarrow \qquad \lambda_0 \sim \mathrm{Re}^{-3/4}\,L, \tag{9.15}$$

where Re is the Reynolds number associated with the largest eddies (comparable to the Reynolds number associated with the mean flow to which the stability analysis was applied). Equation (9.15) for the inner scale of the turbulence (that at which viscous dissipation into heat takes place) clearly constitutes a very rough estimate; it may well miss a substantial numerical factor. Indeed, we know that there exists only one mode of instability when $\mathrm{Re} = \mathrm{Re}_{\mathrm{cr}}$; hence the largest eddy must equal the smallest eddy when the Reynolds number is barely critical. This heuristically suggests that a better formula than equation (9.15) would read

$$\lambda_0 \sim \left(\frac{\mathrm{Re}}{\mathrm{Re}_{\mathrm{cr}}}\right)^{-3/4} L, \tag{9.16}$$

where $\mathrm{Re}_{\mathrm{cr}}$ has a typical order of magnitude of 10^2-10^3 for viscous shear flows.

INERTIAL SUBRANGE

From equation (9.16), we see that λ_0 will be much smaller than L if $\mathrm{Re} \gg \mathrm{Re}_{\mathrm{cr}}$. This corresponds to the situation for fully developed turbulence; so we may have an *inertial subrange* of eddy sizes λ for which

$$L \gg \lambda \gg \lambda_0, \tag{9.17}$$

and Kolmogorov's law, equation (9.13), has room to hold. In the inertial subrange, Richardson's ditty helps us to remember the physical situation:

Large eddies have small eddies,
which feed on their vorticity.
Small eddies have smaller eddies,
and so on to viscosity.

THE ROLE OF COMPRESSIBILITY

Compressibility introduces a role for longitudinal motions: in addition to eddies (vortical motions), there can also be sound waves. Indeed, the eddies themselves will usually generate acoustic noise. Lighthill developed the theory for the generation of sound in a weakly turbulent medium possessing a Kolmogorov spectrum of eddies. He assumed very subsonic turbulence, i.e., that the largest eddy-velocity $U \ll$ sound speed a. For homogeneous isotropic turbulence (an acceptable approximation for aircraft noise), Lighthill discovered the dominant acoustic process to be the quadrupole emission of sound waves, which subsequently propagate away from the region of turbulent air. Thus he calculated a rather low efficiency for the generation of acoustic radiation if the turbulence occurs subsonically. Lighthill found the ratio of energy loss per unit mass in sound waves to the energy cascade per unit mass in eddies to be proportional to the fifth power of the turbulent Mach number $M \equiv U/a$:

$$\mathcal{L}_{\text{acoustic}}/\epsilon \sim M^5. \tag{9.18}$$

Application of equation (9.18) to the turbulence in the solar convection zone allows us to estimate the rate at which acoustic waves are generated in the Sun. Goldreich and Kumar have shown, however, that such a calculation probably underestimates the true emissivity of pressure waves, since the energy-carrying eddies probably have a nonisotropic character; so sound may be emitted by monopole, dipole, as well as quadrupole radiation. (But the former two largely cancel.) In any case, Lighthill's formula has been used to estimate how much acoustic heating elevates the temperature of the Sun's upper atmosphere. From such studies, solar astronomers have generally concluded that acoustic effects might be important for heating the lower chromosphere, but cannot explain the observations for the upper chromosphere and corona. The latter regions probably rely on magnetic activity to maintain their thermal structure and dynamics.

Extrapolated to $M \to 1$, equation (9.18) would predict an increasingly important role for acoustic radiation (relative to turbulent cascade) as we approach a situation where the turbulence becomes supersonic (as may be true for many rarefied gases in astrophysics, such as those encountered in various galactic environments and the interstellar medium.) In fact, this (unjustifiable) extrapolation probably severely underestimates the acoustic

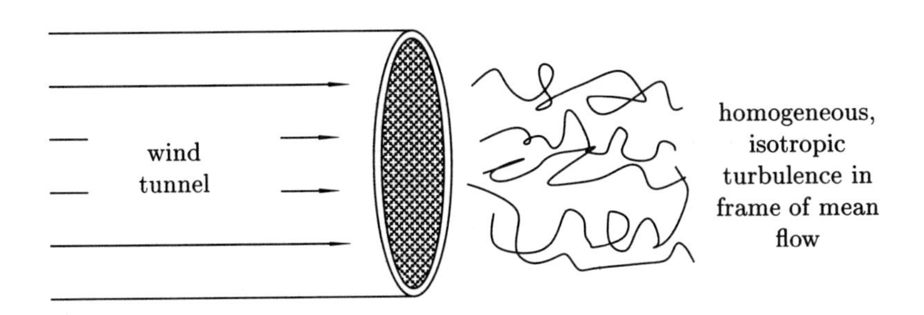

FIGURE 9.6
The laboratory production of the approximate conditions of homogeneous isotropic turbulence must rely on artifices such as placing a fine wire mesh at the inlet of a wind tunnel. Astrophysical phenomena may have greater difficulty in satisfying the conditions of Kolmogorov's theory.

dissipation that would actually take place in a supersonically turbulent fluid. In supersonic turbulence, the generated sound waves would attain highly nonlinear amplitudes (with fluctuating velocities that approach or exceed the speed of sound); so they would quickly steepen and become shock waves (a topic to be discussed in Chapter 15). In a shock wave, the dissipation of kinetic energy of bulk motion into heat occurs in one *jump*, from the largest scales to the smallest, without having to go through an intermediary cascade. Hence supersonic turbulence would require an enormous energy source to maintain it, *unless* the fluid can somehow avoid the generation of strong shock waves. As we shall discuss in Chapter 14 (see also Problem Set 4), an increasing number of astronomers believe that magnetic fields are the key to avoiding such shock waves.

JUPITER'S RED SPOT

Finally, we should ask how the picture of a turbulent cascade might change if we drop the assumption of homogeneous, isotropic turbulence. Such turbulence (in the frame of the mean flow) can be simulated experimentally by placing a uniform grid of wires in the flowing air of a wind tunnel; however, real astronomical objects would probably have difficulty recreating such an idealized situation (Figure 9.6).

In particular, atmospheric storms in the zones and belts of Jupiter's atmosphere (regions of strong shear flow driven by the thermal circulation inside the planet) have a vortical character. The Red Spot forms the most famous of these storms; and it has long posed a puzzle how such a large eddy could manage to persist over the centuries since its initial discovery by Cassini. How does the Red Spot avoid the natural degradation implied by the ditty: "Large whirls have small whirls, etc."? After all, the natural lifetime for a large eddy in this picture equals the overturn timescale L/U, i.e., days, not centuries. Spacecraft missions (principally the two Voyagers) and computer simulations (principally by Marcus and Ingersoll) have revealed a partial answer. The interaction of one shearing layer and another in Jupiter's system of zones and belts apparently sheds vortices containing spin of both signs. If the turbulence in Jupiter's atmosphere behaved as we might have naively expected from the theory of homogeneous, isotropic turbulence outlined two sections earlier, these vortices should randomly merge, cascading their energies into progressively smaller scales. Instead, what is observed, both by the spacecrafts and in the (simplified) numerical simulations, almost defies belief. The Red Spot "swallows" only eddies of its own sign; it "repels" vortices of the opposite sign. In this way, the Red Spot can apparently gain the external sustenance to maintain itself against the natural ebbing of its strength by an internal cascade. The ability of vortices to behave in this counterintuitive fashion may depend critically on the dynamics in Jupiter's atmosphere being essentially two-dimensional in character. (The dynamics of line vortices, free to move in two dimensions, can be reduced to an equivalent N-body problem containing attractive and repulsive forces that depend on the strength of the circulation and the intervortical distance.)

This fascinating fluid behavior is one example of a generic phenomenon that has been termed the physics of "negative viscosity" by the meteorologist Victor Starr. This example, and several others from the field of planetary atmospheres, violates the thermodynamic intuition that energy flow should occur from the orderly to the disorderly. No real violation of the second law of thermodynamics takes place, since there usually exists a source of free energy—the shear system represented by the zones and belts for Jupiter's Red Spot—that can act as a basic driver for the so-called "negative eddy viscosity." The most spectacular example of the phenomena of "negative diffusivity," however, involves not inanimate objects, but living systems. The existence of such examples reminds us how much we have yet to learn about the surprisingly intricate solutions that lie buried in the innocuous-looking equations of fluid mechanics.

10

Mixing-Length Theory of Convection

References: Schwarzschild, *Structure and Evolution of the Stars*, pp. 44–50; Cox and Giuli, *Principles of Stellar Structure*, Chapter 14.

In Chapter 8, we derived the Schwarzschild criterion for convective instability. In this chapter, we consider how to implement practical schemes to take account of the presence of this instability in calculations of stellar (and planetary) models. In other words, we wish to consider the nonlinear resolution of the development of convection, and the consequences of its back reaction on the basic state. No rigorous general procedure starting from first principles (i.e., the equations of fluid mechanics) exists for this problem; this gap forms the major missing element in the modern theory of stellar structure and evolution. Fortunately, a heuristic approach does exist and seems to give reasonable approximate results, at least in comparison with observations. This approach, on which astronomers base practically all their current model computations, relies on Ericka Bohm-Vitense's adaptation to the problem of thermal convection of Prandtl's *mixing-length formulation* for treating transport in a turbulent fluid.

CONVECTION IN INTERIOR ZONES

Before we embark on an exploration of mixing-length theory, we first tackle the easier problem of convection in interior zones, where the convective efficiency may be regarded as much higher than that of diffusive heat-transport mechanisms (radiation and conduction). We shall demonstrate that the nonlinear resolution of the instability results in a spatially constant specific entropy s (see Figure 10.1). For the construction of stellar models, we need to know how to calculate the value of this constant. For interior convection zones, a simple prescription suffices, based on the principle that in the overlying radiative layer, the (radiative plus conductive) energy flux satisfies Fick's law,

$$\mathbf{F}_{\text{rad}} = -\frac{4caT^3}{3\rho\kappa_{\text{eff}}}\boldsymbol{\nabla}T, \tag{10.1}$$

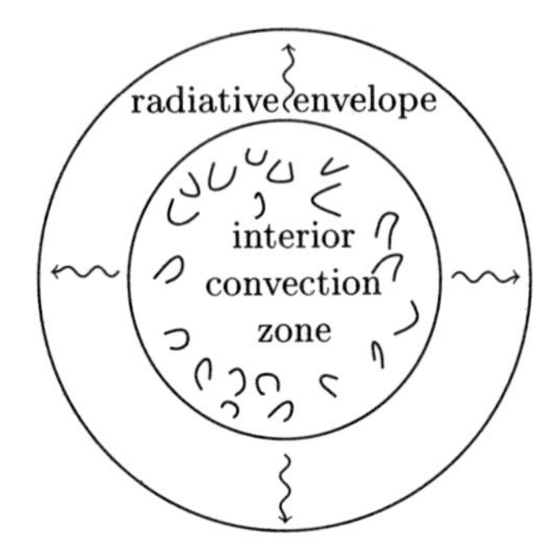

FIGURE 10.1
The practical implementation of the effects of the convective transport of heat is easiest for interior convection zones, where the large convective efficiency causes the specific entropy s to be nearly a spatial constant.

where we have defined an effective opacity κ_{eff} in accordance with equation (5.8). At the convective-radiative interface, the temperature gradient must be adiabatic if s = constant holds inside interior convection zones; thus we are motivated to define a fiducial flux equal to the amount that radiation could carry if s were a constant,

$$\mathbf{F}_s \equiv -\frac{4caT^3}{3\rho\kappa_{\text{eff}}}\left(\frac{\partial T}{\partial P}\right)_s \nabla P, \tag{10.2}$$

where the gradient of the pressure is given in hydrostatic equilibrium by

$$\nabla P = -\rho\nabla\mathcal{V}. \tag{10.3}$$

We then postulate that the convective flux equals whatever it must in order to satisfy $\mathbf{F}_{\text{conv}} = \mathbf{F}_{\text{tot}} - \mathbf{F}_{\text{rad}}$, consistent with a value of s that satisfies $\mathbf{F}_{\text{rad}} = \mathbf{F}_s$ at the convective-radiative interface.

CONCEPT OF MIXING LENGTH

The procedure given in the previous section fails for surface convection zones where the mean free path for photons becomes long enough for radiative transport to offer serious competition to convection. In such zones, we may no longer assume that s = constant right up to the top of the convection zone. Worse, without the condition $\mathbf{F}_{\text{rad}} = \mathbf{F}_s$ at the top, we

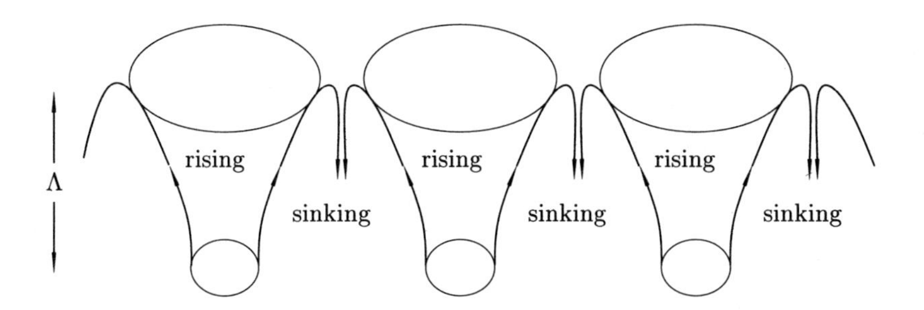

FIGURE 10.2
The convective mixing length Λ probably has a magnitude comparable to a pressure-scale height H, because ascending blobs initially occupying about half the area of any spherical shell cannot rise more than about one H before they expand so much that they must dissolve into the surroundings.

have no way to calculate s in the deeper layers, where it does have a constant value. Mixing-length theory attempts to fill the gap by heuristically treating the motion and thermodynamics of an average convective element in the transition layers.

The central idea is that such an element is driven by an unbalanced buoyancy force that moves it through a distance Λ, called the mixing length, before the element dissolves and joins the ambient medium. The principal uncertainty in the formalism is over how to assign a value for Λ. Astrophysical convention expresses this assignment in terms of a ratio of mixing length to pressure-scale height α:

$$\Lambda \equiv \alpha H \qquad \text{where} \qquad H \equiv P/|\nabla P|, \tag{10.4}$$

with α usually chosen to be a number between 1 and 2. This range of choices, dictated by empirical comparisons with the calculated state of our Sun after 4.5×10^9 yr of evolution, yields a structure for the solar convection zone in good agreement also with the results of helioseismology. Numbers in the range 1 to 2 make physical sense, since ascending blobs which initially occupy, say, half the area of a spherical surface, on rising one or two pressure-scale heights, would have filled the entire spherical surface. After dissolution and equilibration with the ambient medium, the fluid then becomes candidate material for sinking convective elements (Figure 10.2).

CONVECTIVE FLUX

Let us now consider the energy flux carried by convective elements. We suppose that we can represent the excess flux solely in terms of an excess enthalpy Δh carried upward in rising currents,

$$\mathbf{F}_{\text{conv}} = \rho(\Delta h)\mathbf{v}, \tag{10.5}$$

where $\mathbf{v}$ is the average velocity of the rising parcels. (In theory, we should also take into consideration the kinetic energy of the rising motions, but such niceties only complicate what can, at best, be regarded as order-of-magnitude arguments.) Rising parcels contain only half the total material, but sinking parcels carrying a comparable *deficit* of specific enthalpy *downward* make an equal contribution to the outward flux; so we give the right-hand side of equation (10.5) a coefficient equal to 1 instead of 1/2. To compute the average enthalpy excess Δh, we assume that rising currents have an internal temperature T' that differs from the ambient temperature T,

$$\Delta h = c_P \Delta T, \tag{10.6}$$

where ΔT is the temperature excess of the rising blob compared to its surroundings:

$$\Delta T \equiv T' - T = \frac{\Lambda}{2}(\nabla T' - \nabla T) \cdot \hat{\mathbf{n}}. \tag{10.7}$$

In equation (10.6) $c_P \equiv (\partial h/\partial T)_P = T(\partial s/\partial T)_P$ equals the specific heat at constant pressure P (that of the ambient medium with which the convective element exists in pressure equilibrium), and $\hat{\mathbf{n}}$ equals a unit normal in the direction of mean motion, opposite to the direction of the local gravity $\mathbf{g} = -\nabla \mathcal{V}$:

$$\hat{\mathbf{n}} \equiv -\mathbf{g}/|\mathbf{g}|. \tag{10.8}$$

We have also introduced a factor 1/2 in equation (10.6) to incorporate the idea that only over half of the mixing length does the parcel have its full enthalpy difference from its surroundings.

To compute $\mathbf{v} = v\hat{\mathbf{n}}$, we equate the specific kinetic energy of the convective element equal to the work done per unit mass by the buoyancy force,

$$\frac{1}{2}(2v)^2 = (\Delta\rho/\rho)|\mathbf{g}|(\Lambda/2), \tag{10.9}$$

where $\Delta\rho$ equals the density deficit of a rising blob,

$$\Delta\rho = \left(\frac{\partial\rho}{\partial T}\right)_P \Delta T, \tag{10.10}$$

with ΔT given by equation (10.7). The factor 2 in the term $(2v)^2$ in equation (10.9) has been inserted (somewhat arbitrarily) to account for the fact that the maximum kinetic energy attained by the blob (equal to the buoyancy work) is associated with a speed twice that of the average value, v. In equation (10.10), we have assumed that material in the convection zone has a uniform chemical composition (well mixed by the presence of the convection itself). If this is not true, i.e., if there exists a chemical composition gradient in the zone (due to thermonuclear processing, for example), the formula for the density excess must include an additional factor $\mathcal{C}$ on the right-hand side of equation (10.10),

$$\mathcal{C} \equiv 1 - \left(\frac{\partial \ln \mu}{\partial \ln T}\right)_P, \tag{10.11}$$

where μ is the mean molecular weight in the ambient medium. In any case, if we collect terms, we obtain

$$v = \frac{1}{2}\left|\frac{\mathcal{C}}{\rho}\left(\frac{\partial \rho}{\partial T}\right)_P \mathbf{g}\Lambda\Delta T\right|^{1/2}. \tag{10.12}$$

Notice that ΔT as given by equation (10.7) contains one factor of Λ; so v is linearly proportional to the mixing length Λ. The formula (10.12) fails to hold if the implied convective speed should exceed the adiabatic speed of sound $a_s \equiv (\partial P/\partial \rho)_s^{1/2}$ since pressure equilibrium with the surroundings could no longer be maintained. In such cases, astronomers usually adopt the substitution $v = a_s$.

Substituting $\mathbf{v} = v\hat{\mathbf{n}}$, with v given by equation (10.12), into equation (10.5), we get

$$\mathbf{F}_{\text{conv}} = \frac{1}{4\sqrt{2}}\rho c_P \left|\frac{\mathcal{C}}{\rho}\left(\frac{\partial \rho}{\partial T}\right)_P \mathbf{g}\right|^{1/2} |\nabla T - \nabla T'|^{3/2}\Lambda^2\hat{\mathbf{n}}. \tag{10.13}$$

This equation yields the convective flux $\mathbf{F}_{\text{conv}}$ if we know the temperature gradients of the ambient medium ∇T and the temperature gradient of the average rising convective element $\nabla T'$. We now consider how to obtain the latter quantities.

EFFICIENT CONVECTION

In the deep interior, we can make the simple assumption that convective blobs move adiabatically, i.e.,

$$\nabla T' = \left(\frac{\partial T}{\partial P}\right)_s \nabla P, \tag{10.14}$$

where ∇P is given by equation (10.3). We express the temperature gradient of the ambient medium as

$$\nabla T = \left(\frac{\partial T}{\partial P}\right)_s \nabla P + \left(\frac{\partial T}{\partial s}\right)_P \nabla s, \tag{10.15}$$

with $(\partial T/\partial s)_P = T c_P^{-1}$. If chemical composition gradients are important, we should include an additional term $(\partial T/\partial \mu)_P \nabla \mu$ on the right-hand side of equation (10.15); in what follows, we ignore this effect and set $\mathcal{C} = 1$ in equation (10.13). The substitution of equations (10.14) and (10.15) into equation (10.13) now yields

$$\mathbf{F}_{\text{conv}} = F_0 (P|\nabla s|/c_P|\nabla P|)^{3/2}\hat{\mathbf{n}}, \tag{10.16}$$

with F_0 being a coefficient having the units of an energy flux,

$$F_0 \equiv \frac{\alpha^2}{4\sqrt{2}}\rho c_P T \left|\frac{P}{\rho}\frac{T}{\rho}\left(\frac{\partial \rho}{\partial T}\right)_P\right|^{1/2}, \tag{10.17}$$

where we have used equation (10.4) to eliminate Λ.

For a perfect gas, the coefficient F_0 has the order of magnitude

$$F_0 \sim \rho c_P T a_s,$$

where a_s is the adiabatic sound speed. Moreover, in the deep interior of a star, we may compare the energy flux F_{conv} that needs to be carried by convection [from $\mathbf{F}_{\text{conv}} = \mathbf{F}_{\text{tot}} - \mathbf{F}_{\text{rad}}$ when $\mathbf{F}_{\text{rad}}$ is given by equation (10.1)], with the flux that fluid motion can carry if matter were to move at the speed of sound a_s. The ratio provides a very small parameter:

$$\delta \equiv F_{\text{conv}}/\rho c_P T a_s. \tag{10.18}$$

For example, in a region where $T \sim 10^6\,\text{K}$, $\rho \sim 1\,\text{g}\,\text{cm}^{-3}$, and $F_{\text{conv}} \sim L_\odot/4\pi R_\odot^2 \sim 6\times 10^{10}\,\text{erg}\,\text{cm}^{-2}\,\text{s}^{-1}$, we have $\delta \sim 10^{-11}$.

Equation (10.16) then implies that the variations of the specific entropy s with variations of the pressure P forms a very small fraction $\delta^{2/3}$ of the natural value c_P/P:

$$\frac{ds}{dP} \sim \delta^{2/3}\frac{c_P}{P}.$$

In regions where $\delta \ll 1$, therefore, we may adopt the zeroth-order approximation, $s =$ constant, which provides a justification for the adiabatic assumption made in Chapter 8. A similar order-of-magnitude estimate yields for equation (10.12),

$$v \sim \delta^{1/3} a_s,$$

which demonstrates that the needed convective speeds represent a small fraction of the sound speed in regions where the convection is efficient, $\delta \ll 1$. This justifies our ignoring any contribution that the presence of thermally driven currents might make toward a "turbulent pressure."

INEFFICIENT CONVECTION

Clearly, the assumption (10.14) breaks down in layers where $\delta \gtrsim 1$. This happens if stellar convection zones extend to the the subphotospheric and photospheric layers. For example, in regions where $T \sim 10^4\,\mathrm{K}$, $\rho \sim 10^{-8}\,\mathrm{gm\,cm^{-3}}$, and $F_{\mathrm{conv}} \sim 6 \times 10^{10}\,\mathrm{erg\,cm^{-2}\,s^{-1}}$, we have $\delta \sim 1$. In such regions, rising convective elements radiate to the surroundings a significant fraction of their excess heat content before dissolving. Let us estimate the rate of (optically thick) radiant energy loss L_{loss} by approximating a convective element as a sphere of diameter Λ (a cube would have the same ratio of volume to surface area $= \Lambda/6$):

$$L_{\mathrm{loss}} \sim \frac{4acT^3}{3\rho\kappa_{\mathrm{eff}}}\frac{\Delta T}{\Lambda/2}4\pi(\Lambda/2)^2.$$

Multiplying L_{loss} by the turnover time scale Λ/v, we obtain the total energy lost as

$$\Delta E_{\mathrm{loss}} = L_{\mathrm{loss}}\Lambda/v. \tag{10.19}$$

If we (somewhat arbitrarily) take the maximum temperature difference between the blob and its surroundings as $8\Delta T/3$ (to get numerical agreement with Bohm-Vitense), the maximum excess energy contained by a rising convective element before it dissolves equals

$$\Delta E_{\mathrm{max}} = \frac{4\pi}{3}\left(\frac{\Lambda}{2}\right)^3 \rho c_P \frac{8}{3}\Delta T. \tag{10.20}$$

Collecting terms, we may define the ratio of ΔE_{max} to ΔE_{loss} as the convective efficiency Γ:

$$\Gamma \equiv \frac{\Delta E_{\mathrm{max}}}{\Delta E_{\mathrm{loss}}} = \frac{c_P \kappa_{\mathrm{eff}} \rho^2 \Lambda v}{6caT^3}.$$

If we substitute equation (10.12) for v, with $\mathcal{C}$ taken to equal 1, we obtain the expression:

$$\Gamma = \left(\frac{F_0 \rho \kappa_{\mathrm{eff}} P}{3caT^4|\boldsymbol{\nabla} P|}\right)\left(\frac{P|\boldsymbol{\nabla} T - \boldsymbol{\nabla} T'|}{T|\boldsymbol{\nabla} P|}\right)^{1/2}, \tag{10.21}$$

where F_0 is defined by equation (10.17). The first term in the parenthesis on the right-hand side of equation (10.21) represents, up to a numerical coefficient of order unity, the ratio of F_0 to F_{rad}. This term is $\gg 1$ in regions of efficient convection.

To make use of equation (10.21), we need another expression for Γ. This second expression, the thermodynamic efficiency, we may write down intuitively as the ratio of the actual temperature difference between the convective element and the ambient medium to the temperature difference that would have been present if the ambient medium had an adiabatic temperature gradient:

$$\Gamma = \frac{\Delta E_{\rm max}}{\Delta E_{\rm loss}} = \frac{|\nabla T - \nabla T'|}{|\nabla T' - (\partial T/\partial P)_s \nabla P|}. \tag{10.22}$$

In regions where the convective efficiency is large, $\Gamma \gg 1$, we get $\nabla T' \approx (\partial T/\partial P)_s \nabla P$. In regions where the convective efficiency is small, $\Gamma \ll 1$, we get $\nabla T' \approx \nabla T$.

For more general regions, if we equate equations (10.21) and (10.22), we get a cubic equation to solve for $\nabla T'$ in terms of ∇T. The latter is then determined by the condition that we want the sum of the radiative and convective fluxes to equal the total flux,

$$\mathbf{F}_{\rm rad} + \mathbf{F}_{\rm conv} = \mathbf{F}_{\rm tot}, \tag{10.23}$$

where $\mathbf{F}_{\rm tot}$ must satisfy a total heat equation (see Chapter 4). For convection in deep zones, we recover our earlier analysis (to factors of order unity), providing the ultimate justification for the identification (10.22). The real utility of the formalism comes with application to convection in the outer layers of a star. See pp. 311–321 of Cox and Giuli for details.

PART II

WAVES, SHOCKS, AND FRONTS

11

Spiral Density Waves: Dispersion Relation

Reference: G. Bertin, C. C. Lin, S. Lowe, and R. Thurstans, 1989, *Ap. J.*, **338**, 78 and 104.

With this chapter, we begin Part II of this Volume, the study of waves, shocks, and fronts. Many different kinds of waves exist in nature, and selecting examples to study is a matter of taste. At small amplitudes, however, certain features are common to waves of all types: characterization in terms of a dispersion relationship, transport of important properties at the group velocity, etc. We choose to begin our discussion with a type of self-gravitating disturbance—spiral density waves—peculiar to rapidly rotating, flattened, astronomical systems, but we could easily have chosen some other example—e.g., pulsational modes in spherical gas masses—to illustrate the general principles.

The theory of spiral density waves, invented in the 1960s to explain the spiral structure of disk galaxies, has also found application in disturbances excited by resonant interactions with orbiting satellites in the rings of Saturn. Furthermore, people have speculated on its possible importance for the disks believed to be present around young stars and active galactic nuclei.

The development of the original theory took place for a disk composed of encounterless stars under the assumption that disk stars (seen and unseen) made up the vast bulk of a spiral galaxy. Few people realized in the early 1960s (a) the importance of dark matter for explaining the rotation curves of spiral galaxies, and (b) the role of molecular hydrogen as the predominant form of interstellar matter in the inner parts of most spiral galaxies. Thus many of the results described here in a gas-dynamical treatment were actually first derived in the much harder stellar-dynamical context. Only gradually did astronomers realize the direct relevance of a fluid treatment of the problem even for the issue of galactic spiral structure (as contrasted, for example, with the disks around young stars, which are truly gaseous.) As we shall see, the theory of small-amplitude density waves in a gaseous disk is relatively simple, especially if we adopt the simplifying assumption of tightly wrapped spiral waves (the WKBJ approximation). Our development below will not try to incorporate the most general treatment, but will follow a route that makes transparent the most interesting physical interpretations.

GOVERNING EQUATIONS

For simplicity, we assume a disk of infinitesimal thickness in which the volume density, written in cylindrical coordinates (ϖ, φ, z), has the form

$$\rho(\varpi, \varphi, z, t) = \sigma(\varpi, \varphi, t)\delta(z), \tag{11.1}$$

where σ is the local surface-mass density and $\delta(z)$ is the Dirac delta function. We define the vertically integrated pressure as

$$\Pi \equiv \int_{-\infty}^{+\infty} a^2 \rho \, dz, \tag{11.2}$$

where a is the thermal speed (or one-dimensional velocity dispersion) in the midplane of the disk. We denote u as the ϖ-component of the fluid velocity, and j as the z-component of the specific angular momentum. Integrated over all z, the dynamical equations for the disk now read:

$$\frac{\partial \sigma}{\partial t} + \frac{1}{\varpi}\frac{\partial}{\partial \varpi}(\varpi \sigma u) + \frac{1}{\varpi^2}\frac{\partial}{\partial \varphi}(\sigma j) = 0, \tag{11.3}$$

$$\sigma\left(\frac{\partial u}{\partial t} + u\frac{\partial u}{\partial \varpi} + \frac{j}{\varpi^2}\frac{\partial u}{\partial \varphi} - \frac{j^2}{\varpi^3}\right) = -\frac{\partial \Pi}{\partial \varpi} - \sigma\left(\frac{\partial \mathcal{V}}{\partial \varpi}\right)_{z=0}, \tag{11.4}$$

$$\sigma\left(\frac{\partial j}{\partial t} + u\frac{\partial j}{\partial \varpi} + \frac{j}{\varpi^2}\frac{\partial j}{\partial \varphi}\right) = -\frac{\partial \Pi}{\partial \varphi} - \sigma\left(\frac{\partial \mathcal{V}}{\partial \varphi}\right)_{z=0}. \tag{11.5}$$

In equations (11.4) and (11.5), $\mathcal{V}(\varpi, \varphi, z, t)$ equals the gravitational potential in three dimensions and satisfies Poisson's equation:

$$\frac{1}{\varpi}\frac{\partial}{\partial \varpi}\left(\varpi\frac{\partial \mathcal{V}}{\partial \varpi}\right) + \frac{1}{\varpi^2}\frac{\partial^2 \mathcal{V}}{\partial \varphi^2} + \frac{\partial^2 \mathcal{V}}{\partial z^2} = 4\pi G\left[\sigma(\varpi, \varphi, t)\delta(z) + \rho_{\text{ext}}(\varpi, z)\right], \tag{11.6}$$

where ρ_{ext} is the (axisymmetric and time-independent) volume density of material not associated with the disk proper (e.g., the bulge and halo of a galaxy, or the central star for a nebular disk). To close the above set of equations, we need a prescription for calculating the vertically integrated pressure Π. We should derive Π, in principle, from the internal energy equation, but for simplicity in what follows, we shall assume that Π equals a prescribed function of σ, in which case we may write

$$\nabla \Pi = -\sigma \nabla h \qquad \text{where} \qquad h = \int \frac{d\Pi}{\sigma}. \tag{11.7}$$

BASIC STATE

We adopt an axisymmetric disk as the equilibrium state, in which the surface density equals $\sigma_0(\varpi)$, the sound speed equals $a_0(\varpi)$, and the fluid velocity corresponds to differential rotation about the z axis with angular speed $\Omega(\varpi)$. With this assumption, the only nontrivial part of the dynamical equations reads

$$\varpi\Omega^2(\varpi) = \frac{a_0^2}{\sigma_0}\frac{d\sigma_0}{d\varpi} + \left(\frac{\partial \mathcal{V}_0}{\partial \varpi}\right)_{z=0},$$

where $a_0^2 \equiv \Pi'(\sigma_0)$ and $\mathcal{V}_0(\varpi, z)$ satisfies the zeroth-order Poisson equation,

$$\frac{1}{\varpi}\frac{\partial}{\partial \varpi}\left(\varpi\frac{\partial \mathcal{V}_0}{\partial \varpi}\right) + \frac{\partial^2 \mathcal{V}_0}{\partial z^2} = 4\pi G\left[\sigma_0(\varpi)\delta(z) + \rho_{\text{ext}}(\varpi, z)\right].$$

LINEARIZED PERTURBATION EQUATIONS

We now assume perturbations about this equilibrium state of the form:

$$\sigma = \sigma_0(\varpi) + \sigma_1(\varpi, \varphi, t), \tag{11.8}$$

$$u = 0 + u_1(\varpi, \varphi, t), \tag{11.9}$$

$$j = \varpi^2\Omega(\varpi) + j_1(\varpi, \varphi, t), \tag{11.10}$$

$$\mathcal{V} = \mathcal{V}_0(\varpi, z) + \mathcal{V}_1(\varpi, \varphi, z, t). \tag{11.11}$$

We take the perturbations to have small (infinitesimal) amplitude and drop quantities that enter in the governing equations of any order higher than linear in the perturbation quantities. If we then subtract from the governing equations the zeroth-order set given in the previous section, we obtain the *linearized perturbation equations*:

$$\frac{\partial \sigma_1}{\partial t} + \frac{1}{\varpi}\frac{\partial}{\partial \varpi}(\varpi\sigma_0 u_1) + \frac{\partial}{\partial \varphi}(\sigma_1\Omega + \sigma_0 j_1/\varpi^2) = 0, \tag{11.12}$$

$$\frac{\partial u_1}{\partial t} + \Omega\frac{\partial u_1}{\partial \varphi} - 2\Omega\frac{j_1}{\varpi} = -\frac{\partial h_1}{\partial \varpi} - \left(\frac{\partial \mathcal{V}_1}{\partial \varpi}\right)_{z=0}, \tag{11.13}$$

$$\frac{\partial j_1}{\partial t} + \frac{\kappa^2}{2\Omega}\varpi u_1 + \Omega\frac{\partial j_1}{\partial \varphi} = -\frac{\partial h_1}{\partial \varphi} - \left(\frac{\partial \mathcal{V}_1}{\partial \varphi}\right)_{z=0}, \tag{11.14}$$

$$\frac{1}{\varpi}\frac{\partial}{\partial \varpi}\left(\varpi\frac{\partial \mathcal{V}_1}{\partial \varpi}\right) + \frac{1}{\varpi^2}\frac{\partial^2 \mathcal{V}_1}{\partial \varphi^2} + \frac{\partial^2 \mathcal{V}_1}{\partial z^2} = 4\pi G\sigma_1\delta(z), \tag{11.15}$$

where κ is the epicyclic frequency, whose square is given by [see equation (8.9)]:

$$\kappa^2 \equiv \frac{1}{\varpi^3}\frac{d}{d\varpi}\left[\left(\varpi^2\Omega\right)^2\right], \tag{11.16}$$

and where equation (11.7), with $h = h(\sigma)$ and $\sigma = \sigma_0 + \sigma_1$, allows us to write

$$h_1 = h'(\sigma_0)\sigma_1 = a_0^2\frac{\sigma_1}{\sigma_0}, \qquad \text{with} \qquad a_0^2 \equiv \frac{d\Pi}{d\sigma}(\sigma_0) \tag{11.17}$$

being some function of ϖ.

FOURIER DECOMPOSITION

The coefficients for the linear perturbation equations (11.12)–(11.15) do not depend on φ or t; therefore, if we apply homogeneous boundary conditions to this set, we may look for solutions which have the Fourier decomposition:

$$\begin{pmatrix} \sigma_1 \\ u_1 \\ j_1 \\ \mathcal{V}_1 \end{pmatrix} = \mathrm{Re}\left[\begin{pmatrix} S(\varpi) \\ U(\varpi) \\ J(\varpi) \\ V(\varpi, z) \end{pmatrix} e^{i(\omega t - m\varphi)}\right], \tag{11.18}$$

where ω is a (complex) mode frequency and m is an integer (taken to be positive by convention). In equation (11.18), the symbol Re implies that we should take the real part of whatever follows the symbol inside the square brackets.

The substitution of equation (11.18) into equations (11.12)–(11.15) results in the following set:

$$i(\omega - m\Omega)S + \frac{1}{\varpi}\frac{d}{d\varpi}(\varpi\sigma_0 U) - im\sigma_0\frac{J}{\varpi^2} = 0, \tag{11.19}$$

$$i(\omega - m\Omega)U - 2\Omega\frac{J}{\varpi} = -\frac{d}{d\varpi}\left(a_0^2\frac{S}{\sigma_0}\right) - \left(\frac{\partial V}{\partial \varpi}\right)_{z=0}, \tag{11.20}$$

$$\frac{1}{\varpi}\frac{\partial}{\partial\varpi}\left(\varpi\frac{\partial V}{\partial\varpi}\right) - \frac{m^2}{\varpi^2}V + \frac{\partial^2 V}{\partial z^2} = 4\pi G S(\varpi)\delta(z). \tag{11.21}$$

To this set, we must add appropriate boundary conditions, for example, the condition that the radial velocity and angular momentum perturbations vanish at the origin,

$$U = 0, \qquad J = 0 \qquad \text{at} \qquad \varpi = 0, \tag{11.22}$$

and that the perturbation surface density and gravitational potential vanish at infinity,

$$S, V \to 0 \qquad \text{as} \qquad \varpi, (\varpi^2 + z^2)^{1/2} \to \infty. \tag{11.23}$$

Other boundary conditions are possible (e.g., outgoing waves) and have been used, appropriate to the physical situation envisaged by the investigator. [Ultimately, boundary conditions represent the scientist's prejudice (intuition is a more polite word) concerning the nature of the world outside of where he or she possesses equations as guides for behavior.] The importance for us here does not lie so much in the specific form of the boundary conditions as in their being *homogeneous.*

Since the governing equations and boundary conditions are both homogeneous, we should treat ω as an eigenvalue of the (global normal mode) problem if we hope to obtain nontrivial solutions. For the discussion of this chapter and the next, however, we forego application of the boundary conditions to start with, and we analyze the problem of locally propagating waves within the context of a WKBJ approximation. Once we have understood the problem of propagating (spiral) waves with given (real) ω, we stand in good position to take into account the boundary conditions for the normal-modes problem in terms of a superposition of inwardly and outwardly propagating waves (to give a "standing pattern"). Such an asymptotic analysis (including the incorporation of boundary conditions) does not provide as much accuracy as, say, a direct numerical attack on the linearized perturbation equations (11.12)–(11.15), but it yields much more insight into what is happening physically.

WKBJ APPROXIMATION AND GEOMETRIC INTERPRETATIONS

We may write the radial dependence of any (complex) perturbation quantity in terms of an amplitude (modulus) and phase (argument). For example, we may write the gravitational potential computed in the plane $z = 0$ as

$$V(\varpi, z = 0) = A(\varpi)e^{i\Phi(\varpi)}, \tag{11.24}$$

where we take A and Φ to be real. The WKBJ approximation corresponds to the assumption that the phase $\Phi(\varpi)$ varies rapidly in comparison with the amplitude $A(\varpi)$:

$$\left|\frac{d\Phi}{d\varpi}\right| \gg \left|\frac{1}{A}\frac{dA}{d\varpi}\right|. \tag{11.25}$$

Since the logarithmic derivative of A with respect to ϖ will usually turn out to be of order $1/\varpi$, equation (11.25) amounts, in practice, to the assumption,

$$|k|\varpi \gg 1, \tag{11.26}$$

where k is the (negative) of the radial wave number:

$$k \equiv \frac{d\Phi}{d\varpi}. \tag{11.27}$$

In hindsight, we would have fared better to have defined the total phase variation that enters as the product e^{if} differently from

$$f(\varpi, \varphi, t) \equiv \omega t - m\varphi + \Phi(\varpi), \tag{11.28}$$

so that the terms Φ and $m\varphi$, representing the radial and azimuthal variations, both carried a minus sign in their coefficients. As it is, much of the literature follows the original choice made by Lin and Shu, equation (11.28), and we shall continue with their convention, with the anticipation that it trades inconvenient minus signs at one place for inconvenient ones somewhere else.

We now consider the geometric meaning of the WKBJ approximation (11.26). In order to obtain a consistent approximation for equations (11.12)–(11.15), we easily verify that every other perturbation quantity must also possess, to lowest order, the same radial-phase variation $\Phi(\varpi)$, although they may have different (slowly varying) amplitude dependences. (We also allow the amplitudes of other quantities to be complex, so that they need not have zero phase difference with respect to the gravitational potential.) Thus, the surface-density perturbation has the form,

$$S(\varpi) = \hat{S}(\varpi)e^{i\Phi(\varpi)}. \tag{11.29}$$

The full variation of the perturbation-surface density reads

$$\sigma_1(\varpi, \varphi, t) = \mathrm{Re}\left\{\hat{S}(\varpi)e^{i[\omega t - m\varphi + \Phi(\varpi)]}\right\}. \tag{11.30}$$

Since the variation of $\hat{S}(\varpi)$ is negligible (by assumption) to that of $\Phi(\varpi)$, the locus of perturbation-surface density maxima lies on curves where

$$f(\varpi, \varphi, t) \equiv \omega t - m\varphi + \Phi(\varpi) = 2p\pi \qquad \text{with} \qquad p = 0, 1, \dots, m-1. \tag{11.31}$$

At any given t, say, $t = 0$, the locus for any specific choice of the integer p, say, $p = 0$, traces out a spiral (see Figure 11.1):

$$\varphi = \frac{1}{m}\Phi(\varpi).$$

The spiral leads or trails the disk rotation (toward positive φ) depending on whether $k \equiv \Phi'(\varpi)$ is positive or negative:

$$k > 0 \qquad \text{for a leading spiral,} \qquad \text{and} \qquad k < 0 \qquad \text{for a trailing spiral.}$$

For $p \neq 0$, we get another spiral rotated through an angle $\Delta\varphi$ equal to p/m of 2π. The collection of p's from $p = 0$ to $p = m - 1$ gives, therefore, m equidistant spiral arms. (The choices $p = m$, $m + 1$, ... repeat the choices

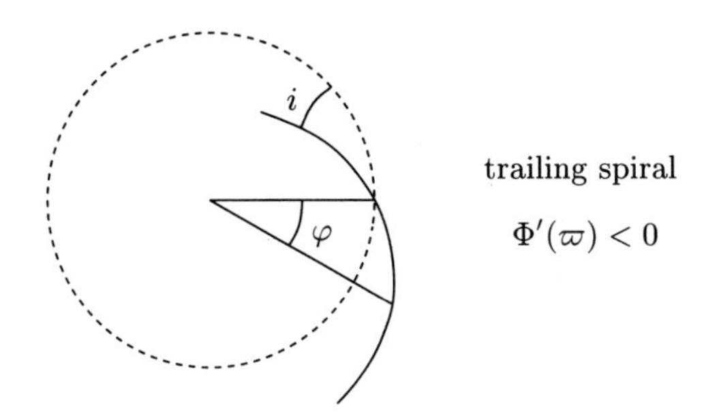

FIGURE 11.1
The locus of the maximum of the perturbational surface density for a spiral density wave. The spiral trails in the sense of rotation if $\Phi'(\varpi) < 0$, and it leads if $\Phi'(\varpi) > 0$.

$m = 0, 1, \ldots$, and therefore yield nothing new.) We obtain the tilt angle i of each spiral arm with respect to the circular direction through the ratio $d\varpi$ to $\varpi d\varphi$ as we follow a curve of constant phase f (with t held fixed):

$$\tan i = \frac{1}{\varpi}\left(\frac{\partial \varpi}{\partial \varphi}\right)_{f,t} = \frac{m}{\varpi}\left(\frac{d\Phi}{d\varpi}\right)^{-1} = m/k\varpi.$$

Hence, for m of order unity, the WKBJ condition, equation (11.26), amounts to the assumption that the spiral waves are *tightly wrapped*, $\tan i \ll 1$.

As time proceeds, the spiral pattern formed by the locus, $f(\varpi, \varphi, t) =$ constant, remains the same except for a rigid rotation at a *pattern speed*,

$$\Omega_{\rm p} \equiv \left(\frac{\partial \varphi}{\partial t}\right)_{f,\varpi} = \omega/m, \tag{11.32}$$

where we have assumed that ω is real. If ω has a complex value, then $\Omega_{\rm p} = \mathrm{Re}(\omega)/m$. Since astrophysical disks rotate differentially, with $\Omega(\varpi)$ decreasing for increasing values of ϖ, the pattern speed (for a direct mode in which $\omega > 0$) will usually rotate slower than the matter in the inner parts of the disk, and it will rotate faster than the matter in the outer parts of the disk. We denote the radius where the matter and pattern rotate at the same angular speed as the *corotation radius*, $\varpi_{\rm CR}$. Its value equals the root of the equation:

$$\Omega(\varpi_{\rm CR}) = \Omega_{\rm p}. \tag{11.33}$$

As we shall see in Chapter 12, the corotation radius plays an important role in possible amplification mechanisms for the spontaneous excitation of spiral density modes.

ASYMPTOTIC SOLUTION OF POISSON'S EQUATION

The WKBJ approximation for the theory of spiral density waves is useful because it reduces Poisson's equation (11.21) to an algebraic relation (see below). The alternative would be to integrate equation (11.21) via the Green's function technique, obtaining the usual integral representation,

$$V(\varpi, z=0) = -2\pi G \int_0^\infty K_m(\varpi, \varpi') S(\varpi')\, \varpi' d\varpi', \tag{11.34}$$

where $K_m(\varpi, \varpi')$ is the Poisson kernel,

$$K_m(\varpi, \varpi') \equiv \frac{1}{\pi} \int_0^\pi \frac{\cos m\varphi\, d\varphi}{(\varpi^2 + \varpi'^2 - 2\varpi\varpi' \cos\varphi)^{1/2}}. \tag{11.35}$$

Although the Poisson kernel can be expressed in terms of complete elliptic integrals, the integral representation (11.34) poses considerable difficulties for physical interpretation (as contrasted with numerical computation) since it makes the fundamental governing set, equations (11.19)–(11.21) and (11.34), a system of *integro-differential* equations, rather than a set of ODEs (or better yet, a set of algebraic relations).

To obtain the asymptotic solution to Poisson's equation (11.21) when the approximation of equation (11.26) holds, we use the rule that each differentiation of perturbation quantities with respect to ϖ (or z, as we shall see) brings down a large factor of k. Thus, to lowest asymptotic order for large $|k|\varpi$, equation (11.21) has the form

$$\frac{\partial^2 V}{\partial \varpi^2} + \frac{\partial^2 V}{\partial z^2} = 4\pi G S \delta(z). \tag{11.36}$$

To dispense with the Dirac delta function, integrate equation (11.36) from $z = 0^-$ to 0^+, making use of the fact that the z component of the gravitational field can suffer a jump (reverse directions) across the plane $z = 0$, but the ϖ component must remain continuous:

$$\left[\frac{\partial V}{\partial z}\right]_{z=0^-}^{0^+} = 4\pi G S. \tag{11.37}$$

From the symmetry of the problem, we see that V must be a function of $|z|$ rather than z itself; therefore we may write equation (11.37) as

$$S = \frac{1}{2\pi G} \left(\frac{\partial V}{\partial |z|}\right)_{|z|=0^+}, \tag{11.38}$$

which allows us to obtain $S(\varpi)$ *locally* from $V(\varpi, |z|)$ if we can just evaluate its $|z|$ derivative in the limit $|z| \to 0$. [From the standpoint of the integral representation, equation (11.34), such a local relation exists between the surface density and the gravity only for a wavy disturbance where distant excesses and deficits of density cancel in their contribution to the gravitational potential. See Chapter 19 of Volume I for a brief discussion of the method of *stationary phase.*]

To obtain the functional form of $V(\varpi, |z|)$ off the midplane, notice that equation (11.36) reads, for $|z| \neq 0$,

$$\frac{\partial^2 V}{\partial \varpi^2} + \frac{\partial^2 V}{\partial |z|^2} = 0, \tag{11.39}$$

which is Laplace's equation in *rectilinear* form. (When spirals are tightly wrapped, we do not see the radius of curvature of the disk to lowest order of approximation.) The theory of complex variables yields as the most general solution of equation (11.39),

$$V = F(\varpi \pm i|z|), \tag{11.40}$$

where F is an arbitrary function of $\varpi \pm i|z|$. To determine F, as well as the choice of sign in $\pm i|z|$, we first note that in the plane of the disk, we wish V to have the form given by equation (11.24):

$$V = A(\varpi)e^{i\Phi(\varpi)} \qquad \text{for} \qquad z = 0.$$

By analytic continuation, then, equation (11.40) requires for arbitrary $|z|$

$$V = A(\varpi + is|z|)e^{i\Phi(\varpi + is|z|)},$$

where the sign of $s \equiv \pm 1$ remains to be determined. The primary contribution to the variation of V comes from the term $e^{i\Phi}$. Expanding $\Phi(\varpi + is|z|)$ in a Taylor series for small $|z|$,

$$\Phi(\varpi + is|z|) = \Phi(\varpi) + is|z|\Phi'(\varpi) + \cdots,$$

and requiring that V decay as we leave the midplane, we obtain the identification

$$s = \pm 1 = \operatorname{sgn}(k),$$

where k is positive for leading waves and negative for trailing waves. Collecting results, we now obtain for equation (11.38),

$$S(\varpi) = \pm \frac{i}{2\pi G} F'(\varpi) \approx -\frac{|k|V}{2\pi G}, \tag{11.41}$$

the desired local relation between surface density and gravitational potential. The negative sign in equation (11.41) implies that spiral potential minima correspond to spiral density maxima; i.e., the excess density in a spiral arm forms a local potential well.

With the derivation of equation (11.41), we no longer need to consider the value of V off the plane of the disk. In what follows, we write $V(\varpi)$ as a shorthand notation for $V(\varpi, z = 0)$.

ASYMPTOTIC THEORY OF SPIRAL DENSITY WAVES

Consider now the substitution of the WKBJ forms,

$$\begin{bmatrix} S(\varpi) \\ U(\varpi) \\ J(\varpi) \\ V(\varpi) \end{bmatrix} = \begin{bmatrix} \hat{S}(\varpi) \\ \hat{U}(\varpi) \\ \hat{J}(\varpi) \\ A(\varpi) \end{bmatrix} e^{i\Phi(\varpi)}, \tag{11.42}$$

into the ODEs (11.19)–(11.21). We invoke the rule that differentiation of $e^{i\Phi}$ increases the order of the term by one factor of $|k|\varpi$ in comparison with differentiation of anything else, and we use equation (11.41) to show that S by itself is (in some sense) one order higher than U, J, or V. On the other hand, as we have already discussed in Chapter 7, the sound speed a_0 in thin disks is small compared to $\varpi\Omega$. To judge $|k|^{-1}$ relative to a_0, we assume that $1/|k|\varpi$ and $a_0/\varpi\Omega$ have the same degree of smallness (to be justified *a posteriori* as the physically relevant regime). With this ordering, we obtain the following set of simultaneous equations as the dominant terms for equations (11.19)–(11.21):

$$\begin{bmatrix} i(\omega - m\Omega) & ik\sigma_0 & 0 \\ ika_0^2/\sigma_0 & i(\omega - m\Omega) & -2\Omega/\varpi \\ 0 & (\kappa^2/2\Omega)\varpi & i(\omega - m\Omega) \end{bmatrix} \begin{pmatrix} S \\ U \\ J \end{pmatrix} = \begin{pmatrix} 0 \\ -ikV \\ 0 \end{pmatrix}. \tag{11.43}$$

Solving equation (11.43) for the surface density S that responds dynamically to an imposed spiral gravitational field V, we get

$$S = -\frac{k^2\sigma_0 V}{\kappa^2 - (\omega - m\Omega)^2 + k^2 a_0^2}. \tag{11.44}$$

Notice that the density response is 180° out of phase with the imposed potential if the denominator is positive. If we equate the density response (11.44) and the imposed density (11.41) needed to maintain (through Poisson's equation) the same potential V, we obtain the *WKBJ dispersion relationship* for spiral density waves,

$$(\omega - m\Omega)^2 = \kappa^2 + k^2 a_0^2 - 2\pi G|k|\sigma_0, \tag{11.45}$$

that gives the absolute value of the radial wave number $|k|$ in terms of the wave frequency ω and the local equilibrium properties of the self-gravitating disk: $\Omega(\varpi)$, $\kappa(\varpi)$, $a_0(\varpi)$, and $\sigma_0(\varpi)$.

AXISYMMETRIC STABILITY CRITERION

We now analyze the implications contained in equation (11.45). We begin with the case of axisymmetric disturbances, $m = 0$:

$$\omega^2 = \kappa^2 + k^2 a_0^2 - 2\pi G|k|\sigma_0. \tag{11.46}$$

We perform a local stability analysis (okay for axisymmetric disturbances that cannot propagate to the boundaries of our system) by restricting k to real values and examining the consequences for ω. Clearly, equation (11.46) requires ω^2 to be real if k is real. Positive values of ω^2 correspond to oscillating rings (stable disturbances), whereas negative values of ω^2 correspond to exponentially growing rings (unstable disturbances). (The disturbances for $\omega^2 < 0$ come in pairs: one of which grows, the other of which decays.) In this analysis, differential rotation, as represented by the term κ^2 (Rayleigh's discriminant; see Chapter 8), contributes positively to ω^2 and therefore constitutes a stabilizing influence. Basically, the tendency to conserve angular momentum throws matter back out that tries to collect into rings. Pressure, as represented by the term $k^2 a_0^2$, also constitutes a stabilizing influence because the overpressure of dense regions tends to push matter away from condensing rings. Self-gravity, as represented by the term $-2\pi G|k|\sigma_0$, constitutes a destabilizing influence because the attraction of a dense condensation for itself tends to make such a region even denser.

Notice that, unlike the Jeans problem [cf. equation (8.18) and Problem Set 3], the self-gravitation term enters into the dispersion relation (11.45) with a (linear) power of $|k|$ *intermediate* between that of two stabilizing influences ($|k|^0$ for differential rotation and $|k|^2$ for pressure). This fact represents an important difference between the (two-dimensional) self-gravity of a disk and that of a homogeneous medium, which is three-dimensional. As a consequence, the combination of differential rotation, which can suppress axisymmetric disturbances of a long spatial scale (small $|k|$), and pressure, which can suppress disturbances of a short spatial scale (large $|k|$), acting in consort may be able to prevent the tendency to clump into rings on all scales. To derive the conditions needed to achieve this feat, we examine the state of marginal stability: $\omega^2 = 0$. (Since ω^2 is real, the transition from ω^2 negative to ω^2 positive must be made through $\omega^2 = 0$, a property of the modal behavior of many dissipationless systems, called the *principle of exchange of stabilities.* Here this principle turns out to apply to axisymmetric disturbances, $m = 0$, but not to non-axisymmetric ones, $m \neq 0$.) If we set $\omega^2 = 0$ in equation (11.46), we get a relationship between the required a_0 and $|k|$ that we may write in a nondimensional form by dividing through by κ^2,

$$1 + \frac{Q^2}{4}\left(\frac{|k|}{k_T}\right)^2 - \frac{|k|}{k_T} = 0, \tag{11.47}$$

where we have chosen to measure the wave number $|k|$ in units of the Toomre wave number scale,

$$k_T \equiv \frac{\kappa^2}{2\pi G\sigma_0}, \tag{11.48}$$

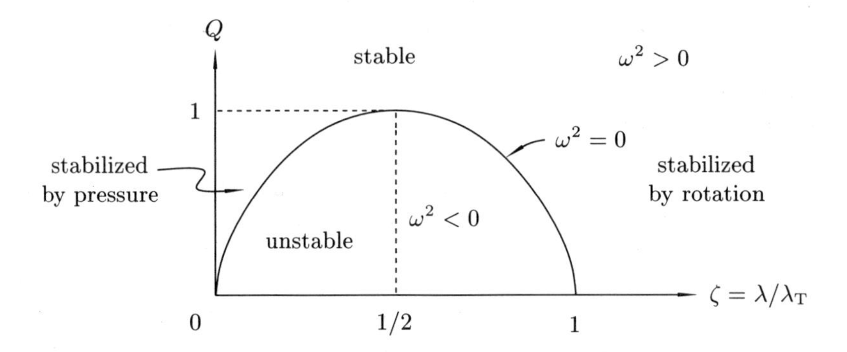

FIGURE 11.2
Stability diagram for clumping into axisymmetric rings by a self-gravitating, differentially rotating, gaseous disk. Short-wavelength disturbances are stabilized by the effects of pressure; long-wavelength, by the effects of rotation. Above a critical value of unity for the Toomre parameter Q, no local axisymmetric disturbance of any scale can grow.

and we have defined Q as (the fluid analog of) Toomre's stability parameter:

$$Q \equiv 2k_T a_0/\kappa = \kappa a_0/\pi G\sigma_0. \tag{11.49}$$

If we solve equation (11.47) for Q in terms of the dimensionless wavelength $\lambda/\lambda_T \equiv k_T/|k| \equiv \zeta$: we obtain the condition of marginal stability as

$$Q = 2[\zeta(1-\zeta)]^{1/2}.$$

This result can be plotted as the curve shown in Figure 11.2, which has the following interpretation.

When the combination of basic state parameters represented by equation (11.49) has a value $Q < 1$, then there exist unstable disturbances possessing an intermediate range of wavelengths that are too short to be stabilized by rotation (κ) and too long to be stabilized by pressure (or velocity dispersion a_0). The most vulnerable wavelength corresponds to $\zeta = 1/2$. On the other hand, when $Q > 1$, axisymmetric disturbances of all wavelength scales are stabilized by the combination of rotation and pressure, or more accurately, by the combination κa_0 compared to the self-gravitation of the disk as measured by $\pi G\sigma_0$. Thus, Toomre's criterion reads

$$Q \geq 1 \qquad \text{for local axisymmetric stability} \tag{11.50}$$

of a self-gravitating disk. Numerical calculations and computer experiments show that if Toomre's stability criterion is satisfied at all ϖ, then the disk is also globally stable to all axisymmetric *modes.*

If we calculate the combination $\kappa a_0/\pi G\sigma_0$ for the local solar neighborhood in the Galaxy, taking the epicylic frequency κ to equal $32\,\text{km}\,\text{s}^{-1}\,\text{kpc}^{-1}$, the surface density (stars plus gas) to equal $60\,M_\odot\,\text{pc}^{-2}$, and the dispersion speed (averaged over stars and gas) to equal $30\,\text{km}\,\text{s}^{-1}$, we get (with $G = 4.298\,M_\odot^{-1}\,\text{pc}^2\,\text{km}^2\,\text{s}^{-2}\,\text{kpc}^{-1}$) $Q = 1.2$, with a large enough uncertainty to make marginal stability, $Q = 1$, a distinct possibility. This coincidence between the conditions actually prevailing locally in the Galaxy and those required for marginal axisymmetric stability has prompted speculation concerning how such a state might have arisen naturally because of the action of past instabilities in the disk of stars, or because of some "self-regulation" by the interstellar gas.

NONAXISYMMETRIC WAVES AND LINDBLAD RESONANCES

Suppose a disk has a certain Q distribution with radius ϖ that makes it stable with respect to all axisymmetric disturbances. How would nonaxisymmetric disturbances, or spiral density waves, behave in such a disk? Begin by defining the dimensionless combination,

$$\nu \equiv (\omega - m\Omega)/\kappa, \tag{11.51}$$

which represents the wave frequency Doppler-shifted relative to the motion of the matter in a pattern with m arms, $(\omega - m\Omega)$, measured in units of the natural frequency κ. If we now solve the dispersion relationship (11.45) as a quadratic equation for $|k|$, we obtain

$$\frac{|k|}{k_\text{T}} = \frac{2}{Q^2}\{1 \pm [1 - Q^2(1 - \nu^2)]^{1/2}\}. \tag{11.52}$$

The upper sign choice, which gives a larger root for $|k|$, yields what is known as *short* waves; the lower sign choice, which gives a smaller root for $|k|$, yields *long* waves. Pressure plays a more important role than rotation as a restoring force for short waves; the reverse is true for long waves.

Each value of the absolute magnitude $|k|$ also has two possibilities for the sign of k (positive for leading, negative for trailing); hence, in all, for given wave frequency ω and disk location ϖ (given ν), there can exist a maximum of four kinds of spiral density waves: short-trailing (ST), long-trailing (LT), short-leading (SL), and long-leading (LL). Notice that the long waves become infinitely long ($|k| = 0$) at a location where $\nu = \pm 1$. These are called the *outer* and *inner Lindblad resonances* (abbreviated ILR and OLR), and they occur at radii where the matter experiences the wave

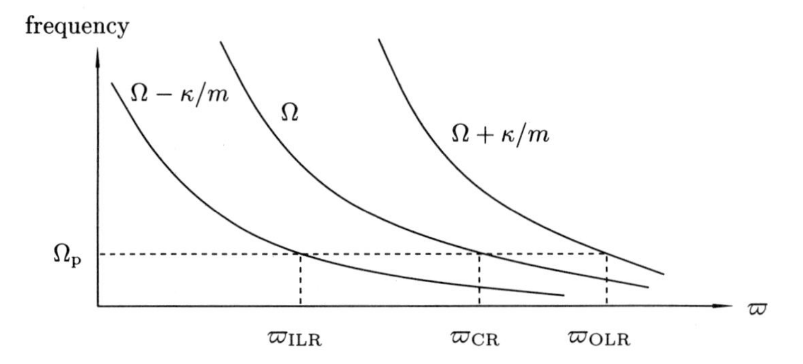

FIGURE 11.3
The radial locations of the inner Lindblad resonance, $\varpi_{\rm ILR}$, the corotation resonance, $\varpi_{\rm CR}$, and the outer Lindblad resonance, $\varpi_{\rm OLR}$, can be found graphically by examining the radius-angular frequency plane. The resonances occur where a horizontal line $\Omega_{\rm p}$ = constant intersects the curves describing $\Omega(\varpi) - \kappa(\varpi)/m$, $\Omega(\varpi)$, and $\Omega(\varpi) + \kappa(\varpi)/m$. Since κ usually is bounded below by Ω (Keplerian value) and above by 2Ω (uniform-rotation value), $m = 1$ (one-armed spiral or eccentric) and $m = 2$ (two-armed spiral or barred) disturbances may have no inner Lindblad resonance if $\Omega_{\rm p}$ has a sufficiently large positive value.

at a Doppler-shifted frequency $(\omega - m\Omega)$ equal to $\pm$ times the natural frequency κ. In other words, the radii, $\varpi_{\rm ILR}$ and $\varpi_{\rm OLR}$, can be found from the roots of the equations

$$\Omega(\varpi_{\rm ILR}) - \frac{1}{m}\kappa(\varpi_{\rm ILR}) = \Omega_{\rm p}, \tag{11.53}$$

$$\Omega(\varpi_{\rm OLR}) + \frac{1}{m}\kappa(\varpi_{\rm OLR}) = \Omega_{\rm p}, \tag{11.54}$$

just as the corotation radius, $\varpi_{\rm CR}$, could be found as the root of the equation $\Omega(\varpi_{\rm CR}) = \Omega_{\rm p}$. The diagram in Figure 11.3 illustrates the situation graphically.

The Lindblad resonances have the following significance for long spiral waves. A position where $|k|$ attains a zero corresponds to a *turning point* of the wave-propagation problem (in the sense of quantum mechanics). At such a turning point, the waves change from being able to propagate (k real) to becoming evanescent (k imaginary). The simple asymptotic analysis breaks down near such turning points, and one must return to the full equations for a deeper treatment (for example, to obtain the WKBJ "connection formulae" that tell us how such waves tunnel or reflect). If the curves $\Omega \pm \kappa/m$ vary mononically with ϖ, then long waves can propagate only in the region between the inner and outer Lindblad resonances; this region, $\varpi_{\rm ILR} < \varpi < \varpi_{\rm OLR}$ is referred to by Lin and Shu as the *principal range*. Goldreich and Tremaine showed that forcing by a rotating bar-like potential or by an orbiting satellite can excite the emission of long spiral waves from the Lindblad resonances. Essentially, the smoothly varying gravitational force associated with such "external influences" can couple well to wavy disturbances only where the waves are very long ($|k| = 0$). The formal violation of the WKBJ condition, $|k|\varpi \gg 1$, when $k = 0$ has raised objections to the concept in the galactic context, but there is no question that this phenomenon occurs in planetary rings. In particular, over 50 resonantly excited waves have been found in the Voyager data of Saturn's rings, and the analysis of these waves provide to date the best diagnostics of the physical conditions in this system.

In a fluid treatment, the Lindblad resonances hold no special significance for short spiral waves; they can propagate right through the ILR and OLR, maintaining a finite value for $|k|$. In a stellar disk, Lin and Shu showed that short waves acquire an *infinite* value for $|k|$. Such behavior also signals a breakdown of the naive WKBJ theory, and Mark showed that a more sophisticated analysis gives *Landau absorption* of short (stellar) density waves at the positions of the Lindblad resonances. Gas and stars also have distinct behaviors at the corotation resonance, but we shall not pursue this topic here.

Q BARRIERS

Another interesting property possessed by density waves concerns the refraction of long waves into short waves (or vice versa) at *Q barriers*. A Q barrier arises whenever the quantity inside the square root in equation (11.52) vanishes, i.e., where

$$\nu^2 = 1 - \frac{1}{Q^2}. \tag{11.55}$$

If we were to imagine lowering Q from a uniformly high value throughout the disk to lower values, the position of the Q barriers would gradually move from the Lindblad resonances ($\nu = \pm\, 1$) for $Q \gg 1$, toward the corotation circle ($\nu = 0$) as $Q \to 1$. For $Q \gg 1$, therefore, we have an exclusion of long waves from almost the entire principal range; whereas , for $Q \gtrsim 1$, they can propagate everywhere except in a narrow annulus surrounding the corotation circle. (For disks where the radial distribution of Q is not uniform, the situation becomes more complicated, with the possibility of multiple zones of evanescence and propagation.)

At a radial position where equation (11.55) holds, the WKBJ dispersion relation (11.52) states that long and short waves merge together in having a common value for $|k|$. Since, as we shall show in Chapter 12, long and short waves (or a given leading or trailing type) have opposite senses of (group) propagation, this behavior suggests (see below) that a long wave propagating into a Q barrier will refract into a short wave, and vice versa. This process multiplies the richness of possibilities for establishing standing-wave patterns ("normal modes") in self-gravitating disks well beyond the simple case if we had only one type of wave (apart from the sign of k), which might reflect (leading into trailing or vice versa) off disk edges and/or propagate through the center. Many different modal cycles have been studied in the literature; we cannot enumerate all of them. Instead, in Chapter 12, we shall focus on the more interesting question of *wave amplification mechanisms* if feedback cycles do exist in a disk.

LOWEST-ORDER WAVE EQUATION

At various turning points of the types discussed so far, the naive WKBJ approximation breaks down, and we must return to the original differential equations to analyze the problem. A simplification in the same spirit as our current asymptotic approach invokes the following procedure. We recognize each factor of ik in the reduced algebraic equations (11.41) and (11.43) as resulting from a differentiation of a perturbation variable by ϖ. Thus the original governing ODEs must have lowest-order asymptotic approximations whose equivalents to equations (11.41) and (11.44) we may write as

$$\frac{dV}{d\varpi} = \mp i2\pi GS,$$

$$\left[-a_0^2\frac{d^2}{d\varpi^2} + \kappa^2 - (\omega - m\Omega)^2\right] S = \sigma_0\frac{d^2V}{d\varpi^2}.$$

In other words, S satisfies the wave equation,

$$L_\pm S = 0,$$

where L_+ and L_- are the second-order differential operators,

$$L_\pm \equiv -a_0^2\frac{d^2}{d\varpi^2} \pm i2\pi G\sigma_0\frac{d}{d\varpi} + \kappa^2 - (\omega - m\Omega)^2, \tag{11.56}$$

with the upper sign choice applying to leading spiral waves, and the lower to trailing.

Except for the term due to self-gravity,

$$\pm\, i2\pi G\sigma_0\frac{d}{d\varpi}, \tag{11.57}$$

and the quadratic dependence on the eigenvalue ω, L_+ and L_- have the same formal structure as the operator for the time-independent Schrödinger equation in one spatial dimension. This property of $L_\pm$ allows us to borrow many of the techniques developed in courses on quantum mechanics (see in particular Problem Set 6 of Volume I). For example, in the problem of *refraction* at a Q barrier (change in magnitude of k but not of sign), we can first factor out the part of the radial phase common to short and long waves:

$$S \equiv \mathcal{S} \exp\left(\pm i \int \frac{\pi G \sigma_0}{a_0^2}\, d\varpi\right).$$

The transformation reduces the governing ODE for $\mathcal{S}$ to normal form, which reads to lowest asymptotic order,

$$\left[-a_0^2 \frac{d^2}{d\varpi^2} + \kappa^2(1 - \nu^2 - Q^{-2})\right] \mathcal{S} = 0,$$

with a reduced wavenumber that vanishes at a Q barrier.

The *reflection* of long leading and trailing waves at Lindblad resonances also involves a gradual transformation of a wavenumber through zero (now the entire k) and can be handled via *two* coupled ODEs that have L_+ and L_-, respectively, as their governing operators. In contrast, the coupling of leading and trailing waves by *nonadiabatic* means (finite changes of k in magnitude and sign, say, across the corotation circle) cannot be treated by the present methods but requires a different technique (e.g., adopting the model of the so-called "sheared sheet"). For further details, we refer the reader to the astrophysical literature.

12

Spiral Density Waves: Group Propagation and Amplification

References: A. Toomre, 1969, *Ap. J.*, **158**, 899; J. W.-K. Mark, 1974, *Ap. J.*, **193**, 539; R. Drury, 1980, *M.N.R.A.S.*, **193**, 337.

In Chapter 11, we derived the dispersion relationship for a tightly wrapped spiral wave. In other words, to lowest asymptotic order in a WKBJ approximation, we obtained the behavior of the phase variation of the wave. To next order, we expect to obtain the amplitude variation. In this chapter, we derive the physical interpretation of this amplitude variation in terms of the propagation of wave energy and angular momentum at the group velocity of the wave. We then indicate schematically how these ideas relate to mechanisms for the amplification of a system of spiral density waves that interact across corotation, and how feedback cycles can lead to the spontaneous growth of spiral modes in differentially rotating, self-gravitating disks.

AMPLITUDE VARIATION

We choose to regard the central results of Chapter 11, equations (11.41) and (11.44), as yielding the density response (from the dynamical equations) in terms of the imposed density (from Poisson's equation) through the relation

$$S_{\text{response}} = D S_{\text{imposed}}, \tag{12.1}$$

where D is the function (related to the concept of a "gravitational dielectric")

$$D(\omega, k, m, \varpi) \equiv \frac{2\pi G|k|\sigma_0}{\kappa^2 - (\omega - m\Omega)^2 + k^2 a_0^2}. \tag{12.2}$$

For a self-consistent wave, $S_{\text{response}} = S_{\text{imposed}}$, and we recover the WKBJ dispersion relationship [cf. equation (11.45)]:

$$D(\omega, k, m, \varpi) = 1, \tag{12.3}$$

which yields the (negative of the) radial wave number k once we are given the (real) wave frequency ω, the azimuthal integer m, and the radial position ϖ.

Equation (12.2) results from our adoption of a fluid model for the dynamics of the disk, but a completely analogous result with a different functional form for D holds in the stellar dynamical case (C. C. Lin and F. H. Shu, 1966, *Proc. Nat. Acad. Sci.*, **55**, 229). If we carry out a formal expansion of the governing equations in powers of $1/|k|\varpi$, the next order of approximation yields a relation for the radial variation of the amplitude A of the gravitational potential that may be written as

$$\varpi A^2 k \left(\frac{\partial D}{\partial k}\right)_{\omega,m,\varpi} = \text{constant}, \tag{12.4}$$

for either a stellar disk (F. H. Shu, 1970, *Ap. J.*, **160**, 99) or a gaseous one which suffers only adiabatic variations (P. Goldreich and S. Tremaine, 1979, *Ap. J.*, **233**, 857). An alternative derivation of the same result follows by applying Whitham's elegant variational formulation in terms of an averaged Lagrangian (R. L. Dewar, 1972, *Ap. J.*, **174**, 301). For details, see the cited references.

WAVE ENERGY AND ANGULAR MOMENTUM

Equation (12.4) has an important interpretation in terms of the propagation of wave energy and angular momentum. We may identify the energy and angular momentum of a spiral density wave with potential amplitude A as the difference in those quantities in the disk with and without the presence of the wave, i.e., when $A \neq 0$ as compared to when $A = 0$. To calculate the densities (per unit area) of the wave energy and angular momentum, we could either carry out (a not very rewarding) direct calculation (which requires a second-order WKBJ analysis to get the nonvanishing contribution), or we can follow Toomre and adopt the following computational trick. We pretend that the disk contains no disturbance at $t = -\infty$, and we force the growth of a spiral wave by slowly turning on a (fictitious) external spiral gravitational potential of the form:

$$\mathcal{V}_{\text{ext}} \equiv \text{Re}\left\{A_{\text{ext}} e^{i[\omega t - m\varphi + \Phi(\varpi)]} e^{\gamma t}\right\}, \tag{12.5}$$

where γ represents a small positive growth rate that we shall let go to zero at the end of our analysis.

We easily verify that the imposition of the additional gravitational influence, equation (12.5), changes equation (11.44) for the zeroth-order density response so that its analog of the full variation (which includes a factor $e^{\gamma t}$ apart from the usual WKBJ terms) reads

$$\sigma_1 = \text{Re}\left[\frac{-k^2\sigma_0(\mathcal{V}_1 + \mathcal{V}_{\text{ext}})}{\kappa^2 - (\omega - i\gamma - m\Omega)^2 + k^2 a_0^2}\right]. \tag{12.6}$$

Notice that the presence of the slight exponential growth modifies the wave frequency ω in the denominator to $\omega - i\gamma$. On the other hand, the relation between the disk density σ_1 and its associated potential through Poisson's equation remains as derived in equation (11.41):

$$\sigma_1 = \text{Re}\left[-\frac{|k|\mathcal{V}_1}{2\pi G}\right]. \tag{12.7}$$

If we now equate equations (12.6) and (12.7) and solve for the external potential $\mathcal{V}_{\text{ext}}$ used to excite the spiral wave of the desired properties, we get

$$\mathcal{V}_{\text{ext}} = \text{Re}\left\{[D(\omega - i\gamma, k, m, \varpi) - 1]\,\mathcal{V}_1\right\}, \tag{12.8}$$

where D is the function given by equation (12.2), except that we replace ω by $\omega - i\gamma$. Only in the limit $\gamma \to 0$, does equation (12.3) hold, i.e., do we recover the dispersion relation for self-sustained spiral density waves (no $\mathcal{V}_{\text{ext}}$ needed).

We let the external forcing act on the disk from time $-\infty$ to t (when the internal wave amplitude has grown to the desired amplitude A), after which we switch $\mathcal{V}_{\text{ext}}$ off. The total amount of angular momentum J_{wave} added to the disk by this process equals the integral of the torque exerted by the external potential, which we may calculate as

$$J_{\text{wave}} = \int_{-\infty}^{t} dt \int_0^{\infty} \varpi d\varpi \oint \sigma_1 \left(-\frac{\partial \mathcal{V}_{\text{ext}}}{\partial \varphi}\right) d\varphi. \tag{12.9}$$

For small γ when $D \approx 1$, we may write equations (12.7) and (12.8) as

$$\sigma_1 = -\frac{|k|A}{2\pi G}\cos[\omega t - m\varphi + \Phi(\varpi)]e^{\gamma t},$$

$$-\frac{\partial \mathcal{V}_{\text{ext}}}{\partial \varphi} = m\gamma \left(\frac{\partial D}{\partial \omega}\right)_{k,m,\varpi} A\cos[\omega t - m\varphi + \Phi(\varpi)]e^{\gamma t},$$

where $(\partial D/\partial\omega)_{k,m,\varpi}$ is evaluated with its first argument taken to be the real wave frequency ω. Integration over φ now yields for equation (12.9),

$$J_{\text{wave}} = 2\pi \int_0^{\infty} \varpi d\varpi \,\frac{m|k|A^2}{4\pi G}\frac{\partial D}{\partial \omega} \int_{-\infty}^{t} \gamma e^{2\gamma t}\, dt, \tag{12.10}$$

where we have switched the order of the ϖ and t integrations, and where we implicitly understand that partial derivatives of D with respect to any of its arguments are to be taken with the other arguments held fixed. The t integration is trivially performed, and if we then take the limit $\gamma \to 0$, we obtain

$$J_{\text{wave}} = \int_0^\infty \left(\frac{m|k|A^2}{8\pi G} \frac{\partial D}{\partial \omega} \right) 2\pi \varpi d\varpi. \tag{12.11}$$

Equation (12.11) suggests that we identify the angular momentum density (per unit area when averaged over the angular direction) of the wave as the following quadratic function of the wave amplitude A:

$$\mathcal{J}_{\text{wave}} = \frac{m|k|A^2}{8\pi G} \frac{\partial D}{\partial \omega}. \tag{12.12}$$

In a similar manner, we may obtain the energy contained in the wave, E_{wave}, by calculating the amount of work performed by the external gravitational potential in establishing the disturbance. For forces derivable as the negative gradient of a time-dependent scalar potential $\mathcal{V}_{\text{ext}}$, the net rate of doing work per unit mass can be shown by elementary means to be given by $\partial \mathcal{V}_{\text{ext}}/\partial t$. Thus,

$$E_{\text{wave}} = \int_{-\infty}^t dt \int_0^\infty \varpi d\varpi \oint \sigma_1 \frac{\partial \mathcal{V}_{\text{ext}}}{\partial t} d\varphi. \tag{12.13}$$

Differentiation with respect to t brings down a factor ω instead of the m we get when we take the negative derivative with respect to φ; hence, if we compare the above equation to equation (12.9), we can write—by analogy with equation (12.12) for the wave angular-momentum density—the corresponding expression for the wave energy density:

$$\mathcal{E}_{\text{wave}} = \frac{\omega|k|A^2}{8\pi G} \frac{\partial D}{\partial \omega}. \tag{12.14}$$

Notice, therefore, that the energy and angular-momentum densities of the wave have a constant proportionality between them,

$$\mathcal{E}_{\text{wave}} = \Omega_{\text{p}} \mathcal{J}_{\text{wave}}, \tag{12.15}$$

a relationship that Kalnajs (private communication) first showed to hold for the *total* wave energy and angular momentum, independently of the validity of the WKBJ approximation.

CHANGE IN SIGN ACROSS COROTATION

Equations (12.12) and (12.14) have the important properties that $\mathcal{J}_{\text{wave}}$ and $\mathcal{E}_{\text{wave}}$ carry their signs in the term $\partial D/\partial \omega$. If we differentiate equation (12.2) with respect to ω, we get

$$\frac{\partial D}{\partial \omega} = \frac{4\pi G|k|\sigma_0(\omega - m\Omega)}{[\kappa^2 - (\omega - m\Omega)^2 + k^2 a_0^2]^2}, \tag{12.16}$$

which switches sign across where $\omega - m\Omega = 0$. Thus, the wave has positive energy and angular-momentum densities outside corotation ($\Omega_p > \Omega$); negative, inside ($\Omega_p < \Omega$). In other words, the disk contains an excess of angular momentum (and energy) where the spiral pattern rotates faster than the matter, and it contains a deficit where the spiral pattern rotates slower than the matter. This fact has two important implications for astrophysical applications.

First, if we have a spiral pattern (rotating in a direct sense) that contains a corotation circle, then the dissipation of that wave by nonadiabatic or frictional effects in the disk would release a negative amount of angular momentum inside corotation and a positive amount outside. The net result would cause the matter on the inside to move inward (accretion); on the outside, to move outward (excretion); i.e., the disk would spread through any secular interaction of the spiral density wave and the basic state, just as it does in models of viscous accretion disks. The difference, however, would be that the transport of angular momentum and energy by waves would not depend on local gradients (e.g., it would not be proportional to the local shear $\varpi d\Omega/d\varpi$).

Second, if the waves can interact across corotation, then we have the possibility of the amplification of the wave system in time on both sides (giving rise to a global instability for the problem of nonaxisymmetric modes). The wave amplitude inside corotation could grow, making $\mathcal{J}_{\text{wave}}$ and $\mathcal{E}_{\text{wave}}$ more and more negative, and the wave amplitude outside corotation could also grow, making $\mathcal{J}_{\text{wave}}$ and $\mathcal{E}_{\text{wave}}$ more and more positive, in such a way that the sums of the inside and outside remain conserved (equal to zero if the disturbance grows spontaneously out of noise). In what follows, we consider how such a mechanism for apparently getting something for nothing would work physically.

GROUP VELOCITY

Physicists usually approach the concept of the *group velocity* $\mathbf{c}_g$ from the point of view of superimposing monochromatic waves (slightly different ω's and $\mathbf{k}$'s) to form a packet (Figure 12.1). The envelope of the group then propagates at a velocity,

$$\mathbf{c}_g = \frac{\partial \omega}{\partial \mathbf{k}}, \tag{12.17}$$

where we evaluate the partial derivative by solving the local dispersion relation, $D(\omega, \mathbf{k}, \mathbf{x}) = 1$, for $\omega(\mathbf{k}, \mathbf{x})$.

Whitham (see, e.g., his book, *Linear and Nonlinear Waves*) adopts an alternative viewpoint. He chooses to regard the equation $D(\omega, \mathbf{k}, \mathbf{x}) = 1$ as a relationship between the temporal and spatial derivatives of the total wave phase,

$$f(\mathbf{x}, t) = \omega t - \phi(\mathbf{x}), \tag{12.18}$$

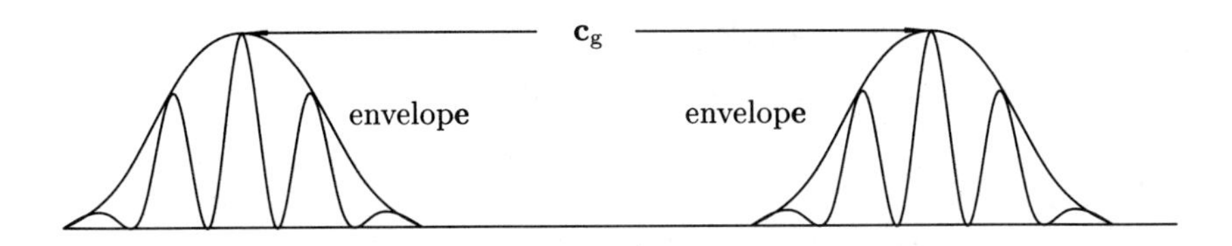

FIGURE 12.1
A wave packet with a well-defined envelope travels at the group velocity $\mathbf{c}_g = \partial\omega/\partial\mathbf{k}$.

when we take the disturbances to have a WKBJ variation $\propto e^{if}$, with

$$\omega = \frac{\partial f}{\partial t}, \qquad \text{and} \qquad \mathbf{k} = -\frac{\partial f}{\partial \mathbf{x}} = \frac{\partial \phi}{\partial \mathbf{x}}. \tag{12.19}$$

Thus Whitham regards the local dispersion relationship as a PDE for f having the functional form,

$$D\left(\frac{\partial f}{\partial t}, -\frac{\partial f}{\partial \mathbf{x}}, \mathbf{x}\right) = 1. \tag{12.20}$$

Since the dispersion relationship usually involves a nonlinear relationship between the wave number $\mathbf{k}$ and the wave frequency ω, no general methods exist to directly solve equation (12.20). Consider, however, differentiating equation (12.20) once in time, holding $\mathbf{x}$ fixed; we get

$$\frac{\partial D}{\partial \omega}\frac{\partial^2 f}{\partial t^2} - \frac{\partial D}{\partial \mathbf{k}} \cdot \frac{\partial^2 f}{\partial t \partial \mathbf{x}} = 0.$$

We may write $\partial^2 f/\partial t^2$ as $\partial\omega/\partial t$, and $\partial^2 f/\partial t\partial\mathbf{x}$ as $\partial\omega/\partial\mathbf{x}$, and obtain a quasilinear PDE of first order for ω:

$$\frac{\partial D}{\partial \omega}\frac{\partial \omega}{\partial t} - \frac{\partial D}{\partial \mathbf{k}} \cdot \frac{\partial \omega}{\partial \mathbf{x}} = 0. \tag{12.21}$$

We term equation (12.21) *quasilinear* because the coefficients $\partial D/\partial\omega$ and $\partial D/\partial\mathbf{k}$ depend only on ω and not on its derivatives in time or space; thus, although equation (12.21) has a nonlinear dependence on ω, the highest (first) order *derivatives* of ω enter linearly.

The *method of characteristics* for quasilinear PDEs of first order in a single dependent variable (see Chapter 13) states that the solution to equation (12.21) can be generated from the family of solutions associated with the ODEs

$$\frac{dt}{\partial D/\partial \omega} = \frac{d\mathbf{x}}{-\partial D/\partial \mathbf{k}} = \frac{d\omega}{0} \tag{12.22}$$

that we get by equating the differential of the independent variables (t and $\mathbf{x}$) over their coefficients ($\partial D/\partial \omega$ and $\partial D/\partial \mathbf{k}$) to the differential of the dependent variable (ω) over whatever appears on the right-hand side (zero if the PDE is homogeneous). Do not worry about formal details like dividing one vector by another, or dividing by zero, because what we really mean by equation (12.22) is the statement

$$\frac{d\omega}{dt} = 0 \qquad \text{on the characteristic} \qquad \frac{d\mathbf{x}}{dt} = -\frac{\partial D/\partial \mathbf{k}}{\partial D/\partial \omega} = \frac{\partial \omega}{\partial \mathbf{k}}. \tag{12.23}$$

In other words, the group velocity, as defined by equation (12.17), refers to the trajectory the solution must follow if we are to maintain a constant wave frequency ω.

Similarly, if we differentiate equation (12.20) once with respect to $\mathbf{x}$, holding t fixed, we obtain

$$-\frac{\partial D}{\partial \omega}\frac{\partial \mathbf{k}}{\partial t} + \frac{\partial D}{\partial \mathbf{k}} \cdot \frac{\partial \mathbf{k}}{\partial \mathbf{x}} = -\frac{\partial D}{\partial \mathbf{x}},$$

which has the associated characteristics,

$$\frac{d\mathbf{k}}{dt} = \frac{\partial D/\partial \mathbf{x}}{\partial D/\partial \omega} \qquad \text{on the trajectory} \qquad \frac{d\mathbf{x}}{dt} = -\frac{\partial D/\partial \mathbf{k}}{\partial D/\partial \omega} = \frac{\partial \omega}{\partial \mathbf{k}}. \tag{12.24}$$

Equation (12.24) tells us how the local wave number varies with changes in position as we follow a trajectory moving at the group velocity $\mathbf{c}_{\mathrm{g}}$.

RELATIONSHIP BETWEEN QUANTUM MECHANICS AND CLASSICAL MECHANICS

For the reader who may feel uncomfortable about the formal manipulations of the last section, we provide a more familiar context for an exposition: the demonstration that the transition from quantum mechanics to classical mechanics involves the same principles. Consider Schrödinger's equation for a particle,

$$-\frac{\hbar^2}{2m}\nabla^2\psi + V(\mathbf{x})\psi = i\hbar\frac{\partial \psi}{\partial t}. \tag{12.25}$$

The classical limit corresponds to looking for a WKBJ solution of the form

$$\psi = A(\mathbf{x}, t)e^{i\hbar^{-1}S(\mathbf{x},t)}, \tag{12.26}$$

where we may assume that the phase $f \equiv \hbar^{-1}S$ varies rapidly in comparison with the amplitude A. (This corresponds to the statement that the classical action S contains many units of $\hbar$.) The substitution of equation (12.26) into equation (12.25) yields, to the lowest order for small $\hbar$,

$$\frac{1}{2m}\left(\left|\frac{\partial S}{\partial \mathbf{x}}\right|\right)^2 + V(\mathbf{x}) = -\frac{\partial S}{\partial t},$$

which represents the Hamilton-Jacobi equation (see Goldstein, *Classical Mechanics*),

$$\frac{\partial S}{\partial t} + H\left(\frac{\partial S}{\partial \mathbf{x}}, \mathbf{x}\right) = 0, \tag{12.27}$$

where $H(\mathbf{p}, \mathbf{x})$ equals the particle Hamiltonian,

$$H = \frac{1}{2m}|\mathbf{p}|^2 + V(\mathbf{x}). \tag{12.28}$$

The Hamilton-Jacobi equation (12.27) represents a nonlinear PDE for the action function $S(\mathbf{x}, t)$ in the same sense that the dispersion relation (12.20) represents a nonlinear PDE for the wave phase $f(\mathbf{x}, t)$. Indeed, the formal solution of equation (12.20) to obtain

$$\frac{\partial f}{\partial t} = F\left(\frac{\partial f}{\partial \mathbf{x}}, \mathbf{x}\right),$$

would make the analogy complete, provided we identify f as being proportional to S, and F as being proportional to $-H$. We use, therefore, a similar trick to derive a quasilinear PDE of first order for the Hamilton-Jacobi equation (12.27) that we used for equation (12.20). Take the first derivative of equation (12.27) with respect to $\mathbf{x}$, and use the definition $\mathbf{p} = \partial S/\partial \mathbf{x}$ to rewrite $\partial^2 S/\partial \mathbf{x}\partial t$ as $\partial \mathbf{p}/\partial t$:

$$\frac{\partial \mathbf{p}}{\partial t} + \frac{\partial H}{\partial \mathbf{p}} \cdot \frac{\partial \mathbf{p}}{\partial \mathbf{x}} = -\frac{\partial H}{\partial \mathbf{x}},$$

which forms a quasilinear PDE of first order for $\mathbf{p}$. The characteristics associated with this PDE read

$$\frac{dt}{1} = \frac{d\mathbf{x}}{\partial H/\partial \mathbf{p}} = \frac{d\mathbf{p}}{-\partial H/\partial \mathbf{x}}; \tag{12.29}$$

i.e., the solution of the Hamilton-Jacobi PDE is equivalent to the solution of the set of ODEs,

$$\frac{d\mathbf{x}}{dt} = \frac{\partial H}{\partial \mathbf{p}}, \qquad \frac{d\mathbf{p}}{dt} = -\frac{\partial H}{\partial \mathbf{x}}, \tag{12.30}$$

which are *Hamilton's equations.* From this discussion, we see that the classical (eikonal) approximation to quantum mechanics corresponds to ray tracing as a substitute for the solution of the full wave equation. What classical mechanics identifies as the velocity of a particle amounts, in this description, to the group velocity associated with a wave packet. (Q.E.D.)

APPLICATION TO SPIRAL DENSITY WAVES

Toomre applied the considerations of two sections ago to the dispersion relation for spiral density waves [actually, the stellar-dynamical analog to equation (11.45)], and obtained, for the component of the group velocity in the ϖ-direction,

$$c_{g\varpi} = -\frac{\partial\omega}{\partial k} = \text{sgn}(k)\frac{2(\pi G\sigma_0 - |k|a_0^2)}{(\omega - m\Omega)}, \tag{12.31}$$

where the minus sign in the first equality comes from our having defined k as the negative of the radial wave number. Short spiral waves have $|k| \geq \pi G\sigma_0/a_0^2$, whereas long spiral waves have $|k| \leq \pi G\sigma_0/a_0^2$ [see equation (11.52)]. Trailing spiral waves correspond to k negative; leading spiral waves, to k positive. Thus, short trailing spiral waves propagate away from corotation: inward for $(\omega - m\Omega) < 0$; outward for $(\omega - m\Omega) > 0$. A flip in sign is introduced for each change: long for short, or leading for trailing.

In a similar manner, since ω and m enter the dispersion relationship only in the combination $(\omega - m\Omega)$, the group velocity in the φ-direction reads

$$c_{g\varphi} = \frac{\partial\omega}{\partial(m/\varpi)} = \varpi\Omega. \tag{12.32}$$

Equation (12.32) shows that spiral density waves propagate in the circular direction at the rotation speed of the matter. Because the rotational motion does not carry density waves to a quantitatively different portion of the disk, however, the φ component of the group velocity has less significance than the radial component.

CONSERVATION OF WAVE ENERGY AND ANGULAR MOMENTUM

With the expressions,

$$c_{g\varpi} = \frac{\partial D/\partial k}{\partial D/\partial\omega}, \qquad \mathcal{J}_{\text{wave}} = \frac{m|k|A^2}{8\pi G}\frac{\partial D}{\partial\omega}, \qquad \mathcal{E}_{\text{wave}} = \frac{\omega|k|A^2}{8\pi G}\frac{\partial D}{\partial\omega},$$

we may express the amplitude relation, equation (12.4), in the suggestive forms,

$$2\pi\varpi c_{g\varpi}\mathcal{J}_{\text{wave}} = \text{constant}, \qquad \text{or} \qquad 2\pi\varpi c_{g\varpi}\mathcal{E}_{\text{wave}} = \text{constant}, \tag{12.33}$$

which express the constancy of the flux of wave angular momentum and energy across each circle of circumference $2\pi\varpi$. Clearly, equation (12.33) must represent the steady-state (real ω) persistent wavetrain expression of more general conservation relations which read

$$\frac{\partial}{\partial t}\begin{pmatrix} \mathcal{J}_{\text{wave}} \\ \mathcal{E}_{\text{wave}} \end{pmatrix} + \frac{1}{\varpi}\frac{\partial}{\partial \varpi}\left[\varpi c_{g\varpi}\begin{pmatrix} \mathcal{J}_{\text{wave}} \\ \mathcal{E}_{\text{wave}} \end{pmatrix}\right] = 0. \tag{12.34}$$

Conservation of total angular momentum and energy always occurs, of course, for any closed system. Equation (12.34) implies that they are conserved *separately* for the waves and the axisymmetric basic state. Without the inclusion of dissipative effects, the transport of angular momentum and energy through the disk by spiral density waves leaves no permanent mark on the properties of the underlying disk; the disk suffers only a temporary jiggling during the passage of a wave packet of finite extent in spacetime. Dissipative effects (e.g., radiative losses or shocks) would change these conclusions, leading to an inward drift of material inside corotation and an outward drift outside corotation.

THE WASER MECHANISM

Mark, and later, Lin and Lau, considered the following question: What would happen to a long trailing wave carrying -1 unit of angular momentum toward the corotation circle from deeper inside? For the special case $Q = 1$, (where long and short waves have the same wavelength at corotation), they found that a short trailing wave carrying $+1$ unit of angular momentum outward gets transmitted across corotation to the outside, whereas another short trailing wave carrying -2 unit of angular momentum gets refracted that propagates back toward the inside (Figure 12.2). The process can be studied mathematically by analyzing the properties of the wave operator (11.56) near corotation, but we omit such an analysis here and merely discuss the physical results.

The *over-refraction* (-1 to -2) occurs because of the need to conserve wave angular-momentum luminosity: the transmission of $+1$ unit of angular momentum outward (of a short trailing wave) beyond corotation must be balanced, inside of corotation, by the propagation outward of -1 unit of angular momentum (original long trailing wave) and the propagation inward of -2 unit of angular momentum (refracted short trailing wave). In other words, if outside of corotation we have $+1$ flux of angular momentum, inside we must also have a flux equal to $-1 - (-2) = +1$.

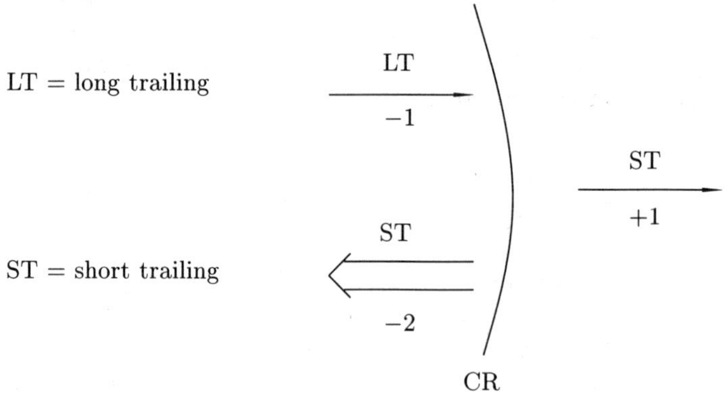

LT = long trailing

LT

-1

ST

$+1$

ST

ST = short trailing

-2

CR

FIGURE 12.2
WASER amplification. The simplest example occurs if no Q barrier exists at corotation (i.e., if $Q = 1$ at CR). In this case, when a long-trailing (LT) spiral wave carrying -1 unit of angular momentum is incident from the inside on the corotation circe, it transmits a short-trailing (ST) wave carrying $+1$ unit of angular momentum outward, and over-refracts into another short-trailing wave carrying -2 units of angular momentum inward. In the process, amplification by a factor of 2 has occurred for the waves inside corotation.

If Q exceeds unity, then a Q-barrier surrounds the corotation circle through which spiral density waves need to tunnel. This lowers the amplitude of the transmitted wave, and the amount of the "over-refraction" as well. If the incident wave carries -1 unit of angular momentum and the transmitted wave carries $+\varepsilon$, the refracted wave carries $-(1+\varepsilon)$. The multiplicative factor (in energy and angular momentum) is now $(1+\varepsilon)$. This general process of the "wave amplification by the stimulated emission of radiation" is called the WASER mechanism when it involves the over-refraction of long waves into short waves near the corotation circle. A feedback cycle could result if the inwardly propagating short trailing spiral waves turn, say, at an inner Q-barrier associated with the existence of a galactic bulge, into an outwardly propagating long trailing spiral wave. The latter reinforces the original incident wave if the pattern speed Ω_{p} satisfies a certain eigenvalue (quantum) relation. We would then have a growing *normal mode*, which amplifies (in the square of the wave amplitude) by a factor $(1+\varepsilon)$ per feedback cycle. A net growth would occur if the natural dissipation in the system (ignored in our treatment of the fluid as adiabatic) reduces the (square of the) amplitude by a smaller factor per wave cycle (Figure 12.3).

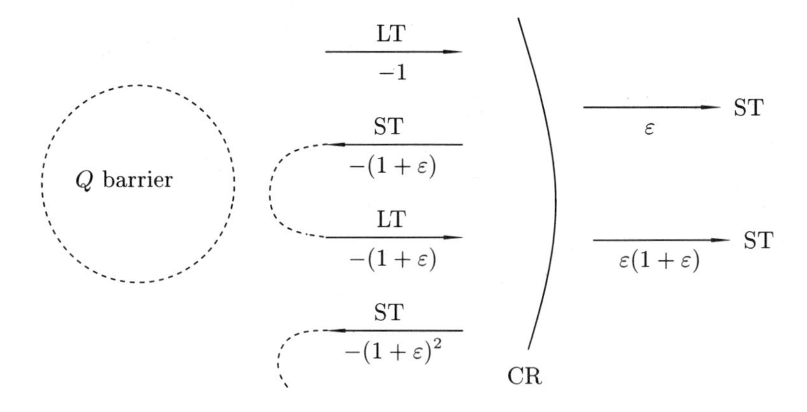

FIGURE 12.3
A growing inner-cavity mode can be set up using only trailing spiral waves if Q barriers exist both near the central parts of a galaxy (because of the existence of large velocity dispersions due to a central bulge) and near the corotation circle (the WASER amplification factor ε will be small unless Q nearly equals unity here). The three-wave cycle works schematically as follows. Begin (arbitrarily) with a LT wave carrying -1 unit of angular momentum, which over-refracts at the CR circle to give a ST wave carrying $-(1+\varepsilon)$ units of angular momentum inward. The ST wave carrying $+\varepsilon$ unit of angular momentum that tunnels through to the outer cavity is irrelevant to our further discussion if it does not reflect or refract back toward the CR circle. (In the formal treatment, people often apply outgoing radiation boundary conditions to make sure that it disappears.) In the absence of wave dissipation, the ST wave propagating toward the central bulge eventually refracts into a LT wave carrying $-(1+\varepsilon)$ units of angular momentum outward. At the CR circle, the LT wave again over-refracts to yield a ST wave, now carrying $-(1+\varepsilon)^2$ units of angular momentum inward, etc. The whole process yields a coherent mode provided we choose the real part of the eigenfrequency ω (i.e., determine the pattern speed Ω_{p}) so that one round-trip cycle contains $(2n+1)$ "wavelengths," where $n = 0, 1, 2, \ldots$. When one eliminates the wave number difference between long and short waves, the problem in the WKBJ limit bears almost an exact analogy with the quantum-mechanical problem of the bound states of an anharmonic oscillator (see Problem Set 6 of Volume I). The main difference here is that ω can have an imaginary part given by the WASER amplification occurring across the CR circle.

The WASER mechanism will also work if the incident trailing long wave comes from outside corotation, transmitting a short trailing wave inside of corotation, and refracting into another short trailing wave that propagates outward outside of corotation. We call any mode established this way (by other feedback cycles) an *outer cavity* mode, and we distinguish it from the *inner cavity* mode depicted earlier in which the important wave interactions take place inside corotation.

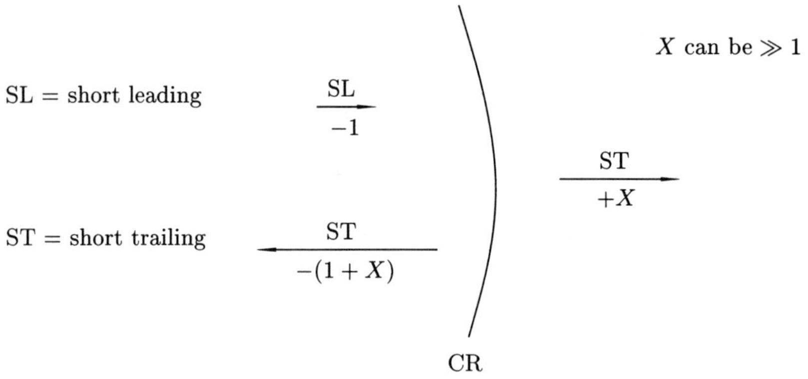

FIGURE 12.4
SWING amplification. The SWING mechanism occurs if a wave cycle involves leading as well as trailing spiral waves. When a Q barrier rises sufficiently quickly compared to a wavelength around the CR circle, the incidence of a short-leading (SL) wave from the inside can lead not only to the refraction of a long-leading (LL) wave traveling inwards (ignored above), but also to the reflection of a short-trailing (ST) wave. The latter *nonadiabatic* (i.e., non-WKBJ) process, involving the transmission of $+X$ units of ST wave into the outer cavity, and the over-reflection of $-(1+X)$ units of ST wave, is illustrated here. Detailed calculations show that the over-reflection factor X (in wave angular-momentum density) can much exceed unity even when Q at CR implies local stability by a fair margin (e.g., $Q = 1.5$); for this reason, SWING amplification, when it occurs, can often dominate over WASER amplification. (The existence of a shear-rate parameter apart from Q governing the SWING mechanism complicates a detailed discussion.) Inner-cavity cycles involving SWING amplification to give growing modes can occur if the central regions feed back the inwardly propagating ST wave (even in part) as an outwardly propagating SL wave. Barred spiral galaxies, among others, may owe their existence to this process.

THE SWING MECHANISM

If leading waves enter anywhere in the feedback cycle (e.g., by propagation in a disk through the galactic center), another amplification mechanism exists that was first studied in a time-dependent context by P. Goldreich and D. Lynden-Bell (1965, *M.N.R.A.S.*, **130**, 125), and elaborated upon by Toomre and other workers (notably, by Lin and his colleagues, and by Goldreich and Tremaine in the context of the "over-reflection" picture). Over-reflection of incident short leading waves into short trailing waves can occur if it is accompanied by the transmission of short trailing waves across corotation (Figure 12.4).

In disks where the dimensionless combination

$$\frac{2\pi G\sigma_0}{r\kappa^2}$$

has appreciable magnitude at corotation, wave amplification by the SWING mechanism (as the turning of short leading waves into short trailing waves is called) can greatly dominate amplification by the WASER mechanism. On the other hand, in systems where the disk mass is only a fraction of the other components (bulge-halo) that support the rotation curve (i.e., enters in the determination of κ^2), the WASER mechanism can underlie the growth of the most important spiral modes (tightly wrapped spiral patterns), as long as Q at corotation does not greatly exceed 1.

SUMMARY

The study of spiral density waves has revealed that self-gravitating differentially rotating disks have access to a surprisingly diverse set of growing normal modes. The existence of differential rotation in any object implies a potential source of free energy to tap to feed into growing disturbances. Without self-gravity as a mediating agent, however, the tendency to conserve angular momentum apparently proves too strong an obstacle to allow actual energy release (at least, on interesting time scales). There now exists many examples that indicate the importance of this mechanism for helping to create structure in otherwise featureless disks. Whether it also competes with anomalous viscosity mechanisms in the basic problem of disk accretion and secular evolution remains to be seen.

13

Method of Characteristics

References: Garabedian, *Partial Differential Equations*; Press *et al.*, *Numerical Recipes*.

In Chapter 12 we introduced the concept of the method of characteristics for solving quasilinear PDEs of first order in one dependent variable. In this chapter we wish to extend this method further to examine the general circumstances under which we can reduce the solution of a PDE to the solution of a set of ODEs. This examination will naturally lead us to a discussion of the classification of a second-order PDE in one dependent variable and two independent variables into *elliptic*, *parabolic*, and *hyperbolic* types. We then digress to consider practical computational procedures for obtaining numerical solutions of such equations. Applied to the problem of steady, barotropic, fluid flow in two spatial dimensions, this knowledge requires, as we shall see in Chapter 14, a fundamental distinction between subsonic and supersonic flow, and gives rise to the notion of the unavoidability of shock waves for certain types of supersonic flow.

PRESSURELESS MOTION OF A FLUID IN A GRAVITATIONAL FIELD

As a simple example to motivate the subsequent discussion, we consider the one-dimensional time-dependent motion of a pressureless fluid subject to a given gravitational field $g(x,t)$:

$$\frac{\partial u}{\partial t} + u\frac{\partial u}{\partial x} = g. \tag{13.1}$$

If we were to plot the solution of this PDE, $u = U(x,t)$, we would have a surface in (t, x, u) space, as depicted schematically in Figure 13.1.

According to our rule of Chapter 12 (set differentials of independent variables over coefficients equal to differential of dependent variable over the

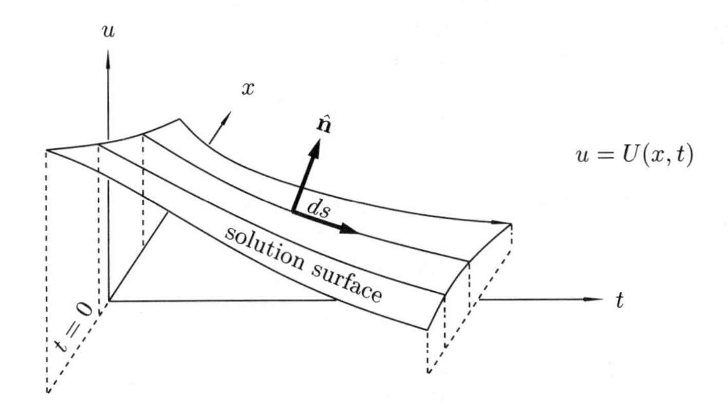

FIGURE 13.1
The schematic solution surface for a quasilinear PDE of first order in one dependent variable, u, and two independent variables, x and t. The vector $\hat{\mathbf{n}}$ describes the local unit vector to the solution surface, $u = U(x,t)$, and $\mathbf{ds}$ describes a line element that follows a characteristic curve embedded in the solution surface.

right-hand side), the shape of this surface can be generated by considering the characteristic equations associated with equation (13.1):

$$\frac{dt}{1} = \frac{dx}{u} = \frac{du}{g}, \tag{13.2}$$

i.e.,

$$\frac{du}{dt} = g \qquad \text{on the trajectory} \qquad \frac{dx}{dt} = u, \tag{13.3}$$

which are the one-dimensional equations of motion for a point particle in a gravitational field g. For each choice of initial conditions $x = x_0$ and $u = u_0$ at $t = 0$, we would get a unique curve in (t, x, u) space, as indicated by the lines embedded in the surface drawn in Figure 13.1. The collection of all such curves, corresponding to the specification of $u(x, 0)$ for all x at $t = 0$, then maps out the solution surface, $u(x,t)$. This then defines the sense in which the solution of the set of characteristic ODEs (13.3) is equivalent to the solution of the original PDE (13.1).

The characteristics (13.2) have the following geometric relationship to the original PDE (13.1). We may think of the solution surface, $u = U(x,t)$, as being defined by the zero of a function,

$$f(t, x, u) \equiv U(x,t) - u = 0. \tag{13.4}$$

In the space of rectilinear coordinates (t, x, u), the "gradient" of a function $f(t, x, u)$ forms a vector $(\partial f/\partial t, \partial f/\partial x, \partial f/\partial u)$ which is perpendicular to any surface of constant f, in particular to the surface $f = 0$. In other words, the vector

$$\mathbf{n} \equiv \left(\frac{\partial U}{\partial t}, \frac{\partial U}{\partial x}, -1\right) \tag{13.5}$$

is normal to the solution surface $u = U(x, t)$. In this notation, the PDE (13.1) has the expression,

$$\mathbf{n} \cdot \mathbf{C} = 0, \tag{13.6}$$

where $\mathbf{C}$ is the vector formed by the coefficients and the right-hand side of the problem:

$$\mathbf{C} \equiv (1, u, g). \tag{13.7}$$

Equation (13.6) states that the vector $\mathbf{n}$ and $\mathbf{C}$ are perpendicular to each other; thus, the vector $\mathbf{C}$ must lie locally in the plane of the solution surface $u = U(x, t)$. Hence, a local differential $\mathbf{ds} \equiv (dt, dx, du)$ parallel to $\mathbf{C}$ will define a curve, called the *characteristic trajectory*, that lies in the solution surface. This requires the ratio of the components of $\mathbf{ds}$ to $\mathbf{C}$ to have the same proportionality, i.e.,

$$\frac{dt}{1} = \frac{dx}{u} = \frac{du}{g},$$

which reproduces equation (13.2). (Q.E.D.)

A simple generalization of the above consideration exists for pressureless motion in three spatial dimensions,

$$\frac{\partial \mathbf{u}}{\partial t} + \mathbf{u} \cdot \frac{\partial \mathbf{u}}{\partial \mathbf{x}} = \mathbf{g}, \tag{13.8}$$

which has the associated characteristics

$$\frac{d\mathbf{u}}{dt} = \mathbf{g} \qquad \text{on the trajectory} \qquad \frac{d\mathbf{x}}{dt} = \mathbf{u}. \tag{13.9}$$

The generalization to more than two independent variables $(t, \mathbf{x})$ turns out to be trivial in this case, but the procedure for writing down the characteristics by inspection works here for more than one dependent variable $\mathbf{u}$ only because the x, y, z components of equation (13.8) do not couple the first derivatives of u_x, u_y, u_z. Nevertheless, we anticipate that the methods of linear algebra should be able to disentangle even coupled PDEs as long as they remain *quasilinear* (linear in the highest-order derivatives).

Equations (13.9) represent the three-dimensional equations of motion of a particle in a gravitational field $\mathbf{g}$. The individual orbits can be pieced together to give the solution of a pressureless fluid. Indeed, the pressureless equations of motion form a viable zeroth approximation for hypersonic flow. The following physical issue now arises. The orbits of a collection of point particles may cross without difficulty if the particles have zero cross section. However, the elements of a fluid can not cross without producing violent interactions. Clearly, the characteristic equations (13.9) can not, by themselves, guarantee the noncrossing of pressureless streamlines. Would inclusion of the effects of pressure automatically prevent such difficulties?

STEADY, INVISCID, IRROTATIONAL FLOW OF A BAROTROPIC GAS IN TWO SPATIAL DIMENSIONS

When we include the effects of pressure P, the simple technique described above fails if we regard P as a dependent variable to be found as part of the solution to the problem, rather than as an imposed source of external force. In this case, the solution and trajectory characteristics no longer separate so cleanly, and we have to do something more sophisticated. We illustrate the situation with the problem of the steady ($\partial/\partial t = 0$), inviscid ($\overleftrightarrow{\pi} = 0$), irrotational ($\nabla \times \mathbf{u} = 0$) flow of an barotropic [$P = P(\rho)$] gas in two dimensions (x and y). For steady flow, the equation of continuity reads

$$\nabla \cdot (\rho \mathbf{u}) = 0,$$

which, in two rectilinear dimensions with $\mathbf{u} = [u(x,y), v(x,y)]$, can be written as

$$u\frac{\partial \rho}{\partial x} + v\frac{\partial \rho}{\partial y} = -\rho\left(\frac{\partial u}{\partial x} + \frac{\partial v}{\partial y}\right). \tag{13.10}$$

If the flow originates from a region of uniform flow, Kelvin's circulation theorem (see Chapter 6) allows us to adopt the irrotational assumption:

$$\frac{\partial v}{\partial x} - \frac{\partial u}{\partial y} = 0. \tag{13.11}$$

Finally, the barotropic relation $P = P(\rho)$ gives Bernoulli's theorem (see Chapter 6):

$$\left(u\frac{\partial}{\partial x} + v\frac{\partial}{\partial y}\right)\left[\frac{1}{2}(u^2+v^2) + \mathcal{V} + \int \frac{dP}{\rho}\right] = 0. \tag{13.12}$$

If we use

$$\left(u\frac{\partial}{\partial x} + v\frac{\partial}{\partial y}\right)P = a^2\left(u\frac{\partial}{\partial x} + v\frac{\partial}{\partial y}\right)\rho, \qquad \text{where} \qquad a^2 = \frac{dP}{d\rho},$$

together with equation (13.10) to eliminate the derivatives of ρ, we get

$$(u^2-a^2)\frac{\partial u}{\partial x} + uv\left(\frac{\partial u}{\partial y} + \frac{\partial v}{\partial x}\right) + (v^2-a^2)\frac{\partial v}{\partial y} = -\left(u\frac{\partial \mathcal{V}}{\partial x} + v\frac{\partial \mathcal{V}}{\partial y}\right). \tag{13.13}$$

Equations (13.13) and (13.11) form a complete quasilinear set to solve for u and v if we know the variation of a^2. We can learn it most easily for an isothermal gas, in which case a^2 = constant $\equiv a_0^2$, an assumption that we shall adopt later to simplify the resultant algebra. For now, however, we continue to allow arbitrary variations of a^2.

SECOND-ORDER PDE FOR VELOCITY POTENTIAL

Because $\nabla \times \mathbf{u} = 0$, we find it instructive to introduce a velocity potential Φ such that $\mathbf{u} = \nabla\Phi$. In two dimensions, we have

$$u = \frac{\partial \Phi}{\partial x}, \qquad v = \frac{\partial \Phi}{\partial y}, \tag{13.14}$$

which provides an automatic solution of equation (13.11). The substitution of equation (13.14) into equation (13.13) produces a second-order PDE for Φ:

$$A\frac{\partial^2 \Phi}{\partial x^2} + 2B\frac{\partial^2 \Phi}{\partial x \partial y} + C\frac{\partial^2 \Phi}{\partial y^2} = D, \tag{13.15}$$

where the coefficients and the inhomogeneous term are given by

$$A \equiv (u^2 - a^2), \quad B \equiv uv, \quad C \equiv (v^2 - a^2), \quad D \equiv -\left(u\frac{\partial}{\partial x} + v\frac{\partial}{\partial y}\right)\mathcal{V}. \tag{13.16}$$

GENERAL PROBLEM IN TWO INDEPENDENT VARIABLES

If A, B, C, and D had arbitrary dependences on x, y, $u = \partial\Phi/\partial x$, and $v = \partial\Phi/\partial y$, equation (13.15) would give the most general form possible for a quasilinear PDE in a single dependent variable, Φ, and two independent variables, x and y. The equation is considered quasilinear because the second-order derivatives—the highest in the problem—enter linearly in the PDE, but the coefficients and the inhomogeneous term may depend nonlinearly on Φ and/or its lower-order derivatives, as well as on x and y. Thus, en route to our desired derivation, we have incidentally proven that such a second-order PDE can always be reduced to a simultaneous set of two quasilinear PDEs of first-order:

$$\frac{\partial v}{\partial x} - \frac{\partial u}{\partial y} = 0, \qquad A\frac{\partial u}{\partial x} + B\left(\frac{\partial u}{\partial y} + \frac{\partial v}{\partial x}\right) + C\frac{\partial v}{\partial y} = D, \tag{13.17}$$

where $u \equiv \partial\Phi/\partial x$ and $v \equiv \partial\Phi/\partial y$. This example, in turn, represents a special case of the general problem of n-coupled PDEs of first order in two independent variables,

$$X_{ik}\frac{\partial u_k}{\partial x} + Y_{ik}\frac{\partial u_k}{\partial y} = H_i, \qquad i = 1, 2, \ldots, n, \tag{13.18}$$

where we have adopted Einstein's summation convention with respect to the repeated index k (with the summation extending over 1 to n).

Motivated by the theory for equation (13.1), we would like to find the n-characteristic directions $\{ds_i^{(j)}\}_{i,j=1}^n$ such that ds_i (we suppress the different characteristics label j henceforth) times equation (13.18), summed over i, gives a sum of total derivatives of u_k on the left-hand side. In other words, we want to find ds_i so that

$$ds_i X_{ik} = L_k dx, \qquad ds_i Y_{ik} = L_k dy, \tag{13.19}$$

for some $\{L_k\}_{k=1}^n$. If we can find such ds_i, then ds_i times equation (13.18) gives

$$L_k du_k = H_i ds_i \qquad \text{where} \qquad du_k = \frac{\partial u_k}{\partial x}dx + \frac{\partial u_k}{\partial y}dy. \tag{13.20}$$

Equation (13.19) implies that appropriate ds_i can be found providing

$$ds_i\left(X_{ik}dy - Y_{ik}dx\right) = 0, \qquad k = 1, 2, \ldots, n, \tag{13.21}$$

for some dx and dy.

Regarded as a simultaneous set of linear homogeneous equations for $\{ds_i\}_{i=1}^n$, we know from linear algebra that nontrivial solutions can be found for ds_i if and only if

$$\det\left(X_{ik}dy - Y_{ik}dx\right) = 0. \tag{13.22}$$

In other words, we must find the relation between dx and dy such that if we form the matrix where the individual elements equal the coefficients of the partial derivatives of x times dy minus the coefficients of the partial derivatives of y times dx, its resultant determinant equals zero.

CLASSIFICATION OF SINGLE PDE OF SECOND ORDER

If we apply equation (13.22) to equation (13.17), which is equivalent to the single quasilinear PDE (13.15) of second order, we get

$$\begin{vmatrix} dx & dy \\ Ady - Bdx & Bdy - Cdx \end{vmatrix} = 0. \tag{13.23}$$

This gives the requirement

$$-A(dy)^2 + 2Bdxdy - C(dx)^2 = 0,$$

which has the two associated trajectory characteristics

$$\left(\frac{dy}{dx}\right)_\pm = \frac{1}{A}\left[B \pm (B^2 - AC)^{1/2}\right]. \tag{13.24}$$

We thus have a classification dependent on the sign of the combination $B^2 - AC$, where $2B$ is the coefficient of the mixed derivative of the second order PDE (13.15) and A and C are, respectively, the coefficients of the second-order derivative in x and y. If $B^2 - AC > 0$, there are two real characteristics, and we say that the corresponding second-order PDE is *hyperbolic*. If $B^2 - AC = 0$, the two characteristics degenerate into one, and we say that the corresponding second-order PDE is *parabolic*. If $B^2 - AC < 0$, then there are no real characteristics, and we say that the corresponding second-order PDE is *elliptic*.

EXAMPLES OF HYPERBOLIC, PARABOLIC, AND ELLIPTIC PDES OF SECOND ORDER

Undergraduate physics supplies the following (linear) prototypes for the different kinds of PDE of second order. The linear wave equation, with a constant propagation speed c, constitutes the prototype for a hyperbolic PDE:

$$\frac{\partial^2 \phi}{\partial t^2} - c^2 \frac{\partial^2 \phi}{\partial x^2} = 0. \tag{13.25}$$

In this equation $A = 1$, $B = 0$, $C = -c^2$; so $B^2 - AC > 0$, and the two real characteristics read $dx/dt = \pm c$. In other words, $x \mp ct =$ constant give the two trajectory characteristics. Indeed, the transformation to characteristic variables, $\xi \equiv x + ct$ and $\eta \equiv x - ct$ yields for equation (13.25) another prototypical form for a hyperbolic PDE,

$$4\frac{\partial^2 \phi}{\partial \xi \partial \eta} = 0,$$

where A and C now both equal zero, but $B = 2$. The trajectory characteristics read $d\eta/d\xi = 0$ or ∞, i.e., $\eta =$ constant or $\xi =$ constant. We leave it as an exercise for the reader to show that the solution characteristics read

$$dv = 0 \quad \text{on} \quad d\eta = 0, \quad \text{and} \quad du = 0 \quad \text{on} \quad d\xi = 0,$$

where $u \equiv \partial\phi/\partial\xi$, and $v \equiv \partial\phi/\partial\eta$. In other words, we have v constant on $\eta =$ constant, i.e., $v = f(\eta) = \partial\phi/\partial\eta$, where f is an arbitrary function. Similarly, we have $u = g(\xi) = \partial\phi/\partial\xi$. Since ξ and η are independent variables, if we integrate either equation for ϕ, we would obtain the most general solution as

$$\phi = F(\eta) + G(\xi) = F(x - ct) + G(x + ct), \tag{13.26}$$

where F and G are the indefinite integrals of f and g with respect to their natural arguments. This solution can also be derived, of course, by more elementary means.

The time-dependent linear heat-diffusion equation constitutes the prototype for a parabolic PDE:

$$\chi \frac{\partial^2 T}{\partial x^2} = \frac{\partial T}{\partial t}, \tag{13.27}$$

where χ equals the thermal diffusivity. The coefficients read $A = \chi$, $B = 0$, $C = 0$, and $D = \partial T/\partial t \equiv v$. Hence, $B^2 - AC = 0$, and the governing PDE is parabolic, with one real trajectory characteristic, $dt/dx = 0$, i.e., lines of constant t. (In other words, although the solution of the time-dependent heat-diffusion equation cannot be reduced to a set of ODEs, it can be solved in a natural way by marching forward from one value of time to another.)

Laplace's equation constitutes the prototype for an elliptic PDE:

$$\frac{\partial^2 \phi}{\partial x^2} + \frac{\partial^2 \phi}{\partial y^2} = 0. \tag{13.28}$$

In this equation, $A = 1$, $B = 0$, $C = 1$, so $B^2 - AC < 0$, and there are no real characteristics. Laplace's equation cannot be reduced to a set of ordinary differential equations in real variables, nor can it be regarded as naturally solved by marching in any combination of x and y.

BOUNDARY CONDITIONS FOR WELL-POSED PROBLEMS

The discussion of the previous section puts us in a good position to consider the imposition of proper boundary conditions for each type of second-order PDE. Since the linear wave equation (13.25) has two derivatives in t and two in x, we expect that it needs two initial conditions and two boundary conditions. For example, to have a well-posed problem for the oscillation of a string, we need to specify both the initial position and the velocity of the string,

$$\phi(x, 0) = Y(x), \qquad \frac{\partial \phi}{\partial t}(x, 0) = V(x), \tag{13.29}$$

where $Y(x)$ and $V(x)$ are specified functions. In addition, we usually impose one boundary condition each on the two ends of the string, say, $x = 0$ and $x = L$ for all time $t > 0$; for example, we could pin down the position of the string on one end and let it move on the other (e.g., if someone waves the string at $x = L$):

$$\phi(0, t) = 0, \qquad \phi(L, t) = \Psi_L(t), \tag{13.30}$$

with $\dot{\Psi}_L(0) = V(L)$ for self-consistency. We depict in Figure 13.2 the generic situation for a well-posed problem involving hyperbolic PDEs.

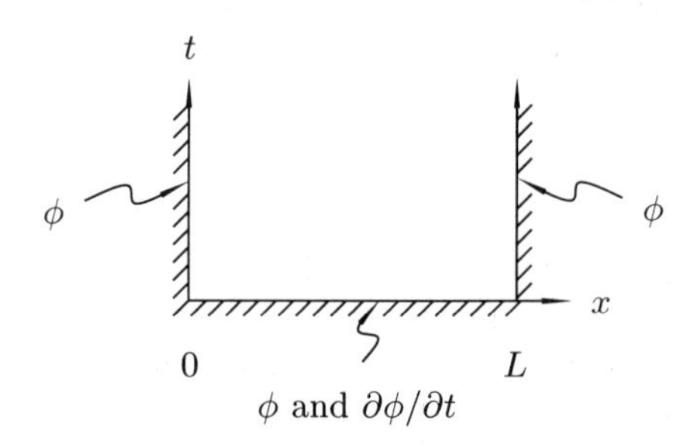

FIGURE 13.2
Natural boundary and initial conditions for the time-dependent wave equation in one spatial dimension (a hyperbolic quasilinear PDE of second order in both space and time).

The time-dependent heat-diffusion equation (13.27) has one derivative in time t and two derivatives in x. Thus, in order to have a well-posed problem for it, we need one initial condition,

$$T(x,0) = \Theta(x), \tag{13.31}$$

and two boundary conditions, for example, the specification of the value of the temperature at one boundary and of the flux (i.e., the spatial derivative of T) at the other,

$$T(0,t) = T_0(t), \qquad \frac{\partial T}{\partial x}(L,t) = F_L(t). \tag{13.32}$$

The situation is depicted generically in Figure 13.3.

Laplace's equation (13.28) has two derivatives in both x and y; consequently, to have a well-posed problem for it, we need a boundary condition at each end of a rectangular configuration. We may specify at each edge either the function ϕ or its normal derivative, but not both simultaneously. The situation is depicted schematically in Figure 13.4.

This analysis lends insight into the usual statement, which holds that for an elliptic PDE we must specify either the function (Dirichlet boundary conditions) or its normal derivative (Neumann boundary conditions), but not both, on all closed bounding surfaces of the problem (see Figure 13.5).

Naively, we might have thought that the specification of only one condition (ϕ or $\partial\phi/\partial n$) violates the intuitive requirements of a second-order (partial) differential equation, but now we see that the *closed* surface aspect of the problem really means that we are specifying one condition each

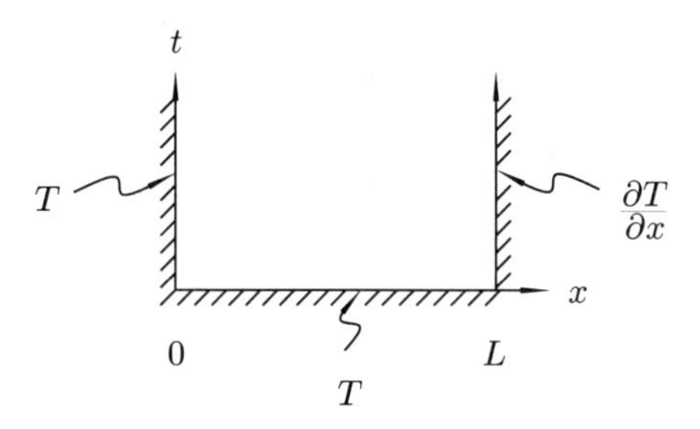

FIGURE 13.3
Natural boundary and initial conditions for the time-dependent equation of heat conduction in one spatial dimension (a parabolic quasilinear PDE of second order in space and first order in time).

on the "opposite" sides that face each other. In other words, Figure 13.5 for a smooth closed curve does not differ topologically from Figure 13.4 for rectangular boundaries, when we found the specification of ϕ or $\partial\phi/\partial n$ on each of the edges quite natural.

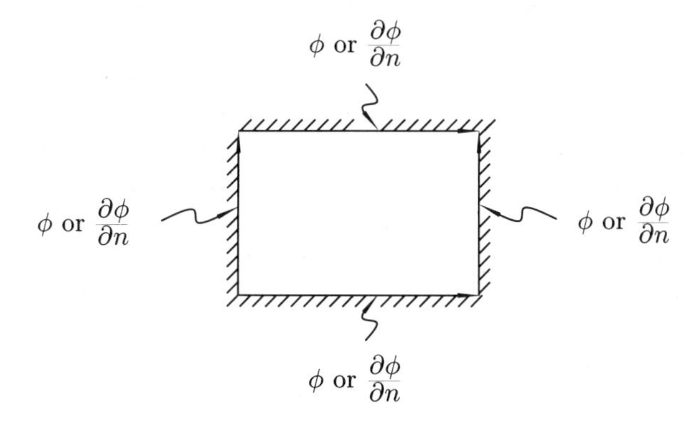

FIGURE 13.4
Natural boundary conditions for Laplace's equation (or, more generally, Poisson's equation) in two spatial dimensions (an elliptic quasilinear PDE of second order in two space variables).

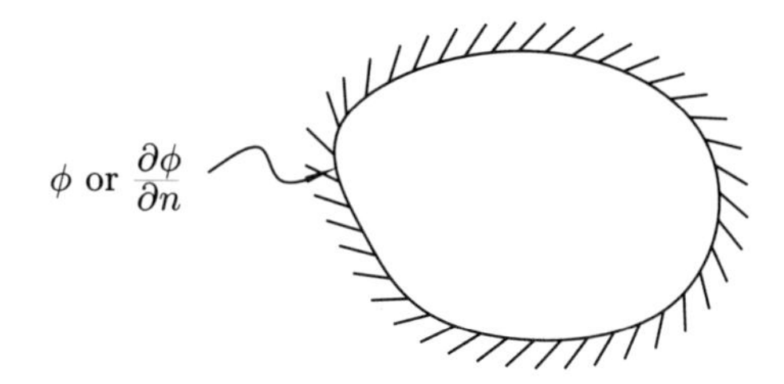

FIGURE 13.5
The topology of a closed rounded surface does not differ from the rectangular boundaries of Figure 13.4.

DIFFERENCE BETWEEN HYPERBOLIC AND ELLIPTIC CASES

The alert reader may have noticed that the natural specification of "boundary conditions" differed for the hyperbolic and elliptic problems (wave equation and Laplace's equation). For the hyperbolic problem, we found it natural to specify two conditions, the values of ϕ *and* its "normal" derivative $\partial\phi/\partial t$ on the *same* time "edge" ($t = 0$). For the elliptic problem, on the other hand, we found it natural to specify the necessary two conditions in the form of one each (either ϕ or $\partial\phi/\partial n$) on *opposite* space edges. Why can we not specify both conditions for the elliptic problem on one edge, and let the solution become whatever it will at the other edge (in the same way that we allow the solution for the wave equation to develop at future times t to whatever it wants, given the two initial conditions)? The reader might be tempted to respond that there exists an innate difference between space and time, and between boundary conditions and initial conditions, but this answer begs the question (apart from getting into a sticky situation with respect to relativity theory). Moreover, in other problems (e.g., steady fluid flow), it might not be so obvious which variable represents the timelike direction, and which the spacelike dimension. A correct answer should respond to the mathematical issue as well as to the physical one.

The mathematical answer involves, in essence, a question of numerical stability. We have noted elsewhere that the elliptic operator,

$$\frac{\partial^2}{\partial x^2} + \frac{\partial^2}{\partial y^2}, \tag{13.33}$$

has the property that homogeneous solutions oscillatory in x, say, $e^{\pm ikx}$, are exponential in y, $e^{\pm ky}$. Consider what this implies if we try to specify both $\phi = \Phi(x)$ and $\partial\phi/\partial y = \Psi(x)$ on the x-axis, $y = 0$, and to march the solution to large values of positive y. Small oscillatory errors in the specification of $\Phi(x)$ and $\Psi(x)$ then propagate in the y direction to becomes exponentially large errors in y. (The errors initially oscillate in x because we usually pin down the values at a discrete number of points, and the errors in between can not wander off indefinitely in the x direction.) Thus, we regard as ill-posed the numerical problem with both the function and its normal derivative specified as "initial" data on one edge for an elliptic PDE. The problem remains ill-posed in a mathematical sense even if we could find the solution by analytical means (where there exists no question of numerical errors) because small differences in the imposition of the boundary values for ϕ and $\partial\phi/\partial y$ on $y = 0$ would lead to vastly different solutions. Thus, we generally seek to pose boundary-value problems for elliptic PDEs so as to pin down the solution on a closed boundary (i.e., in positive and negative directions for both x and y), so that we do not get solutions in the interior which have an exponential sensitivity to the boundary conditions, with an exponent that diverges as the inverse of the grid spacing.

The above discussion has the simple corollary that hyperbolic PDEs are well-posed as initial-value problems. The prototypical hyperbolic operator,

$$\frac{\partial^2}{\partial t^2} - c^2\frac{\partial^2}{\partial x^2}, \tag{13.34}$$

has the property that homogeneous solutions oscillatory in x, $e^{\pm ikx}$, are also oscillatory in t, $e^{\pm ickt}$. Thus, any oscillatory numerical error in the specification of $\phi = \Phi(x)$ and $\partial\phi/\partial t = \Psi(x)$ at $t = 0$ remains bounded as the solution progresses in t.

RELAXATION TECHNIQUE FOR SOLVING ELLIPTIC PDES

In Chapter 14, we shall consider numerical methods for solving hyperbolic PDEs. In this chapter, we wish to indicate briefly some practical schemes for the numerical solution of elliptic and parabolic PDEs. Our examples, based on Laplace's equation and the time-dependent heat-diffusion equation, are linear, and thus could also be attacked by the specialized techniques—e.g., separation of variables, Laplace transforms, etc.—that one learns in undergraduate physics and mathematics courses. The finite difference techniques we describe here, however, with simple adaptation, work equally well for linear and nonlinear problems; they are also relatively fast to implement on modern computers.

We begin with Laplace's equation. Consider the problem with a rectangular boundary, on which, for simplicity of exposition, we impose the value of the function ϕ. Divide the computational space into a rectangular grid evenly spaced, again, for simplicity, in x and y. Let $\phi_{i,j}$ denote the value

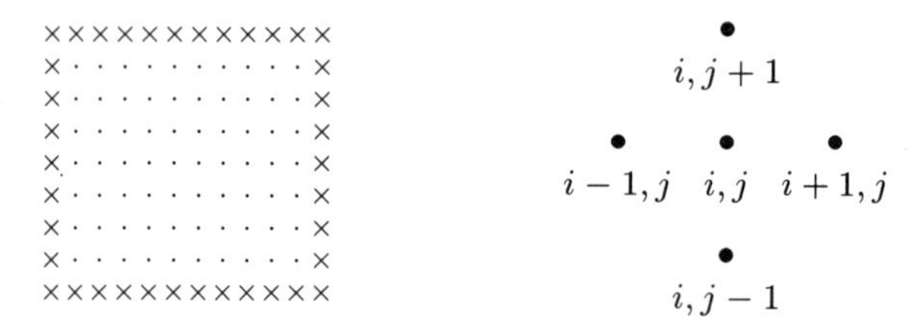

FIGURE 13.6
Finite-difference techniques for solving elliptic PDEs divide up the computational grid into a mesh of interior points (dots) surrounded by boundary points (crosses). Iterative (relaxation) methods obtain improved values for the dependent variable by using the finite-difference representation of the (second-order) PDE to relate an interior point (i, j) to its immediate neighbors: $(i-1, j)$, $(i+1, j)$, $(i, j-1)$, and $(i, j+1)$. Boundary points have their values specified by the boundary conditions.

of $\phi(x_i, y_j)$ on our grid, and approximate the second order derivatives by the finite differences:

$$\begin{aligned}
\frac{\partial^2 \phi}{\partial x^2}(x_i, y_j) &\approx \frac{1}{h^2}(\phi_{i+1,j} - 2\phi_{i,j} + \phi_{i-1,j}), \\
\frac{\partial^2 \phi}{\partial y^2}(x_i, y_j) &\approx \frac{1}{h^2}(\phi_{i,j+1} - 2\phi_{i,j} + \phi_{i,j-1}),
\end{aligned} \tag{13.35}$$

where h represents the grid spacing, and the formulae are accurate to second order for small h. The finite-difference representation for equation (13.28) may now be solved formally for the central value, $\phi_{i,j}$, in terms of ϕ evaluated at neighboring grid points:

$$\phi_{i,j} = \frac{1}{4}(\phi_{i+1,j} + \phi_{i-1,j} + \phi_{i,j+1} + \phi_{i,j-1}). \tag{13.36}$$

Equation (13.36) has a simple geometric interpretation for a solution by repeated iteration, namely, that the inversion of the Laplace operator involves the rule that we should (repeatedly) average the values of ϕ on neighboring grid points (see Figure 13.6).

The averaging procedure applies only to *interior* grid points (dot's in Figure 13.6), because they reside where the PDE (and our finite-difference representation for it) holds. The PDE does not hold on the boundary points (X's in the above figure), but boundary values enter the problem when we average interior grid points that have boundary points as a nearest neighbor. With repeated iteration, we easily see how the propagation (with iteration, not with time) of the boundary values for ϕ into the interior takes place. With an arbitrarily guessed first iterate (usually the solution for some previous problem), we generally do not have a converged solution until after many iterations have passed. By that point, the information contained in the boundary values will have had an opportunity to "diffuse" into the interior. Thus, unlike the localized influence exerted by boundary conditions in hyperbolic PDEs, where a given point in space has to wait in time before the influence arrives (at speed c) from a distant location in space, boundary values in elliptic PDEs have a global effect, with every interior point influenced by every boundary point. The mathematical technique of passing this influence by repeated averagings, equation (13.36), resembles a "random walk" in which there is a 1/4 chance (in two dimensions) of going east or west or north or south; hence, this method has the name of "relaxation technique."

Relaxation techniques often benefit from the procedures of "under-relaxation" or "over-relaxation." Let a superscript n indicate the n-th iterate. To get the $(n+1)$-st iterate, instead of straightforwardly taking the full value suggested by equation (13.36),

$$\phi_{i,j}^{*} = \frac{1}{4}[\phi_{i+1,j}^{(n)} + \phi_{i-1,j}^{(n)} + \phi_{i,j+1}^{(n)} + \phi_{i,j-1}^{(n)}], \tag{13.37}$$

we can sometimes benefit by taking a linear combination of the provisional value (13.37) and the n-th iterate,

$$\phi_{i,j}^{(n+1)} = \theta\phi_{i,j}^{*} + (1-\theta)\phi_{i,j}^{(n)}, \tag{13.38}$$

where θ is a pure number called the *relaxation parameter* (see Problem Set 1 of Volume I). If θ lies between 0 and 1, we term the procedure *under-relaxation*; if θ exceeds 1, we term it *over-relaxation*; and if θ is less than zero, we prefer back iterates to present ones. How should we choose θ? Trial-and-error to find what works (i.e., gives convergence) forms one possible procedure. A more systematic approach chooses θ so as to minimize some measure of the total error (e.g., sum of the squares of the value at each grid point when the current iterate is substituted back into the original difference equation). The latter technique has the advantage that it guarantees a gradual lowering of the total error in the solution of the finite-difference equation, since the worst that one can do is to choose the previous iterate as the present iterate (i.e., choose $\theta = 0$).

Always proceeding in a direction that lowers the error, however, presents the danger—particularly in nonlinear problems—that successive iterates will get stuck in a local minimum, rather than proceed to find a true global minimum. A cure exists in the procedure called *annealing*. The name comes from analogy with metallurgy, where occasional large changes in the heat treatment of metals can help the material to find a good state of near-minimum energy. Large temperature perturbations help to bounce the system over the potential barriers of local minimums where it might otherwise get stuck. Once the state of absolute-minimum energy has been approximately approached, however, one wants to reduce the frequency and intensity of the temperature "shocks." In an analogous manner, it may help occasionally to introduce a random numerical perturbation in the iterative solution of equations, even at the expense of temporarily increasing the error made in the "solution." Methods of annealing in numerical analysis are in their infancy, and in practice represent more of an art than a science. Nevertheless, the basic principle is sound, and the student would be well advised to keep the idea in mind when faced with an otherwise intractable problem.

EXPLICIT AND IMPLICIT TECHNIQUES FOR SOLVING PARABOLIC PDES

Consider next the solution of the time-dependent heat-diffusion equation (13.27). If we denote time levels by superscripts and spatial positions by subscripts, we may write the finite-difference approximation to equation (13.27) as

$$\frac{\chi}{(\Delta x)^2}\left[T_{i+1}^{(j)} - 2T_i^{(j)} + T_{i-1}^{(j)}\right] = \frac{1}{\Delta t}\left[T_i^{(j+1)} - T_i^{(j)}\right], \tag{13.39}$$

where Δx and Δt are the intervals between successive spatial and temporal points. On the right-hand side of equation (13.39), we actually had a choice of forming the forward difference in time, $T_i^{(j+1)} - T_i^{(j)}$, as we did, or the backward difference, $T_i^{(j)} - T_i^{(j-1)}$. The two choices lead, respectively, to *explicit* and *implicit* methods for solving the problem. In the explicit method, we basically differentiate in space and integrate in time. In the implicit method, we essentially do the opposite: differentiate in time and integrate in space. The two methods have their own advantages and disadvantages, as we now proceed to describe.

We begin with the explicit method. Equation (13.39) can most naturally be regarded as an equation to find the function at the next time level $j+1$, when we know it spatially at time j. Thus, if we solve for $T_i^{(j+1)}$, we get

$$T_i^{(j+1)} = T_i^{(j)} + \frac{\chi \Delta t}{(\Delta x)^2}\left[T_{i+1}^{(j)} - 2T_i^{(j)} + T_{i-1}^{(j)}\right]. \tag{13.40}$$

Since we know the value of T at all spatial grid points (including the boundaries) for time j, equation (13.40) allows us to march forward one step in time to obtain explicitly the solution at time $j+1$. In practice, the marching can not be done with arbitrarily large Δt, because we need to allow the information (temperature) time to diffuse from one grid point to another. For the procedure to have numerical stability, we can show (see any book on numerical analysis of PDEs) that our time step for an explicit solution is restricted to values:

$$\Delta t \lesssim (\Delta x)^2/2\chi, \tag{13.41}$$

a condition that we could have guessed on the basis of physical intuition (often the best guide, because a formal numerical stability analysis can prove difficult or impossible in complicated situations).

Notice that the explicit method easily generalizes to multiple spatial dimensions,

$$\chi \nabla^2 T = \frac{\partial T}{\partial t}, \tag{13.42}$$

since the formation of spatial derivatives in additional coordinates does not pose any special problems. Indeed, the explicit technique comes fully into its own in multiple spatial dimensions because, apart from memory demands (more grid points in three dimensions than in one), the three-dimensional problem does not pose any more programming or calculational difficulties than the one-dimensional one. (One does not have to invert very large matrices, as one has to do, for example, with the implicit techniques to be described below.) Notice also that if the boundary conditions contain no explicit time-dependence, then the solution of equation (13.42) asymptotically goes to steady state in the limit $t \to 0$ and satisfies Laplace's equation:

$$\nabla^2 T = 0.$$

Thus, occasionally, people will artificially convert elliptic PDEs to parabolic PDEs (by adding a bogus term containing a first time-derivative), which they then proceed to solve explicitly, evolving the solution to steady state to recover the solution for the original problem. We may view this trick as a physically motivated iteration technique, with the time variable replacing a mathematical iteration parameter j.

The need to limit the time steps to satisfy a Courant-like condition, equation (13.41), constitutes the principal disadvantage of the explicit method, which is otherwise simple and direct. Fast computers have vitiated this disadvantage to a large extent; so explicit techniques are now often attractive. For problems containing appreciable spatial structure (so that we need to have small values for Δx) and involving very long evolutionary times in comparison with diffusion time scales (e.g., stellar evolution theory, where we do not want to be limited to taking very small Δt's), however, we may find it desirable to adopt an implicit technique.

We can easily invent at least two versions of an implicit method. In the first, we use the backward time difference instead of the forward difference on the right-hand side of equation (13.39). We then transpose all terms involving the time level j to the left-hand side, and all terms involving $j-1$ to the right-hand side, and obtain

$$T_{i+1}^{(j)} - 2(1+\eta)T_i^{(j)} + T_{i-1}^{(j)} = -\eta T_i^{(j-1)}, \tag{13.43}$$

where η equals the dimensionless combination

$$\eta \equiv \frac{(\Delta x)^2}{2\chi \Delta t}. \tag{13.44}$$

We now regard equation (13.43) as a matrix equation (in the spatial index) for $T^{(j)}$, with the right-hand side as a known quantity. The difference equation applies only to interior grid points; i.e., we have two fewer equations than the full run of the spatial index $i = 1$ to N. The two missing pieces of data are supplied by the two-point boundary conditions, which we take, for sake of definiteness, to be (the finite-difference versions of) equation (13.32):

$$T_1^{(j)} = T_0, \qquad 3T_N^{(j)} - 4T_{N-1}^{(j)} + T_{N-2}^{(j)} = (2\Delta x)F_N. \tag{13.45}$$

To retain the second-order accuracy of the finite-difference representation (13.43) of the PDE (13.42), we have used a more complicated expression than the first-order accurate $(T_N^{(j)} - T_{N-1}^{(j)})/\Delta x$ to represent $\partial T/\partial x$ in the second of the boundary conditions in equation (13.32). Because we deal with a boundary point rather than an interior point, we cannot simply center the spatial derivative to get a second-order expression, but must use the two grid points to the interior side of the boundary point. To deduce the coefficients for $f'(x)$ in terms of the values $f(x)$, $f(x-h)$, and $f(x-2h)$, expand the latter two in Taylor series about x, including terms of order h^2. If one then takes 4 times the $f(x-h)$ equation and subtracts it from the $f(x-2h)$ equation, one sees that the terms proportional to $f''(x)h^2$ cancel. Rearranging terms to solve for $f'(x)$ then yields the derivation of the coefficients in equation (13.45). At some basic level, this procedure underlies many of the high-order interpolation, extrapolation, differentiation, and integration schemes that one finds in books on numerical analysis.

Equations (13.43) and (13.45) may be written in the matrix form:

$$\begin{bmatrix} 1 & 0 & 0 & 0 & \dots & 0 & 0 & 0 \\ 1 & -2(1+\eta) & 1 & 0 & \dots & 0 & 0 & 0 \\ 0 & 1 & -2(1+\eta) & 1 & \dots & 0 & 0 & 0 \\ \vdots & \vdots & \vdots & \vdots & & \vdots & \vdots & \vdots \\ 0 & 0 & 0 & 0 & \dots & 1 & -2(1+\eta) & 1 \\ 0 & 0 & 0 & 0 & \dots & 1 & -4 & 3 \end{bmatrix} \begin{bmatrix} T_1^{(j)} \\ T_2^{(j)} \\ T_3^{(3)} \\ \vdots \\ T_{N-1}^{(j)} \\ T_N^{(j)} \end{bmatrix}$$

$$= \begin{bmatrix} T_0 \\ -2\eta T_2^{(j-1)} \\ -2\eta T_3^{(j-1)} \\ \vdots \\ -2\eta T_{N-1}^{(j-1)} \\ (2\Delta x)F_N \end{bmatrix}. \tag{13.46}$$

If χ is a constant (so that η is a constant), then equation (13.46) forms a linear inhomogeneous matrix equation that we may invert to find $\{T_i^{(j)}\}_{i=1}^N$. The solution is greatly simplified if we first subtract the next-to-last row from the last (boundary) row. This operation removes the entry that prevents the coefficient matrix from taking a tridiagonal form. With a tridiagonal matrix, the inversion procedure has a straightforward solution (see Mihalas, *Stellar Atmospheres*, pp. 156–158). Problems with multiple spatial dimensions can be formulated in terms of tridiagonal *block* matrices, which also have an easy inversion process.

If χ does not equal a constant, but depends on T itself, say, we need additional steps to find the solution if we wish to use current values (at time level j) in the evaluation of χ. A linearization procedure—Taylor expansion about a guessed solution—followed by repeated matrix inversions until one converges onto the correct solution by a Newton-Raphson technique is popular in stellar interiors calculations, where it is known by the name of the *Henyey technique.* Alternatively, we may be satisfied with evaluating χ at past time levels (e.g., $j-1$), in which case the problem is still linear, but with variable (known) coefficients. A single matrix inversion then suffices for each new time step.

For an explicit procedure, we are restricted to time steps that correspond to $\eta \gtrsim 1$ [see equation (13.41)]. No such restriction for numerical stability applies to the implicit scheme described above; indeed, taking very large time steps so that $\eta \ll 1$ formally looks good for equation (13.46). Mnemonic: When Δt is small, you want to multiply by it (explicit method); when Δt is large, you want to divide by it (implicit method). Common sense dictates, however, that we should not take time steps that begin to approach the system diffusion time L^2/χ, where L equals the total dimension of the system. Nevertheless, the potential gain in time-step requirements over an explicit method approaches a factor $(L/\Delta x)^2 = N^2$, which is considerable if we need to use a large number N of spatial grid points. Offsetting this gain in general, however, is the need to invert a different matrix many instances per time step if the problem is nonlinear to begin with. The latter problem will not pose a great difficulty if the governing matrix equations can be made tridiagonal, or if the machine architecture has multiple parallel processors.

Finally, we note that the matrix inversion method can also be adapted to solve elliptic PDE problems. For parabolic PDEs, the inversion of large matrices can be avoided for problems containing spatial variations in only a single dimension. In the case of a single space variable, we may form the time derivative of T by finite-differencing *two* past solutions:

$$\frac{\partial T}{\partial t}(x_i, t_j) \approx \frac{1}{\Delta t}\left[T_i^{(j-1)} - T_i^{(j-2)}\right] \equiv Q^{(j)}(x_i) \qquad (13.47)$$

so that the right-hand side looks like a completely known source term, Q. The heat-diffusion equation now looks like an ODE in x,

$$\chi \frac{\partial^2 T}{\partial x^2} = Q, \tag{13.48}$$

where time t simply enters as a parameter in the problem. The spatial integration of equation (13.48) can be performed with very high accuracy (e.g., with Runge-Kutta techniques or with one of the other schemes described in the book *Numerical Recipes* by Press *et al.*), independent of whether χ has an additional dependence on T or not. Of course, the implementation of two-point boundary conditions (one at $x = 0$, the other at $x = L$) does not have a convenient solution in spatial integration methods that essentially march from one boundary to another. In the *shooting method*, one makes an arbitrary guess for either the function or its derivative (whichever is missing at the first boundary); integrate by marching to the other boundary; fail to satisfy the imposed boundary condition there; go back to the original end and make another guess; and iterate (typically, by Newton's method for finding roots) until both boundary conditions are satisfied.

Each of the methods described above has its virtues and its faults. A little thought ahead of time as to which one (or which combination) is most suitable for attacking the particular problem at hand will usually save a lot of wasted programming effort. Numerical analysis possesses no *failsafe* recipe for success.

14

Steady Supersonic Flow

Reference: Courant and Friedrichs, *Supersonic Flow and Shock Waves.*

In this chapter we apply the method of characteristics to the quasilinear PDEs that appear in problems of fluid flow. We begin by consolidating the results derived in Chapter 13. For steady compressible flow in two spatial dimensions, we then find that the governing equations are elliptic or hyperbolic depending on whether the flow occurs subsonically or supersonically. From this point of view, the supersonic flow of gas past bodies of various shapes differs fundamentally from subsonic flow past the same bodies in that no warning exists upstream of the presence of the obstacle. As we shall see, this leads to the formal appearance of double-valued fluid properties when obstacles force the flow to decelerate or to turn into itself. The resolution of this contradictory behavior necessitates the introduction of discontinuous jumps, or shock waves, a topic that we shall pursue in Chapters 15 and 16.

TRAJECTORY AND SOLUTION CHARACTERISTICS

According to the discussion of Chapter 13, the equations governing the steady, irrotational flow of a barotropic fluid in two spatial dimensions read

$$\frac{\partial v}{\partial x} - \frac{\partial u}{\partial y} = 0, \qquad A\frac{\partial u}{\partial x} + B\left(\frac{\partial u}{\partial y} + \frac{\partial v}{\partial x}\right) + C\frac{\partial v}{\partial y} = D, \tag{14.1}$$

where A, B, C, and D are given by equations (13.16). The introduction of the velocity potential Φ via equation (13.14) transforms the equations (14.1) to equation (13.15). For arbitrary functional dependences of A, B, C, and D on x, y, Φ, u, $\equiv \partial\Phi/\partial x$, and $v = \partial\Phi/\partial y$, our discussion below applies therefore to the most general form for a quasilinear PDE of second order in two independent variables.

The eigenvector ds_i needed to find characteristics satisfies the matrix equation,

$$\begin{pmatrix} ds_1 & ds_2 \end{pmatrix} \begin{pmatrix} dx & dy \\ Ady - Bdx & Bdy - Cdx \end{pmatrix} = \begin{pmatrix} 0 & 0 \end{pmatrix}, \tag{14.2}$$

which has the trajectory characteristics

$$\frac{dy}{dx} = \frac{1}{A}[B \pm (B^2 - AC)^{1/2}], \tag{14.3}$$

and the corresponding eigenvector components

$$\begin{pmatrix} ds_1 & ds_2 \end{pmatrix} \begin{bmatrix} 1 \\ \pm(B^2 - AC)^{1/2} \end{bmatrix} = 0, \qquad i.e., \quad ds_1 = \mp(B^2 - AC)^{1/2} ds_2. \tag{14.4}$$

Equation (13.19), $L_k dx = ds_i X_{ik}$, now implies

$$L_1 dx = ds_1 \cdot 0 + ds_2 \cdot A = A ds_2,$$
$$L_2 dx = ds_1 \cdot 1 + ds_2 \cdot B = [B \mp (B^2 - AC)^{1/2}] ds_2.$$

Hence, if we divide one relation by the other, we get

$$\frac{L_2}{L_1} = \frac{1}{A}[B \mp (B^2 - AC)^{1/2}]. \tag{14.5}$$

We obtain the solution characteristics from equation (13.20), $L_k du_k = H_i ds_i$, i.e.,

$$L_1 du + L_2 dv = 0 \cdot ds_1 + D ds_2 = D\frac{L_1}{A} dx. \tag{14.6}$$

Let us define

$$\lambda_\pm \equiv \frac{1}{A}[B \pm (B^2 - AC)^{1/2}]; \tag{14.7}$$

then we may summarize by writing,

$$du + \lambda_\mp dv = \frac{D}{A} dx \qquad \text{on} \qquad dy = \lambda_\pm dx, \tag{14.8}$$

as the characteristic ODEs associated with the PDEs (14.1).

DISTINCTION BETWEEN SUBSONIC AND SUPERSONIC FLOW

We now specialize the discussion to our original flow problem. With A, B, and C given by equations (13.16), we obtain

$$B^2 - AC = a^2(u^2 + v^2 - a^2) = a^2(q^2 - a^2),$$

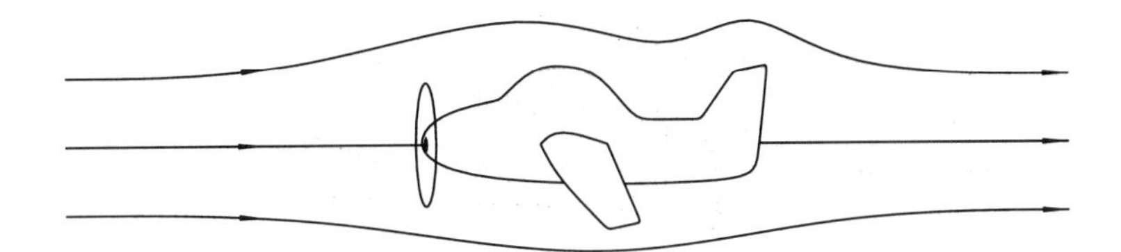

FIGURE 14.1
Subsonic flow about an obstructing body occurs smoothly because the upstream gas can be forewarned of the presence of the object.

where $q \equiv (u^2 + v^2)^{1/2}$ is the total fluid speed. Hence, $B^2 - AC$ is positive or negative depending on whether q^2 exceeds a^2 or not; i.e., the governing PDEs are *hyperbolic for supersonic flow* and *elliptic for subsonic flow.*

This conclusion has the following far-reaching consequence. Consider the subsonic flight of a prop-driven airplane. The air flow in steady state has the appearance shown in Figure 14.1 when viewed from the frame of rest of the plane.

Because the governing equations are elliptic, the influence of the boundary conditions that apply on the surface of the airplane has opportunity (in steady state) to propagate throughout all of space. Thus, the air incident on the nose of the plane can be forewarned of the presence of the plane, and it can be brought smoothly to a stop at a stagnation point, past where it slowly creeps around the plane, eventually to leave through another stagnation point at the tail. Similarly, a buildup of pressure gradients near the plane can push the air on neighboring streamlines to pass smoothly around the plane. (We ignore here the issue of turbulent boundary layers.)

Consider now the supersonic flight of a jet (Figure 14.2).

Because the governing equations are now hyperbolic, the boundary conditions generated by the surface of the airplane has a limited regime of influence, governed by how far sound waves can travel while being swept downstream by the supersonic flow. In particular, no information concerning the presence of the airplane can reach onrushing air directly upstream of the plane. On the other hand, air incident on the nose of the plane can not penetrate its surface; the flow must ultimately be brought to rest at a stagnation point as before. How does this deceleration occur if the far-upstream gas cannot be forewarned that it has to stop? Almost all the deceleration must occur in one jolt, a nearly discontinuous jump in a *bow shock*, which brings the supersonically flowing air (relative to the plane) to subsonic speeds, from where it and the neighboring streamlines can flow

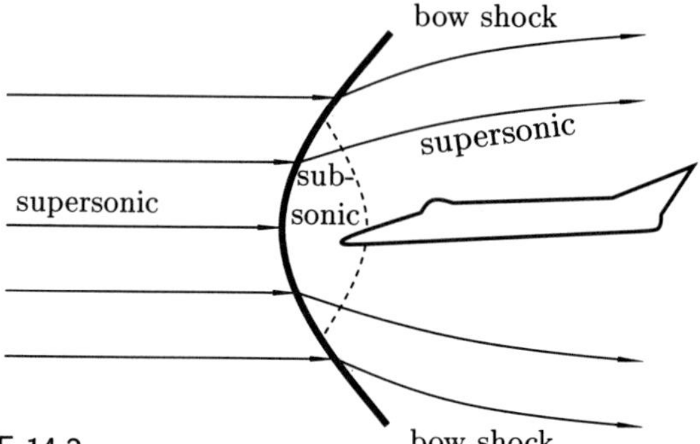

FIGURE 14.2
Supersonic flow about an obstructing body involves the production of shock waves because the upstream gas cannot be forewarned of the presence of the object.

smoothly around the plane. In Chapter 15, we shall examine the details of how shock waves work; here we concentrate on computing the conditions which inevitably bring about such a dilemma for the incident supersonic flow.

MACH CHARACTERISTICS

We begin by giving a geometric interpretation for the trajectory characteristics, which we choose to write in the form that directly emerges from setting the determinant of the coefficient matrix in equation (14.2) equal to 0:

$$(u^2 - a^2)dy^2 - 2uv\,dxdy + (v^2 - a^2)dx^2 = 0.$$

Rearranging terms, we obtain

$$a^2(dx^2 + dy^2) = (vdx - udy)^2. \tag{14.9}$$

This form suggests that we may find it profitable to transform to polar coordinates for both the velocity components,

$$u = q\cos\vartheta, \qquad v = q\sin\vartheta, \tag{14.10}$$

and for the displacement along a trajectory characteristic,

$$dx = \cos\chi\,ds, \qquad dy = \sin\chi\,ds. \tag{14.11}$$

Pictorially, the transformation looks as drawn in Figure 14.3.

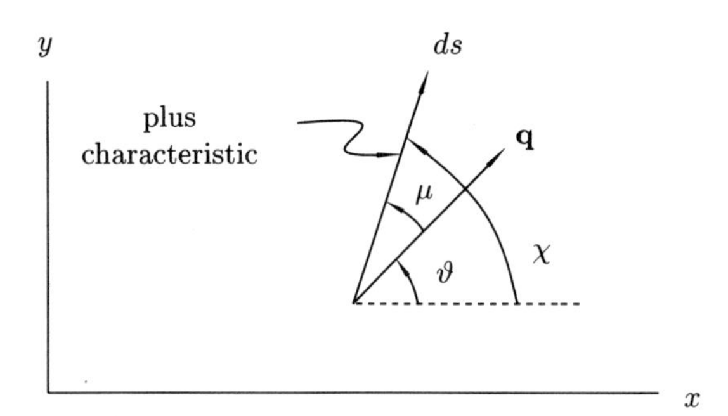

FIGURE 14.3
The geometry of streamlines and characteristics for two-dimensional, steady, supersonic flow, where μ is the Mach angle defined by $\sin\mu = a/q$.

Let us further define the Mach angle μ through

$$\sin\mu \equiv \frac{a}{q} \equiv \frac{1}{M}, \tag{14.12}$$

where M equals the Mach number. Notice that μ is real if $M > 1$, an assumption that we adopt from here on. We can now transform equation (14.9) to read

$$\sin^2\mu\, ds^2 = (\sin\vartheta\cos\chi - \cos\vartheta\sin\chi)^2\, ds^2.$$

Canceling the common term ds^2 and making use of the trigonometric formula for the sine of the difference of two angles, $\vartheta - \chi$, we may express the above relation as

$$\sin^2\mu = \sin^2(\vartheta - \chi) \qquad \Rightarrow \qquad \chi = \vartheta \pm \mu. \tag{14.13}$$

Written in the form of equation (14.13), the trajectory characteristics are known as *Mach characteristics.* They have the geometric interpretation shown in Figure 14.4, as the loci of sound waves emitted by a hypothetical source moving at the local fluid speed when supersonic motion sweeps the acoustic disturbances steadily downstream.

In three dimensions, we would have gotten a Mach cone, which plays a role in information propagation in gas dynamics like that of the light cone in the theory of special relativity. In particular, only matter lying within the Mach cone can be influenced by hydrodynamical effects emanating from the point of origin of the acoustic disturbance (Figure 14.4).

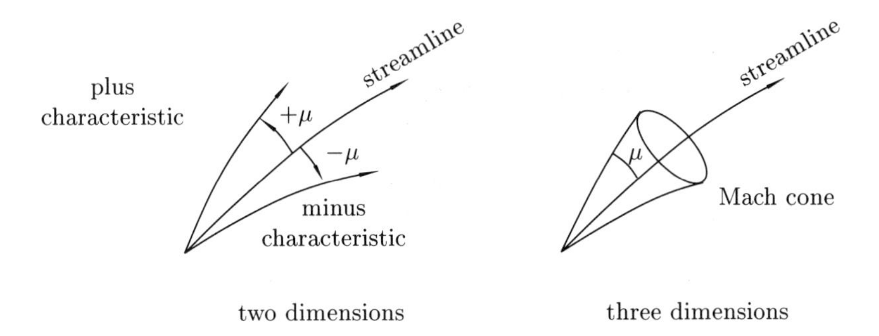

FIGURE 14.4
In two dimensions, a streamline in steady supersonic flow bisects a plus characteristic and a minus characteristic, making angles of $\pm\mu$ with these characteristics. In three dimensions, a Mach cone, carrying acoustic information to other parts of the flow, emanates from each point of a streamline. The opening angle of the cone is again the Mach angle $\mu = \arcsin(a/q)$.

RIEMANN INVARIANTS

Having obtained the geometric interpretation of the trajectory characteristics, we now move to consider the information contained in the solution characteristics. To simplify the algebra, we specialize to the following circumstances: (a) We suppose the flow occurs isothermally, so that the sound speed everywhere equals the same constant a. (Courant and Friedrichs work out the algebra for isentropic flow, a case more relevant to terrestrial applications than isothermal flow.) (b) We ignore accelerations produced by gravity, so that we have no inhomogeneous term, $D = 0$. In these circumstances, equations (14.8) become

$$du + \tan(\vartheta \mp \mu)dv = 0 \qquad \text{on} \qquad \left(\frac{dy}{dx}\right)_{\pm} = \lambda_{\pm} = \tan(\vartheta \pm \mu). \qquad (14.14)$$

Using equation (14.10), which we choose to write as

$$u = a\cos\vartheta/\sin\mu, \qquad v = a\sin\vartheta/\sin\mu,$$

we may rewrite the first expression in equation (14.14), after dividing out a factor of $a/\sin^2\mu$, as

$$-(\sin\vartheta\sin\mu\, d\vartheta + \cos\vartheta\cos\mu\, d\mu)$$
$$+\left(\frac{\sin\vartheta\cos\mu \mp \cos\vartheta\sin\mu}{\cos\vartheta\cos\mu \pm \sin\vartheta\cos\mu}\right)(\cos\vartheta\sin\mu\, d\vartheta - \sin\vartheta\cos\mu\, d\mu) = 0.$$

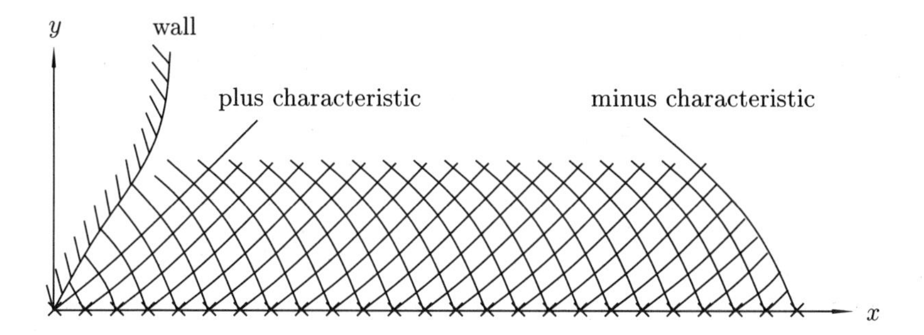

FIGURE 14.5
When added to boundary conditions, a mesh of plus and minus characteristics yields the complete solution for the two-dimensional problem of steady supersonic flow (see text for details).

Multiplying through by the denominator, we get, upon collecting terms,

$$\sin^2\mu\, d\vartheta \pm \cos^2\mu\, d\mu = 0 \qquad \Rightarrow \qquad d\vartheta \pm \cot^2\mu\, d\mu = 0.$$

The last equation can be integrated to give (what are called in other contexts) the *Riemann invariants*:

$$\vartheta \mp (\mu + \cot\mu) = \text{constant} \qquad \text{on} \qquad \frac{dy}{dx} = \tan(\vartheta \pm \mu). \tag{14.15}$$

SOLUTION OF FLOW PROBLEM BY THE METHOD OF CHARACTERISTICS

Equation (14.15) allows us to reduce the fluid flow problem to a solution of a problem in differential geometry, i.e., we can deduce much of what happens physically simply by drawing some pictures. Consider an "initial-value" problem of steady supersonic flow where we are given initial data on the x-axis, with boundary conditions imposed at some left- and right-hand boundaries (Figure 14.5).

From the "initial" data, we know the value of everything on the x-axis; in particular, we know the values of the angle ϑ that a streamline makes locally with respect to the x-axis and the Mach angle μ at which sound would propagate with respect to the streamline if an acoustic disturbance were to be emitted there. Thus, if we divide the x-axis up into a number of small grid points, we can construct the slopes of the plus and minus characteristics, $dy/dx = \tan(\vartheta \pm \mu)$, that emanate from the x-axis. For a given streamline which bisects its two characteristics, the plus characteristic bends more to the right; the minus, more to the left. A plus characteristic carries the information that the Riemann invariant, $\vartheta - (\mu + \cot\mu)$, retains the same value as it had on the position along the x-axis from where the plus characteristic originated. This does not suffice to determine either ϑ or μ individually, but only that particular combination represented by the Riemann invariant. A minus characteristic carries the information that the other Riemann invariant, $\vartheta + (\mu + \cot\mu)$, retains the same value as it had at the position on the x-axis from which that minus characteristic emanated. Hence, where a plus and a minus characteristic intersect, we know both combinations, $\vartheta \mp (\mu + \cot\mu)$, and thus can determine the values of ϑ and μ independently at the intersection point. From the first generation of intersections (upper vertices of triangles in the figure), we obtain therefore a row of points on which we again know everything. We may regard this row as a new set of "initial data," from which we may carry the solution another step forward by marching.

What happens when we reach the left-hand boundary? If the left-hand boundary represents the surface of a solid body, we don't have points to the left of the boundary (inside the body) that would give a plus characteristic. However, we do have the boundary condition that the streamline must be tangent to a solid body (no normal velocity), i.e., $\vartheta = \vartheta_{\text{w}}$, where ϑ_{w} is the local slope angle of the given wall. This knowledge, plus the value of $\vartheta + (\mu + \cot\mu)$ carried to the same point by a minus characteristic, serves to determine ϑ and μ individually. With ϑ and μ known, we can construct the plus characteristic that emanates from the left-hand boundary, and continue with our interior solution. The same argument holds for the right-hand boundary, except we must reverse the roles of the plus and minus characteristics (Figure 14.6).

What do we do if the left-hand boundary represents a *free boundary*, whose position we do not know in advance, but whose pressure is given by some external medium? Again, we don't have a plus characteristic, but we do have the boundary condition that $P = \rho a^2 = P_{\text{ext}}$. We also know from Bernoulli's theorem applied to an isothermal gas that

$$\frac{1}{2}q^2 + a^2 \ln\rho = \text{constant on a streamline}, \tag{14.16}$$

where the relevant streamline is the one that makes an angle $-\mu$ with respect to the minus characteristic that comes to the boundary from the fluid interior. Knowing $\rho = P_{\text{ext}}/a^2$ on a free boundary determines the magnitude of $\mathbf{q}$ from Bernoulli's theorem. Given q, we may determine μ from $\sin\mu = a/q$. We may then find ϑ from the value of $\vartheta + (\mu + \cot\mu)$ brought to the boundary point by the minus characteristic. Thus we can again derive both μ and ϑ at the boundary point; the latter provides the

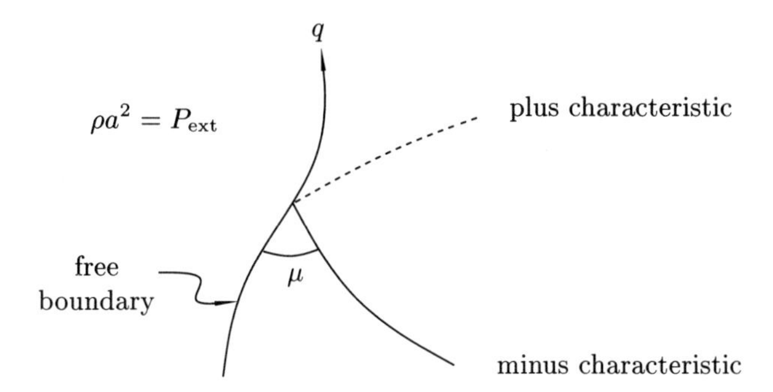

FIGURE 14.6
Pressure equilibrium is maintained across a free boundary, and the latter must also correspond to a streamline. This knowledge suffices both to fix the shape of the free boundary and to obtain the emergence of a plus characteristic (for the disappearance of each minus characteristic into a left-hand boundary) that interacts with the rest of the flow.

local inclination of the free surface with respect to the x-axis. (By definition, a free surface must correspond to a streamline, so that the flow occurs locally parallel to such a surface.) This inclination allows an extension of the free surface whose shape prior to that point we have mapped out by previous construction. Marching of the solution at interior points now continues as before.

EXPANSIVE FLOWS AROUND CURVED WALLS

A special, but important, situation exists that greatly simplifies the calculations outlined in the previous section. This situation arises when we have a problem involving uniform flow incident on some curved surface, as illustrated in Figure 14.7.

The minus characteristics (the light lines going from top left to bottom right) originate from a region of uniform flow; thus the Riemann invariant associated with them satisfies

$$\vartheta + (\mu + \cot\mu) = \text{same constant everywhere.} \tag{14.17}$$

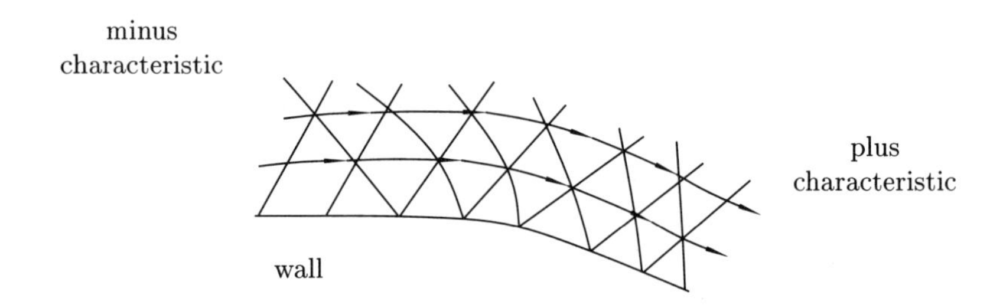

FIGURE 14.7
The expansion associated with flow out of a corner, where the fluid turns away from itself.

The plus characteristics originate from the wall; thus once these characteristics begin to come from the curved part of the wall, the Riemann invariant associated with the plus characteristics satisfies

$$\vartheta - (\mu + \cot \mu) = \text{different constants on different plus characteristics.} \tag{14.18}$$

On a given plus characteristic, both combinations of $\vartheta \pm (\mu + \cot \mu)$ are constant; thus ϑ and μ maintain the same values on a plus characteristic, and these characteristics must form straight lines. The slopes of the lines will generally be different for two plus characteristics that emanate from different parts of the curved wall, but each plus characteristic will trace a straight line.

A minus characteristic as it propagates from top left to bottom right intersects different plus characteristics. As long as these intersections involve only plus characteristics from the straight part of the wall (where the flow is uniform), the value of $\vartheta - (\mu + \cot \mu)$ brought to the intersection point by the plus characteristic will not vary, and the minus characteristic will also be a straight line. However, as soon as a minus characteristic begins to intersect plus characteristics that emanate from the curved part of the wall, the values of $\vartheta - (\mu + \cot \mu)$ brought to the intersections by different plus characteristics will begin to differ, and the trajectory of the minus characteristic will begin to curve. This curvature is not important to follow, however, because we already know the information [equation (14.17)] carried by *any* minus characteristic, independently of how it arrived at the point in question.

A flow situation that depends on the variation of only one of the Riemann invariants, equation (14.18), with the other being constant, equation

(14.17), goes by the name of a *simple wave.* For steady flow in multiple spatial dimensions, these disturbances form the nonlinear analogs of "standing waves" (in the frame of rest of the obstacle or wall). In Chapter 15, we shall consider simple waves in a time-dependent context, which constitute direct nonlinear analogs of linear waves that propagate to the right or to the left as functions (not necessarily sinusoids) of the combination of variables $(x - ct)$ or $(x + ct)$.

Let us consider the result of carrying out the geometric construction outlined in Figure 14.7. As the wall curves away from the flow, the value of ϑ on each successive plus characteristic (which equals the ϑ of the part of the wall from which it emanates) must change from 0 to more negative values. The plus characteristics begin then to bend more and more downward, in the direction of the curving wall. The streamlines that bisect the plus and minus characteristics follow this trend, and in this fashion, they receive the information to follow the bend of the curve of the wall. Streamlines close to the wall receive this information earlier; those far from the wall, later; but eventually the whole flow is turned parallel to the downsloping straight portion of the wall. Streamlines initially close together are spread farther apart by the process; thus the process of a flow turning away from itself must *expand* the fluid. (The corresponding simple wave is an *expansion wave.*) Since expanding a supersonic flow *accelerates* the flow (see Chapter 6), the flow must speed up upon turning such a bend.

The power for speeding up the flow comes from the internal energy of the fluid, which is potentially infinite if we assume that external agents exist for maintaining the fluid always at the same temperature T. For an ideal fluid with no external heating or cooling agents, the flow would occur adiabatically instead of isothermally, and the fluid would have limited capability of following a bend that curves away from the flow. If the bend occurs through too large an angle, the fluid would separate from the wall and develop a free surface that expands trying to fill in the vacuum that forms between the wall and the free surface (Figure 14.8).

An isothermal gas has no such difficulties, because the heat function $a^2 \ln \rho$ in Bernoulli's theorem (14.16) permits some of the gas to acquire very high speeds if we lower the density ρ sufficiently. Thus, an isothermal gas will (perhaps artificially) never allow the development of a free surface bounded by a vacuum.

From a coarse point of view, every bend seen from sufficiently afar looks like a sharp corner. The mathematical problem becomes especially simple for flow around a sharp corner (Figure 14.9).

In this case, a (Prandtl-Meyer) "fan" of plus characteristics emanates from the single point of the corner. The slopes of these characteristics take all intermediate values from the straight line appropriate for uniform flow parallel to the leading straight wall to that appropriate for uniform flow parallel to the trailing straight wall. The interaction of the expansion fan

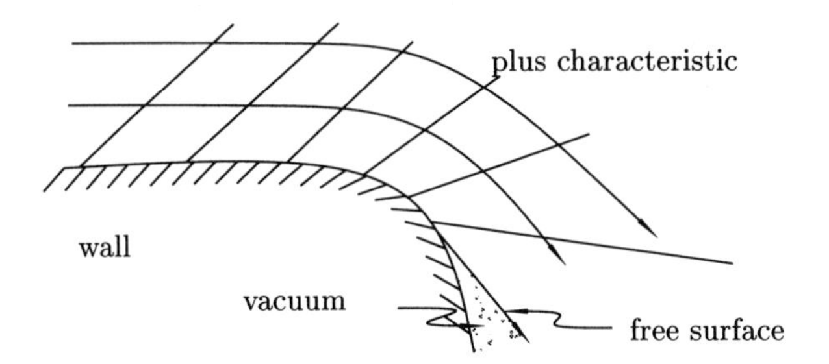

FIGURE 14.8
For turn out of a greatly rounded corner, the fluid may not be able to follow the wall, and a free surface will form that is separated from the wall by a vacuum. Such a vacuum cannot form if the flow is isothermal, but a polytropic gas with $\gamma > 1$, where the sound speed $\propto \rho^{(\gamma-1)/2}$ goes to zero in the limit $\rho \to 0$, has more limited ability to expand into a vacuum.

with the minus characteristics (not drawn for sake of clarity) turns the flow from one uniform state to another uniform state, as indicated schematically by the arrowed streamlines.

COMPRESSIVE FLOWS INTO CURVED SURFACES

A crucial difference exists between turning a flow out of a corner and turning into one. In the former case, the process is expansive, and the flow remains smooth; in the latter, the process is compressive, and the flow tends

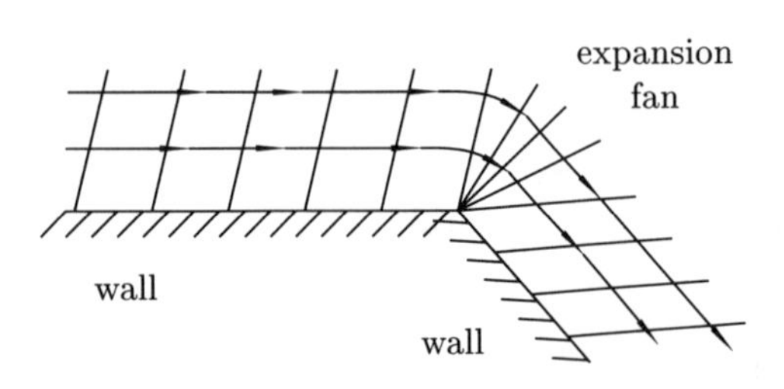

FIGURE 14.9
Supersonic turn out of a sharp corner occurs via a Prandtl-Meyer expansion fan.

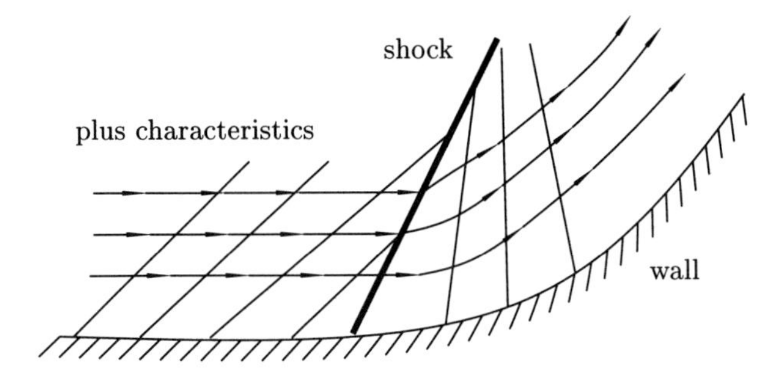

FIGURE 14.10
Supersonic flow into a corner, where the fluid turns into itself, involves the crossing of characteristics of the same family and the production of a shock wave.

to develop shocks. To see this, consider first the problem of supersonic flow into a rounded corner (Figure 14.10).

By arguments similar to those produced in the last section, we easily see that the same family of plus characteristics tend to intersect *each other*. In the case of expansion around the corner, the intersections occur inside the wall (i.e., not part of the fluid); for the present example, the intersections occur as part of the fluid flow. For a gently rounded corner, two plus characteristics will intersect at a finite angle at a finite distance from the wall. At an intersection point, there formally exist *two* values of $\vartheta - (\mu + \cot \mu)$ carried into the intersection between two plus characteristics, in addition to the one value of $\vartheta + (\mu + \cot \mu)$ carried there by a minus characteristic (not shown). Thus, the fluid velocity (and density, etc.) formally tries to become double-valued. One part of the fluid tries to penetrate into another part; this can not happen without the introduction of strong viscous forces, which have been neglected so far in our analysis.

As we shall see in Chapter 15, the competition between the resultant viscous forces and the inertia (and pressure) of the fluid leads to the production of a shock wave that resolves the conflict for the flow. Mathematically, the solution by the method of characteristics breaks down when characteristics of the same family cross. The plus characteristics enter the shock front and merge (to define, in a manner of speaking, the locus of the shock discontinuity). The minus characteristics enter the shock front and re-emerge on the downstream side, but the Riemann invariant that they carry, $\vartheta + (\mu + \cot \mu)$ for an isothermal gas, will generally suffer a jump in value as they cross the shock. (We defer the derivation of the jump conditions for an "isothermal shock" to Chapter 16.) Intuitively, we expect that

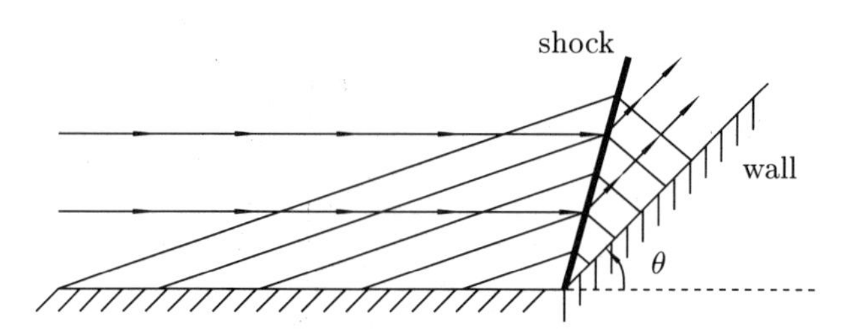

FIGURE 14.11
Supersonic flow into a sharp corner will produce an attached shock, if the exterior angle θ formed by the two walls is not too large.

the amount of this jump will be small if the shock wave is weak, so that we may "patch up" a solution. Courant and Friedrichs show that for weak shocks, we may locate the shock front (the difficult part of the problem) so that it bisects the intersecting (plus) characteristics of the same family. The minus characteristics carry their Riemann invariant across the shock virtually intact, and the totality of jump conditions for an isothermal shock then provide the additional information needed to generate a postshock set of "initial data" from which we can continue the downstream part of the supersonic flow by marching.

By taking the limit of a rounded corner, we can see that a sharp corner will produce an attached shock (Figure 14.11).

We now have the reverse situation from an expansion fan; all the plus characteristics "intermediate" in slope between the precorner value and the postcorner value compress (because the range is *negative*) into a single compromise inclination that defines the shock front, which radiates in a straight line from the corner. The shock front now transforms one uniform flow (upstream) into another uniform flow (downstream) in a *single discontinuous jump* (in the approximation that the mean free path in the fluid equals zero). The streamlines are brought closer by the process; so the shock is *compressive*, and the flow must decelerate. The transformation can be calculated in this idealized case for (oblique) shocks of arbitrary strength (corners of arbitrary angles) once we have derived the appropriate jump conditions (Chapters 15 and 16).

DIFFICULTY WITH STEADY SUPERSONIC TURBULENCE

We now see the problem with the notion of steady supersonic turbulence as presented sometimes in a purely gas-dynamical context to explain the line widths of spectroscopic measurements of certain astronomical objects. The word "turbulence" invokes the image of many swirling eddies (see Chapter 9). An eddy corresponds to fluid turning into itself. In contrast to gravity, gas-dynamical forces cannot cause a fluid to turn into itself without creating shockwaves. Shockwaves involve strong dissipation, transforming bulk motion into heat, which in many astrophysical environments will be carried away by radiation (see Chapter 16). The time for an eddy of typical size λ and velocity U to turn into itself is λ/U, much shorter than the time to cross the system L/U if $\lambda \ll L$. Thus, unless there exists a powerful input of random kinetic energy, supersonic turbulence will decay quickly.

Supersonic turbulence differs from subsonic turbulence in the crucial sense that the latter induces relatively little compression. As a consequence, little energy leaks from the system in the form of acoustic waves (review Chapters 6 and 9). The main drain of energy in subsonic turbulence comes via a turbulent cascade from large eddies to small ones, and "so on to viscosity." No such queue awaits supersonic turbulence. The separation from the largest-scale differential motions to heat can be spanned in one jump, via a shockwave. Under these circumstances, we have no rational reason to expect a Kolmogorov spectrum of eddies.

How, then, do we resolve the astronomical puzzle of ubiquitous large line widths? The inclusion of magnetic fields may help tremendously in this regard, since they make the system much more elastic. More specifically, a flow may be supersonic yet still be slower than the other wave speeds that characterize a magnetized fluid (see Chapter 22). As a consequence, the fluid may be "forewarned" before it has to turn a corner, and thereby it can accomplish the cornering smoothly. We shall see, moreover, that bent magnetic fields exert restoring forces that oppose strong curvature. Thus, when the fluid moves more slowly than the relevant wave speed, the undulations take on a more wavelike than eddylike character. Supersonic turbulence in a magnetized medium is therefore more likely to comprise many MHD waves than many eddies.

SUPERSONIC FLOW PAST OBJECTS OF ARBITRARY SHAPE

We can now piece together, at least qualitatively, what supersonic flow will look like past an object of arbitrary shape. The front of a sharp-nosed jet looks like a wedge. Symmetry of the upstream conditions makes supersonic flow past a wedge look like two halves of uniform flow into a corner; thus we get two attached shocks (Figure 14.12).

If the flow speed is too close to being sonic for the given opening angle of the wedge, the shock wave actually forms ahead of the nose, and we get a detached bow shock. The same happens if the airplane has a rounded nose (Figure 14.13).

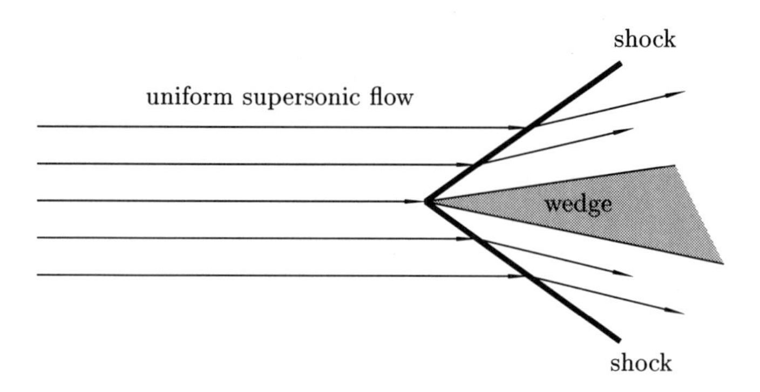

FIGURE 14.12
Supersonic flow past an infinite wedge with a small opening angle occurs via two attached shocks that transform regions of one uniform flow to regions of another uniform flow.

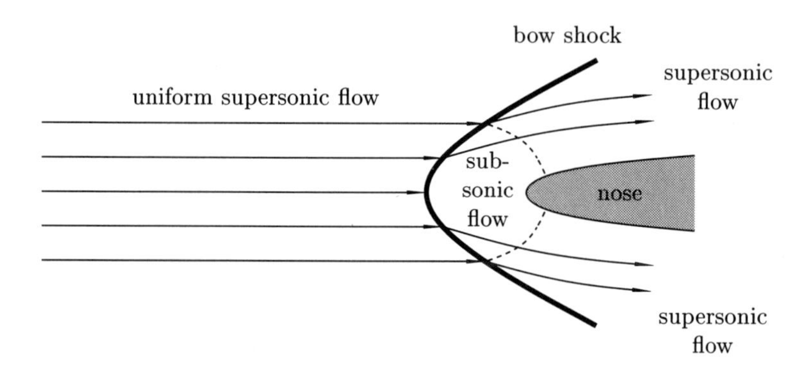

FIGURE 14.13
Supersonic flow past an object with a blunt nose produces a detached bow shock. The flow downstream from the part of the shock front where the flow enters and exits nearly perpendicularly is decelerated to subsonic values, whereas the flow downstream from the parts of the shock front where the flow enters more obliquely is left at supersonic speeds. The transition from subsonic speeds back to supersonic speeds as the gas leaves the area of the nose occurs essentially by expansion out of a corner, like nozzle flow (see Chapter 6).

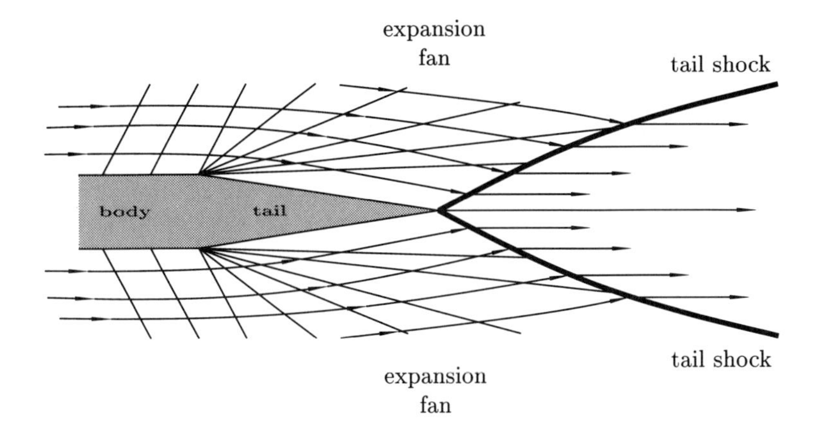

FIGURE 14.14
The convergence of the two supersonic flows on opposite sides of a body as that body comes to an end generates tail shocks.

In this case, behind the standing bow shock and just ahead of the nose, there exists a region of subsonic flow which can bring one of the streamlines to a stagnation point, and push the other streamlines smoothly out of the way of the nose. The combination of the governing PDE being hyperbolic in some part of the flow and elliptic in another part causes considerable mathematical difficulties, but the physics remains clear. The high pressure of the (strongly shocked) subsonic region eventually causes this gas to make a sonic transition, accelerating it back up to supersonic speeds. As the air turns out of the rounded portion where the nose joins to the fuselage, the dynamics looks like supersonic flow out of a corner, and we get an expansion fan, which accelerates the gas further.

The fuselage of any finite plane eventually tapers to an end at the tail. The convergence of the two supersonic flows from the top and bottom of the plane looks like supersonic flow into a corner, and we get two tail shocks. These turn the two flows coming off the top and bottom of the aircraft parallel to one another (Figure 14.14). (If the two flows do not merge well, we might also get a turbulent wake.)

By the same token any protuberence on the airplane's body will form a local shock wave if it causes the supersonic flow to turn into itself, but will form an expansion fan if it causes the flow to turn away from itself. Since finite protuberences will generally have both types of corners (as does the whole airplane), shock waves and expansion waves (i.e., the compressive and

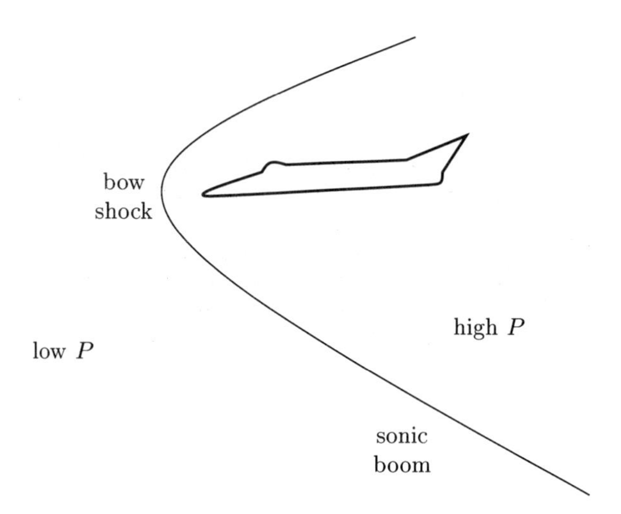

FIGURE 14.15
The bow shock pushed ahead of a supersonically flying aircraft loses strength as we move farther away from the plane. Eventually, the shock front makes an angle with respect to an entering streamline that begins to approach the Mach angle, and the shock wave degenerates to a small amplitude acoustic wave.

expansive parts of the flow) will generally interact with each other via the propagating characteristics. The expansional motion will weaken the shock waves, so that, far from the airplane, all shock fronts acquire an angle that approaches the Mach angle of the flow; i.e., crossing characteristics of the same family progressively intersect each other at shallower and shallower angles. This causes the shock wave to degenerate eventually to one of very small amplitude (a linear sound wave). To experience a "sonic boom," one has to be fairly close to a supersonically flying jet; otherwise, one just hears the gentle rumble of aircraft noise (Figure 14.15).

15

Steepening of Acoustic Waves Into Shock Waves

Reference: Whitham, *Linear and Nonlinear Waves.*

If nature used equations to move matter and radiation around (which I doubt), the method of characteristics—with its emphasis on the propagation of information at the natural speeds of the problem—would probably represent the way she would do it. Yet in our study of steady flow in a fluid, we found the characteristics to be real only for supersonic flow. For subsonic flow, the governing equations are elliptic, and the range of influence of any one part of the flow on the rest extends infinitely in all directions. How does this "action at a distance" jibe with modern perceptions that all physical influences have only a finite propagation speed?

The answer to this apparent paradox lies in the fact that we have assumed *steady state* conditions; i.e., we have allowed the fluid infinite duration to communicate with itself. If we consider instead the development of the flow in *time*, as we shall proceed to do below, we find the governing equations always to be hyperbolic, independent of whether the flow occurs supersonically or subsonically. The associated characteristics, which do not equal their counterparts for the steady flow problem, are always real. Physically, the difference arises because acoustic disturbances are always swept ahead by the passage of time, but they are swept downstream by steady flow only if that flow occurs supersonically (Figure 15.1).

SMALL-AMPLITUDE ACOUSTIC WAVES

In what follows we wish to consider the propagation of acoustic disturbances in an isentropic ideal gas. For an isentropic fluid, s = constant, implying pressure variations $P \propto \rho^\gamma$ with variations in density ρ. Thus, the gradient of the pressure can be calculated as

$$\nabla P = \left(\frac{\partial P}{\partial \rho}\right)_s \nabla \rho = a_s^2 \nabla \rho, \tag{15.1}$$

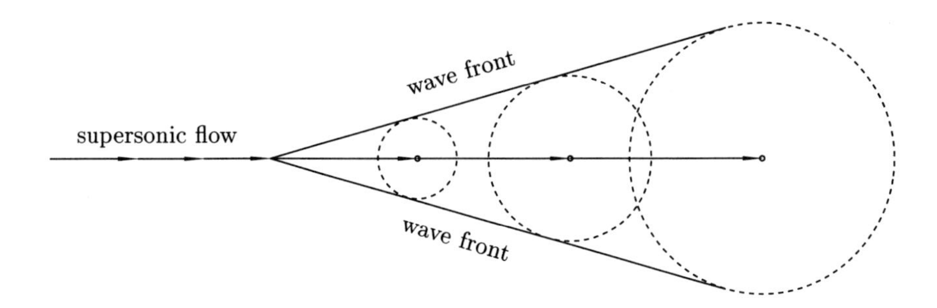

FIGURE 15.1
A small disturbance that occurs at a certain point in a gas in steady supersonic flow propagates away from that point (with respect to the moving fluid) at the speed of sound. The envelope of the wave front associated with this propagation forms a Mach cone with apex at the original point. In other words, knowledge of the original disturbance is swept downstream with the flow and cannot make its way upstream to the incident supersonic gas.

where a_s is the adiabatic speed of sound:

$$a_s = a_{s0}\left(\frac{\rho}{\rho_0}\right)^{(\gamma-1)/2}, \tag{15.2}$$

with ρ_0 and a_{s0} being the uniform density and acoustic speed of the static ambient background and $\gamma = c_P/c_v = 5/3$ for a perfect monatomic gas.

If we assume the disturbances to have infinitesimal amplitude, the linearized equations of motion in one dimension give rise to the following perturbation equations:

$$\frac{\partial \rho_1}{\partial t} + \rho_0 \frac{\partial u_1}{\partial x} = 0, \tag{15.3}$$

$$\rho_0 \frac{\partial u_1}{\partial t} = -a_{s0}^2 \frac{\partial \rho_1}{\partial x}. \tag{15.4}$$

If we subtract the space derivative of the second equation from the time derivative of the first, we get the homogeneous wave equation

$$\frac{\partial^2 \rho_1}{\partial t^2} - a_{s0}^2 \frac{\partial^2 \rho_1}{\partial x^2} = 0, \tag{15.5}$$

which has as its most general solution

$$\rho_1 = f(x - a_{s0}t) + g(x + a_{s0}t). \tag{15.6}$$

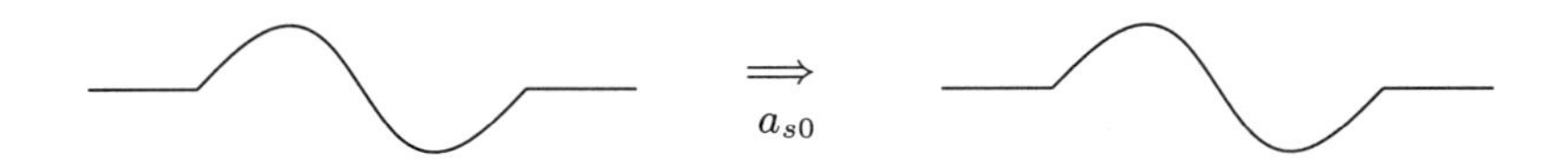

FIGURE 15.2
In the absence of dissipation and spatial inhomogeneities (or dispersion), the waveform of a disturbance governed by a linear wave equation maintains its size and shape forever, apart from propagation at a constant wave speed.

The corresponding solution for the velocity perturbations reads

$$u_1 = \frac{a_{s0}}{\rho_0}\left[f(x - a_{s0}t) - g(x + a_{s0}t)\right]. \tag{15.7}$$

Notice that the individual terms maintain *forever* their original waveforms, $f(x)$ or $g(x)$, apart from a propagation to the right or left at the constant speed a_{s0} (Figure 15.2).

Waves can propagate without changing shape if their governing PDE is linear. Although linear equations are a basic feature in some parts of physics (e.g., Maxwell's equations in a vacuum; Schrödinger's equation in an external potential), the equations of fluid mechanics are fundamentally nonlinear, unless they have been forced to take a linear form by fiat [as we have done in equation (15.5)]. The interesting question then arises, can acoustic waves of finite amplitude also propagate without changes of the waveform? The answer, we shall find below, is no. Acoustic waves having finite amplitude of *any waveform* must always steepen (if we ignore the action of viscosity), until eventually they develop such steep wavefronts that they become shock waves. (Almost all nonlinear waves have a tendency to steepen. However, some dispersive waves, whose different Fourier components propagate at different speeds when they have small amplitudes, have special finite-amplitude solutions, called *solitons* or *solitary waves*, in which the nonlinear steepening tendency can be exactly balanced by the dispersive tendency in such a way that the soliton maintains its shape as it propagates. Acoustic waves, which are not dispersive at small amplitudes, do not have such solitary-wave solutions; to pursue this topic further, see Whitham's book.)

UNSTEADY FLOW OF AN ISENTROPIC IDEAL GAS IN ONE DIMENSION

Without the small-disturbance assumption, the full equations read

$$\frac{1}{\rho}\left(\frac{\partial \rho}{\partial t} + u\frac{\partial \rho}{\partial x}\right) + \frac{\partial u}{\partial x} = 0, \tag{15.8}$$

$$\frac{\partial u}{\partial t} + u\frac{\partial u}{\partial x} = -\frac{a_s^2}{\rho}\frac{\partial \rho}{\partial x}, \tag{15.9}$$

where we have made use of equation (15.1) to eliminate the pressure gradient. We now use equation (15.2) to eliminate ρ in favor of a_s (which we write henceforth as a for simplicity of notation):

$$\frac{d\rho}{\rho} = \frac{2}{\gamma - 1}\frac{da}{a}.$$

Equations (15.8) and (15.9) now become

$$\frac{\partial}{\partial t}\left(\frac{2}{\gamma-1}a\right) + u\frac{\partial}{\partial x}\left(\frac{2}{\gamma-1}a\right) + a\frac{\partial u}{\partial x} = 0, \tag{15.10}$$

$$\frac{\partial u}{\partial t} + u\frac{\partial u}{\partial x} + a\frac{\partial}{\partial x}\left(\frac{2}{\gamma-1}a\right) = 0. \tag{15.11}$$

The coupled set of quasilinear PDEs of first order, equations (15.10) and (15.11), could be analyzed formally by the method of characteristics that we developed in Chapters 13 and 14. A faster route to the same answer follows the blazed trail by Riemann, who adopted the simple trick of adding and subtracting equations (15.10) and (15.11) to obtain

$$\left[\frac{\partial}{\partial t} + (u+a)\frac{\partial}{\partial x}\right]\left(u + \frac{2}{\gamma-1}a\right) = 0, \tag{15.12}$$

$$\left[\frac{\partial}{\partial t} + (u-a)\frac{\partial}{\partial x}\right]\left(u - \frac{2}{\gamma-1}a\right) = 0. \tag{15.13}$$

Now equations (15.12) or (15.13) individually represent a quasilinear PDE of first order in a single dependent variable Q or R, where

$$Q \equiv u + \frac{2}{\gamma-1}a, \tag{15.14}$$

$$R \equiv u - \frac{2}{\gamma-1}a. \tag{15.15}$$

The method of characteristics for such equations yields the ODEs

$$\frac{dt}{1} = \frac{dx}{u+a} = \frac{dQ}{0}, \tag{15.16}$$

$$\frac{dt}{1} = \frac{dx}{u-a} = \frac{dR}{0}. \tag{15.17}$$

In other words, we have the Riemann invariants,

$$Q \equiv u + \frac{2}{\gamma-1}a = \text{constant} \qquad \text{on the plus characteristic} \qquad \frac{dx}{dt} = u + a, \tag{15.18}$$

$$R \equiv u - \frac{2}{\gamma - 1}a = \text{constant} \qquad \text{on the minus characteristic} \qquad \frac{dx}{dt} = u - a. \tag{15.19}$$

The characteristic velocities $dx/dt = u \pm a$ equal the velocities at which (nonlinear) acoustic disturbances propagate at speed a to the right or to the left with respect to the fluid motion u. In contrast with the infinitesimal-amplitude analysis, we do not here assume u to be negligible compared to a, nor do we approximate a by its undisturbed value a_0 ($\equiv a_{s0}$). For the finite-amplitude case, what remains conserved by propagation at the two characteristic speeds is not the shape of the initial waveform, but the Riemann invariants, Q and R.

PROPAGATION AND STEEPENING OF A SIMPLE WAVE

Consider the situation of a simple wave, in which R on left-propagating minus characteristics is a strict constant, but Q is a different constant on different right-propagating plus characteristics.

$$R = u - \frac{2}{\gamma - 1}a = \text{the same constant for all spacetime}, \tag{15.20}$$

$$Q = u + \frac{2}{\gamma - 1}a = \text{different constants on different plus characteristics.} \tag{15.21}$$

The physical situation, shown in Figure 15.3, arises because the minus characteristics all originate from a static region of uniform properties ahead of the wave on the right before the simple wave (propagating from left to right) has had a chance to disturb the medium.

By arguments similar to those used in Chapter 14, we easily show that the minus characteristics form curved trajectories in spacetime, whereas the plus characteristics yield straight lines. We already know the information, equation (15.20), carried by any minus characteristic; so we need not concern ourselves with plotting any of them. Indeed, knowing that all minus characteristics come from a static region of uniform properties allows us to identify the constant in equation (15.20):

$$R \equiv u - \frac{2}{\gamma - 1}a = -\frac{2}{\gamma - 1}a_0. \tag{15.22}$$

At time $t = 0$ we have assumed a sinusoidal profile that possesses a density ρ equal to the undisturbed value ρ_0 at three points, marked A, B, C. The disturbance has vanishing amplitude at these points, and therefore plus characteristics emanating from these points must propagate at the undisturbed sound speed a_0, making three parallel lines in space time. Consider a part of the wave that has a density excess; by equation (15.2) it has a

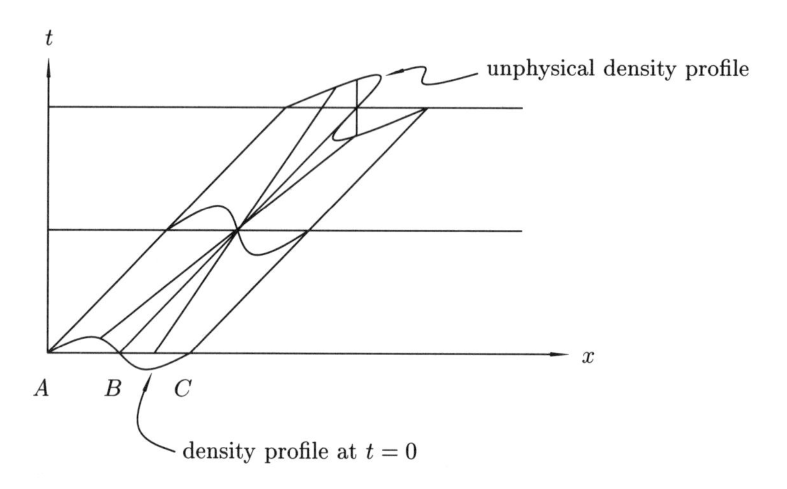

FIGURE 15.3
An acoustic wave of finite amplitude, even if it starts with a perfect sinusoidal shape and propagates in an undisturbed medium of exactly uniform properties, would inevitably steepen in its waveform.

sound speed $a > a_0$; therefore this effect tends to make a plus characteristic emanating from this part of the wave travel at a speed greater than average. There also exists the term u in the characteristic speed, $u + a$, which, by equation (15.22) $u = 2(a - a_0)/(\gamma - 1)$, is positive for parts of the wave that have density excesses, since $a > a_0$, and this just amplifies the effect. Now, consider a part of the wave that has a density deficit, where equations (15.2) and (15.22) imply $a < a_0$ and $u = 2(a - a_0)/(\gamma - 1) < 0$. Thus plus characteristics emanating from regions containing an initial density deficit travel slower than average. The net result is that the crest of the nonlinear wave tends to catch up with the trough, and the wave profile steepens. Nonlinear acoustic waves cannot maintain a constant waveform.

Eventually, if we were to follow this analysis far enough, we would find plus characteristics (belonging the same family) trying to cross. When this happens, the flow attempts to take on multiple values for the density, velocity, etc. This can not happen, of course, and the steepening must be halted by viscous forces (and have its thermal energy redistributed by heat conduction), so that the wave profile cannot steepen beyond the point where it suffers essentially one discontinuous jump (Figure 15.4).

At this stage, we have a shock wave, and although the characteristic analysis begins to break down, we describe what happens subsequently in

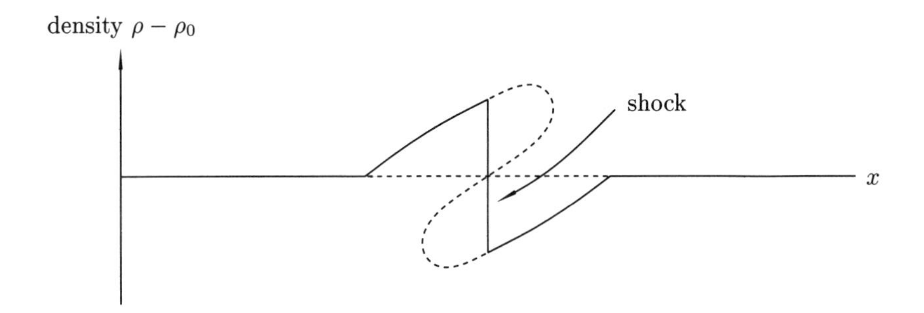

FIGURE 15.4
The tendency for nonlinearities to steepen the wave profile, which would produce multiple values for fluid properties such as gas density and velocity, must be eventually offset by the onset of strong viscous forces. The balance of the viscous forces and the steepening tendency mediates a shock, which is approximated in ideal fluid flow as a discontinuous jump of gas properties across the front.

a qualitative fashion so that we may form a relatively complete physical picture for the phenomenon. (See Whitham or Landau and Lifshitz for quantitative calculations.)

The high-density and high-pressure region just behind the shock wave must continue to push the front ahead faster than the undisturbed speed of sound a_0, so that eventually the shock front completely overruns the trough of the wave, and begins to propagate at supersonic speeds through the undisturbed medium. In the meantime, the tail of the wave, whose end propagates only at speed a_0, begins to lag behind the front, so that the wave profile gets stretched out along its length (Figure 15.5).

The redistribution of the density and pressure excess to more and more fluid must damp the disturbance, so that the height of the discontinuity (the strength of the shock) progressively weakens with time. Eventually, the wave will dissipate itself and become a small amplitude acoustic disturbance. Unlike a steady shock wave driven by a constant source of momentum and energy input (e.g., an airplane), in the current problem, we have the resources of only the initial input to draw on, and these resources must inevitably degrade once viscosity and heat conduction come into play.

THE STRUCTURE OF VISCOUS SHOCKS

In this section we wish to examine the structure of viscous shocks. We will verify first that the deceleration and compression of the fluid (in a frame

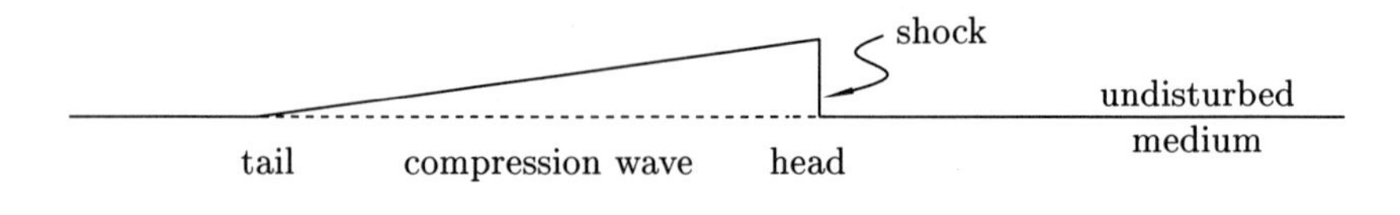

FIGURE 15.5
Since a shock front propagates at supersonic speeds with respect to the gas ahead of the front, the front eventually overruns the trough of a wave that initially had one cycle of a sinusoid, and the front increasingly also leaves behind the part of the wave that forms its tail. Thus, after shock formation, the initial single-sinusoid wave (indeed, any waveform consisting of no more than a single compression peak) increasingly takes on the shape of a triangular waveform.

which travels with the shock front) occur in a very thin layer. To the extent that the details of this transition do not interest us, we may approximate the entire process as a single discontinuous jump. We have as another goal, therefore, the derivation of the shock jump conditions.

Because the shock transition occurs in a thin layer, the time required for fluid to traverse the layer will be small compared to the time for the exterior conditions to change appreciably. In this case, if we transform to a frame which moves at the local velocity of the shock front, we may approximate the flow to be steady and to vary appreciably along only the one dimension perpendicular to the surface of the shock front. In addition, we assume that we may ignore any variation of the large-scale gravitational force field across the thin layer, as well as any radiative losses that might occur because the gas gets strongly heated behind the shock. (We shall treat radiative shocks in Chapter 16.) For sake of definiteness, we could pretend that we are blowing up the sharp part of the steepened wave profile of the previous section for detailed examination. In the frame of the wave front, preshock gas in region 1 rushes toward the front from the right (at minus the shock wave speed in the frame used in the previous section) and suffers a sudden compression and deceleration to become postshock gas of region 2 (Figure 15.6).

On the length scale of interest for investigating the shock layer, we can not see any of the smooth variations in the structure of region 1 (undisturbed gas) or of region 2 (gas inside the compression wave). In this approximation (which can be made mathematically rigorous by the method of matched asymptotic expansions), we may regard 1 and 2 as two regions of uniform flow, and the shock layer as the mechanism by which the fluid

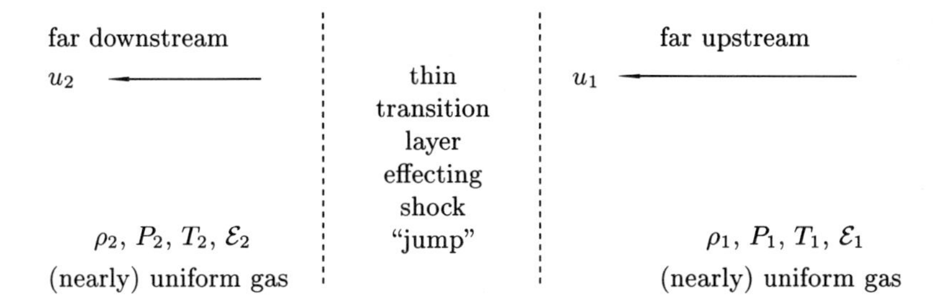

FIGURE 15.6
Schematic representation of the narrow transition region in a shockwave whereby upstream gas properties get transformed to downstream gas properties.

effects a transformation from 1 to 2. From this point of view, the macroscopic flow environment which brought about the production of the shock does not matter; once the shock exists, the structure would be the same if some completely different set of circumstances had led to the same upstream conditions 1. Thus, for example, our analysis would also apply to the part of the standing bow shock illustrated in Figure 15.7 that develops when supersonic gas flows past a blunt body.

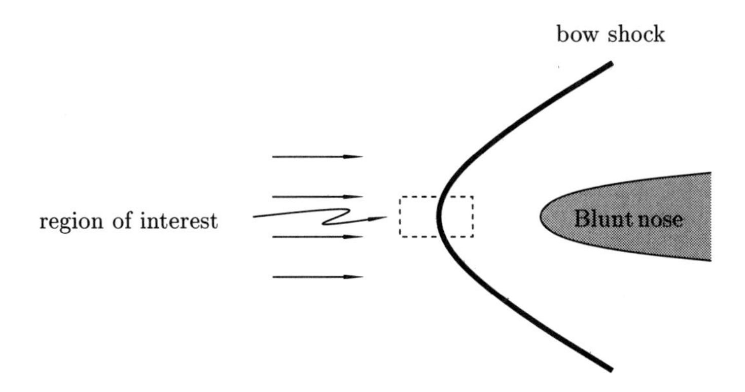

FIGURE 15.7
The part of the bow shock near the nose of a blunt object has the geometry of a normal shock.

We begin by writing the steady-state fluid equations with variations in one spatial dimension in their conservative form when we ignore any external sources for changing the mass, momentum, or energy content of the gas [see equations (4.16), (4.19), and (4.21)]:

$$\frac{d}{dx}(\rho u) = 0, \tag{15.23}$$

$$\frac{d}{dx}\left(\rho u^2 + P - \frac{4}{3}\mu\frac{du}{dx}\right) = 0, \tag{15.24}$$

$$\frac{d}{dx}\left[\rho\left(\frac{1}{2}u^2 + \mathcal{E}\right)u + \left(P - \frac{4}{3}\mu\frac{du}{dx}\right)u - \mathcal{K}\frac{dT}{dx}\right] = 0, \tag{15.25}$$

where we have written the xx-component of the viscous stress tensor as [cf. equation (4.28) with zero bulk viscosity]

$$\pi_{xx} = \frac{4}{3}\mu\frac{du}{dx}. \tag{15.26}$$

Integration of equations (15.23)–(15.25) gives the constancy of the fluxes of mass, momentum, and energy:

$$\rho u = \text{constant}, \tag{15.27}$$

$$\rho u^2 + P - \frac{4}{3}\mu\frac{du}{dx} = \text{constant}, \tag{15.28}$$

$$\rho u\left(\frac{1}{2}u^2 + \mathcal{E} + \frac{P}{\rho}\right) - \frac{4}{3}\mu u\frac{du}{dx} - \mathcal{K}\frac{dT}{dx} = \text{constant}. \tag{15.29}$$

In the shock transition layer (see Figure 15.8), we expect the frictional momentum flux $-(4/3)\mu du/dx$ to have an order of magnitude comparable to that of any other term in equation (15.28), for example, the momentum flux advected through the layer, ρu^2 (which diminishes as the gas suffers rapid deceleration). This requirement allows us to estimate the shock thickness, Δx, over which a jump in fluid velocity Δu is made. (In Problem Set 3, you are asked to perform a detailed integration to make a more quantitative analysis.) If $(4/3)\mu du/dx \sim \rho\nu\Delta u/\Delta x$ is to be comparable to ρu^2, we require

$$\Delta x \sim \nu\Delta u/u^2.$$

For a strong shock Δu is comparable to u itself, whereas u jumps (as we shall see) from a supersonic value to a subsonic one; so an estimate $u \sim v_T$ represents a good compromise. On the other hand, for a neutral gas, the kinematic viscosity satisfies, $\nu \sim \ell v_T$. Collecting expressions, we estimate that the thickness of the transition layer in a strong shock,

$$\Delta x \sim \ell,$$

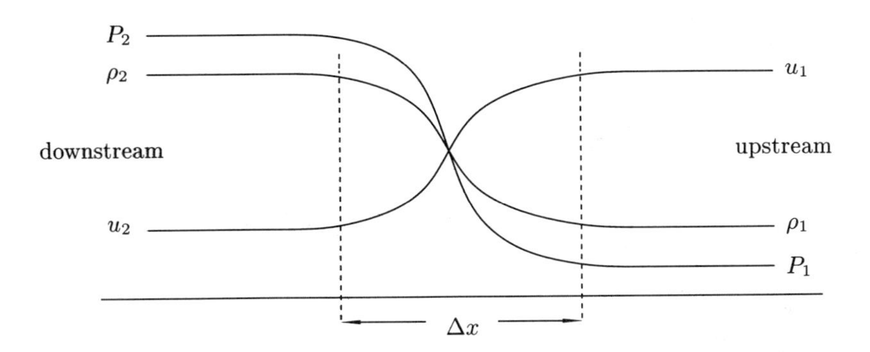

FIGURE 15.8
Across a viscous shock, the pressure and density increase and the velocity decreases as the gas flows from the upstream state to the downstream state. The transition is made in a characteristic distance Δx that equals a few mean free paths ℓ for the elastic scattering of the gas particles.

occupies a distance comparable to only a single mean free path ℓ for elastic collisions!

We can get the same conclusion from equation (15.29). A shock wave takes preshock gas at a low temperature and transforms it to postshock gas at a high temperature. In the transition layer, the heat flux by thermal conduction $\mathcal{K}dT/dx$ must become comparable to the enthalpy flux ρhu, where

$$h \equiv \mathcal{E} + \frac{P}{\rho}, \tag{15.30}$$

is the specific enthalpy, as defined in thermodynamics. (This quantity equals $\int dP/\rho$ for a barotropic fluid, which we have also denoted by the symbol h, only if we restrict ourselves to an isentropic fluid where $ds = 0$.) In order for $\mathcal{K}dT/dx \sim \mathcal{K}\Delta T/\Delta x$ to be comparable to $\rho hu \sim nkTu$, we require

$$\Delta x \sim \mathcal{K}\Delta T/nkTu.$$

For a strong shock, $\Delta T \sim T$, while $u \sim v_T$. On the other hand, for a neutral gas, the heat conduction coefficient satisfies $\mathcal{K} \sim c_v\mu \sim knv_T\ell$, so that we again get

$$\Delta x \sim \ell.$$

Of course, when gases begin to acquire structure on the scale of a few mean free paths, we must question whether a fluid treatment remains viable, or whether we need a kinetic analysis based on a solution of the

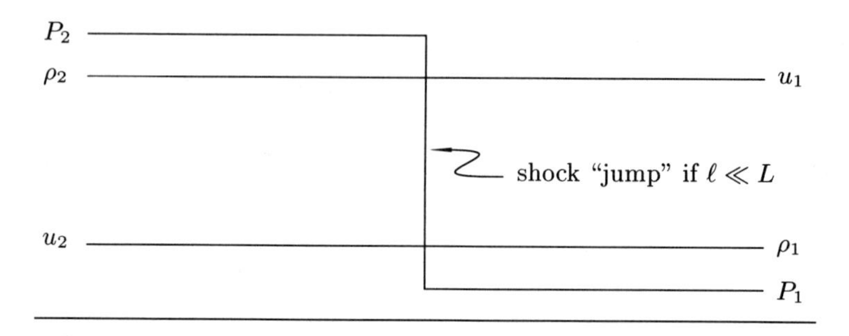

FIGURE 15.9
On macroscopic scales, shock transitions may be approximated as single discontinuous jumps.

Boltzmann transport equation. From a kinetic viewpoint, a strong shock wave consists of two interpenetrating Maxwellians (gas in LTE moving at some high upstream mean speed slamming into gas in LTE moving at some low downstream speed). The gas in the interaction layer is unlikely to have a good representation in terms of an expansion about LTE (the basis of the Chapman-Enskog procedure by which we obtained the Navier-Stokes equations from the moment equations of the Boltzmann equation—see Chapter 3). Nevertheless, detailed calculation using the Boltzmann equation *does* show that, apart from details on the upstream side, the viscous shock structure given by a fluid analysis yields a reasonable representation of the actual state of affairs (Problem Set 3). In particular, the transition layer thickness in a strong shock really does occupy only a few mean free paths for elastic scattering. (That's all it takes for elastic collisions to transform one state of local thermodynamic equilibrium to another.) If we ignore structure on the scale of ℓ, we may approximate a shock transition as a discontinuous jump (Figure 15.9).

RANKINE-HUGONIOT JUMP CONDITIONS

On a scale much larger than a mean free path ℓ, we may ignore the diffusive terms proportional to μ and $\mathcal{K}$ in equations (15.28) and (15.29). This allows us to calculate the "constants" on the right-hand sides of equations (15.27)–(15.29) as equal to the sum of the rest of the terms evaluated *either* far upstream (region 1) or far downstream (region 2) from the transition layer. In other words, the net transition satisfies the *jump conditions*:

$$\rho_2 u_2 = \rho_1 u_1, \tag{15.31}$$

$$\rho_2 u_2^2 + P_2 = \rho_1 u_1^2 + P_1, \tag{15.32}$$

$$\frac{1}{2}u_2^2 + h_2 = \frac{1}{2}u_1^2 + h_1. \tag{15.33}$$

In equation (15.33), we have used equation (15.31) to eliminate a factor of $\rho_2 u_2 = \rho_1 u_1$. Equations (15.31)–(15.33) give the downstream conditions (2) if we know the upstream ones (1).

On both sides of equation (15.33), h equals the specific enthalpy, which for a perfect gas satisfies the constitutive relations

$$h = \frac{\gamma}{\gamma - 1}\frac{P}{\rho} = \frac{\gamma}{\gamma - 1}\frac{kT}{m}. \tag{15.34}$$

Far upstream (on a scale of ℓ) and far downstream from the shock front, we may use thermodynamic relations of the type given by equation (15.34) because the gas exists in LTE (at least insofar as its translational degrees of freedom are concerned; see Chapter 9 of Volume I).

There are many different ways to algebraically solve the (Rankine-Hugoniot) jump conditions (15.31)–(15.33) supplemented by the constitutive relations (15.34). The most useful of these relations may be the expressions that give the ratios $\rho_2/\rho_1 = u_1/u_2$, P_2/P_1, and T_2/T_1 in terms of the upstream Mach number $M_1 \equiv u_1/a_s$, where $a_s^2 \equiv \gamma P/\rho$. The quantity M_1 is known as the Mach number of the shock. (Recall that we are in the frame of the shock front; so the shock front moves at velocity $-u_1$ with respect to the upstream gas that is often regarded by a laboratory observer as being at rest). For a perfect gas, we find

$$\frac{\rho_2}{\rho_1} = \frac{(\gamma+1)M_1^2}{(\gamma+1)+(\gamma-1)(M_1^2-1)} = \frac{u_2}{u_1}, \tag{15.35}$$

$$\frac{P_2}{P_1} = \frac{(\gamma+1)+2\gamma(M_1^2-1)}{(\gamma+1)}, \tag{15.36}$$

$$\frac{T_2}{T_1} = \frac{[(\gamma+1)+2\gamma(M_1^2-1)][(\gamma+1)+(\gamma-1)(M_1^2-1)]}{(\gamma+1)^2 M_1^2}. \tag{15.37}$$

Notice that $P_2 \geq P_1$, $\rho_2 \geq \rho_1$ ($u_2 \leq u_1$), and $T_2 \geq T_1$ if $M_1 \geq 1$ (supersonic upstream) with equality if $M_1 = 1$ (no shock at all). In the limit of a very strong shock, $M_1 \to \infty$, the density jump is bounded by a finite value, $(\gamma+1)/(\gamma-1)$, which equals 4 if $\gamma = 5/3$. In the same limit, the pressure and temperature jumps have no bound. In any case the deceleration of a gas from supersonic speeds to subsonic values in a shock produces compression and heating, outcomes which have observational consequences in astronomical objects if the process results in enhanced radiation. (It is not directly obvious from the above relations that $M_2 < 1$ when $M_1 > 1$, but Problem Set 3 shows it to be true.)

The equations (15.35)–(15.37) formally allow "expansive shocks" in which $M_1 < 1$ and $M_2 > 1$, because the jump conditions themselves do not prevent a reversal of roles for regions 1 and 2. Rarefaction shocks, in which subsonically moving hot gas suddenly expands and accelerates to become supersonically moving cool gas, do not arise in nature. It is possible to show that the entropy jump, $s_2 - s_1$, has positive values for compressive shocks, and negative ones for rarefaction shocks. The second law of thermodynamics forbids internal processes by which a gas in one uniform state can spontaneously lower its specific entropy to become gas in another uniform state. In other words, viscosity can transform the energy of bulk motion into heat, but it can not do the reverse. From another point of view—that which began this chapter—nonlinear compression waves steepen into shocks; expansion waves remain smooth as they propagate. For these reasons, the word "shocks" always refers to "compressive shocks," and we never need consider adding either modifier, "expansive" or "compressive."

Finally, notice that viscosity and heat conduction are the microscopic mechanisms which counteract the nonlinear tendency for compressive waves to get ever steeper. The balance between spreading by diffusion and steepening by nonlinearity gives the wave a semi-permanent profile (that of a shock wave). Yet, despite this fundamental role for viscosity and conductivity, the end values of quantities like s, ρ, P, T, etc., at 1 and 2 do not depend on any of the detailed properties of the transport coefficients μ and $\mathcal{K}$. From a thermodynamic point of view, quantities like s, etc. represent *state variables*; they cannot depend on the path of the transformation taken to reach state 2 from state 1. What happens on a mechanistic level is that the shock thickness *automatically adjusts* itself to whatever value is necessary so that the level of viscosity and heat conduction available can balance the nonlinear steepening tendency. [Notice the counterintuitive result: The smaller the viscosity, the thinner the shock transition layer (if a shock arises at all).] The end states of the transformation are completely determined by the joint conservation of mass, momentum, and energy.

ARTIFICIAL VISCOSITY

Because the magnitude of the viscous term does not affect the net shock-jump conditions, many numerical schemes implicitly or explicitly incorporate the trick of *artificial viscosity* for halting the ever-growing steepening tendency produced by nonlinear effects, thereby gaining the automatic insertion of shock waves wherever they are needed (in time-dependent calculations). The central idea involves the introduction of a numerical viscous term large enough to spread out shock transitions over a few zones of the computational grid, making the inclusion of shock waves resolvable by a finite-difference calculation. As long as the grid size remains small compared to interesting macroscopic lengths (essential for reliable results), no harm results from the smearing of the shock wave over a few zones instead of a few mean free paths. However, do not blindly adopt artificial viscosity schemes for astrophysical applications that involve radiative transfer. If you spread out shocks over more than a fraction of a photon mean free path, you may obtain spurious flows of radiant energy (wrong direction as well as wrong magnitude). Fortunately, photon mean free paths usually greatly exceed particle mean free paths (review Chapter 1), so one can often achieve adequate numerical resolution for the former even if one cannot for the latter.

16

Oblique Shocks and Optically Thin Radiative Layers

References: Landau and Lifshitz, *Fluid Mechanics*, pp. 333–337.
G. B. Field, J. Rather, P. Aannestad, and S. Orzsag, 1968, *Ap. J.*, **151**, 953.

At the end of Chapter 15 we derived the so-called Rankine-Hugoniot jump conditions for a *normal shock* in which the upstream fluid encounters the shock front in a direction perpendicular to the face of the front. In this chapter we wish to consider oblique shocks, for which the incident angle does not equal 90° (Figure 16.1).

Shock waves of all types, oblique or normal, heat the affected gas. In many astrophysical situations, the rise in temperature produces a significant increase in radiation. Shock emission, particularly in spectral lines, can provide an invaluable diagnostic for the physical state of the system; this motivates us to examine the gas dynamics of radiative shocks in the second part of this chapter. We specialize to the case of radiation under optically thin conditions (every photon generated escapes from the system), a good approximation when the shock occurs in the rarefied regions of interstellar space or external galaxies. We defer to Chapter 19 the examination of the problem of radiative losses from shocks in which preshock and/or postshock gas may be optically thick to the emitted photons in the continuum.

OBLIQUE SHOCKS

To derive the jump conditions for oblique shocks we start with the equations of fluid mechanics in their conservation form (see Chapter 5), which take the generic form

$$\frac{\partial q}{\partial t} + \nabla \cdot \mathbf{f} = Q, \tag{16.1}$$

where q is the density of some quantity, $\mathbf{f}$ is the associated flux, and Q equals the difference between volumetric sources and sinks. We wish to

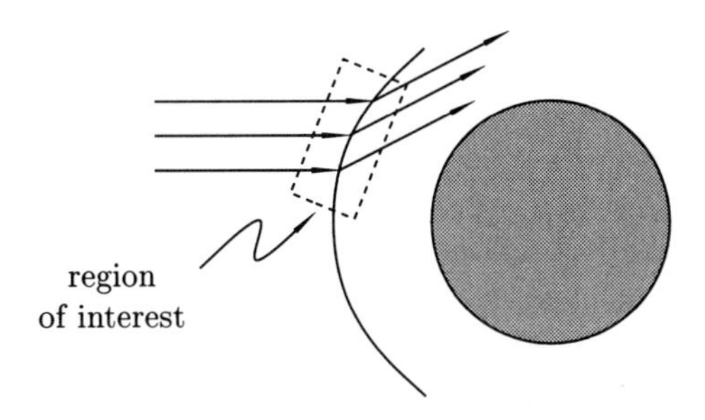

FIGURE 16.1
The deflection of streamlines not at normal incidence to a solid body occurs by oblique shocks.

apply equation (16.1) in a frame which moves with the instantaneous shock speed. Denoting measurements in this frame with a prime, we write

$$\nabla \cdot \mathbf{f}' = Q' - \frac{\partial q'}{\partial t}, \tag{16.2}$$

which we wish to integrate across the surface of the shock front.

To perform the integration, we find it convenient first to construct a primed coordinate system in such a way that n describes a Cartesian coordinate that increases in the direction normal to local shock front. Decompose the incident velocity vector $\mathbf{u}_1'$ into components $u_\perp$ and $u_\parallel$ perpendicular and parallel to the shock front, and let s represent another Cartesian coordinate that increases in the direction of $u_\parallel$. The symmetry of the problem guarantees then that the local flow occurs in the (n, s)-plane, as depicted in Figure 16.2.

Discarding primes for a cleaner notation, we may now write equation (16.2) as

$$\frac{\partial f_\perp}{\partial n} = Q - \frac{\partial q}{\partial t} - \frac{\partial f_\parallel}{\partial s}. \tag{16.3}$$

Integrate over n from n_1 to n_2, just ahead and behind the shock front, holding s fixed:

$$[f_\perp]_1^2 = \int_{n_1}^{n_2} \left(Q - \frac{\partial q}{\partial t} - \frac{\partial f_\parallel}{\partial s} \right) dn. \tag{16.4}$$

Since $n_2 - n_1 \sim \ell$, a mean free path for elastic scattering, and since the term $(Q - \partial q/\partial t - \partial f_\parallel/\partial s)$ is bounded inside the shock layer, we have, in the limit that $\ell \to 0$,

$$[f_\perp]_1^2 = 0, \tag{16.5}$$

which gives the desired generic jump condition.

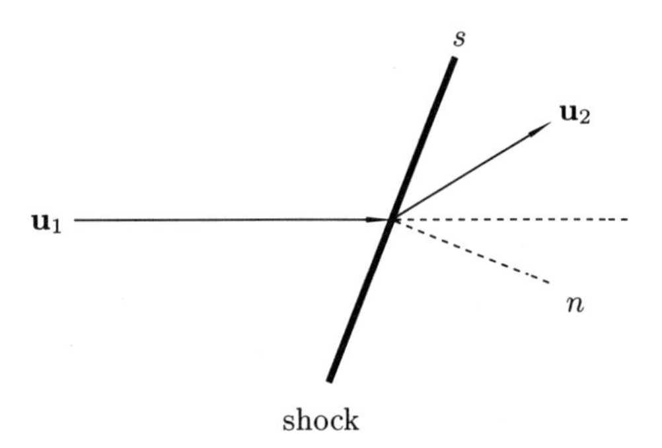

FIGURE 16.2
The geometry of an oblique shock in the plane of the incident velocity vector and the unit normal to the shock front.

JUMP CONDITIONS

The Rankine-Hugoniot jump conditions for oblique shocks can now be obtained by writing down the inviscid fluid equations in the shock frame:

$$\frac{\partial \rho}{\partial t} + \frac{\partial}{\partial n}(\rho u_\perp) + \frac{\partial}{\partial s}(\rho u_\parallel) = 0, \tag{16.6}$$

$$\frac{\partial}{\partial t}(\rho u_\perp) + \frac{\partial}{\partial n}(\rho u_\perp u_\perp + P) + \frac{\partial}{\partial s}(\rho u_\perp u_\parallel) = \rho g_\perp, \tag{16.7}$$

$$\frac{\partial}{\partial t}(\rho u_\parallel) + \frac{\partial}{\partial n}(\rho u_\parallel u_\perp) + \frac{\partial}{\partial s}(\rho u_\parallel u_\parallel + P) = \rho g_\parallel, \tag{16.8}$$

$$\frac{\partial}{\partial t}\left[\rho\left(\frac{1}{2}u_\perp^2 + \frac{1}{2}u_\parallel^2 + \mathcal{E}\right)\right] + \frac{\partial}{\partial n}\left[\rho\left(\frac{1}{2}u_\perp^2 + \frac{1}{2}u_\parallel^2 + h\right)u_\perp\right] \tag{16.9}$$

$$+ \frac{\partial}{\partial s}\left[\rho\left(\frac{1}{2}u_\perp^2 + \frac{1}{2}u_\parallel^2 + h\right)u_\parallel\right] = -\rho\mathcal{L}. \tag{16.10}$$

By inspection, equation (16.5) yields for the above set:

$$[\rho u_\perp]_1^2 = 0, \tag{16.11}$$

$$\left[\rho u_\perp^2 + P\right]_1^2 = 0, \tag{16.12}$$

$$\left[\rho u_\perp u_\parallel\right]_1^2 \quad \Rightarrow \quad \left[u_\parallel\right]_1^2 = 0, \tag{16.13}$$

$$\left[\rho u_\perp\left(\frac{1}{2}u_\perp^2 + \frac{1}{2}u_\parallel^2 + h\right)\right]_1^2 = 0 \quad \Rightarrow \quad \left[\frac{1}{2}u_\perp^2 + h\right]_1^2 = 0. \tag{16.14}$$

In equation (16.13) we have used the constancy of $\rho u_\perp$ to deduce that $u_\parallel$ is conserved across the shock front. In equation (16.14) we have likewise used the constancy of $\rho u_\perp$ and $u_\parallel$ to deduce the conservation of $u_\perp^2/2 + h$, where h equals the specific enthalpy, $\mathcal{E} + P/\rho$, of the gas.

The constancy of $u_\parallel$ implies that translation at a velocity $u_\parallel$ parallel to the surface of the shock front would allow us to transform an oblique shock geometry to a normal shock geometry. For curved shocks, however, like those that arise in bow shocks around blunt bodies, we would need a locally different Galilean transformation for different portions of the shock surface, which would prove impractical in actual calculations. Nevertheless, Galilean invariance of the dynamics explains why the jump conditions for the perpendicular components of the flow remain the same as their normal shock counterparts. [Compare equations (16.11), (16.12), and (16.14) with equations (15.31), (15.32), and (15.33).]

SHOCK POLAR

If the shock geometry were specified in advance, then equations (16.11)–(16.14) would uniquely determine the downstream state given the upstream state. In many situations, however, the location and shape of the shock surface are not given *a priori*, but must be deduced as part of an overall solution. As the simplest example of such a situation, consider for concreteness the problem of supersonic flow past a wedge of known opening angle. We know that the shock which develops in front of the wedge must redirect the postshock flow to occur parallel to the faces of the wedge. Above the upper face, let the downstream flow make an angle ψ with respect to the upstream velocity vector $\mathbf{u}_1$ (Figure 16.3). What angle ϕ does the shock front need to make with respect to $\mathbf{u}_1$ to allow the required turn?

With the introduced variables, equation (16.13) now implies

$$u_1 \cos\phi = u_2 \cos(\phi - \psi). \tag{16.15}$$

According to equation (15.35), the other three relations, (16.11), (16.12), and (16.14), condense to the statement

$$\frac{u_{\perp 1}}{u_{\perp 2}} = \frac{(\gamma+1)M_{\perp 1}^2}{(\gamma+1) + (\gamma-1)(M_{\perp 1}^2 - 1)} = \frac{(\gamma+1)M_{\perp 1}^2}{2 + (\gamma-1)M_{\perp 1}^2}.$$

Taking the reciprocal of the above equation, we get

$$\frac{u_2 \sin(\phi - \psi)}{u_1 \sin\phi} = \frac{2a_1^2}{(\gamma+1)u_1^2 \sin^2\phi} + \frac{\gamma-1}{\gamma+1}. \tag{16.16}$$

We wish to eliminate ϕ from equations (16.15) and (16.16). To do this, we perform the indicated trigonometric expansions to obtain:

$$u_1 \cos\phi = u_2 \cos\phi \cos\psi + u_2 \sin\phi \sin\psi, \tag{16.17}$$

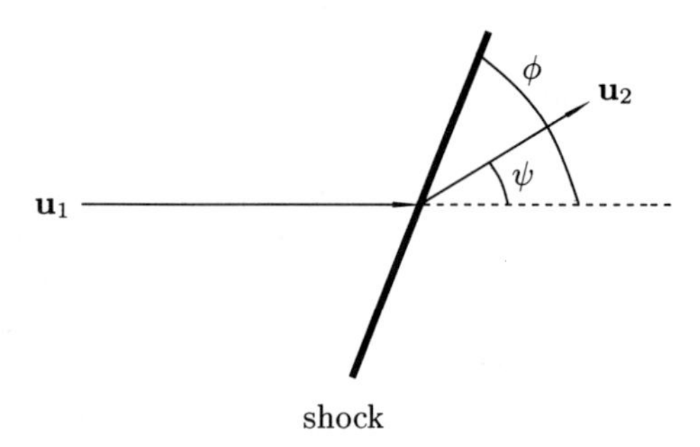

FIGURE 16.3
The geometry for an oblique shock in terms of the angle ψ through which the velocity vector turns after passage through the front, and the angle ϕ (to be determined) that the shock front makes with respect to the incident velocity in the flow plane.

$$\frac{u_2}{u_1}\cos\psi - \frac{u_2}{u_1}\cot\phi\sin\psi = \frac{2a_1^2}{(\gamma+1)u_1^2\sin^2\phi} + \frac{\gamma-1}{\gamma+1}. \tag{16.18}$$

Equation (16.17) implies

$$\cot\phi = \frac{u_2\sin\psi}{u_1 - u_2\cos\psi}; \tag{16.19}$$

i.e.,

$$\sin^2\phi = \frac{(u_1-u_2\cos\psi)^2}{u_1^2 - 2u_1u_2\cos\psi + u_2^2} \qquad \Rightarrow \qquad \frac{1}{\sin^2\phi} = 1 + \frac{u_2^2\sin^2\psi}{(u_1-u_2\cos\psi)^2}. \tag{16.20}$$

Substitution of equations (16.19) and (16.20) into equation (16.18) now gives the desired relation

$$u_2^2\sin^2\psi = (u_1-u_2\cos\psi)^2\left\{\frac{u_1u_2\cos\psi - c_*^2}{c_*^2 + [2/(\gamma+1)]u_1^2 - u_1u_2\cos\psi}\right\}, \tag{16.21}$$

where we have written the combination $(\gamma-1)u_1^2 + 2a_1^2$ as $(\gamma+1)c_*^2$, with the critical velocity c_* being defined through

$$\frac{1}{2}u_1^2 + h_1 = \left(\frac{\gamma+1}{\gamma-1}\right)\frac{c_*^2}{2} \qquad \text{with} \qquad h_1 = \frac{\gamma P_1}{(\gamma-1)\rho_1} = \frac{a_1^2}{\gamma-1} \tag{16.22}$$

for a perfect gas.

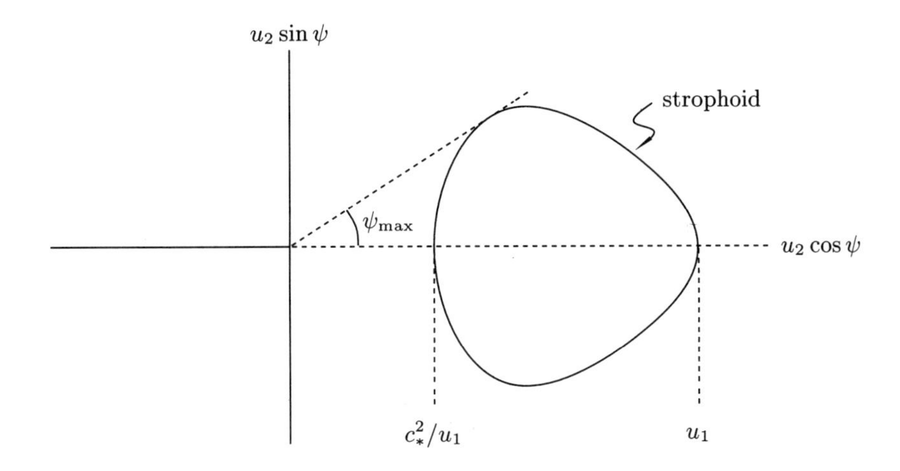

FIGURE 16.4
The shock polar (see text for description).

Equation (16.21), the so-called *shock polar*, gives $u_2 \sin\psi$ in terms of $u_2 \cos\psi$ once u_1 and c_* are given. Figure 16.4 shows a plot of this relation.

The angle $\psi_{\max}$ represents the maximum angle through which the flow can be deflected by an oblique shock. It is a monotonically increasing function of $M_1 = u_1/a_1$, and it has the limiting values

$$\psi_{\max} \to \frac{8\sqrt{2}}{3[3(\gamma-1)]^{1/2}}(M_1-1)^{3/2} \to 0 \qquad \text{as} \qquad M_1 \to 1, \tag{16.23}$$

$$\psi_{\max} \to \arcsin(1/\gamma) \qquad \text{as} \qquad M_1 \to \infty. \tag{16.24}$$

Equation (16.23) implies that a body with a sharp nose cannot have an attached shock if it flies too slowly; a detached bow shock must form to turn the flow through the requisite wedge angle (Figure 16.5).

Likewise, equation (16.24) implies that a body with a nose more blunt than $\arcsin(1/\gamma)$ must also have a detached bow shock, no matter how fast it flies (Figure 16.6).

In both cases, the deceleration of the flow to subsonic speeds at the head of the bow shock generates the high pressures needed to push the gas through angles large enough to allow it to slip past the flying body.

Notice that the shock polar implies that if we wish to turn the flow

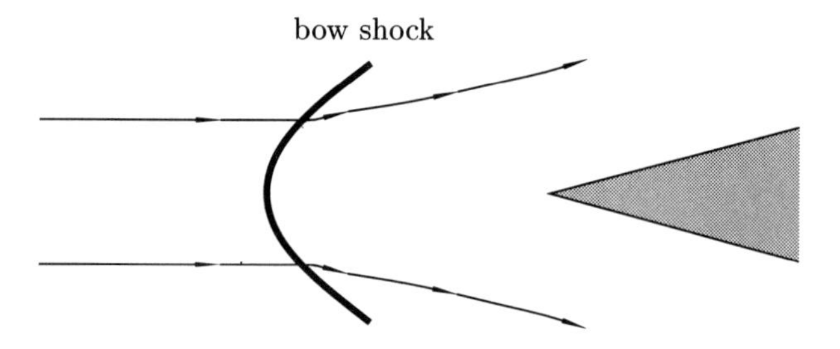

FIGURE 16.5
Schematic illustration that a body with a sharp nose which flies supersonically but too slowly will have a detached bow shock.

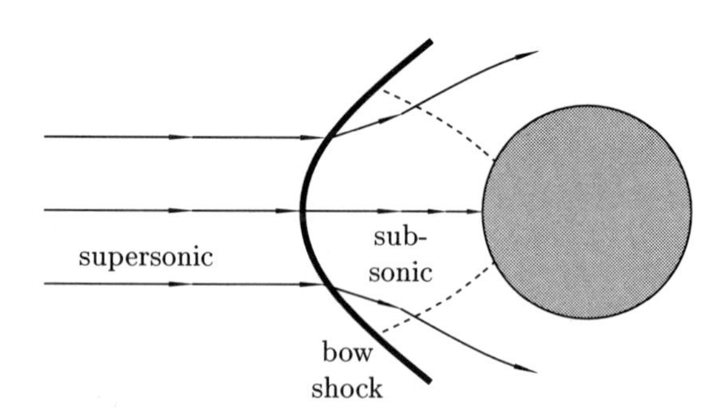

FIGURE 16.6
Schematic illustration that a body with a blunt nose flying supersonically at any speed through a fluid will always have a detached bow shock.

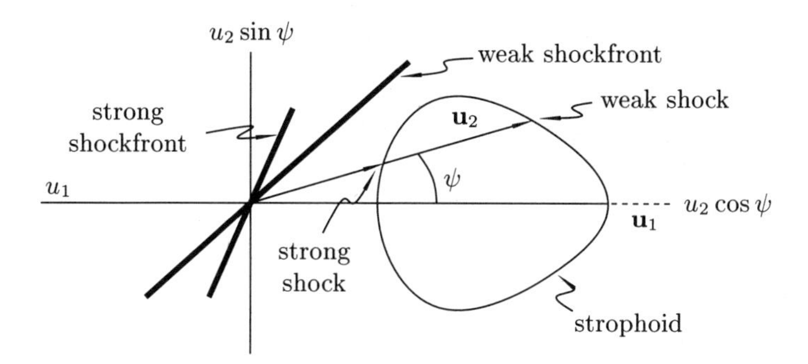

FIGURE 16.7
Two ways exist to turn supersonic flow through a given angle ψ: a strong, nearly perpendicular shock that leads to a smaller downstream velocity; and a weak, more oblique shock that leads to a larger downstream velocity.

through an angle smaller than ψ_{max}, we have two ways to do it: by a strong shock, or by a weak one (Figure 16.7).

For an infinite wedge with known angles, then, there exist *two* possible shocks that will allow supersonic incident gas to turn parallel to a given face. Which one would nature choose? Since infinite wedges are hard to come by, we cannot call on wind-tunnel experiments for an answer to this problem. Construct instead the following thought experiment. Imagine a wedge which begins as a completely flattened sheet oriented edge-on to the flow. Such a configuration offers no obstacle to the flow, so the "shock" that arises has infinitesimal strength, i.e., we have the transition: $u_2 \cos\psi = u_1$ and $u_2 \sin\psi = 0$. As we open up the wedge gradually, the velocity vector $(u_2 \cos\psi, u_2 \sin\psi)$ moves from the back of the "strophoid" along the weak shock branch until the turn angle reaches ψ_{max}. Further opening of the wedge angle produces a detached shock; so the strong shock branch becomes irrelevant for this idealized problem.

For bodies of finite size, conditions downstream from the flow affect which choice ultimately gets made. Moreover, although the quantity $u^2/2 + h$ remains conserved across viscous shocks of all strengths, Bernoulli's constant,

$$\mathcal{B} \equiv \frac{1}{2}u^2 + \int \frac{dP}{\rho} + \mathcal{V},$$

suffers a jump because $\int dP/\rho$ does not equal the thermodynamic enthalpy h for nonadiabatic processes, i.e., when s is not conserved. Thus, $\mathcal{B}$ suffers jumps across shocks, and the jumps will be of different magnitudes across shocks of different strengths, as happens for a curved front. As previewed in Chapter 6, such curved shocks can generate vorticity in gas whose preshock flow was irrotational.

Let me here comment on the term "adiabatic." The word as commonly applied by astronomers to nonradiative shock waves is a pernicious misnomer and should be avoided if possible. Adiabatic does not mean "energy-conserving," but "entropy-conserving." Entropy represents precisely the quantity that is not preserved in a strong shock. A strong shock wave sends a violent *irreversible* perturbation through the system; in no sense can such a process be regarded as "adiabatic."

RADIATIVE SHOCKS

To discuss the physics (rather than the geometry) of radiative shocks, let us revert to the case of a normal shock (with the understanding that a simple Galilean transformation allows us to consider oblique shocks, at least for unmagnetized media). We further assume that the radiation occurs under optically thin conditions, so that we may use a net cooling function $\mathcal{L}(\rho, T)$ that depends only on the local density ρ and temperature T. Finally, we assume that the cooling length is long compared to a particle mean free path ℓ, but is short compared to the scale over which external forces have an effect on the flow. In this case, we anticipate the problem to have the following features (Figure 16.8).

Cold rarefied gas moving rapidly and initially in thermal equilibrium, $\mathcal{L} = 0$, enters from the left. For one reason or another, this gas shocks viscously at position 1. The immediate downstream conditions at 2 are given by the Rankine-Hugoniot jump conditions, and they yield hot gas moving subsonically with respect to the viscous shock front. The density at 2 has, at most, been compressed by a factor of $(\gamma + 1)/(\gamma - 1)$, but the temperature has increased by a large factor typically. These circumstances throw the gas badly out of thermal equilibrium, so that $\mathcal{L}$ exceeds 0 by a substantial margin. The subsequent radiative cooling of the gas causes it to compress (because of the ambient pressure), and the increased density leads, for a given mass flux ρu, to smaller values of the flow velocity u. By point 3, the gas has cooled enough as to come back into thermal equilibrium, $\mathcal{L} = 0$, at a much higher density ρ_3.

For steady flow, the governing equations read

$$\rho u = \text{constant}, \tag{16.25}$$

$$(\rho u)u + P = \text{constant}, \tag{16.26}$$

$$\rho u \frac{d\mathcal{E}}{dx} = -P\frac{du}{dx} - \rho\mathcal{L}, \tag{16.27}$$

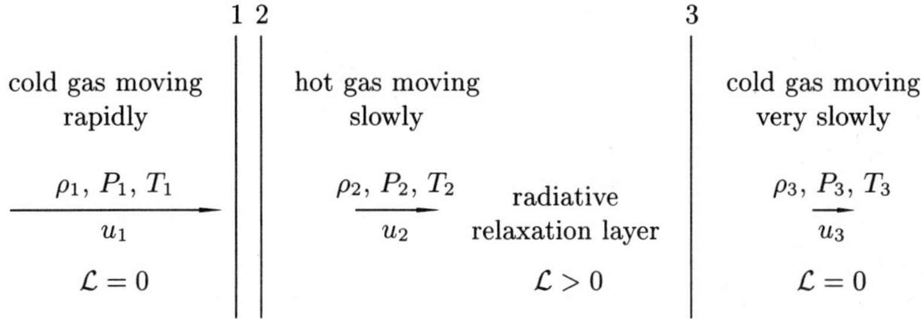

1 2 3

cold gas moving rapidly

ρ_1, P_1, T_1

u_1

$\mathcal{L} = 0$

hot gas moving slowly

ρ_2, P_2, T_2

u_2

radiative relaxation layer

$\mathcal{L} > 0$

cold gas moving very slowly

ρ_3, P_3, T_3

u_3

$\mathcal{L} = 0$

FIGURE 16.8
Schematic regions of interest in a radiative shock. Ahead of a normal viscous shock front lies cold, rapidly moving gas (in the frame of rest of the shock front). A sudden deceleration, compression, and heating of the gas occurs in the viscous transition layer, 1 to 2. Downstream from the viscous layer, the shocked gas is thrown badly out of thermal equilibrium and radiates profusely. The radiative cooling, assumed here to occur under optically thin conditions, lowers the temperature until the gas eventually reattains radiative balance at point 3.

where equation (16.27) is the internal-energy equation for this problem. For a perfect gas (we ignore the effects of ionizations or chemical reactions),

$$\mathcal{E} = \frac{1}{(\gamma - 1)} \frac{P}{\rho}. \tag{16.28}$$

The substitution of equation (16.28) into equation (16.27), with the constant ρu moved inside the derivative on the left-hand side, yields

$$\frac{1}{(\gamma-1)} \frac{d}{dx}(Pu) = -P\frac{du}{dx} - \rho\mathcal{L} \quad \Rightarrow \quad \frac{u}{(\gamma-1)} \frac{dP}{dx} + \frac{\gamma}{(\gamma-1)} P \frac{du}{dx} = -\rho\mathcal{L}.$$

Equations (16.25) and (16.26) allow us to write $dP/dx = -\rho u du/dx$; consequently, we obtain

$$\frac{1}{(\gamma-1)} (a_s^2 - u^2) \frac{du}{dx} = -\mathcal{L}(\rho, T), \tag{16.29}$$

where we have defined the square of the adiabatic speed of sound as $a_s^2 \equiv \gamma P/\rho$.

Equation (16.29) gives an ODE for u to be solved in conjunction with equations (16.25) and (16.26) when P and T are related through the equation of state,

$$P = \frac{\rho}{m} kT. \tag{16.30}$$

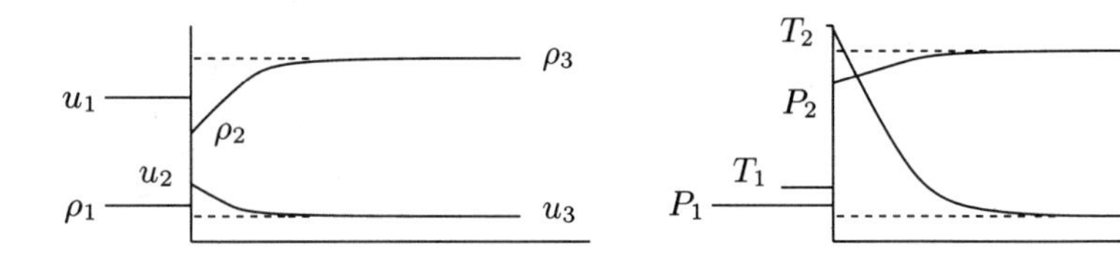

FIGURE 16.9
Schematic depiction of the variation of the velocity and density (first panel), and of the pressure and temperature (second panel), in a radiative shock.

We use as initial conditions the post-viscous-shock values of ρ_2, P_2, T_2, u_2. Since $\mathcal{L} > 0$ and $u < a_{s2}$ (see Problem Set 3), equation (16.29) implies that u will decrease with increasing x inside the radiative relaxation layer, as advertised in our preliminary discussion. The decrease in u, with ρu remaining constant, means, through equation (16.26), that the pressure will rise slightly in the cooling process. The rise in P is small because most of the momentum flux after the viscous shock already resides in the contribution from random motions. The slight increase in P, coupled with larger increases in ρ, leads via the equation of state (16.30) to lower values of T. This generally decreases the net cooling rate, and eventually $\mathcal{L}$ asymptotically approaches 0 (Figure 16.9).

In some situations, the compression can throw the gas into a thermally unstable region, with the consequence that the final equilibrium state has much cooler temperatures T_3 than the initial state T_1. Such shock-induced changes of thermal phase have been studied in a variety of astronomical contexts (e.g., in "galactic shocks"), but they probably represent the exceptional case rather than the rule. In many circumstances, the gas returns to a temperature T_3 that does not differ very much from the initial equilibrium value T_1, although the final equilibrium density ρ_3 can greatly exceed the initial value ρ_1. In any case, if we are not interested in the structure of the radiative relaxation layer, we may effect a total jump, from 1 to 3, by invoking the jump conditions

$$\rho_3 u_3 = \rho_1 u_1, \tag{16.31}$$

$$\rho_3 u_3^2 + P_3 = \rho_1 u_1^2 + P_1, \tag{16.32}$$

with

$$\mathcal{L}(\rho_3, T_3) = 0, \qquad \text{and} \qquad P_3 = \rho_3 k T_3 / m. \tag{16.33}$$

Equations (16.31)–(16.33) completely specify the downstream variables at 3 when the upstream ones at 1 are known.

Equations (16.33) can often be approximated by the isothermal jump condition

$$T_3 = T_1. \tag{16.34}$$

With a constant isothermal sound speed,

$$a_T = (kT/m)^{1/2}, \tag{16.35}$$

we may easily solve the remaining relations to obtain the so-called "isothermal shock" jump conditions,

$$u_3 u_1 = a_T^2 \qquad \text{with} \qquad \frac{\rho_3}{\rho_1} = \left(\frac{u_1}{a_T}\right)^2. \tag{16.36}$$

The second relation of equation (16.36) states that the density contrast in an isothermal shock equals the square of the upstream isothermal Mach number. Unlike the limiting value $(\gamma+1)/(\gamma-1)$ for a nonradiative shock, the density contrast for an isothermal shock can reach arbitrarily high values. Very strong density compressions constitute the chief distinguishing feature of radiative shocks compared to their nonradiative counterparts. This phenomenon, for example, gives rise to the so-called dense-shell formation stage of the late-time evolution of supernova remnants (see, e.g., the review of L. Woltjer, 1972, *Ann. Rev. Astr. Ap.*, **10**, 129). In realistic circumstances, however, compression to very high densities may be limited by magnetic field effects, a topic that we shall discuss when we come to magnetohydrodynamics.

17

Blast Waves and Supernova Remnants

References: Landau and Lifshitz, *Fluid Mechanics*, pp. 392–396; J. P. Ostriker and C. F. McKee, 1988, *Rev. Mod. Phys.*, **60**, no. 1.

As part of the atomic-weapons research in the West and East, Taylor and Sedov independently developed the theory of strong point-like explosions in a uniform quiescent medium. Shklovskii applied the theory to explain the early stages of the evolution of a supernova remnant; later, others considered realistic estimates of the effects of radiation in limiting the late-time validity of the blast-wave solution. In this chapter we examine the Taylor-Sedov solution, not only with an eye to astrophysical applications, but also for its interest as an example of the mathematical method of *similarity analysis.*

Similarity arguments go deeper than dimensional analysis. The latter states that if one nondimensionalizes any set of equations and boundary conditions properly, one will obtain a minimum number of dimensionless parameters on which the solution of the problem depends. Different physical problems with the same values of the dimensionless parameters will then have solutions that look the same except for a change in scale.

A self-similar flow, on the other hand, represents the case when the *same* flow at any location and time looks the same as it did at a different location and an earlier time. (For steady-state problems, the self-similarity can occur on different scales in physical space.) In a given situation, the initial and/or boundary conditions may not be compatible with an exactly self-similar flow. Thus a similarity solution may become applicable only asymptotically in time or in space as one gets away from the influence of said initial and boundary conditions.

SIMILARITY BASIS FOR BLAST-WAVE PROBLEM

Consider a point release of an enormous amount of energy E into a static medium of homogeneous density ρ_1 (the value that enters the shock on the side of the undisturbed medium; see Figure 17.1).

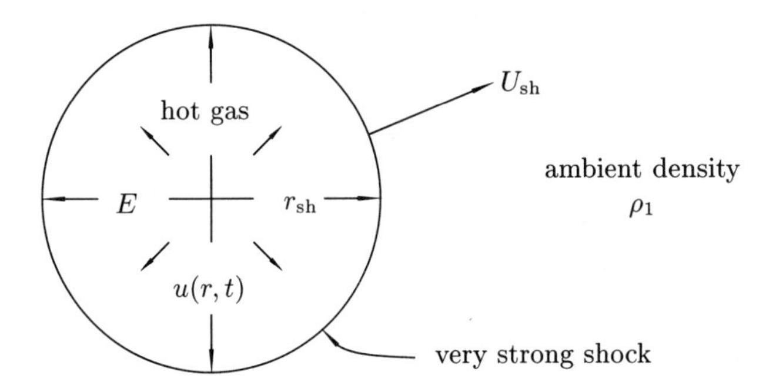

FIGURE 17.1
The point release of a large amount of energy E creates a spherical blast wave that propagates at speed $U_{sh}(t)$ through an ambient medium of uniform density ρ_1.

This picture approximates the state of a supernova remnant when the mass swept up by the outwardly propagating shock wave much exceeds the mass of the initial ejecta. In addition, we shall be concerned with that stage in which the total radiated energy has not yet begun to approach E, and the ram pressure $\rho_1 U_{sh}^2$ associated with the matter that enters the shock wave much exceeds the thermal pressure P_1 of the undisturbed medium. Under these circumstances, the strong explosion produces a *blast wave.* When the relevant assumptions hold, astronomers say that the flow exists in the *energy-conserving phase* of the evolution of supernova remnants.

The problem so posed must possess a well-behaved solution that evolves in spacetime (r, t). Any such solution must ultimately be expressible in terms of dimensionless variables as arguments for various mathematical functions. We cannot take, for example, the sine of $r = 2\,\text{pc}$; there must occur some length scale into which we can divide r. However, in the approximation where we consider the undisturbed pressure P_1 to be indistinguishable from zero, we have only two dimensional parameters left in the problem, E and ρ_1. With an energy E and a density ρ_1 alone, we cannot form a quantity with the dimension of a length. To get a length we must use the only other available quantity, the time t since the initial explosion. In other words, wherever r enters as part of the dimensionless argument of a function, so must some product of powers of the *parameters* E and ρ_1, as well as a power of t. The relevant dimensionless variable therefore has the form

$$\xi \equiv r t^{\ell} \rho_1^m E^n,$$

where ℓ, m, and n are pure numbers that remain to be evaluated. We want the left-hand side to be dimensionless. The right-hand side has the dimensions

$$LT^{\ell}M^{m}L^{-3m}M^{n}L^{2n}T^{-2n},$$

where L, T, and M are, respectively, a length, a time, and a mass. For this expression to be dimensionless, we require

$$1-3m+2n=0, \qquad \ell-2n=0, \qquad m+n=0,$$

which represent three equations in three unknowns that can be solved to obtain

$$\ell=-2/5, \qquad m=1/5, \qquad n=-1/5.$$

Hence, the desired similarity variable reads

$$\xi = r(\rho_1/Et^2)^{1/5}. \tag{17.1}$$

When properly nondimensionalized, the governing equations will depend on r and t only through the combination ξ (see below). In particular, we expect the position of the shock wave to correspond to some fixed value of ξ, say ξ_0. Thus,

$$r_{\rm sh}(t) = \xi_0(Et^2/\rho_1)^{1/5}. \tag{17.2}$$

We also expect, in the absence of good arguments to the contrary, that ξ_0 will turn out to be a number of order unity. (For $\gamma = 5/3$, $\xi_0 = 1.17$ according to the detailed solution to be presented in the following sections.) Without doing any work much harder than dimensional analysis, then, we may recover the velocity of the shock wave as

$$U_{\rm sh} = \frac{dr_{\rm sh}}{dt} = \frac{2}{5}\frac{r_{\rm sh}}{t} = \frac{2}{5}\xi_0(E/\rho_1 t^3)^{1/5}; \tag{17.3}$$

i.e., $U_{\rm sh} \propto t^{-3/5} \propto r_{\rm sh}^{-3/2}$, and the shock will weaken as it propagates outward in time. (A useful mnemonic device to remember this result is $\rho_1 r_{\rm sh}^3 U_{\rm sh}^2 =$ constant; i.e., the blast wave conserves energy.)

SOME TYPICAL NUMBERS

To obtain an order-of-magnitude feeling for application to supernova remnants, we suppose $E = 10^{51}$ erg (e.g., ejection of $1\,M_\odot$ at $10^4\,{\rm km\,s^{-1}}$), $\rho_1 = 2\times10^{-24}\,{\rm g\,cm^{-3}}$, then equations (17.2) and (17.3) lead to Table 17.1. Since the shock wave cannot move much faster than the original ejecta, $10{,}000\,{\rm km\,s^{-1}}$, we should not use the part of the table much before $t =$ 100 yr. The same conclusion follows by noting that the swept-up mass, $\rho_1(4\pi r_{\rm sh}^3/3)$, equals the mass of the original ejecta, $1\,M_\odot$, when $r_{\rm sh} = 2$ pc.

Table 17.1
Typical values for supernova remnants.

t (yr)	1	10	100	1,000	10,000	100,000
$r_{\rm sh}$ (pc)	0.315	0.791	1.99	4.99	12.5	31.5
$U_{\rm sh}$ (km s^{-1})	124,000	31,000	7,820	1,970	494	124

Thus we expect the blast-wave approximation to apply only for t appreciably greater than 100 yr.

For $\gamma = 5/3$, the postshock temperature equals $T_2 = 3m_2U_{\rm sh}^2/16k$, where m_2 is the mean particle mass behind the shock. (We assume that the energy required for the shock to ionize the medium [if the preshock medium consists of neutral gas] is negligible compared to the thermal content otherwise imparted to the postshock gas.) With $m_2 = 1 \times 10^{-24}$ g (completely ionized gas of cosmic abundance), Table 17.1 implies that $T_2 \sim 3 \times 10^6$ K at $t \sim 10^4$ yr. Such a supernova remnant should radiate strongly in X-rays, a prediction verified by direct observations of supernova remnants with sizes $r_{\rm sh} \sim 10$ pc. At a time $t \sim 10^5$ yr, $T_2 \sim 2 \times 10^5$ K, when the cooling function (see Figure 17.2) would have a value

$$\rho\mathcal{L} \sim 10^{-21} - 10^{-22}\,\text{erg}\,\text{cm}^{-3}\,\text{s}^{-1}. \tag{17.4}$$

Thus, the total energy radiated in time $t \sim 10^5$ yr will roughly equal

$$(\rho\mathcal{L})t\left(\frac{4\pi}{3}r_{\rm sh}^3\right) \sim 10^{51} - 10^{52}\,\text{erg},$$

the same order as the original explosion energy E. Consequently, we expect the blast-wave solution to break down because of radiative losses for t of order 10^5 yr or greater.

Finally, we should ask whether we should have expected a shock wave to form in the first place. An ejection velocity of 10^4 km s^{-1} corresponds to the supernova throwing out a hail of 2 MeV protons. Let ℓ equal the stopping length of 2 MeV protons in a medium containing 1 H atom per cm^{-3}. The hydrogen ionization cross-section $\sigma_{\rm ion}$ for 2 MeV protons $\sim 10^{-17}$ cm^2, and $\sim$ 50 eV are lost per ionization. Therefore, the stopping length ℓ for a 2 MeV proton can be estimated as

$$\left(\frac{2\text{ MeV}}{50\text{ eV}}\right)\frac{1}{n_1\sigma_{\rm ion}} \sim 10^3\text{ pc}.$$

If this flight took place in a straight line, it would greatly exceed any macroscopic length scale of interest, and we would not have been justified

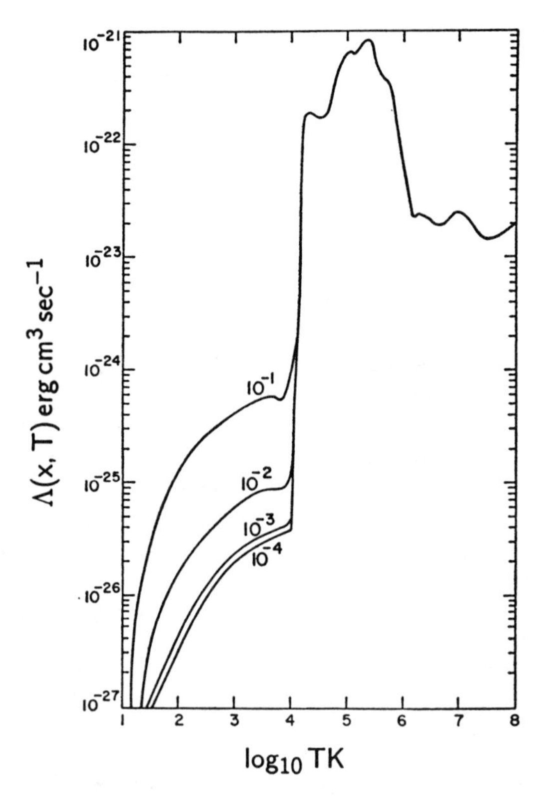

FIGURE 17.2
The cooling function $\Lambda(x, T)$, in units of cm^3 s^{-1} [when multiplied by the square of the number density, $\Lambda(x, T)$ equals the volumetric radiative loss rate that we usually denote by the symbol Λ in this text], as a function of the electron fraction x and the gas temperature T. (From A. Dalgarno and R. McCray, 1972, *Ann. Rev. Astr. Ap.*, **10**, p. 383.)

in adopting a fluid approximation. On the other hand, if magnetic fields of strengths $\sim 3\,\mu G$ exist in the general interstellar medium, the radius of gyration of MeV protons $\sim 10^{11}$ cm would tie the streaming protons to the ambient fluid much better than estimated by our stopping-length calculation. The remaining issue concerns whether the protons might not freely stream along the direction of the ambient magnetic field. Calculations by Wentzel, Kulsrud and Pierce, and others indicate that the amount of relative streaming would be limited by the resonant excitation of Alfven waves to speeds less than $\sim 50\,\mathrm{km\,s^{-1}}$. As a consequence, astronomers generally believe that a fluid treatment of supernova remnant evolution does have rough validity; however, subtle differences may still reside in the problem. In particular, workers in the subject of "collisionless shocks" [where the "jumps" are mediated by (chaotic) magnetic fields] believe that a significant part of the shock-deposited energy may end up not as heat in the bulk of the fluid, but rather as relativistic kinetic energy of a fraction of the charged particles, which become cosmic rays. The acceleration of cosmic rays in collisionless shocks constitutes a variation of a mechanism first described by Fermi. In what follows, we ignore this part of the story and focus on what happens to the bulk fluid behind the shock front of the blast wave (see, however, Chapter 28).

TAYLOR-SEDOV SOLUTION FOR INTERIOR OF A BLAST WAVE

In a frame fixed with respect to the center of the explosion, the Rankine-Hugoniot jump conditions in the limit of a very strong shock can be expressed as

$$\rho_2 = \left(\frac{\gamma+1}{\gamma-1}\right)\rho_1, \qquad u_2 = \frac{2}{(\gamma+1)}U_{\text{sh}}, \qquad P_2 = \frac{2}{(\gamma+1)}\rho_1 U_{\text{sh}}^2. \tag{17.5}$$

If we know r_{sh} and $U_{\text{sh}} = \dot{r}_{\text{sh}}$, equations (17.5) yield the physical conditions just inside the blast wave. For the interior flow, we have

$$\frac{\partial \rho}{\partial t} + \frac{1}{r^2}\frac{\partial}{\partial r}(r^2 \rho u) = 0, \tag{17.6}$$

$$\frac{\partial u}{\partial t} + u\frac{\partial u}{\partial r} = -\frac{1}{\rho}\frac{\partial P}{\partial r}, \tag{17.7}$$

$$\frac{\partial}{\partial t}\left[\rho\left(\mathcal{E} + \frac{1}{2}u^2\right)\right] + \frac{1}{r^2}\frac{\partial}{\partial r}\left[r^2 \rho u\left(\mathcal{E} + \frac{P}{\rho} + \frac{1}{2}u^2\right)\right] = 0, \tag{17.8}$$

with the specific internal energy

$$\mathcal{E} = \frac{P}{(\gamma-1)\rho}$$

for an ideal gas. Instead of the energy equation (17.8), we could also use the entropy equation,

$$\left(\frac{\partial}{\partial t} + u\frac{\partial}{\partial r}\right)\ln\left(P\rho^{-\gamma}\right) = 0, \tag{17.9}$$

which states that the pressure P of a given fluid element compresses adiabatically as $K\rho^{\gamma}$ once it crosses the shock front, but different fluid elements that cross at different times (when the shocks have different strengths) would generally have different adiabatic "constants" K.

We define the dimensionless similarity variable ξ through equation (17.1), and we look for solutions of equations (17.6)–(17.8) having the self-similar form:

$$\rho(r,t) = \left(\frac{\gamma+1}{\gamma-1}\right)\rho_1\alpha(\xi) = \rho_2\alpha(\xi), \tag{17.10}$$

$$u(r,t) = \frac{4}{5(\gamma+1)}\frac{r}{t}v(\xi) = \frac{2}{(\gamma+1)}\frac{U_{\rm sh}}{r_{\rm sh}}rv(\xi) = u_2\frac{r}{r_{\rm sh}}v(\xi), \tag{17.11}$$

$$P(r,t) = \frac{8}{25(\gamma+1)}\rho_1\left(\frac{r}{t}\right)^2 p(\xi) = P_2\left(\frac{r}{r_{\rm sh}}\right)^2 p(\xi), \tag{17.12}$$

where the coefficients in front of the reduced density α, velocity v, and pressure p have been chosen so that these quantities are normalized behind the shock,

$$\alpha(\xi_0) = v(\xi_0) = p(\xi_0) = 1. \tag{17.13}$$

To obtain the condition that determines the dimensionless position of the shock front, ξ_0, we integrate equation (17.8) over all space, noting that $ru = 0$ both at $r = 0$ and $r = \infty$:

$$0 = \frac{d}{dt}\int_0^\infty \rho\left(\mathcal{E} + \frac{1}{2}u^2\right)4\pi r^2\,dr = \frac{d}{dt}\int_0^{r_{\rm sh}(t)}\left[\frac{P}{(\gamma-1)} + \frac{1}{2}\rho u^2\right]4\pi r^2\,dr$$
$$+ \frac{d}{dt}\int_{r_{\rm sh}(t)}^\infty \frac{P_1}{(\gamma-1)}4\pi r^2\,dr. \tag{17.14}$$

The last term equals

$$-\frac{P_1}{(\gamma-1)}4\pi r_{\rm sh}^2\frac{dr_{\rm sh}}{dt},$$

which can be ignored in comparison with the others if $P_1/\rho_1 = kT_1/m_1 \ll U_{\rm sh}^2$, as we have assumed from the outset. [Typically, $(kT_1/m_1)^{1/2} \sim$ 1–10 km s^{-1} in the interstellar medium.] In this strong shock approximation, we may integrate equation (17.14) to obtain

$$\int_0^{r_{\rm sh}(t)}\left[\frac{P}{(\gamma-1)} + \frac{1}{2}\rho u^2\right]4\pi r^2\,dr = E, \tag{17.15}$$

which gives the nondimensional requirement,

$$\frac{32\pi}{25(\gamma^2-1)}\int_0^{\xi_0}\left[p(\xi) + \alpha(\xi)v^2(\xi)\right]\xi^4\,d\xi = 1. \tag{17.16}$$

Notice that

$$(\text{total thermal energy})/E = \frac{32\pi}{25(\gamma^2-1)}\int_0^{\xi_0} p(\xi)\xi^4\,d\xi$$

and

$$(\text{total kinetic energy})/E = \frac{32\pi}{25(\gamma^2-1)} \int_0^{\xi_0} \alpha(\xi) v^2(\xi) \xi^4 \, d\xi$$

are individually independent of time, and are, therefore, separately conserved, as one might have guessed on an intuitive basis.

When differentiating functions of ξ, we note that we can use

$$\frac{\partial}{\partial t} = -\frac{2}{5}\frac{\xi}{t}\frac{d}{d\xi}, \qquad \frac{\partial}{\partial r} = \frac{\xi}{r}\frac{d}{d\xi}.$$

If we now substitute equations (17.10)–(17.12) into equations (17.6)–(17.8), we get, after a little algebra,

$$-\xi\frac{d\alpha}{d\xi} + \frac{2}{(\gamma+1)}\left[3\alpha v + \xi\frac{d}{d\xi}(\alpha v)\right] = 0, \tag{17.17}$$

$$-v - \frac{2}{5}\xi\frac{dv}{d\xi} + \frac{4}{5(\gamma+1)}\left(v^2 + v\xi\frac{dv}{d\xi}\right) = -\frac{2}{5}\left(\frac{\gamma-1}{\gamma+1}\right)\frac{1}{\alpha}\left(2p + \xi\frac{dp}{d\xi}\right), \tag{17.18}$$

$$-2(p+\alpha v^2) - \frac{2}{5}\xi\frac{d}{d\xi}(p+\alpha v^2) + \frac{4}{5(\gamma+1)}\left\{5v(\gamma p + \alpha v^2) + \xi\frac{d}{d\xi}[v(\gamma p + \alpha v^2)]\right\} = 0. \tag{17.19}$$

We need to integrate equations (17.17)–(17.19) from ξ_0 to 0, subject to the normalization conditions (17.13). The value of ξ_0 must satisfy equation (17.16). These requirements serve to define ξ_0, $\alpha(\xi)$, $v(\xi)$, and $p(\xi)$ uniquely. In one of the first numerical integrations carried out with a modern electronic computer, Taylor obtained the complete solution of the blast-wave problem. By examining motion pictures of the Alamogordo atomic blast, he then used equation (17.2) to deduce and publish the energy released in the explosion as $E \sim 10^{21}$ erg, a number that had been considered classified information! Without access to a comparably powerful computer in the Soviet Union, Sedov (amazingly enough) was able to find an *analytical* solution of equations (17.17)–(17.19) in closed form. (The results are quoted—with some misprints—in Landau and Lifshitz.) For $\gamma = 5/3$, the solution looks as drawn in Figure 17.3.

Notice that the interior temperature $T \propto P/\rho$ reaches its highest value (∞) at the center of the blast wave and reaches its minimum at the edge. This behavior occurs because the material near the center passed through the front earlier, when the shock speed was higher (infinitely large in the formal extrapolation to $t \to 0$). As time proceeds, the entire temperature profile drops as $T \propto U_{\text{sh}}^2 \propto t^{-6/5}$. The rapid rise of the function $\rho\mathcal{L}$ as the interior cools then implies that radiative losses eventually become severe in

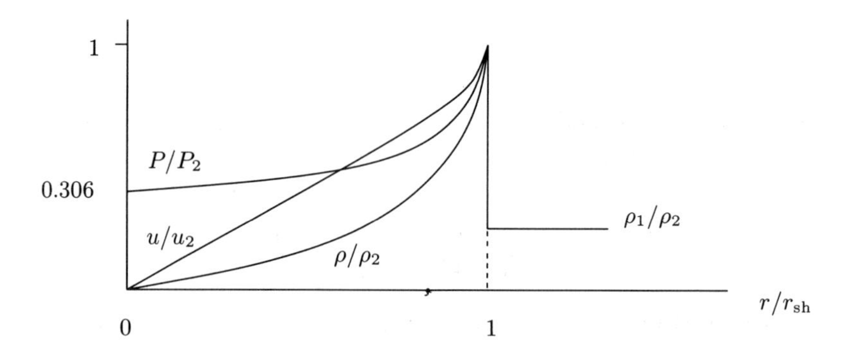

FIGURE 17.3
The blast-wave solution in units of the immediate postshock values for the case $\gamma = 5/3$.

a shell just behind the shock front. When the cooling time $kT/m\mathcal{L}$ there becomes comparable to the flow time, $r_{\rm sh}/U_{\rm sh}$, we expect the supernova remnant to undergo a phase of *dense shell formation* (Figure 17.4).

Past this epoch, the supernova remnant no longer conserves its mechanical and thermal energy, but tends to resemble a "snowplow model" in which ambient matter is swept up by the inertia of a moving shell, with the approximate independent conservation of momentum per unit solid angle toward every direction.

MODIFICATIONS OF THE CLASSICAL PICTURE

The above idealized picture for the evolution of a supernova remnant has undergone modifications in recent years. Theorists have pointed out that the clumpy nature of the actual interstellar medium introduces (at least) two different shock propagation speeds, one through the intercloud medium and one through clouds. Blast waves in a rarefied intercloud medium take longer to turn radiative than if we were to take the undisturbed density ρ_1 as an average of the clouds and the intercloud medium. This effect introduces the possibility that the hot interiors of different supernova remnants might merge before they lose their outward drive via radiative cooling. Such ideas form the basis for models of the interstellar medium based on "tunnel networks" (Cox and Smith) or the existence of "three thermal phases" (McKee and Ostriker). In addition, type II supernovae, arising as they do from short-lived O and B stars recently formed from molecular clouds, are

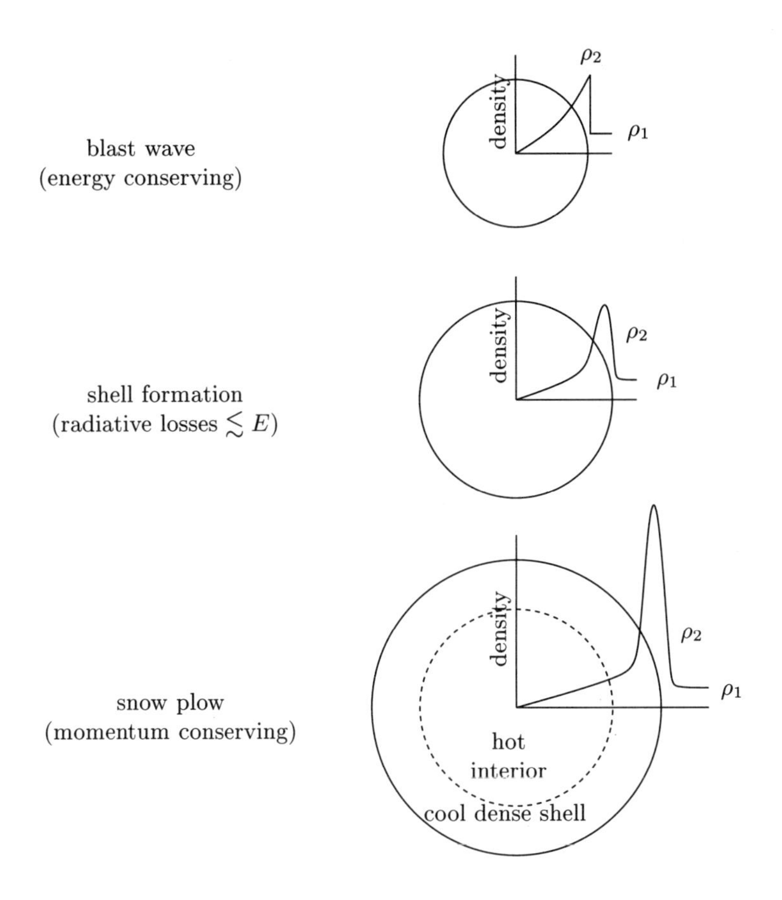

FIGURE 17.4
The transition from the blast-wave phase (energy conserving) to the snow-plow phase (momentum conserving) phase in the evolution of a supernova remnant.

probably highly correlated in space and time. Thus, their overlapping remnants may create huge "superbubbles" that break out of the galactic disk (like "mushroom clouds" in terrestrial nuclear blasts), channeling much of the collective energy into the gaseous halo. (Such superbubbles have been found by radio astronomers both in our own Galaxy and in external spiral galaxies.) Moreover, the hydrodynamic crushing of clouds represents an additional source of energy loss not included in the usual classical theory. Astronomers have also come to realize that presupernova stars blow powerful stellar winds that modify the medium into which the explosion later occurs. This tends to homogenize the preshock medium more than if the supernova explosion were to occur in an undisturbed cloudy interstellar medium, thus restoring some of the validity of the simple assumption. These issues, and others, are comprehensively discussed in the review article by Ostriker and McKee listed at the beginning of this chapter.

18

Gravitational Collapse and Star Formation

References: C. Hayashi 1966, *Ann. Rev. Astr. Ap.*, **4**, 171; R. B. Larson, 1973, *Ann. Rev. Astr. Ap.*, **11**, 219; F. H. Shu, F. C. Adams, and S. Lizano, 1987, *Ann. Rev. Astr. Ap.*, **25**, 23.

Given that stars age and die, and given that young stars still exist in the Galaxy, it is natural to ask how they are born. In the 1950s and 1960s, theoretical astrophysicists attempted to pose this question by asking how stars might contract quasistatically from some larger gaseous configuration to the so-called main sequence, where they acquire central temperatures high enough to ignite hydrogen fusion. This approach descends almost in a direct line from research begun in the nineteenth century with the investigations of Lord Kelvin in England, Hermann Helmholtz in Germany, and Homer Lane in America, and it culminated in the pioneering work of L. Henyey, R. LeLevier, and R. D. Levee (1955, *Pub. Astr. Soc. Pac.*, **67**, 154) and of C. Hayashi, R. Hoshi, and D. Sugimoto (1962, *Prog. Theor. Phys. Suppl.*, **34**, 754).

MAXIMUM SIZE FOR A PRE-MAIN-SEQUENCE STAR

The study of Hayashi and colleagues demonstrated that, independently of how large stars were evolved "backward in time," their photospheres could not drop below a certain minimum temperature that gas opacities (mainly H^-) could maintain in a self-consistent manner. In the Hertzsprung-Russell (H-R) diagram of luminosity plotted against effective temperature, there exists a "forbidden zone" which pre-main-sequence stars (and, as it turns out, also post-main-sequence stars) must remain to the left of (Figure 18.1).

Since their surface temperatures T_{eff} are virtually fixed (for a given value of mass M), giant (pre-main-sequence) stars with large radii R must have large surface luminosities in accordance with the formula

$$L = 4\pi R^2 \sigma T_{\text{eff}}^4. \tag{18.1}$$

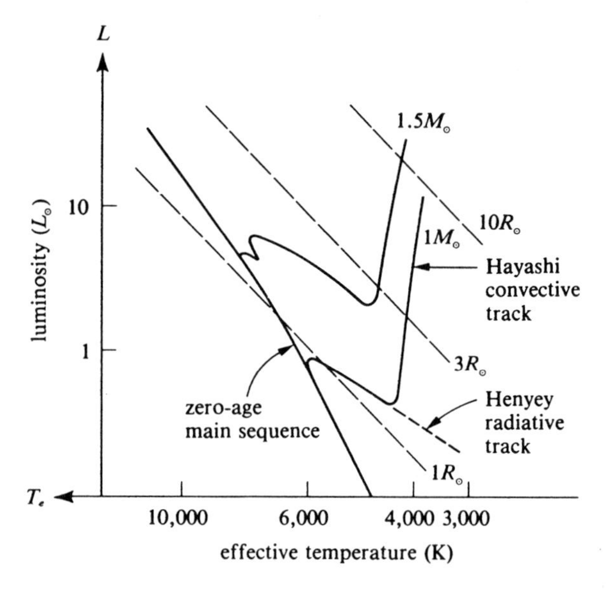

FIGURE 18.1
Theoretical evolutionary tracks for the pre-main-sequence phase of quasi-static contraction. Tracks are shown for a $1\,M_\odot$ and 1.5',$M_\odot$ star. The dashed diagonal lines give the loci of constant radius. The $1.5\,M_\odot$ track has an extra bump just before reaching the main-sequence position that the $1\,M_\odot$ track does not have, because the heavier star must make an extra adjustment to burn hydrogen in equilibrium by the CN cycle.

If this luminosity exceeds what radiative diffusion can carry in the interior (according to the "mass-luminosity" relationship for quasistatic stars), convection must compensate for the difference.

The nearly ideal gas of a fully convective star satisfies a polytropic equation of state, $P \propto \rho^{5/3}$. Lane had shown that such a configuration has a self-gravitational energy content equal to

$$W = -\frac{6}{7}\frac{GM^2}{R}. \tag{18.2}$$

On the other hand, the virial theorem (see below and Chapter 22) states that in quasistatic contraction, half of the gravitational energy released goes into radiation; the other half goes to raising the internal thermal energy content $U = -W/2$. If we set the rate of the radiation loss equal to L, we get the evolutionary equation for homologous gravitational contraction (the quasistatic transformation of one polytrope to another satisfying a law of homology: $r' = \lambda r$ where λ is a constant for each r):

$$L = -\frac{1}{2}\frac{dW}{dt} = -\frac{3}{7}\frac{GM^2}{R^2}\frac{dR}{dt}. \tag{18.3}$$

If we substitute equation (18.1) into equation (18.3), we get, as the characteristic (Kelvin-Helmholtz) time scale associated with quasistatic gravitational contraction,

$$t_{\text{KH}} \equiv -\frac{R}{dR/dt} = \frac{3GM^2/7R}{4\pi R^2 \sigma T_{\text{eff}}^4}. \tag{18.4}$$

For fixed M and T_{eff}, this time scale becomes very short for large values of R. Indeed, if it becomes shorter than the orbit time scale at the surface of the star,

$$t_{\text{dyn}} \equiv \frac{R}{(GM/R)^{1/2}}, \tag{18.5}$$

the quasistatic assumption must break down. Setting t_{KH} equal to t_{dyn}, we obtain an estimate for the maximum size at which a star on the Hayashi track could remain quasistatic as

$$R_{\text{max}} = \left(\frac{3}{28\pi}\right)^{2/9} \frac{G^{1/3}M^{5/9}}{(\sigma T_{\text{eff}}^4)^{2/9}}. \tag{18.6}$$

For example, with $M = 1\ M_{\odot}$ and $T_{\text{eff}} = 4000\,\text{K}$, we obtain $R_{\text{max}} \approx 500$ $R_{\odot}$. Beyond this point, the evolution from interstellar cloud to star must involve dynamical collapse. Gaustad and others reached a complementary conclusion by considering the problem from the other end, namely, the radiative losses suffered by a contracting interstellar cloud. These authors deduced that radiative losses would keep the cloud so nearly isothermal (under optically thin conditions) that a phase of dynamical evolution became inevitable when the dynamical time scale exceeded the thermal time scale.

In fact, following Cameron, Hayashi gave an elegant argument that stars should begin their pre-main-sequence phase with radii at least one order of magnitude smaller than the estimate given in equation (18.6). Condensed to its essentials, Hayashi's argument proceeds as follows. After the dynamic collapse phase is over, we have a hydrostatic (pre-main-sequence) star. This object satisfies the virial theorem,

$$W + 2U = 0. \tag{18.7}$$

Independent of the details of the collapse process, we may also relate the hydrostatic object to the original gas cloud from which it formed by an energy budget equation, which applies after the kinetic energy of the collapse has been thermalized:

$$W + U + I + \Delta = 0, \tag{18.8}$$

where I equals the total energy required to break the molecules in the original cloud first into atoms and later into ions and electrons, and where Δ equals the total energy radiated into space. On the right-hand side of equation (18.8), we have assumed that the initial energy of the greatly distended interstellar cloud may be set to zero in comparison with the much larger value of the final state. If we eliminate U from equation (18.8) by using equation (18.7), we obtain

$$-W = 2(I+\Delta) \qquad \Rightarrow \qquad R = \frac{3}{7}\frac{GM^2}{(I+\Delta)}, \tag{18.9}$$

where we have made use of equation (18.2). Hayashi obtained a maximum size by assuming that no binding energy at all is radiated away, $\Delta = 0$; consequently, his estimate for when stars first appear as quasistatic pre-main-sequence objects reads

$$R_{\rm H} = \frac{3}{7}\frac{GM}{\chi}, \tag{18.10}$$

where we have written $\chi \equiv I/M$ as the dissociation and ionization energy per unit mass for a gas of cosmic composition. For $M = 1\ M_\odot$ and $\chi = 1.6 \times 10^{13}\,\text{erg}\,\text{g}^{-1}$, we get $R_{\rm H} \approx 50\ R_\odot$, about an order of magnitude smaller than $R_{\rm max}$, as advertised. Since $R_{\rm H}$ is less than $R_{\rm max}$, the subsequent contraction would indeed be quasistatic, as verifiable by direct calculation. Hayashi and his colleagues completed the argument by demonstrating that if interstellar clouds were to collapse homologously in a dynamical state, reaching free fall from interstellar dimensions to stellar dimensions, the total amount of energy radiated away Δ would indeed be negligible, justifying the argument that led to equation (18.10) from equation (18.9).

NONHOMOLOGOUS NATURE OF GRAVITATIONAL COLLAPSE

Unfortunately, computer simulations by McNally, by Bodenheimer and Sweigart, and by others demonstrated that the gravitational collapse of a gaseous mass marginally in excess of the Jeans criterion for instability proceeds very nonhomologously, with the dense central regions falling in well before a more extended envelope does so. In a classic numerical study, Larson (1969, *M.N.R.A.S.*, **145**, 271) showed that the process of star formation quickly resolves itself to the problem of the accretion of material by a central hydrostatic object (a "protostar"), which steadily builds up its mass through the addition of mass from an infalling envelope. The time scale for the complete assemblage of the star ($\sim 10^6$ yr typically) then greatly exceeds the value (~ 1 yr) assumed by Hayashi and coworkers. Under these circumstances, the total energy Δ radiated away in the collapse process can not be ignored in comparison with I, and Larson obtained beginning sizes (after the dynamical accretion has halted) for a $1\,M_\odot$ pre-main-sequence star of $R \sim 2\ R_\odot$. In particular, Larson claimed that young stars missed having a completely convective phase of quasistatic contraction altogether, and began their lives as optically visible objects in the H-R diagram on Henyey radiative tracks. (During the infall phase, so much gas and dust surrounds the protostar that almost all emergent stellar photons are degraded into the far infrared. Only after the infall has stopped would the central object be revealed as an optical star.)

Larson's disagreement with the conclusions of Hayashi generated considerable early controversy, and attempts by other groups to reproduce his results sometimes supported his point of view and sometimes Hayashi's. The process of transforming a piece of interstellar cloud to a star involves changes in density of more than twenty orders of magnitude, a computational chore not easy to carry out accurately even in its simplest spherically symmetric context. Eventually, resolution of the controversy turned out to hinge on two issues: (1) whether the final accumulation time scale for a (low-mass) star is more like 10^6 yr or 10^0 yr, and (2) whether Larson's simplified treatment of the strongly radiating accretion shock at the surface of the protostar was justified. We shall examine issue (2) in Chapter 19. For the rest of this chapter, we shall concentrate on issue (1), the naturalness of nonhomologous gravitational collapse for isolated objects.

The numerical calculations contain a clue to the physical basis of the latter behavior. Independent largely of the details of the initial or boundary conditions, the density profile of the envelope acquires power-law forms satisfying $\rho \propto r^{-3/2}$ in the freely falling inner parts, and $\rho \propto r^{-2}$ in the nearly static outer parts. The first law has a simple explanation. For small r, the crossing time $r/|u|$ becomes short compared to the evolutionary time t characterizing the overall problem; consequently, the flow looks locally like steady-state accretion, with a mass accretion rate by the central protostar given by

$$\dot{M} = -4\pi r^2 \rho u. \tag{18.11}$$

If the fluid velocity corresponds to free fall,

$$u = -(2GM/r)^{1/2}, \tag{18.12}$$

where the central mass M equals the time integral of $\dot{M}$ over the evolutionary time t, we get $u \propto r^{-1/2}$ and $\rho \propto r^{-3/2}$ if the changes in M and $\dot{M}$ can be ignored over the flow time $r/|u|$.

Does the behavior $\rho \propto r^{-2}$ at large r have a similarly simple physical explanation? Yes, if we believe P. Bodenheimer and T. Sweigart (1968, *Ap. J.*, **152**, 515). These authors argued that all self-gravitating isothermal gas clouds which have a chance to evolve subsonically before undergoing gravitational collapse (i.e., if they do not lie initially far from hydrostatic balance) tend to develop r^{-2} density profiles in their outer parts. Indeed, we show below that this behavior is a basic property of self-gravitating states of isothermal equilibria.

ISOTHERMAL EQUILBRIA OF SELF-GRAVITATING SPHERES

Force balance for a static gaseous configuration reads

$$\nabla P = -\rho \nabla \mathcal{V}. \tag{18.13}$$

For an ideal gas, $P = a^2\rho$, where $a^2 \equiv kT/m =$ constant if we assume isothermal conditions. With the assumption of spherical symmetry, we may write equation (18.13) in the form

$$\frac{a^2}{\rho}\frac{d\rho}{dr} = -\frac{d\mathcal{V}}{dr} \qquad \Rightarrow \qquad \rho = \rho_0 \exp(-\mathcal{V}/a^2), \tag{18.14}$$

where ρ_0 is the value of the density at which $\mathcal{V}$ has a fiducial value equal to zero. The gravitational potential $\mathcal{V}$ in turn satisfies Poisson's equation,

$$\frac{1}{r^2}\frac{d}{dr}\left(r^2\frac{d\mathcal{V}}{dr}\right) = 4\pi G\rho = 4\pi G\rho_0 \exp(-\mathcal{V}/a^2), \tag{18.15}$$

a nonlinear ODE known as the *Lane-Emden equation for an isothermal sphere.*

The regular solutions of equation (18.15), which satisfy $\mathcal{V} = 0$ and $d\mathcal{V}/dr = 0$ at $r = 0$, have well-known properties that are tabulated, for example, in Chandrasekhar's book *Stellar Structure.* If the system lacks an outer bounding surface (e.g., a change in phase to a different kind of [much hotter] gas), all solutions acquire the asymptotic behavior

$$\rho \to \frac{a^2}{2\pi G r^2} \qquad \text{for} \qquad r \gg a(4\pi G\rho_0)^{-1/2}.$$

In particular, if the density contrast between center and outer edge is very large (as appears now to be empirically true for the molecular cloud cores that represent the sites of low-mass star formation), all such configurations approach the solution given by the singular isothermal sphere (corresponding to $\rho_0 \to \infty$),

$$\rho = a^2/2\pi G r^2 \qquad \text{and} \qquad \frac{d\mathcal{V}}{dr} = \frac{2a^2}{r}, \tag{18.16}$$

which can easily be verified to exactly satisfy equations (18.14) and (18.15).

SELF-SIMILAR COLLAPSE OF THE SINGULAR ISOTHERMAL SPHERE

The gravitational collapse of the singular isothermal sphere, equation (18.16), takes a self-similar form (F. H. Shu, 1977, *Ap. J.*, **214**, 488). Since power laws have no characteristic scale, if the outer boundary condition does not introduce a characteristic radius R_0 or pressure P_{ext}, then the universal gravitational constant G and the isothermal sound speed a form the only dimensional parameters of the problem. From these quantities and the independent variables r and t, we can form the dimensionless similarity variable

$$x \equiv r/at. \tag{18.17}$$

We choose to write the equation of continuity,

$$\frac{\partial \rho}{\partial t} + \frac{1}{r^2}\frac{\partial}{\partial r}(r^2 \rho u) = 0, \tag{18.18}$$

in terms of mass conservation in spherical shells (cf. Chapter 5):

$$\frac{\partial M}{\partial t} + u\frac{\partial M}{\partial r} = 0, \qquad \frac{\partial M}{\partial r} = 4\pi r^2 \rho. \tag{18.19}$$

For ideal isothermal flow, the force equation reads

$$\frac{\partial u}{\partial t} + u\frac{\partial u}{\partial r} = -\frac{a^2}{\rho}\frac{\partial \rho}{\partial r} - \frac{GM}{r^2}. \tag{18.20}$$

We now look for a similarity solution of equations (18.19) and (18.20) in the form

$$\rho(r,t) = \frac{\alpha(x)}{4\pi G t^2}, \qquad M(r,t) = \frac{a^3 t}{G} m(x), \qquad u(r,t) = a v(x). \tag{18.21}$$

The substitution of equations (18.21) into equations (18.19) yields

$$m + (v - x)\frac{dm}{dx} = 0, \qquad \frac{dm}{dx} = x^2 \alpha.$$

The term dm/dx can be eliminated from the above to give

$$m = x^2 \alpha (x - v). \tag{18.22}$$

Some straightforward manipulation now allows us to express equations (18.18) and (18.20) as the coupled set of nonlinear ODEs

$$[(x-v)^2 - 1]\frac{dv}{dx} = \left[(x-v)\alpha - \frac{2}{x}\right](x - v), \tag{18.23}$$

$$[(x-v)^2 - 1]\frac{1}{\alpha}\frac{d\alpha}{dx} = \left[\alpha - \frac{2}{x}(x-v)\right](x - v). \tag{18.24}$$

SOLUTION

The singular isothermal sphere, equation (18.16), corresponds to an exact (static) solution of equations (18.23) and (18.24):

$$v = 0, \qquad \alpha = 2/x^2, \qquad m = 2x, \tag{18.25}$$

which give through equation (18.21) the time-independent forms: $\rho = a^2/2\pi G r^2$, and $M = 2a^2 r/G$. If we use equation (18.25) as the "initial state," we expect the solution for $t > 0$ to look self-similar. (Notice that the "initial state" as $t \to 0^+$ is represented at all r by the behavior at $x \to \infty$.)

The loci of points given by

$$x - v = 1, \qquad \alpha = 2/x, \tag{18.26}$$

represent "critical points" for the flow, in the same sense that the sonic point represents a "critical point" for Bondi flow (refer back to Problem Set 1). To pass through such a critical point smoothly, we would need to expand in a Taylor series about equation (18.26). (For details, refer to Shu, 1977.) Notice, in particular, that the singular solution (18.25) satisfies the critical point condition at the point $x = 1$:

$$v = 0 \qquad \text{and} \qquad \alpha = 2 \qquad \text{at} \qquad x = 1. \tag{18.27}$$

Thus, the solution that evolves continuously from an initial state given by the singular isothermal sphere looks as drawn in Figure 18.2 for the reduced velocity $v(x)$.

PHYSICAL INTERPRETATION

Table 18.1 gives another representation of the solution for the gravitational collapse of a singular isothermal sphere. This solution has the following physical interpretation.

At $t = 0^+$, imagine a perturbation to occur that causes the "inside-out" collapse of a singular isothermal sphere, with the dense central regions forming a hydrostatic protostar whose dimensions are tiny in comparison with those of the infalling envelope and have therefore been approximated in the similarity solution as a single point. For dimensionless radii in the envelope $x > 1$, the configuration remains undisturbed, for two reasons. Hydrodynamic signals have insufficient time to propagate to radii $r > at$, and no changes occur in the gravitational field, because the regions interior to the point in question remain spherically symmetric. Thus, exterior to the head of an expansion wave, $r = at$, the material remains in the hydrostatic

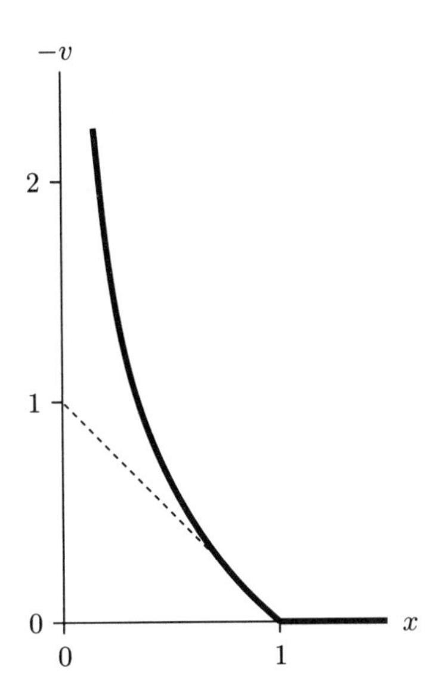

FIGURE 18.2
Similarity solution for the reduced velocity in isothermal flow. The line inclined at $-45°$ with respect to the x-axis gives the locus of the critical points: $-v = 1 - x$. The heavy solid curve gives the collapse solution for an initial state corresponding to a singular isothermal sphere (see also Table 18.1).

Table 18.1
Gravitational Collapse of the Singular Isothermal Sphere.

x	α	$-v$	m	x	α	$-v$	m
0.00...	∞	∞	0.975	0.55 ...	3.66	0.625	1.30
0.05...	71.5	5.44	0.981	0.60...	3.35	0.528	1.36
0.10...	27.8	3.47	0.993	0.65...	3.08	0.442	1.42
0.15...	16.4	2.58	1.01	0.70...	2.86	0.363	1.49
0.20...	11.5	2.05	1.03	0.75...	2.67	0.291	1.56
0.25...	8.76	1.68	1.05	0.80...	2.50	0.225	1.64
0.30...	7.09	1.40	1.08	0.85...	2.35	0.163	1.72
0.35...	5.95	1.18	1.12	0.90...	2.22	0.106	1.81
0.40...	5.14	1.01	1.16	0.95...	2.10	0.051	1.90
0.45...	4.52	0.861	1.20	1.00...	2.00	0.000	2.00
0.50...	4.04	0.735	1.25				

equilibrium of the initial state (18.25). The front at $r = at$ initiates inward motion as the bottom drops out from below in a wave of falling that moves outward at the speed of sound. Inward of the (expansion) wave front, matter accelerates toward the center, reaching asymptotic free-fall speeds given by the formula:

$$v \to -(2m_0/x)^{1/2} \text{ and } \alpha \to (m_0/2x^3)^{1/2} \qquad \text{where} \qquad m \to m_0 \text{ as } x \to 0. \tag{18.28}$$

Numerically, the reduced mass m_0 that has already fallen into the center can be found as

$$m_0 = 0.975, \tag{18.29}$$

and corresponds, via equation (18.21), to a dimensional central mass for the protostar equal to

$$M(0,t) = m_0 a^3 t/G.$$

In other words, the similarity solution yields a central object whose mass increases linearly with time, with a mass infall rate given by

$$\dot{M} = m_0 a^3/G. \tag{18.30}$$

This formula should be compared with the Bondi rate, equation (6.43), appropriate for steady isothermal accretion onto a (fixed) central mass M. At first sight, these two formulae seem in violent disagreement; e.g., the dependence on $a^3 = c_\infty^3$ appears in the numerator of equation (18.30) while it appears in the denominator of equation (6.43). However, if we set $M = \dot{M}t$ by fiat in equation (6.43), and if we identify $4\pi G\rho_\infty$ with the inverse-square of the (free-fall) time, t^{-2}, we see that equations (6.43) and (18.30) give basically the same mass infall rate, within numerical factors of order unity.

For a molecular cloud core with temperature $T = 10\,\text{K}$, $a = 0.19\,\text{km}\,\text{s}^{-1}$, and equation (18.30) yields $\dot{M} = 2 \times 10^{-6}\ M_\odot\,\text{yr}^{-1}$. At such an accretion rate, it would take $5 \times 10^5\,\text{yr}$ to build up a $1\,M_\odot$ star, more in agreement with Larson's point of view than with Hayashi's. [Notice, however, that the similarity solution gives no clue as to why the infall should stop after any particular time, i.e., it contains no characteristic mass scale for (low-mass) star formation, only a characteristic mass-accretion *rate*. The question of what determines stellar mass scales remains one of the intriguing problems of astrophysics; indications exist that the initiation of powerful stellar winds, which ultimately reverse the inflow, plays an important role in the definition of stellar masses.]

In any case, between the head of the wave of infall at $x = 1$ and the tail just outside the origin at $x = 0$, where the material falls the fastest (free fall), the reduced density α drops below the equilibrium values $2/x^2$. When multiplied by x^2 and integrated over $x = 0^+$ to $x = 1$, the reduced density α gives a reduced mass equal to 1.025. This reduced mass added to the reduced mass that has already fallen into the origin, $m_0 = 0.975$, yields a total reduced mass m inside $x = 1$ equal to 2—the original equilibrium value. Thus, at any instant t, about 49% of the mass interior to the head of the expansion wave at $r = at$ has already fallen into the center, and another 51% is on the way down, with each quantity increasing linearly with time. In other words, the self-gravity of the gas interior to the wave of infall maintains an approximately equal importance to the gravity of the star, independent of how large a mass is acquired by the star.

MORE REALISTIC MODELS

The above description clearly makes a number of oversimplifications; the most serious involve the neglect of the effects of rotation and magnetic fields. The idealizations prove useful to telescope the overall problem down to a tractable form, in which we can compute all accompanying physical processes (e.g., the strong radiating shock at the surface of the protostar; see Chapter 19) in an *a priori* fashion, but they ignore structures (e.g., the formation of a nebular disk that surrounds the protostar) that may have important consequences when one attempts to make detailed comparisons of the models with observations. Some success has been achieved in incorporating the effects of rotation as a perturbation calculation to the above treatment (see S. Terebey, F. H. Shu, and P. Cassen, 1984, *Ap. J.*, **286**, 529), but the dynamical effects of interstellar magnetic fields have just begun to be incorporated in a realistic fashion.

19

Radiative Shocks Under Optically Thick Conditions

Reference: Ya. B. Zeldovich and Yu. P. Raizer, 1967, *Physics of Shock Waves and High-Temperature Hydrodynamic Phenomena*, Vol. 2.

In the controversy concerning the sizes of forming stars discussed in Chapter 18, debate arose about the proper treatment of the radiative shock caused by the infalling hypersonic gas, which slams to a virtual halt at the surface of a central protostar (Figure 19.1).

Schematically, we may picture the situation as shown in Figure 19.2.

If we depict radial distance to increase to the right, gas enters the shock at a hypersonic upstream velocity u_1 close to free fall onto a star of radius R. This material decelerates to about a quarter of its value to u_2 behind a thin viscous layer, shocking to a very high post-shock temperature T_2. The gas then radiates the shock-deposited energy until it comes back into radiative equilibrium at temperature T_3.

Using reasoning that Hayashi disputed, Larson claimed [see equation (A2) of Larson, 1969, *M.N.R.A.S.*, **145**, 271] that this temperature T_3 would be bounded by the inequalities

$$0.781[F_{\rm rad}(1)/\sigma]^{1/4} \le T_3 \le [F_{\rm rad}(1)/\sigma]^{1/4}. \qquad (19.1)$$

Because he did not aim for high accuracy, he ended up using the heuristically plausible formula $T_3 = [F_{\rm rad}(1)/\sigma]^{1/4}$ in his actual numerical calculations. The issue has importance because the post-radiative-relaxation temperature T_3 yields one of the surface boundary conditions that we need for a stellar-structure calculation of the underlying protostar. In turn, the protostellar-evolution calculation ultimately determines the value of the stellar radius R that was the main source of disagreement between different groups of workers (review Chapter 18). In this chapter, we discuss the property of radiative shocks with the protostellar problem providing the basic motivation, but our treatment has more general validity (see, for example, the analysis of Zeldovich and Raizer of high-temperature shock waves).

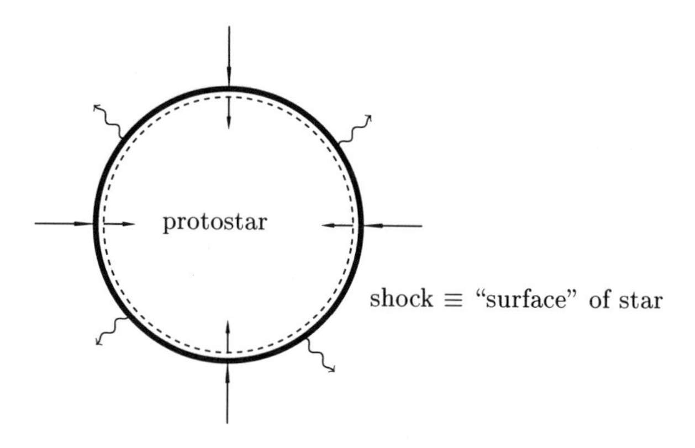

FIGURE 19.1
During the "main accretion phase" in the problem of forming a protostar by spherical collapse, infalling matter is brought nearly to rest on the surface of the star by passage through a standing shockwave. The high-temperature postshock gas cools radiatively, and this radiation adds to the total emerging from the protostar to heat the infalling envelope as well as the ambient molecular cloud core.

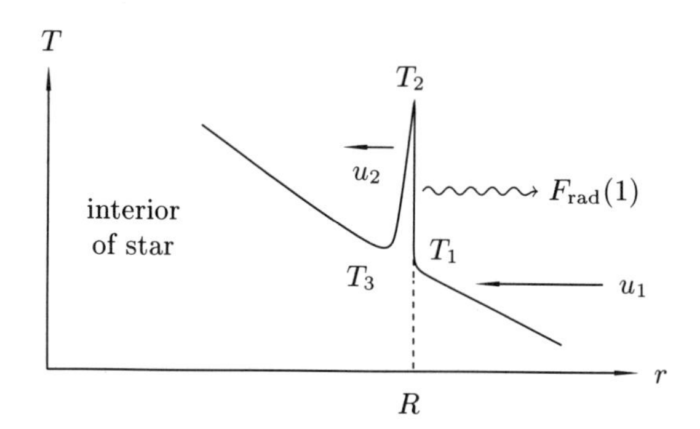

FIGURE 19.2
Schematic depiction of the structure of a radiative accretion shock at the surface of a star.

JUMP CONDITIONS

We begin with the jump conditions that apply across the strong viscous shock: $1 \rightarrow 2$. The viscous layer occupies only a few particle mean free paths for elastic scattering, which will generally be very much smaller than a photon mean free path (Chapter 1). This implies that the quantities describing the radiation field will not change as we cross the viscous shock:

$$[E_{\text{rad}}]_1^2 = [F_{\text{rad}}]_1^2 = [P_{\text{rad}}]_1^2 = 0. \tag{19.2}$$

The viscous jump for the matter field have their usual expressions. Thus, for free-fall upstream motion, $u_1^2/2 \approx GM/R$, the post-viscous-shock temperature T_2 has an energy equivalent,

$$kT_2 \approx \frac{3}{16}\frac{GMm_2}{R}, \tag{19.3}$$

that is more appropriate for the center of a star than for its surface ($\sim 10^6$ K rather than a few times 10^3 K). Thus, the strong viscous shock throws the gas badly out of thermal equilibrium; so the material must radiate intensely to relax back to equilibrium at a temperature T_3.

We assume (as can be verified after the fact) that the gas has sufficient emissivity to keep the transition-layer thickness $\Delta r \equiv r_1 - r_3$ spatially narrow: $\Delta r \ll R$. Thus we adopt a plane-parallel local analysis. We further assume steady flow in a frame that moves with the velocity of the shock front, $v_s = dR/dt$, where we may obtain the latter by time-differencing two successive values of the stellar radii R emergent from the underlying protostar calculation. The conservation of total mass, momentum, and energy contained in both matter and radiation across the shock and radiative relaxation layer then implies

$$\rho_3(u_3 - v_s) = \rho_1(u_1 - v_s), \tag{19.4}$$

$$P_{\text{rad}}(3) + P_3 + \rho_3(u_3 - v_s)^2 = P_{\text{rad}}(1) + P_1 + \rho_1(u_1 - v_s)^2, \tag{19.5}$$

$$\begin{aligned} F_{\text{rad}}(3) &+ \rho_3(u_3 - v_s)\left[h_3 + \frac{1}{2}(u_3 - v_s)^2\right] \\ &= F_{\text{rad}}(1) + \rho_1(u_1 - v_s)\left[h_1 + \frac{1}{2}(u_1 - v_s)^2\right], \end{aligned} \tag{19.6}$$

where P and h are the pressure and specific enthalpy of the gas.

To fix ideas, we provide a concrete framework for equations (19.4)–(19.6), although such an interpretation does not prove essential for the development that follows. In the protostellar problem, we have such a small mean shock speed, $v_s = dR/dt \sim 10^{11.5}\,\text{cm}/10^{12.5}\,\text{s} \sim 10^{-1}\,\text{cm}\,\text{s}^{-1}$, that there exists little difference between the radiative flux in the shock frame and that in the inertial frame. In other words, $F_{\rm rad} = L/4\pi R^2$, where L is the photon luminosity. To the same order of approximation, $\rho_1(u_1 - v_s) \approx -\dot{M}/4\pi R^2$, where $\dot{M}$ equals the mass accretion rate by the star calculated in accordance to the considerations of Chapter 18. With $(u_3 - v_s)^2/2$, h_1, and h_3 all much less than $(u_1 - v_s)^2/2 \approx GM/R$, equation (19.6) then implies that the luminosity ahead of the shock equals the stellar luminosity L_3 plus the release of gravitational binding energy inside the relaxation layer:

$$L_1 \approx L_3 + \frac{GM\dot{M}}{R}. \tag{19.7}$$

This result has intuitive validity if we ignore the thermal content of the gas before and after the relaxation process, and if we neglect any radiative deceleration of the preshock gas that would have led to departures from free-fall motions at radius 1.

RADIATIVE TRANSFER ACROSS THE RELAXATION LAYER

More generally, equations (19.4)–(19.6) provide only three constraints on the matter and radiation fields. To close the set, we need not only additional constitutive relations for the matter (e.g., equations of state, etc.), but also equations for $E_{\rm rad}$, $F_{\rm rad}$, and $P_{\rm rad}$. Two of these relations we may obtain as the steady-state angular moments of the frequency-integrated equation of radiative transfer in plane-parallel geometry (see Volume I and Chapter 4),

$$\frac{dF_{\rm rad}}{dr} = \rho \int_0^\infty j_\nu \, d\nu - c\rho\kappa_{\rm abs} E_{\rm rad}, \tag{19.8}$$

$$\frac{dP_{\rm rad}}{dr} = -\frac{\rho}{c}(\kappa_{\rm abs} + \kappa_{\rm sca}) F_{\rm rad}, \tag{19.9}$$

where $\kappa_{\rm abs}$ and $\kappa_{\rm sca}$ represent appropriate mean absorption and scattering opacities for the frequency distribution of photons present in the ambient radiation field, and j_ν equals the monochromatic emissivity of the matter over all solid angles (assumed isotropic). In writing down equation (19.8), we have assumed that the work done by the radiative force contributes negligibly to the energy imbalance between emission and absorption; we have also ignored the distinction between the radiative flux in an inertial frame and that in the shock frame.

If the emission occurred under LTE conditions, we could use Kirchhoff's law to write $j_\nu = 4\pi\kappa_\nu^{\rm abs} B_\nu(T)$, where $B_\nu(T)$ equals the Planck function at the local matter temperature T. In this case, to a rough order of approximation, we have

$$\int_0^\infty j_\nu \, d\nu = 4\pi \int_0^\infty \kappa_\nu^{\rm abs} B_\nu(T) \, d\nu \approx \kappa_{\rm abs} c a T^4,$$

where κ_{abs} is the same mean opacity that we used in the absorption term in equation (19.8). However, the emission probably does not occur under LTE conditions, nor, for the purposes of deriving the total jump conditions (rather than evaluating the spectral-energy distribution of the emergent radiation), do we need to assume equality of mean emission and absorption opacities. Instead, we *define* a mean emission opacity κ_{em} so that the relation

$$\int_0^\infty j_\nu \, d\nu = \kappa_{\text{em}} caT^4 \tag{19.10}$$

holds *exactly*. For our purposes below, we need only the physically plausible assumption that κ_{em} and κ_{abs} (for energy transfer) have the same *order of magnitude* as the total opacity $\kappa \equiv \kappa_{\text{abs}} + \kappa_{\text{sca}}$ (for momentum transfer).

If we now define the vertical optical depth τ increasing inward so that

$$d\tau = -\rho\kappa \, dr, \tag{19.11}$$

we may schematically integrate equations (19.8) and (19.9) to obtain

$$F_{\text{rad}}(1) - F_{\text{rad}}(3) = \int_{\tau_2}^{\tau_3} c\left(\frac{\kappa_{\text{em}}}{\kappa} aT^4 - \frac{\kappa_{\text{abs}}}{\kappa} E_{\text{rad}}\right) d\tau, \tag{19.12}$$

$$P_{\text{rad}}(1) - P_{\text{rad}}(3) = -\int_{\tau_2}^{\tau_3} \frac{1}{c} F_{\text{rad}} \, d\tau, \tag{19.13}$$

where we have used equation (19.2) to eliminate the contribution from τ_1 to τ_2 to the right-hand sides. (Note the combination $ca = 4\sigma$, where σ is the Stefan-Boltzmann constant.) Equation (19.12) states that the difference of the radiative fluxes at 1 and 3 equals the difference in the integrated contributions of emission and absorption in the radiative relaxation layer. Equation (19.13) states that the difference in the radiative momentum flux (radiation pressure) at 1 and 3 equals the negative of the amount of directed momentum flux absorbed by the matter in the radiative relaxation layer.

At point 3, behind which we assume the protostar to be optically thick, we apply the requirement that we have a radiation field that satisfies Eddington's closure relation and that is well-coupled thermally to the matter:

$$E_{\text{rad}}(3) = 3P_{\text{rad}}(3) = aT_3^4. \tag{19.14}$$

Consider now the implication for equation (19.12). The radiative flux $F_{\text{rad}} \leq cE_{\text{rad}}$ since photons cannot stream on average faster than the speed of light. Moreover, $aT^4 \gg E_{\text{rad}}$ throughout much of the region between 2 and 3 since E_{rad} will have an order-magnitude given by aT_3^4, where $aT_2^4 \gg aT_3^4$ if $T_2 \gg T_3$ (as occurs behind a hypersonic shock). Consequently, the individual terms on the left-hand side of equation (19.12) are much smaller than the first term in the integrand of the right-hand side, which represents an impossible balance unless the range of integration is much less than unity:

$$\tau_3 - \tau_2 \ll 1. \tag{19.15}$$

This conclusion forms one of the central results of our analysis: the radiative relaxation layer is *very optically thin* for a strong radiative shock in which the post-relaxation layers are optically thick. In hindsight, we see that this conclusion must follow if the shock-deposited energy that produces a strong temperature spike does not all flow down the temperature gradient into the star (a potential problem with brute-force numerical calculations that do not succeed in resolving the optically thin nature of the radiative shock).

We can present another way to see the physical reasonableness of the conclusion that the radiative relaxation layer must be very optically thin (in the continuum). If it were not, the shock front would radiate an energy flux comparable to σT_2^4, which would far exceed the rate at which the shock deposits energy per unit area into the region: $\rho_1(u_1 - v_s)(h_1 + u_1^2/2)$. Since this situation cannot hold for very long in practice, the relaxation layer must form an optically thin medium, so that it does not radiate anywhere near its blackbody limit. Note, however, that our argument may break down if the radiative relaxation occurs mainly in a few spectral lines. In that case, the effective emission opacity κ_{em}, defined by equation (19.10), may amount to only a small fraction of the the (continuum) extinction opacity, $\kappa \equiv \kappa_{\text{abs}} + \kappa_{\text{sca}}$, because the integral in equation (19.10) has weight only for a small frequency bandwidth $\Delta\nu$ (in lines) over which appreciable emission takes place. In what follows, we ignore the pitfalls presented by this possibility and continue with our analysis under the assumption that equation (19.15) does hold.

Consider then the implications of equation (19.15) for equation (19.13). Since the individual terms on the left-hand side are at least comparable to if not larger than the integrand F_{rad}/c, the small range of integration implies that the contribution of the right-hand side is negligible to the difference of $P_{\text{rad}}(1)$ and $P_{\text{rad}}(3)$. Hence, we have

$$P_{\text{rad}}(1) = P_{\text{rad}}(3), \tag{19.16}$$

which means that the momentum flux contained in the radiation field and the matter are *separately conserved* across the radiative relaxation layer. (This latter result holds only as long as the radiation momentum flux does not greatly exceed that contained in the matter.) In other words, in an optically thin region where the radiative *energy* flux has an order-unity coupling to the matter field, we should not be surprised to find a correspondingly negligible coupling of *momentum* fluxes. [Consult again the comment in Chapter 1 that "radiation can heat (or cool), but frequently finds it difficult to push."] Equation (19.16) constitutes one of the two radiative jump conditions that we wanted to derive.

The other jump condition depends on whether the preshock region 1 is also optically thick, or is optically thin. If the former, then we have two optically thick regions 1 and 3 that face each other, divided by an optically thin layer $\tau_3 - \tau_2 \ll 1$. We intuitively expect two such regions to come into thermodynamic equilibrium with each other in steady state,

$$T_1 = T_3; \tag{19.17}$$

i.e., the total jump from 1 to 3 occurs with the satisfaction of *isothermal shock jump conditions.* To derive equation (19.17) formally, we note that if region 1 is optically thick, we expect Eddington's closure relation and thermal equilibrium between matter and radiation to hold:

$$E_{\rm rad}(1) = 3P_{\rm rad}(1) = aT_1^4. \tag{19.18}$$

With equations (19.14) and (19.18), equation (19.16) gives equation (19.17). (Q.E.D.) For optically thick preshock conditions, equations (19.14), (19.16), (19.17), and (19.18), coupled with equations (19.4)- (19.6), suffice to yield a complete link between the upstream and downstream (post-relaxation) states.

What do we do if the preshock conditions are thin to optical photons but are thick to the soft X-rays that are emitted directly from the post-viscous-shock regions? (This turns out to be relevant for protostar calculations that involve relatively modest infall rates $\dot{M}$.)

In this case (Figure 19.3), outwardly traveling optical photons can fly freely past radius 1, but X-ray photons get absorbed there and reprocessed into optical photons. Upon re-emission, in an optically thin medium, half of these optical photons would generally fly outward and half inward. The latter get absorbed in the layers below radius 3, and therefore contribute to the heating of region 3. We now compute the equilibrium value for T_3.

For values of $\tau \geq \tau_3$ (settling zone inside the protostar), we assume we may adopt the radiation conduction approximation:

$$F_{\rm rad} = \frac{c}{3}\frac{d}{d\tau}(aT^4). \tag{19.19}$$

In the realistic problem, the accreted matter settling quasistatically into the protostar behind the post-relaxation layer contributes to changes of $F_{\rm rad}$, i.e., the settling zone does not perfectly satisfy a condition of *radiative equilibrium.* However, the effect is not large, and we shall ignore it here to simplify the discussion; in other words, we postulate that $F_{\rm rad} = F_{\rm rad}(3) =$ constant $\equiv F_3$. We may now integrate equation (19.19) to obtain the usual linear relationship between T^4 and τ (see Chapter 4 of Volume I):

$$aT^4 = \frac{3F_3}{c}(\tau - \tau_3) + E_3, \tag{19.20}$$

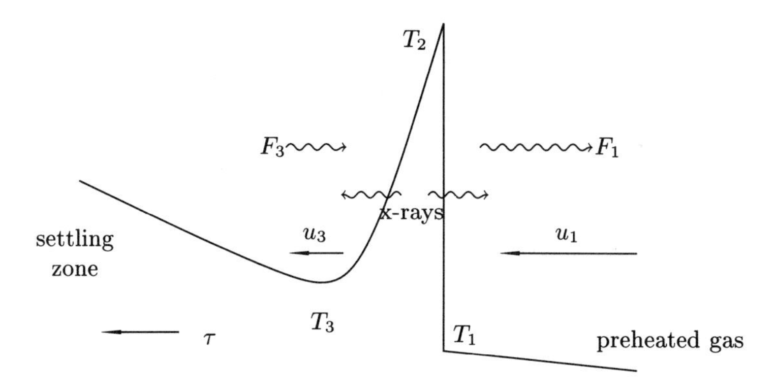

FIGURE 19.3
Radiative processes near a strong accretion shock at the surface of a protostar start with the raising of the gas to a high viscous-postshock temperature T_2. The hot gas then copiously radiates soft x-rays that travel both into and out of the star. The strong radiative cooling brings the gas to a radiative equilibrium value T_3 at the bottom of a radiative-relaxation zone. Because the radiative-relaxation layer is optically thin, the x-rays can fly freely through the zone both into and out of the star. Heating by these soft x-rays help to define the preshock and post-relaxation temperatures, T_1 and T_3. The continued press of new material added on top causes the post-relaxation gas to sink gradually in a settling zone to become part of the body of the growing star.

where we have required that $aT^4 = E_{\text{rad}}(3) \equiv E_3$ at $\tau = \tau_3$ [cf. equation (19.14)].

The LTE source function equals $c/4\pi$ times equation (19.20); hence the outwardly directed specific intensity at point 3 can be obtained as

$$I(\tau_3, \mu) = \int_{\tau_3}^{\infty} \left(\frac{caT^4}{4\pi} \right) e^{-(\tau-\tau_3)/\mu} \frac{d\tau}{\mu} = \frac{3F_3}{4\pi}\mu + \frac{cE_3}{4\pi} \qquad \text{for} \qquad \mu \geq 0, \tag{19.21}$$

where μ is the cosine of the angle of the emergent ray with respect to the radial direction (Figure 19.4).

We evaluate the constant E_3 by requiring that the first angular moment of $I(\tau_3, \mu)$ yield the net flux F_3:

$$F_3 = 2\pi \int_0^1 I(\tau_3, \mu)\mu \, d\mu + 2\pi \int_{-1}^{0} I(\tau_3, \mu)\mu \, d\mu. \tag{19.22}$$

The first integral can be performed by substituting equation (19.21). To compute the second integral, we need to count all the incoming photons.

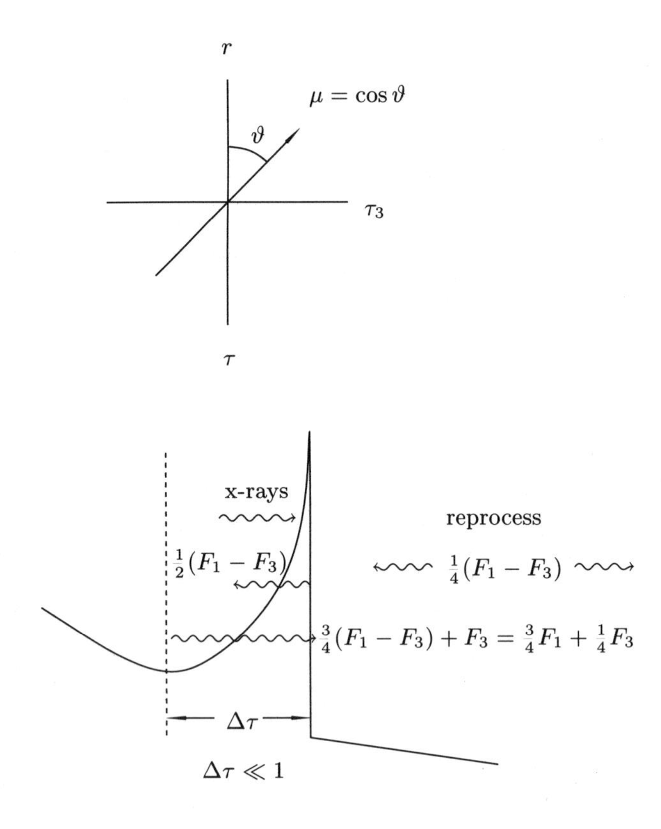

FIGURE 19.4
Geometry and the radiative energy budget for an accretion shock.

The total rate of energy release by shock deposition equals $F_1 - F_3$ (see Figure 19.4).

Half the X-rays so generated between τ_2 and τ_3 fly outward; half inward. The half, $(F_1 - F_3)/2$, in soft X-rays that fly outward will be absorbed by the preshock gas and be reprocessed into optical photons. Half of the latter fly freely outward (into the optically thin preshock region); half inward. Thus, region 3 receives $-(F_1 - F_3)/2$ in soft X-rays from the shock layer, and $-(F_1 - F_3)/4$ from reprocessed optical photons from the preshock layer, giving $-3(F_1 - F_3)/4$ as the contribution of the second integral on the right-hand side of equation (19.22). (We ignore here the possibility of "backwarming" from infrared photons from hot dust that may surround the protostar.) Equation (19.22) now becomes

$$F_3 = \left[\frac{1}{2}F_3 + \frac{1}{4}cE_3\right] - \left[\frac{3}{4}(F_1 - F_3)\right], \tag{19.23}$$

which we may solve for $cE_3/4 = \sigma T_3^4$ to obtain the desired result:

$$T_3 = \left[\frac{1}{\sigma}\left(\frac{3}{4}F_1 - \frac{1}{4}F_3\right)\right]^{1/4}. \tag{19.24}$$

Equation (19.24) replaces equation (19.17) when the preshock conditions are thin to optical photons rather than thick.

Notice that if $(F_1 - F_3) \ll F_3$, so that the shock-deposited release of energy is much less than the intrinsic stellar luminosity (as happens, for example, for high-mass protostars or for very low rates of mass infall), we obtain the standard formula for the boundary temperature of a plane-parallel atmosphere in the Eddington approximation:

$$T_3 = (2)^{-1/4}T_{\text{eff}} \qquad \text{where} \qquad T_{\text{eff}} \equiv (F_3/\sigma)^{1/4} \qquad \text{when} \qquad F_1 = F_3.$$

Notice also that although equation (19.24) does not exactly reproduce Larson's formula, equation (19.1), the discrepancy does not make a crucial difference to his problem. Finally, notice that the flux of photons seen by an outside observer from the radiative relaxation layers equals

$$2\pi \int_0^1 I(\tau_3, \mu)\mu \, d\mu = \frac{1}{2}F_3 + \frac{1}{4}cE_3 = \frac{3}{4}F_1 + \frac{1}{4}F_3.$$

In addition, this observer sees $(F_1 - F_3)/4$ of the reprocessed X-rays that fly outward. In other words, the outside observer sees a sum equal to F_1, the total radiative flux at position 1. Larson's setting of σT^4 in the post-relaxation layer equal to the flux F_1 (a common practice also in X-ray astronomy when dealing with accreting neutron stars) contains a conceptual error, but not a serious numerical one (when the preshock layers are optically thin).

RADIATIVE PRECURSORS

How does the transition occur from the optically thin case, equation (19.24), to the optically thick case, equation (19.17)? A detailed answer to this question involves the problem of *radiative precursors* and would require a close look at the radiative transfer problem in the preshock gas when this region is partially optically thick. Intuitively, we expect the gas to suffer preheating by the emergent radiation from the radiative relaxation layer behind the viscous shock. For a sequence of precursor regions of increasing optical depth (corresponding, e.g., to increasing rates of mass infall $\dot{M}$), we expect the radiative heating to raise the preshock temperature T_1 until it eventually equals the post-relaxation equiibrium value T_3, as predicted by equation (19.17). For the fully self-consistent problem (the evolution of the protostar), the equilibrium temperature T_3 itself may not remain constant for such a sequence of models; indeed, it will rise to whatever value proves necessary so that the entire region, including the radiative precursor, can push out the requisite shock-deposited energy resulting from the conversion of the kinetic energy of infall into heat. Until we get to very strong shocks, the radiative deceleration of gas in the precursor region will usually be a minor effect, and the main issues involved in the precursor problem concern radiative transfer rather than gas dynamics. We defer such discussions, therefore, to other books; the interested reader may wish to consult, for example, the discussion of Zeldovich and Raizer on the issue of "supercritical shocks."

IMPLICATIONS FOR PROTOSTELLAR RADIUS

Using considerations like those discussed above, S. W. Stahler, F. H. Shu, and R. E. Taam (see also K.-H. Winkler and M. J. Newman, 1980, *Ap. J.*, **236**, 201) found that low-mass protostars that accumulate their mass over an interval $\sim 10^5$–10^6 yr lose much more binding energy than is accounted for by Hayashi's original arguments. As a consequence, such young stellar objects have radii equal to several times their main-sequence values. The actual sizes somewhat exceed those estimated by Larson (giving pre-main-sequence stars finite time on the convective tracks of Hayashi) if one properly includes the effects of deuterium burning. Deuterium possesses a store of nuclear energy comparable to the gravitational binding energy of T Tauri stars (pre-main-sequence stars of sunlike masses); consequently, when deuterium undergoes fusion, it tends to thermostat the central stellar temperatures to $\sim 10^6$ K, yielding radii for the corresponding pre-main-sequence stars that amount to several times their main-sequence values. This result, involving as it does the physics of nuclear burning, is relatively insensitive to whether the original mass buildup occurred through spherical infall or through disk accretion. The inferred sizes of the corresponding stars agree reasonably well with observations of the locations in the H-R diagram of the youngest and most luminous T Tauri stars.

STABILITY OF SHOCK LOCATION

In the interim, a new question has arisen concerning the stability of a radiating accretion shock with respect to oscillations of its standoff distance from the surface (r_3) of a star. Using a radiative hydrodynamics code to study spherical accretion onto a white dwarf, W. Langer, G. Chanmugan, and G. Shaviv [1981, *Ap. J. (Letters)*, **245**, L23] found the radiative relaxation region to suffer an oscillatory instability ("overstability"). The physical basis of the instability is as follows. When the shock front moves outward, increasing the relative speed $|u_1 - v_s|$, the gas heats to a higher postshock temperature T_2. If, as a consequence, the cooling time should become longer than the steady-state case, then the standoff distance $(r_2 - r_3)$ increases, maintaining a larger radiative relaxation layer. When the shock front moves inward, the situation reverses. M. Balluch (1991, *Astr. Ap.*, **243**, 205) has claimed that this effect underlies the large oscillations of the shock radius reported by some workers from their hydrodynamic simulations of protostellar collapse. In Balluch's own calculations, however, the oscillation amplitude is relatively small (a few percent) during the main accretion phase; so the mechanism, if operative, probably has little effect on the net evolutionary history of the quasisteady models that assume no oscillations.

R. Chevalier and J. Imamura (1982, *Ap. J.*, **261**, 543) performed a linear stability analysis of the (plane-parallel) situation when the (optically thin) radiation process can be characterized by a cooling function (see Chapter 16), $\rho\mathcal{L} \propto \rho^2 T^\alpha$ (with $\alpha = 0.5$ for bremsstrahlung [or free-free] radiation), and they established that oscillatory instability of the shock location arises when the temperature exponent α is less than a certain critical value (~ 0.4 for the fundamental mode, ~ 0.8 for the first and second overtone, etc.). Replacing $\mathcal{L}$ by $\kappa_{\text{em}} caT^4$ [cf. equation (554)], we see that instability might prevail in the protostellar case if κ_{em} were given (in the LTE approximation), by Kramers' law, $\kappa \propto \rho T^{-7/2}$ (cf. Vol. I); however, no one has yet taken into account the influence of non-LTE effects on the bound-bound and bound-free transitions likely to be of primary importance to the protostellar accretion problem.

20

Ionization Fronts and Expanding H II Regions

References: Spitzer, *Physical Processes in the Interstellar Medium*, Chapter 12; H. W. Yorke, 1986, *Ann. Rev. Astr. Ap.*, **24**, 49.

A landmark in the modern understanding of gaseous nebulae came in 1939 with Bengt Stromgren's study of H II regions surrounding newly born OB stars (see Volume I). However, Stromgren's classic paper analyzing the structure of H II regions concerned itself only with the problem of the ionization balance. Later work by Lyman Spitzer and others showed that the temperatures of H II regions are $T_2 \approx 10^4$ K, much hotter than the temperatures T_1 of the neutral gas (hereafter assumed to be H I for simplicity of discussion) within which H II regions may be embedded. Since the density of the parent gas cloud probably increases, or at best remains constant, as we go closer to the recently formed star, a great pressure imbalance should exist between its Stromgren sphere and the ambient gas. This mechanical imbalance should cause the H II region to expand, lowering the density of the ionized gas until the final, enlarged, H II region comes into pressure equilibrium with its surroundings (or, more likely, until the star explodes in a supernova explosion).

In actual practice, two facts complicate this simple picture: (a) *molecular* hydrogen (H_2) clouds rather than atomic hydrogen (H I) clouds yield the main sites for present-day star formation, and (b) young stars ubiquitously possess powerful winds that help processes such as expanding H II regions sweep away the natal material. Nevertheless, there is historical as well as pedagogical value for us to study the idealized problem of the sudden appearance of a luminous OB star in an infinite static region of pure atomic hydrogen with uniform initial number density n_0. How does the presence of such a powerful source of ionization and heating affect the subsequent evolution of its surroundings?

OUTLINE OF EVENTS

Let $N_u(t)$ represent the number of ultraviolet photons beyond the Lyman limit released per second by the star, which we place at the origin of a spherical coordinate system (r, θ, φ). We idealize the star as having negligible size and gravity on the scales of primary interest. We suppose the ultraviolet output to turn on instantaneously at $t = 0$, and to maintain a constant (main-sequence) value thereafter:

$$N_u(t) = 0 \quad \text{for} \quad t < 0, \quad \text{and} \quad N_u(t) = N_* \quad \text{for} \quad t \geq 0. \tag{20.1}$$

We characterize the responses in density ρ and velocity $\mathbf{u} = u\mathbf{e}_r$ of the surrounding medium as

$$\rho = m_H n_0 \quad \text{for} \quad t < 0, \quad \text{and} \quad \rho = \rho(r,t) \quad \text{for} \quad t \geq 0; \tag{20.2}$$

$$u = 0 \quad \text{for} \quad t < 0, \quad \text{and} \quad u = u(r,t) \quad \text{for} \quad t \geq 0. \tag{20.3}$$

At all stages of the evolution before the star dies, a sharp transition, called an *ionization front*, will generally exist between gas that is nearly completely ionized (H II region) and gas that is nearly completely neutral (H I region), with excess ultraviolet photons (if any) causing the front to encompass ever more neutral gas if the rate of recombinations within the H II region fails to keep up with the rate of production of ionizing radiation by the central star. The characteristics of the motion of such ionization fronts, first elucidated by Franz Kahn, have elements in common with the propagation of flames (combustion and detonation fronts; see Chapter XIV of Landau and Lifshitz, *Fluid Mechanics*); and they largely govern how newly created H II regions will expand into the surrounding medium. In what follows, we shall show that this expansion occurs, under idealized interstellar conditions, in two nearly distinct stages (see Figures 20.1a and 20.1b):

(a) The first stage, lasting a few thousand years, involves the fast outward propagation of a wave of ionization as the young star attempts to ionize the surrounding hydrogen out to an equilibrium radius R_{init} equal to an "initial Stromgren sphere" (refer to Chapter 10 of Volume I):

$$R_{\text{init}} = \left(\frac{3N_*}{4\pi n_0^2 \alpha} \right)^{1/3}. \tag{20.4}$$

In equation (20.4) α is the (case B) recombination coefficient, assumed to be a constant, $\alpha = 2.6 \times 10^{-13}\,\text{cm}^{-3}$ if the temperature of the H II region remains fixed at about 10^4 K. Since the evolution in the first stage produces relatively little actual motion of any gas, n_0 represents both the number density of the surrounding H I region and the electron or proton density of the new H II region. For an O5 star born into a region of density $n_0 \sim 10^2\,\text{cm}^{-3}$ (typical value for dense H I clouds when averaged over large volumes), $R_{\text{init}} \sim 5.6\,\text{pc}$ (see Problem Set 3 of Volume I).

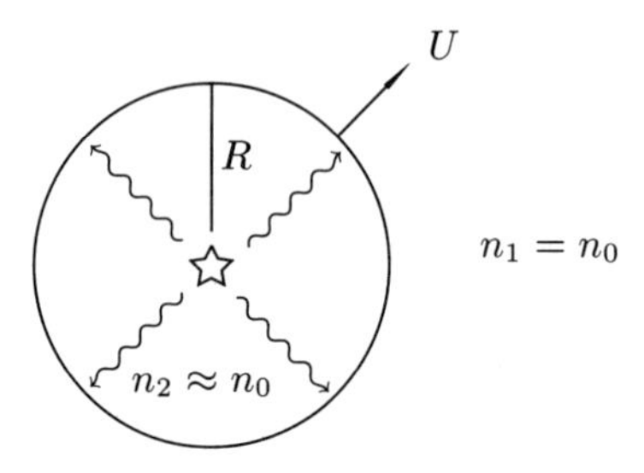

FIGURE 20.1a
In the idealized first stage of the evolution of a newly born H II region, an ionization front expands quickly at velocity U into a static H I cloud of hydrogen density $n_1 = n_0$. This expansion produces little accompanying fluid motion so that the electron or ion density of the H II region behind the ionization front satisfies $n_2 \approx n_0$ also. This stage ends when the radius R of the ionization front approaches the radius of the "initial" Stromgren sphere, R_{init}, defined so that the total rate of recombinations inside a region of uniform electron and proton density n_0 equals the rate of outpouring of ultraviolet photons beyond the Lyman limit from the central star.

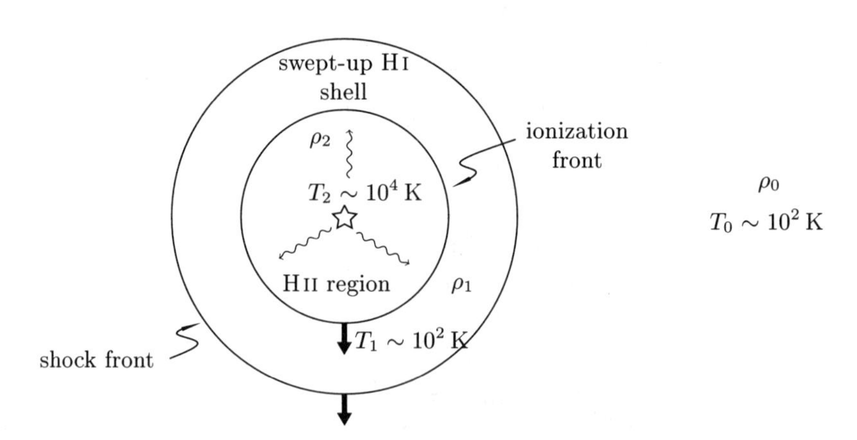

FIGURE 20.1b
In the idealized second stage of the evolution of a newly born H II region, the pressure imbalance of the H II region and the ambient H I gas drives an isothermal (radiating) shock into the atomic gas, sweeping it into a narrow dense shell. On the inner face of the dense shell, an ionization front continues to eat into the H I gas, as long as the interior H II region continues to expand, because the lowering of the electron and ion densities in the ionized gas allows a given stellar ultraviolet output to create a larger Stromgren sphere. This stage ends either when the radius R of the ionization front reaches R_{final}, defined by where the H II region can maintain *both* ionization balance in its interior and mechanical balance with its exterior, or when the central star explodes as a supernova.

(b) The second stage begins when the overpressure of the nascent H II region makes its presence felt dynamically, causing an outward expansion of the ionized gas (and not just the boundary that divides it from its nonionized counterpart). This motion pushes on the ambient medium, shocking it and compressing the swept-up neutral gas into a thin shell that possesses nearly the same pressure as the H II region driving its outward motion. Interior to this shell, the physical expansion of the ionized gas lowers the density of the H II region, slowing down the volumetric rate of recombinations (proportional to the product of electron and proton densities), and allowing the constant rate of output of stellar ultraviolet photons to ionize progressively larger volumes (and masses) of hydrogen gas. Thus, the ionization front separating the ionized hydrogen of the H II region from the neutral hydrogen in the compressed shell will continue to move out, following the path plowed by the preceding shock wave. The shock wave will gradually weaken as expansion saps the strength of the driving H II region, with the overpressure dropping to zero when the H II region reaches the size of a "final Stromgren sphere:"

$$R_{\text{final}} = \left(\frac{3N_*}{4\pi n_{\text{final}}^2 \alpha} \right)^{1/3}. \tag{20.5}$$

The final number density n_{final} of electrons or protons is given by pressure balance between the H II region and the surrounding H I region:

$$2n_{\text{final}} T_2 = n_0 T_1. \tag{20.6}$$

In its "final" state, the H II region possesses both ionization and mechanical balance (as well as thermal equilibrium), and therefore will remain in this state for as long as the star continues to maintain the same ultraviolet luminosity. Comparing equations (20.4) and (20.5), we see that R_{final} is larger than R_{init} by a factor $(2T_2/T_1)^{2/3} \sim 34$ if $T_2 \sim 10^4$ K and $T_1 \sim 10^2$ K. Thus, $R_{\text{final}} \sim 200$ pc, if $R_{\text{init}} \sim 5.6$ pc. Since it takes an H II region 2×10^7 yr to expand at $\sim 10 \text{ km s}^{-1}$ to 200 pc, and since the lifetimes of the most massive stars do not exceed a few times 10^6 yr, the H II region cannnot realistically grow to its "final" equilibrium size, but is probably fated to end inside a supernova remnant.

JUMP CONDITIONS FOR IONIZATION FRONTS

Since the transition from H I region to H II region occurs in a few photon mean free paths for ionization of the neutral medium at the Lyman limit (see Problem Set 3 of Volume I), we may regard an ionization front as a

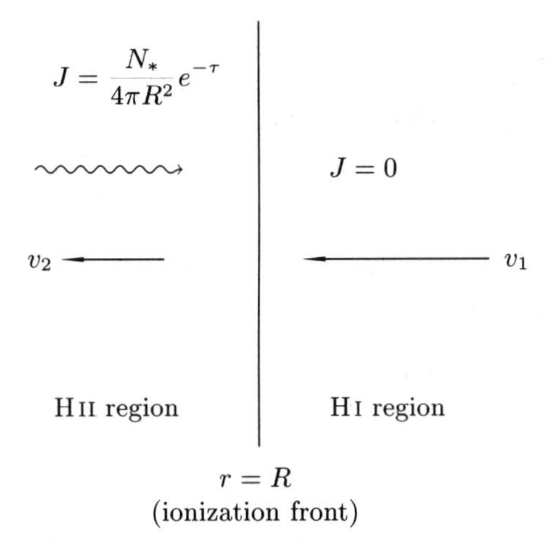

FIGURE 20.2
The gas and radiation flow across an ionization front in the frame of rest of the front. The mass flux of nonionized gas $\rho_1 v_1$ flowing into the front will be large as long as the number flux J of ionizing radiation impingent on the back (downstream) side of the front remains intense.

sharp discontinuity in a fluid treatment. In a frame that moves with the velocity U of the ionization front located at radius R, the local gas flow looks as drawn in Figure 20.2. The situation superficially resembles the case for normal shock waves; there exist, however, two important differences.

(a) Thermal balance, and not the flow dynamics, determines the temperatures T_1 and T_2 on either side of the ionization front. For simplicity, we shall assume that these temperatures remain constant (at $\sim 10^2$ K and 10^4 K, respectively) independent of density changes of the H I and H II regions.

(b) The neutral gas just ahead of the ionization front may or may not have been set into motion by a preceding shock wave. If we denote the velocity of this gas with respect to the star by u_1 (which may be zero), the associated mass-flux through the ionization front, $\rho_1(u_1 - U)$, can not have arbitrary values, but is restricted by the rate at which excess ultraviolet photons reach the front and are able to ionize more neutral material.

To derive the jump conditions for the ionization front, we first note that the equation for electron production reads

$$\frac{\partial n_e}{\partial t} + \nabla \cdot (n_e \mathbf{u}) = \mathcal{I} - \mathcal{R}. \tag{20.7}$$

In equation (20.7) $\mathcal{I}$ and $\mathcal{R}$ are the volumetric rates of ionization and recombination,

$$\mathcal{I} = -\nabla \cdot \mathbf{J}, \tag{20.8}$$

$$\mathcal{R} = \alpha n_e^2, \tag{20.9}$$

where n_e is the electron number density, $\mathbf{J}$ is the ionizing number flux at radius r,

$$\mathbf{J} = J(r)\mathbf{e}_r, \qquad J(r) = \frac{N_* e^{-\tau(r)}}{4\pi r^2}, \tag{20.10}$$

and $\tau(r)$ is the optical depth from the origin to r for absorbing ionizing photons by the small fraction of recombined protons and electrons in the H II region. By equating $\mathcal{I}$ with the negative divergence of the number flux of ionizing photons from the central star in equation (20.8), we have ignored effects due to any diffuse radiation field or finite time of propagation of the ultraviolet light. If we now place ourselves in a frame of reference that moves at the speed U of the ionization front, we easily integrate equation (20.7) across the front at $r = R$ to derive the jump condition,

$$n_{e2}(u_2 - U) + J_2 = n_{e1}(u_1 - U) + J_1 = 0,$$

where the last step comes from setting $n_{e1} = 0$ (prefront medium completely neutral) and $J_1 = 0$ (no ionizing photons beyond the ionization front). With the further identification that the number flux of atomic hydrogen flowing into the front, $\rho_1(u_1 - U)/m_H$, gets completely converted to a number flux of electrons (and protons) flowing away from the front, $n_{e2}(u_2 - U)$, we have the result,

$$\frac{\rho_1}{m_H} v_1 = J_2, \tag{20.11}$$

where we denote inward velocities relative to the motion of the ionization front by the symbol v,

$$v \equiv U - u. \tag{20.12}$$

For given ρ_1 and J_2, equation (20.11) determines $v_1 = U - u_1$. In particular, for given ρ_1, v_1 will be large if J_2 is large (strong ionizing flux illuminating the front, as must be the case, for example, before the ionization front has reached its "initial" Stromgren radius). In contrast, v_1 will be small if J_2 is small, as it must be after the ionization front passes the initial Stromgren sphere. Mass and momentum conservation across the ionization front requires (cf. Chapter 16)

$$\rho_2 v_2 = \rho_1 v_1, \tag{20.13}$$

$$P_2 + \rho_2 v_2^2 = P_1 + \rho_1 v_1^2, \tag{20.14}$$

with P_1 and P_2 given, not by energy conservation, but by the condition that the temperatures of the H I and H II gases are fixed:

$$P_1 = \rho_1 a_1^2, \qquad P_2 = \rho_2 a_2^2, \tag{20.15}$$

where

$$a_1^2 = kT_1/m_{\rm H}, \qquad a_2^2 = 2kT_2/m_{\rm H}. \tag{20.16}$$

The formulae of this section can accommodate a cosmic mix of elements simply by adjusting the isothermal sound speeds in equation (20.16), computed for pure hydrogen, to take into account the actual mean molecular weights.

The substitution of equation (20.15) into equation (20.14) results in the relation,

$$\rho_2(a_2^2 + v_2^2) = \rho_1(a_1^2 + v_1^2).$$

Since equation (20.11) motivates us to think of v_1 as known, we use equation (20.13) to eliminate $v_2^2 = v_1^2(\rho_1/\rho_2)^2$, and we then solve the resulting expression as a quadratic equation for $\rho_2/\rho_1 = v_1/v_2$:

$$\frac{\rho_2}{\rho_1} = \frac{1}{2a_2^2}\left\{(a_1^2 + v_1^2) \pm \left[(a_1^2 + v_1^2)^2 - 4a_2^2 v_1^2\right]^{1/2}\right\} = \frac{v_1}{v_2}. \tag{20.17}$$

Equation (20.16) implies $a_2^2 \gg a_1^2$ by a factor of 200 or more; so the quantity inside the square root,

$$f(v_1^2) \equiv (a_1^2 + v_1^2)^2 - 4a_2^2 v_1^2 = v_1^4 - 2(2a_2^2 - a_1^2)v_1^2 + a_1^4, \tag{20.18}$$

can become negative for intermediate values of v_1^2 neither too small nor too large (see Figure 20.3).

The roots of $f(v_1^2) = 0$ occur at

$$\begin{aligned} v_\pm^2 &\equiv (2a_2^2 - a_1^2) \pm \left[(2a_2^2 - a_1^2)^2 - a_1^4\right]^{1/2} \\ &= (2a_2^2 - a_1^2) \pm 2a_2(a_2^2 - a_1^2)^{1/2} = \left[a_2 \pm (a_2^2 - a_1^2)^{1/2}\right]^2. \end{aligned}$$

We call the larger root $v_+ \equiv v_{\rm R}$, the smaller, $v_- \equiv v_{\rm D}$, with the subscripts denoting "rarefied" and "dense" for reasons that will become apparent shortly:

$$v_{\rm R} = a_2 + (a_2^2 - a_1^2)^{1/2} \approx 2a_2, \qquad v_{\rm D} = a_2 - (a_2^2 - a_1^2)^{1/2} \approx \frac{a_1^2}{2a_2}. \tag{20.19}$$

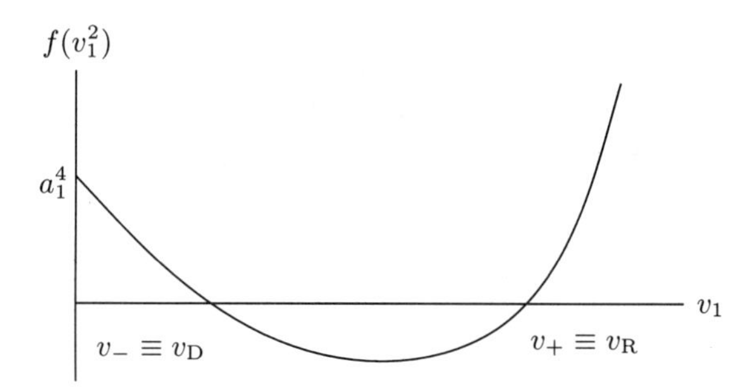

FIGURE 20.3
Schematic plot of the function $f(v_1^2)$ against the prefront velocity v_1. Notice that f dips below zero for v_1 between $v_- \equiv v_D$ and $v_+ \equiv v_R$.

The approximate parts of the expressions in equation (20.19) hold for $a_2^2 \gg a_1^2$. The exact expressions for v_R and v_D satisfy

$$v_R v_D = a_1^2 \qquad \text{and} \qquad v_R + v_D = 2a_2. \tag{20.20}$$

CLASSIFICATION OF IONIZATION FRONTS

We can now write equation (20.17) for the density and velocity jumps across the ionization front as

$$\frac{\rho_2}{\rho_1} = \frac{1}{2a_2^2}\left\{(v_R v_D + v_1^2) \pm \left[(v_1^2 - v_R^2)(v_1^2 - v_D^2)\right]^{1/2}\right\} = \frac{v_1}{v_2}, \tag{20.21}$$

which makes explicit the requirement that v_1 must satisfy either

$$v_1 \geq v_R \qquad \text{or} \qquad v_1 \leq v_D \tag{20.22}$$

for meaningful physical solutions. If $v_1 \geq v_R$, the ionization front is called R-type; if $v_1 \leq v_D$, D-type. When equalities apply, the ionization front is said to be R-critical ($v_1 = v_R$) or D-critical ($v_1 = v_D$). For each noncritical case, R-type or D-type, we have a choice between the plus or minus sign in equation (20.21). We therefore further classify R-fronts and D-fronts as being "weak" or "strong," with the appellation "weak" attached to a small contrast in densities; "strong," to a large contrast. Thus, for R-type fronts, we associate "weak" with the minus sign and "strong" with the plus sign; for D-type, "weak" with the plus sign and"strong" with the minus sign.

Weak and strong branches coalesce when the ionization fronts become R-critical or D-critical.

The reason for the R and D nomenclature now follows. For given ionizing flux J_2 reaching the front, the velocity v_1 in equation (20.11) will be large (larger than $v_{\rm R}$) if ρ_1 is very small, i.e., if the prefront medium is very *rarefied.* Alternatively, v_1 will be large if J_2 is large for fixed ρ_1. We expect R-type ionization fronts to apply, therefore, during the early stages of the evolution of an expanding H II region, before the ionization front has moved to $R_{\rm init}$ and a large ionizing flux from the central star still reaches the front.

On the other hand, for given J_2, v_1 will be small (smaller than $v_{\rm D}$), if ρ_1 is very large, i.e., if the prefront medium is very *dense.* Alternatively, v_1 will be small if J_2 is small for fixed ρ_1. We expect D-type fronts, therefore, during the late stages of the evolution of an H II region, when the ionization front is evolving from $R_{\rm init}$ to $R_{\rm final}$.

Equation (20.17) implies that the product of the "weak" and "strong" roots for v_2 of either R-type or D-type ionization fronts equals a_2^2. Thus, one root must correspond to $v_2 > a_2$; the other, to $v_2 < a_2$. In other words, one root (weak R, strong D) involves supersonic motion of the pos-tionization (H II) region relative to the front; the other root (strong R, weak D), subsonic motion. The applicable sound speed in the present context is the isothermal sound speed, not the adiabatic one. At the critical condition (when the prefront speed v_1 equals $v_{\rm R}$ or $v_{\rm D}$), the "weak" and "strong" roots merge, and the postfront fluid motion v_2 is exactly sonic ($a_2 \approx 13\,{\rm km\,s^{-1}}$). Let us now examine the implications of these properties a little more closely.

FIRST EXPANSION PHASE AND R-TYPE IONIZATION FRONTS

If we plot for a R-type front the density contrast, $\rho_2/\rho_1 = v_1/v_2$, versus the upstream speed in units of the downstream sound speed, v_1/a_2, we get the schematic result indicated in Figure 20.4 (not to scale): For the weakest ionization fronts on the weak-R branch, where $v_1 \gg a_2$ toward the bottom of the diagram, the density contrast,

$$\frac{\rho_2}{\rho_1} \to 1 + \frac{a_2^2}{v_1^2} \approx 1 \qquad \text{for} \qquad v_1 \gg a_2.$$

Equal density ahead and behind the front corresponds to the starting condition for the expanding H II region, right after the star turns on. The ionization front races out at speed U into static matter (so the gas flows inward at speed $v_1 = U$ relative to the front), without changing appreciably the initial density of the medium. In this phase of the evolution, every part of the H II and H I regions has nearly the same mass density $\rho = m_{\rm H} n_0$,

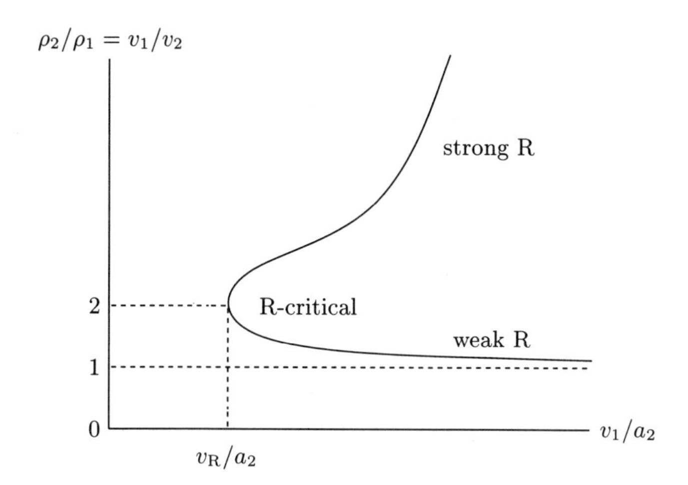

FIGURE 20.4
Schematic plot of $\rho_2/\rho_1 = v_1/v_2$ versus v_1/a_2 for R-type ionization fronts. Physical solutions exist only for $v_1 \geq v_R$. The R-critical case $v_1 = v_R$, where $v_2 = a_2$, separates the weak R branch from the strong R branch. For large v_1 the weak R solution asymptotically approaches a value of unity for $\rho_2/\rho_1 = v_1/v_2$, whereas the strong R solution yields $\rho_1/\rho_2 = v_1/v_2 \to v_1^2/a_2^2$.

although the *number* density of particles increases through ionization by a factor of two as gas flows across the front. The early evolution of an expanding H II region therefore has a simple description, which we now quantify by an analytical calculation.

Since the evolution toward the establishment of the "initial" Stromgren sphere involves little mass motion or change of mass density, charge conservation, or equation (20.7), represents the only nontrivial consideration in the problem. If we integrate equation (20.8) over the volume V of the instantaneous H II region, we have

$$\int \mathcal{I}\, dV = -\int \nabla \cdot \mathbf{J}\, dV = N_*. \tag{20.23}$$

To derive the last result from equation (20.10), we have invoked the divergence theorem to convert the (negative of the) second volume integral to the difference of a surface integral just outside the star (which yields N_* because $\tau = 0$ at the star) and a surface integral just outside the H II region (which yields nothing, because $J_1 = 0$ by assumption). Equation (20.23) states that all the ionizing photons released per second by the star gets used up inside the H II region if we include the ionization front as part of V. If we integrate equation (20.9) over $V = 4\pi R^3/3$, we obtain

$$\int \mathcal{R}\, dV = \alpha n_0^2 \frac{4\pi}{3} R^3, \tag{20.24}$$

where $n_e \approx n_0$ for $dR/dt = v_1 \gg a_2$. Performing the integration of equation (20.7) over $V(t)$ now yields

$$\frac{d}{dt}\left(n_0 \frac{4\pi}{3} R^3 \right) = N_* - \alpha n_0^2 \frac{4\pi}{3} R^3. \tag{20.25}$$

We define $z \equiv (R/R_{\text{init}})^3$ as the fraction of the volume of the initial equilibrium Stromgren sphere occupied by the H II region, with R_{init} given by equation (20.4), and we rewrite equation (20.25) as

$$\frac{dz}{dt} = \alpha n_0 (1-z) \equiv \frac{1}{t_{\mathcal{R}}}(1-z), \tag{20.26}$$

where

$$t_{\mathcal{R}} \equiv \frac{1}{\alpha n_0} \tag{20.27}$$

is the recombination time in a region of electron density n_0. For a temperature $T_2 = 10^4$ K and a density $n_0 = 10^2\ \text{cm}^{-3}$, $t_{\mathcal{R}} = 1.2 \times 10^3$ yr. The solution of equation (20.26), subject to the initial condition $z = 0$ at $t = 0$, reads

$$z = 1 - \exp(-t/t_{\mathcal{R}}); \tag{20.28}$$

i.e., the volume of the H II region relaxes from small values to the "initial" ionization-equilibrium value with a time constant $t_{\mathcal{R}}$. The velocity U of the ionization front can be obtained by differentiating $R = R_{\text{init}} z^{1/3}$, which gives

$$U = \frac{dR}{dt} = R_{\text{init}} \frac{dz/dt}{3z^{2/3}} = \left(\frac{R_{\text{init}}}{3t_{\mathcal{R}}} \right) \frac{(1-z)}{z^{2/3}}, \tag{20.29}$$

For an O5 star, $R_{\text{init}} \approx 5.6$ pc for the values of n_0 and T_2 adopted here (see Problem Set 3 of Volume I), and the combination $R_{\text{init}}/3t_{\mathcal{R}} = 1300\ \text{km s}^{-1}$ is much greater than $a_2 \approx 13\ \text{km s}^{-1}$. For $t \to 0$ ($z \to 0$), U formally $\to \infty$, so equation (20.29) needs modification at small times after the star turns on to account for the finite propagation speed of the ultraviolet photons. A more significant breakdown occurs for large t, when $R \to R_{\text{int}}$ ($z \to 1$) and equation (20.29) implies $U \to 0$. Clearly, the assumption $v_1 = U \gg a_2$ made to derive equation (20.29) must fail before R reaches R_{init}.

As R increases, we slide along the curve in Figure 20.4 from the extreme weak-R branch toward the R-critical point. At the R-critical point, our previous discussion demonstrates that $v_1 = v_{\text{R}} \approx 2a_2$, whereas $v_2 = a_2$.

Thus, the relative motion between prefront gas and postfront gas suffers a discontinuous jump, $v_1 - v_2 \approx a_2$, which is sonic with respect to gas in the H II region. Since the prefront H I gas remains still, the gas of the H II region itself must be itself moving outward at a speed equal to its internal sound speed. Clearly, we can not ignore the effects of fluid motion under these circumstances.

To estimate when the ionization front becomes R-critical, let us use equation (20.29)—which has broken down, but not badly—to get an approximate criterion. If we set U equal to $v_{\rm R} \approx 2a_2$ in equation (20.29), we get the R-critical condition at

$$\frac{(1-z)}{z^{2/3}} \approx \frac{2a_2}{R_{\rm init}/3t_{\mathcal{R}}},$$

i.e., at $(1-z)/z^{2/3} = 26\,{\rm km\,s^{-1}}/1300\,{\rm km\,s^{-1}} = 0.02$, if we use our previous fiducial numbers. When the ionization front occupies a volume within about 2 percent of its initial ionization-equilibrium value, the expansion of the H II region turns fully dynamical. In other words, the first stage of the expansion of our fiducial H II region ends when $t = t_0 = 3.9\ t_{\mathcal{R}} \approx 5 \times 10^3\,{\rm yr}$ after the star turns on. What happens next?

TRANSITION TO D-TYPE IONIZATION FRONTS

Since the expansion velocity of the ionization front, $U = dR/dt$, can not drop below $v_{\rm R}$ as long as the front remains on the R branch, and since a gap between $v_{\rm R}$ and $v_{\rm D}$ exists for allowable values of v_1 (the relative speed between the prefront gas and the front itself), how does the evolution proceed once the ionization front becomes R-critical? Numerical simulations by W. G. Mathews (1965, *Ap. J.*, **142**, 1120) and B. M. Lasker (1966, *Ap. J.*, **143**, 700) provide the answer; for here, let us reason out the answer qualitatively.

A clue comes from the fact that when the R-critical condition is reached, the gas in the H II region just behind the front is moving into the preionization-front medium at a speed equal to $a_2 \gg a_1$. This should drive a shock wave into the preionization-front gas. Before this point, the large pressure discrepancy between the H II region and the H I region ahead of it has no chance to act dynamically, because the ionization front races ahead with speed U so much faster than a pressure wave can catch it. When the ionization front slows down to a speed $v_1 = v_{\rm R} \approx 2a_2$, however, the pressure wave (moving at a speed a_2 on top of the speed $u_2 \approx a_2$ that the H II fluid itself moves) can catch up with the ionization front and overtake it. In so doing, the pressure wave will steepen into a shock wave, thereby compressing the atomic material behind it into a denser state that

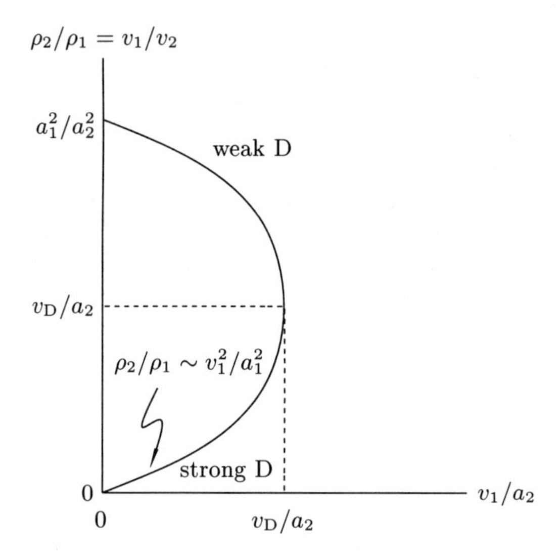

FIGURE 20.5
Schematic plot of $\rho_2/\rho_1 = v_1/v_2$ versus v_1/a_2 for D-type ionization fronts. Physical solutions exist only for $v_1 \leq v_D$. The D-critical case $v_1 = v_D$, where $v_2 = a_2$, separates the weak D branch from the strong D branch. For very small v_1, the weak D solution approaches a value of a_1^2/a_2^2 for $\rho_2/\rho_1 = v_1/v_2$, whereas the strong D solution yields $\rho_2/\rho_1 = v_1/v_2 \to v_1^2/a_1^2$.

the lagging ionization front then has to eat into. Thus, can the R-type front, which has evolved from R-weak to R-critical in the first phase of the expansion, become a D-type front, via the intermediary of a preceding (isothermal) shock? To answer this question, let us consider more closely the property of D-type ionization fronts (see Figure 20.5).

Notice that the weakest of weak-D ionization fronts corresponds to a density discontinuity,

$$\frac{\rho_2}{\rho_1} = \frac{a_1^2}{a_2^2},$$

i.e., to *static pressure equilibrium*, $\rho_2 a_2^2 = \rho_1 a_1^2$, the state that we expect for the final Stromgren sphere. Thus, we anticipate that the evolution in the final stages must involve the propagation of weak D fronts. Since we cannot get to weak D from strong D by a monotonic evolution of v_1, can the second phase of the expansion of an H II region start with a D-critical ionization front? In particular, since we have previously argued that the first phase of the expansion of an H II region ends with an R-critical ionization front, is

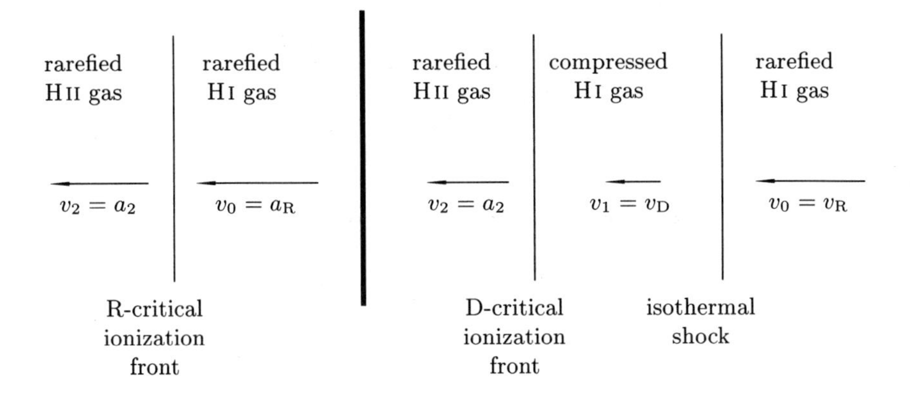

FIGURE 20.6
Two visualizations of the transition between stage 1 and stage 2 of the early evolution of an expanding H II region. In the first visualization, we have a R-critical ionization front where the incoming fluid velocity relative to the front $v_0 = v_R$ and the outgoing velocity $v_2 = a_2$. The transition takes rarefied H I gas directly to rarefied H II gas. In the second visualization, we have rarefied H I gas entering an isothermal shock front at velocity $v_0 = v_R$; the compressed H I gas exits at velocity $v_1 = v_D$ (note $v_0 v_1 = v_R v_D = a_1^2$ for an isothermal shock) and enters a D-critical ionization front, where the gas gets ionized, and becomes rarefied in the process, with an outgoing H II velocity $v_2 = a_2$. The two visualizations have the same net result and are exactly equivalent when the location of the isothermal shock coincides with the location of the D-critical ionization front, a circumstance that defines (in the idealized case) the exact moment of the transition between stage 1 and stage 2 of the overall problem.

there some way to make a transition from a R-critical front to a D-critical front, and thereby make a transition from the first phase of evolution to the second?

The answer to this question is an elegant yes, because it turns out that

> an R-critical ionization front = an isthothermal shock + a D-critical front.

Figure 20.6 illustrates the physical situation.

Proof: An isothermal shock is a viscous shock plus a radiative layer, approximated to be very thin. Chapter 16 showed that, across an isothermal shock, the product of the upstream velocity v_0 of the undisturbed medium (relative to the motion of the shock front) and the downstream velocity of the compressed gas v_1 equals the square of the isothermal sound speed a_1^2:

$$v_0 v_1 = a_1^2 \tag{20.30}$$

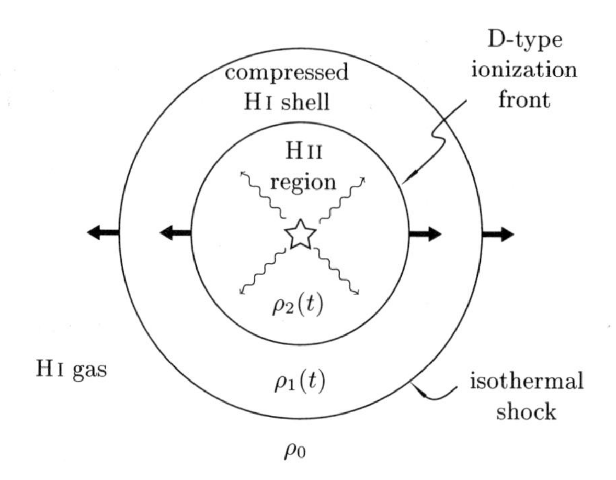

FIGURE 20.7
Schematic depiction of the physical situation during the second stage of the expansion of a newly born H II region. An isothermal shock wave precedes an outwardly expanding (weak D) ionization front. The shock wave sweeps up the ambient atomic gas into a narrow H I shell, whose back side the ultraviolet radiation from the central star ionizes and adds to the material of the growing H II region. In the analytical treatment discussed in the text, we approximate the densities in the compressed H I shell and in the H II region, $\rho_1(t)$ and $\rho_2(t)$, as time-variable but spatially constant.

In the critical state, the shock front moves at the same speed as the ionization front; so the velocites denoted by v's all refer to the same frame. If the further transition of state 1 to state 2 occurs via a D-critical ionization front, then $v_1 = v_{\rm D}$ and $v_2 = a_2$. This implies $v_0 = a_1^2/v_{\rm D} = v_{\rm R}$, since equation (20.20) states $v_{\rm R}v_{\rm D} = a_1^2$. If we think of region 1 as being very thin, we can also think of the net transition from state 0 to state 2 as taking place by an R-type ionization front, with $v_0 = v_{\rm R}$ and $v_2 = a_2$. This demonstrates our assertion that an R-critical front equals an isothermal shock plus a D-critical front. (Q.E.D.)

SECOND PHASE OF EXPANSION OF H II REGIONS

After the transition from R-type to D-type has been made, we have the physical situation depicted in Figure 20.7. How does the further evolution of a weak-D ionization front preceded by an isothermal shock take place? The numerical calculations of Lasker and others show that $\rho_2(r,t)$ and $\rho_1(r,t)$

are nearly independent of r, and that the region 1 (swept-up atomic gas behind a radiative shock) remains fairly thin (between 1 and 10 percent of the radius R_s of the shock front). Physically, ρ_2 remains nearly uniform because the weak-D ionization front moves subsonically with respect to the H II gas, and the pressure of the H II region has a chance to iron out any density variations in that region. The uniformity of ρ_1 results from the shell of swept-up atomic gas being thin. For purposes here, we can follow Spitzer's analytical treatment by assuming that the dynamics of the situation can be well-approximated by simply setting

$$\rho_0 = \text{constant}, \qquad \rho_1 = \rho_1(t), \qquad \rho_2 = \rho_2(t), \tag{20.31}$$

all independent of r. The assumption (20.31) then replaces any detailed consideration of the momentum equation.

We wish to show that the subsequent expansion occurs with velocities comparable to $a_2 \approx 13\,\mathrm{km\,s^{-1}}$ during the main-sequence lifetime of the central star. Because there now exists a shock front in addition to an ionization front, we must introduce notation more complicated than we had to deal with before. Relative to a frame fixed with respect to the star, we denote the velocities of the ionization and shock fronts, respectively, as U_i and U_s; we also call

u_{2i} = H II velocity just inside the ionization front,

u_{1i} = postshock H I velocity just outside the ionization front,

u_{1s} = postshock H I velocity just inside the shock front,

u_{0s} = preshock H I gas just outside the shock front.

We start on the outside and work our way in with a number of jump conditions. Clearly, $u_{0s} = 0$, since we assume that the undisturbed gas lies at rest with respect to the star. The jump conditions for an isothermal shock imply

$$(U_s - u_{0s})(U_s - u_{1s}) = a_1^2, \qquad \frac{\rho_1}{\rho_0} = \frac{(U_s - u_{0s})^2}{a_1^2},$$

which, with $u_{0s} = 0$, yields

$$u_{1s} = U_s - \frac{a_1^2}{U_s}, \qquad \frac{\rho_1}{\rho_0} = \frac{U_s^2}{a_1^2}. \tag{20.32}$$

Equation (20.32) gives $u_{1s}(t)$ and $\rho_1(t)$ if we can determine $U_s(t)$.

The approximations that $\rho_1(t)$ is independent of r and that region 1 is thin imply that $u_{1i} = u_{1s}$ in order to satisfy the equation of continuity; i.e., the back and front of a homogeneous thin shell must move at the same fluid speed. Indeed, the whole shell must move with the speed $u_{1s} = U_s - a_1^2/U_s$.

For a weak-D ionization front, the form of the jump condition (20.17) that applies has the plus sign, with $v_1 \equiv U_{\rm i} - u_{1\rm i}$ and $v_2 \equiv U_{\rm i} - u_{2\rm i}$. Substitution of $u_{1\rm i} = u_{1\rm s}$, together with the expression (20.32) for $u_{1\rm s}$, then gives

$$\frac{a_1 \eta}{U_{\rm i} - u_{2\rm i}} = \frac{a_1^2}{2a_2^2}\xi = \frac{\rho_2}{\rho_1}, \tag{20.33}$$

where we have defined

$$\eta \equiv \frac{U_{\rm i} - U_{\rm s}}{a_1} + \frac{a_1}{U_{\rm s}}. \tag{20.34}$$

$$\xi \equiv 1 + \eta^2 + \left[(1+\eta^2)^2 - \frac{4a_2^2}{a_1^2}\eta^2\right]^{1/2}, \tag{20.35}$$

If the H II region could reach the "final" state, the shock wave would become an acoustic wave of infinitesimal amplitude propagating in the neutral medium at the acoustic speed, $U_{\rm s} = a_1$. Equation (20.32) then shows that such a wave would induce negligible fluid motion, $u_{1\rm s} = 0$, and no density jump, $\rho_1/\rho_0 = 1$, in the neutral gas. The lack of upstream fluid motion would halt the ionization front, $U_{\rm i} = 0$, yielding $\eta = 0$ for equation (20.34). With $\eta = 0$, equation (20.35) gives $\xi = 2$, so the first part of equation (20.33) implies that the H II region becomes static, $u_{2\rm i} = 0$, while the second part of equation (20.33) implies that the H II region reaches pressure equilibrium with its surroundings, $\rho_2 a_2^2 = \rho_1 a_1^2 = \rho_0 a_1^2$.

In actual practice, however, the central star does not live long enough for the shock wave to propagate as a weak pressure pulse well ahead of the ionization front. As long as the shell of neutral gas between the ionization and shock fronts remains thin, we expect the difference of $U_{\rm i}$ and $U_{\rm s}$, both individually of order a_2, to be small. Indeed, we shall show *a posteriori* that $|U_{\rm i} - U_{\rm s}|$ is of order a_1^2/a_2 with $a_1 \ll a_2$, as long as $U_{\rm i} \approx U_{\rm s}$ remains of order a_2 (essentially throughout the life of an O star). As a consequence, η^2 is of order $a_1^2/a_2^2 \ll 1$, and we may approximate ξ by the simpler expression,

$$\xi = 1 + \left[1 - \frac{4a_2^2}{a_1^2}\eta^2\right]^{1/2}. \tag{20.36}$$

To obtain more general solutions than the final state for $U_{\rm i}$ and $U_{\rm s}$ from equations (20.33), (20.34), and (20.36), we first need to find $u_{2\rm i}$. The latter task requires us to examine the structure of the H II region. Inside the H II region, we have assumed that the momentum equation is consistent with the approximation $\rho(r,t) = \rho_2(t)$, independent of r. For the equation of continuity to be satisfied, we require $u(r,t)$ to have a form consistent with

$$\frac{1}{r^2}\frac{\partial}{\partial r}(r^2 u) = -\frac{1}{\rho}\frac{D\rho}{Dt} = -\frac{\dot{\rho}_2}{\rho_2}. \tag{20.37}$$

The right-hand side's being independent of r requires that the the velocity field must correspond to uniform expansion, i.e., satisfy "Hubble's law," $u(r,t) = H(t)r$, where the proportionality factor $H(t)$ is set by requiring $u = u_{2i}$ at $r = R_i(t)$, the instantaneous position of the ionization front. Thus we have the identification

$$u(r,t) = \frac{r}{R_i(t)} u_{2i}(t) \qquad \text{for} \qquad r \leq R_i(t). \tag{20.38}$$

Substitution of equation (20.38) into equation (20.37) yields

$$\frac{3u_{2i}}{R_i} = -\frac{\dot{\rho}_2}{\rho_2}. \tag{20.39}$$

Since ρ_2 is given in terms of ρ_1 by equation (20.33) and ρ_1 can be obtained from equation (20.32), equation (20.39) yields u_{2i} if we know R_i.

To get an equation for R_i, we consider the ionization balance inside the H II region. Since we deal here with a situation where an approximately uniform Stromgren sphere uses up all the ultraviolet photons (its expansion is now pressure driven rather than ionization driven), we have

$$\alpha n_2^2 \frac{4\pi}{3} R_i^3 = N_*.$$

Assuming that N_* and T_2 (on which α depends) are independent of t, we see that the above equation requires

$$\rho_2^2 R_i^3 = \text{constant} \equiv \rho_0^2 R_{\text{init}}^3, \tag{20.40}$$

where we have chosen to evaluate the constant in terms of the conditions applicable to the "initial" Stromgren sphere. Taking the logarithmic derivative of equation (20.40) yields

$$\frac{3}{2}\frac{\dot{R}_i}{R_i} = -\frac{\dot{\rho}_2}{\rho_2}, \tag{20.41}$$

with $\dot{R}_i \equiv U_i$. Comparison of equations (20.39) and (20.41) now implies

$$u_{2i} = \frac{1}{2} U_i; \tag{20.42}$$

i.e., the fluid velocity of the H II region just behind the ionization front equals half the (flame) velocity of the ionization front itself.

The substitution of equation (20.42) into equation (20.33) gives

$$\eta = \frac{a_1}{4a_2^2} U_i \xi, \tag{20.43}$$

which can be put into equation (20.36) to obtain

$$\xi = 1 + \left(1 - \frac{U_\mathrm{i}^2}{4a_2^2}\xi^2\right)^{1/2},$$

an equation for ξ that has as its meaningful solution

$$\xi = \frac{8a_2^2}{4a_2^2 + U_\mathrm{i}^2}. \tag{20.44}$$

Equations (20.34) and (20.33) now become

$$U_\mathrm{i} - U_\mathrm{s} = U_\mathrm{i}\left[\frac{2a_1^2}{4a_2^2 + U_\mathrm{i}^2} - \frac{a_1^2}{U_\mathrm{i}U_\mathrm{s}}\right], \tag{20.45}$$

$$\frac{\rho_2}{\rho_0} = \frac{4U_\mathrm{s}^2}{4a_2^2 + U_\mathrm{i}^2}. \tag{20.46}$$

Equation (20.45) provides an explicit demonstration of the approximation that we used to simplify ξ, namely, that $|U_\mathrm{i} - U_\mathrm{s}|$ is of order a_1^2/a_2 as long as $U_\mathrm{i} \approx U_\mathrm{s}$ remains of order a_2.

We now set $U_\mathrm{s} = U_\mathrm{i} = \dot{R}_\mathrm{i}$ in equation (20.46), and we use equation (20.40) to eliminate ρ_2 to obtain

$$\dot{R}_\mathrm{i} = \frac{2a_2}{[4(R_\mathrm{i}/R_\mathrm{init})^{3/2} - 1]^{1/2}}. \tag{20.47}$$

Notice that $U_\mathrm{i} = \dot{R}_\mathrm{i}$ remains of order a_2 as long as R_i does not expand very far beyond the value of the "initial" Stromgren sphere. Equation (20.47) constitutes an ODE for $R_\mathrm{i}(t)$, which we may solve subject to the "initial" condition $R_\mathrm{i} = R_\mathrm{init}$ at $t = t_0$, the "instant" of transition from stage 1 of the expansion to stage 2. According to our previous fiducial estimate, $t_0 \approx 5 \times 10^3$ yr. To an overall accuracy of about 10 percent, we may ignore the 1 on the right-hand side of equation (20.47) and obtain a rough analytic solution as

$$\frac{R_\mathrm{i}(t)}{R_\mathrm{init}} = \left[1 + \frac{7}{4}\frac{a_2}{R_\mathrm{init}}(t - t_0)\right]^{4/7}. \tag{20.48}$$

For $R_\mathrm{init} = 5.6\,\mathrm{pc}$, $a_2 = 13\,\mathrm{km\,s^{-1}}$, and $t_0 \ll t = 3 \times 10^6\,\mathrm{yr}$, we obtain $R_\mathrm{i}/R_\mathrm{init} \approx 4$ at the end of the main-sequence lifetime of an O5 star, substantially less than the factor of 34 by which the H II region needs to expand in order to reach its "final" Stromgren sphere (where the approximation $U_\mathrm{i} = U_\mathrm{s}$ would fail badly). According to this picture, then, a typical H II region spends most of its life in a fully dynamic state of expansion and ends its existence as part of the debris of a supernova remnant.

COMPARISONS WITH OBSERVATIONS

The solution derived in the previous section makes an interesting prediction concerning the mass of neutral gas swept up into a thin shell by the expansion process. This result obtains independent of the somewhat crude approximation we made to solve equation (20.47). Equation (20.32), which is valid without the approximation, states that the velocity of the shock front relative the gas just behind it in the shell equals

$$U_\mathrm{s} - u_{1\mathrm{s}} = \frac{a_1^2}{U_\mathrm{s}}. \tag{20.49}$$

On the other hand, the thin-shell assumption allows us to approximate $u_{1\mathrm{i}} = u_{1\mathrm{s}}$; thus, if we add equation (20.49) to equation (20.45), also valid without the approximation used to solve equation (20.47), we get the velocity of the ionization front relative to the gas just ahead of it in the shell:

$$U_\mathrm{i} - u_{1\mathrm{i}} = \frac{2a_1^2 U_\mathrm{i}}{4a_2^2 + U_\mathrm{i}^2}. \tag{20.50}$$

The ratio of the velocities in equations (20.49) and (20.50) equals

$$\frac{U_\mathrm{s} - u_{1\mathrm{s}}}{U_\mathrm{i} - u_{1\mathrm{i}}} = \frac{4a_2^2 + U_\mathrm{i}^2}{2U_\mathrm{i}U_\mathrm{s}} \approx \frac{4a_2 + U_\mathrm{i}^2}{2U_\mathrm{i}^2}. \tag{20.51}$$

where we have used $U_\mathrm{s} \approx U_\mathrm{i}$ to write the last equality. Finally, we use equation (20.47) itself, without any approximations, to eliminate $U_\mathrm{i} = \dot{R}_\mathrm{i}$, obtaining for equation (20.51):

$$\frac{U_\mathrm{s} - u_{1\mathrm{s}}}{U_\mathrm{i} - u_{1\mathrm{i}}} = 2\left(\frac{R_\mathrm{i}}{R_\mathrm{init}}\right)^{3/2}. \tag{20.52}$$

The expression on the right-hand side of equation (20.52) always exceeds 2 during the stage of the expansion when $R_\mathrm{i} \geq R_\mathrm{init}$. In the frame of rest of fluid 1, therefore, the shock front outruns the ionization front by more than a factor of two, and the mass in the compressed H I shell should exceed one-half of the total mass swept up by the shock. In particular, sufficiently old objects should contain more mass in the thin H I shells than in the H II regions.

A practical difficulty arises when one tries to compare this prediction with observations. Observations show that many optically visible (and therefore, probably old) H II regions appear to be *density-bounded* (runs out of gas) rather than *ionization-bounded* (runs out of photons). One possible explanation for the effect has been given by J. Franco, G. Tenario-Tagle, and P. Bodenheimer (1990, *Ap. J.*, **349**, 126, and references therein). Stars, and in particular massive stars, tend to form in molecular cloud cores that have fairly steep density gradients. When the density ρ decreases with distance r from the star faster than $r^{-3/2}$, the radius for the "initial" Stromgren sphere can formally become infinite. In that case, stage 1 of our schematic evolutionary classification involves the rapid propagation of an ionization front essentially to infinite distance, so that the entire region becomes photoionized. The resulting pressure imbalance in an isothermal H II region with a steep gradient in density then results in a outwardly propagating pressure wave that quickly steepens into a shockwave. This shockwave can set the ionized gas behind it into expansional motion at speeds considerably in excess of a_2, the exact enhancement depending on the numerical value of the exponent in the power law for the density decline with radius r. (An analytic theory for the process can be developed using similarity methods.) If the ambient density distribution is not spherically symmetric, but has a steeper decline in some directions than in others, the fast flow of ionized gas can "pop out" preferentially in those directions, much as a cork driven by the overpressure of champagne pops out in the direction of least resistance. Most of the old H II regions in the Galaxy appear to be better described by such "champagne flows" than by the more classical picture developed in this chapter. However, even this description contains oversimplification, inasmuch as spatially resolved maps of nearby H II regions like the Orion nebula show that much of the internal velocity field has a chaotic or turbulent character, rather than one of overall expansion. On the scale of a few parsecs, we probably cannot realistically ignore the complex, clumpy, and possibly fractal structure of the molecular cloud medium from which high-mass stars are actually born.

What about young H II regions? The youngest H II regions are too deeply embedded in gas and dust to be seen optically. When they possess sufficiently high emission measures, however (see Volume I for the definition of the "emission measure"), they can be found by astronomers through their radio free-free emission. Many H II regions discovered this way turn out to be "compact" or "ultracompact" H II regions, with sizes inferred to be less than 0.1 to 0.01 pc. D. B. Wood and E. B. Churchwell (1989, *Ap. J. Suppl.*, **69**, 831) estimate that 10–20 percent of all the H II regions in the Galaxy are of the compact or ultracompact variety. If we assume that most H II regions live for a few times 10^6 yr, the lifetimes of compact H II regions must exceed a few times 10^5 yr. Such a long lifetime poses a severe problem for the theory of expanding H II regions. To have the observed emission measures, the densities (and therefore the pressures) of compact H II regions have to be very high, much too large probably to be confined by any reasonable combination of ambient gas pressure and magnetic stresses. But the time to expand 0.1 pc at a speed of $10\,\mathrm{km\,s^{-1}}$ is only 10^4 yr, and the dynamical lifetimes of ultracompact H II regions should be an order of magnitude shorter still. No definitive answer yet exists for this puzzle. Proposals include the possibility that some ultracompact H II regions (especially those having a "cometary" shape) might be "ram-pressure confined" if the exciting stars, coupled with their protostellar winds, move with sufficient velocity through the ambient molecular cloud. Another promising possibility is that, like young stars of lower luminosities, massive stars are born with surrounding disks. When observed at low angular resolution, the photoevaporation of such a disk by the ionizing photons from the star may then drive a dense and continuous H II flow from a small area that may mimic the appearance of a compact or ultracompact H II region. The reservoir of matter in the disk need typically amount to only a few solar masses to explain the apparent longevity of the phenomenon.

PART III

MAGNETOHYDRODYNAMICS AND PLASMA PHYSICS

21

Magnetohydrodynamics

Reference: Jackson, *Classical Electrodynamics*, pp. 309–315.

With this chapter, we begin Part III of this Volume, the discussion of magnetohydrodynamics (MHD) and plasma physics. The name "magnetohydrodynamics" contains the subject's main concern: the dynamics of electrically conducting fluids in the presence of magnetic fields. Why this emphasis on magnetic fields? Why not also electric fields? Indeed, doesn't relativity theory teach us that we should make no absolute distinction between electric and magnetic fields? What appears as one to an observer will appear as the other according to a second observer in motion with respect to the first. Yet, despite this formal equivalence between electric and magnetic fields, the astronomer may be forgiven if he or she prefers to calculate all quantities in a particular reference frame, that in which the material sources for the electromagnetic fields have nonrelativistic motions (or equivalently, slow macroscopic time dependences). In such a frame, electric and magnetic fields do *not* generally play equal roles, because there exists a fundamental asymmetry in Maxwell's equations concerning their material *sources.*

Electric charges constitute a practical source for electric fields; magnetic charges (monopoles) do not constitute the same for magnetic fields. Although it is formally possible to construct a theory of electromagnetism that accords complete symmetry between electric and magnetic charges, in point of fact, all empirical evidence points to the conclusion that free magnetic charges are effectively absent in the present universe. (Magnetic monopoles do make a natural entrance in the formalisms of some quantum field theories, and cosmologists then invoke the phenomenon of *cosmic inflation* to explain away the apparent dearth of such particles in the present universe.) Consider the practical difference caused by this basic fact.

Large-scale electric fields are rare in the cosmos because free electric charges of both signs exist which can flow to short out such fields should they develop. Only in contrivances such as batteries can we easily create nonzero electric fields. Even in such devices, the flow of the negative charges to the anode and the positive charges to the cathode will inexorably cause the battery to "run down." Similarly, any natural batteries that might have existed initially to generate substantial potential drops in the universe must have long since largely run down. (In making such categorical statements, we must concede occasional exceptions, such as newly born pulsars.) As a general rule, then, macroscopic electric fields exist in the present universe only where there are *time-varying* magnetic fields [that then give rise to (time-varying) electric fields through Faraday's law of induction].

In contrast, no free magnetic charges exist to short out cosmic magnetic fields. The universe can not easily rid itself of magnetic fields once these get generated through the action of electric *currents.* Although negative and positive charges occur with equal abundance in almost every macroscopic volume of space, they often do not move identically. In particular, even a slight drift of electrons relative to ions can yield astrophysically powerful electric currents. In the magnetohydrodynamic regime, where we concern ourselves with slow macroscopic changes, such *conduction currents* generally vastly dominate time-varying electric fields ("displacement currents") as sources for cosmical magnetic fields. We now proceed to give mathematical expression to these physical ideas.

THE FOUR APPROXIMATIONS OF MAGNETOHYDRODYNAMICS

In this section, we wish to derive the MHD approximations that allow us to collapse the equations of Maxwell's theory of electrodynamics down to a more manageable set:

$$\nabla \cdot \mathbf{E} = 4\pi\rho_{\text{e}}, \tag{21.1}$$

$$\nabla \times \mathbf{E} = -\frac{1}{c}\frac{\partial \mathbf{B}}{\partial t}, \tag{21.2}$$

$$\nabla \cdot \mathbf{B} = 0, \tag{21.3}$$

$$\nabla \times \mathbf{B} = \frac{4\pi}{c}\mathbf{j}_{\text{e}} + \frac{1}{c}\frac{\partial \mathbf{E}}{\partial t}, \tag{21.4}$$

where ρ_{e} and $\mathbf{j}_{\text{e}}$ are, respectively, the electric charge and current densities. Because the divergence of the curl of any vector quantity equals zero, the divergence of Ampere's law, as modified by Maxwell to include the "displacement current" $c^{-1}\partial\mathbf{E}/\partial t$, equation (21.4), combined with the time derivative of Coulomb's law, equation (21.1), yields the statement of the conservation of electric charge (discovered by Benjamin Franklin):

$$\frac{\partial \rho_{\text{e}}}{\partial t} + \nabla \cdot \mathbf{j}_{\text{e}} = 0. \tag{21.5}$$

Similarly, if we take the divergence of Faraday's law of induction, equation (21.2), we obtain

$$\frac{\partial}{\partial t}(\nabla \cdot \mathbf{B}) = 0,$$

which demonstrates that Gilbert's discovery of the absence of magnetic monopoles, equation (21.3), will hold for all times if it holds at any initial instant throughout space.

To simplify the set, we proceed in four steps. First, we suppose slow time variations so that the displacement current $c^{-1}\partial \mathbf{E}/\partial t$ has negligible magnitude in comparison with either $\nabla \times \mathbf{B}$ or $(4\pi/c)\mathbf{j}_e$ in equation (21.4). At first sight, we might think that we should then also be able to discard the term $-c^{-1}\partial \mathbf{B}/\partial t$ in equation (21.2), but a little reflection demonstrates the illegitimacy of such a treatment, since we would then have nothing else with which to compare $\nabla \times \mathbf{E}$. As we shall directly verify shortly, what must happen instead is that $\mathbf{E}$ has an order of magnitude (in the laboratory frame) which equals a factor of u/c times $\mathbf{B}$, and that $c^{-1}\partial/\partial t$ operating on either $\mathbf{E}$ or $\mathbf{B}$ essentially scales as u/c times the spatial derivatives of the same quantities. Hence, for small u/c, we may ignore the term $c^{-1}\partial \mathbf{E}/\partial t$ in comparison with $\nabla \times \mathbf{B}$ in equation (21.4), giving the original form of Ampere's law:

$$\nabla \times \mathbf{B} = \frac{4\pi}{c}\mathbf{j}_e. \tag{21.6}$$

Equation (21.6) contains the approximation that electric currents effectively constitute the only source for magnetic fields in our problem. The second step of our analysis involves, then, a realization of how small drift velocities between electrons and ions can get and still supply enough current to explain typically observed magnetic fields in the cosmos. To make this estimate, we adopt a model of a medium containing an overall charge balance between free electrons, with number density n_e, and positive ions, with number density n_i,

$$\rho_e \equiv Zen_i - en_e = 0, \tag{21.7}$$

where Ze equals the (average) charge of each ion. The gas may also contain an electrically neutral component of atoms or molecules which contribute to the overall mass density ρ, but not to the charge density ρ_e. The condition of charge neutrality does not prevent the medium from possessing electromagnetic properties, in particular, from carrying an electric current $\mathbf{j}_e$, if the mean velocity of electrons $\mathbf{u}_e$ differs from that of the ions $\mathbf{u}_i$,

$$\mathbf{j}_e = Zen_i\mathbf{u}_i - en_e\mathbf{u}_e = -en_e\mathbf{v}'_e, \tag{21.8}$$

where we have made use of equation (21.7) to eliminate Zn_i and where we have defined the electron velocity relative to the ions as

$$\mathbf{v}'_e \equiv \mathbf{u}_e - \mathbf{u}_i. \tag{21.9}$$

As a numerical example, we consider the case of solar magnetic fields, which are believed to be generated in an outer convection zone of depth $L \sim 2 \times 10^{10}$ cm, containing on average an electron density $n_e \sim 10^{23}\ \text{cm}^{-3}$. The drift velocity of electrons relative to ions needed to support magnetic fields of even the greatest observed strengths, $B \sim 10^3$ G, can then be estimated from equations (21.6) and (21.9) as

$$v_e \sim cB/4\pi e n_e L \sim 10^{-12}\ \text{cm s}^{-1}.$$

This represents an absurdly small velocity difference; the formula can withstand many orders of magnitude changes in the variables of the problem, n_e, B, and L, and we would still come to the conclusion that—except for the single purpose of calculating the conduction current—the practical difference between the mean motions of electrons and ions are usually negligible in astrophysically realistic circumstances. This greatly simplifies the resultant dynamical problem, because we need not write down separate equations of motion for the electrons and the ions. Indeed, in single-component MHD, once we have calculated the conduction current (see below), we do not even make a distinction between the motions of electrons ions, and neutrals, but assume that collisions occur often enough to mechanically couple the three constituents of the gas to a well-defined mean fluid velocity $\mathbf{u}$. (In a later chapter on *ambipolar diffusion*, we relax this assumption enough to allow a slip between neutrals and the charged plasma.)

For theoretical calculations involving equations (21.6) and (21.9), we must use the laws of dynamics to discover a formula for the drift velocity of electrons relative to the ions, $\mathbf{v}_e'$. Equivalently, we may state the third step of our list of chores as the calculation of the conduction current $\mathbf{j}_e$ in terms of the other variables of the problem. This turns out in practice to mean the derivation of *Ohm's law* for this problem.

In the frame of rest of the ions (which we denote by primes), the equation of motion for a fluid of (cold) electrons can be written as

$$m_e \frac{d\mathbf{v}_e'}{dt} = -e\left(\mathbf{E}' + \frac{\mathbf{v}_e'}{c} \times \mathbf{B}'\right) - m_e \nu_c \mathbf{v}_e' + \text{gravitational and inertial terms}, \tag{21.10}$$

where we have included the possibility of inertial terms because the ion rest frame will generally not correspond to an inertial frame of reference, and where ν_c equals the mean collision frequency associated with momentum transfer ("drag") between the electrons and ions. For an incompletely ionized gas, we should also include a term on the right-hand side equal to the mean drag force exerted by the neutrals on the electron,

$$-m_e \nu_c^{(n)} (\mathbf{u}_e - \mathbf{u}_n),$$

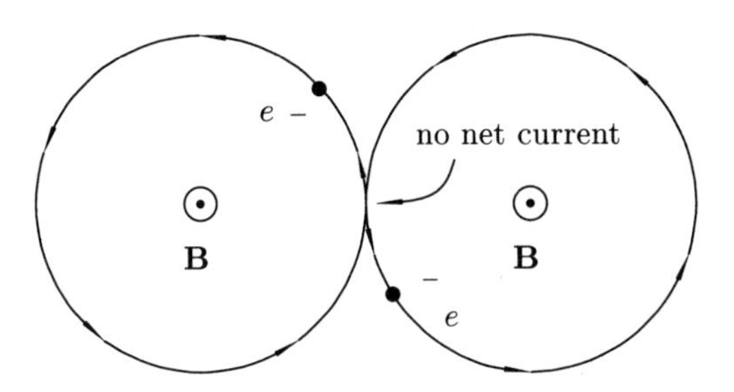

FIGURE 21.1
In the MHD region, where we assume collisions among the charged species will balance the electromagnetic forces, we can treat the effects of gyromotion somewhat cavalierly. Even when the collision frequency does not greatly exceed the gyrofrequency, however, the gyromotions of the collection of electrons (or ions) tend to cancel in their net contribution to carrying a current as long as we can ignore small-scale spatial gradients of the magnetic field. In Chapter 30, we shall consider the plasma effects that arise when these assumptions do not hold.

where $\mathbf{u}_{\rm n}$ is the fluid velocity of neutrals. In the present simplified treatment, where we wish to focus on the difference between the motion of electrons and ions, we ignore such added sophistication.

In equation (21.10), we furthermore suppose that we may ignore the inertia of the electron; i.e., we assume that the average electron quickly reaches a balance between the drag and electromagnetic forces in which we may drop the left-hand side and the terms attributed schematically to gravity and inertia. This implicitly assumes that the electron gyromotion (generally quite fast compared to the macroscopic bulk motions) does not contribute on average to the conduction current. Figure 21.1 illustrates that this follows intuitively if the collection of electrons resides in a spatially quasi-uniform region.

Strong spatial gradients in the magnetic field strength can make a difference to this conclusion, but we shall defer discussion of the phenomenon to Chapter 30. For now, we merely throw out in equation (21.10) the part of the Lorentz force $(\mathbf{v}'_{\rm e}/c) \times \mathbf{B}'$ that contributes to the gyromotion. Problem Set 4 demonstrates that the term adds nothing qualitatively new if it too is included in the balance with the frictional force.

With the above approximations, equation (21.10) yields the terminal velocity,

$$\mathbf{v}_\mathrm{e}' = -e\mathbf{E}'/m_\mathrm{e}\nu_\mathrm{c}. \tag{21.11}$$

The conduction current in the ion rest frame now becomes

$$\mathbf{j}_\mathrm{e}' = -en_\mathrm{e}\mathbf{v}_\mathrm{e}' = \sigma\mathbf{E}', \tag{21.12}$$

which represents Ohm's law in the frame of rest of the ions. The proportionality constant between the conduction current and the electric field equals the *electrical conductivity* σ, given by

$$\sigma = n_\mathrm{e}e^2/m_\mathrm{e}\nu_\mathrm{c}. \tag{21.13}$$

The fourth and final step of our analysis involves the nonrelativistic transformation relations between the ion rest frame and the laboratory rest frame. Galilean invariance for relative velocities gives $\mathbf{j}_\mathrm{e}' = \mathbf{j}_\mathrm{e}$, whereas the Lorentz transformations for the electric and magnetic fields, specialized to the case $u_\mathrm{i}/c \ll 1$, yields

$$\mathbf{B}' = \mathbf{B}, \tag{21.14}$$

$$\mathbf{E}' = \mathbf{E} + \frac{\mathbf{u}_\mathrm{i}}{c} \times \mathbf{B}. \tag{21.15}$$

To remember the second term in equation (21.15), consider the following mnemonic. If we have only a magnetic field $\mathbf{B}$ in the laboratory frame, but no electric field $\mathbf{E} = 0$, we know that, apart from a trivial uniform translation down the length of $\mathbf{B}$, a single ion will gyrate with angular velocity $\boldsymbol{\omega}_\mathrm{Li} = -Ze\mathbf{B}/m_\mathrm{i}c$ and circular velocity $\mathbf{v}_\mathrm{i} = \boldsymbol{\omega}_\mathrm{Li} \times \mathbf{r}_\mathrm{i}$. In the rotating frame, the ion feels a centrifugal force equal to $-m_\mathrm{i}\boldsymbol{\omega}_\mathrm{Li} \times (\boldsymbol{\omega}_\mathrm{Li} \times \mathbf{r}_\mathrm{i}) = -Ze(\mathbf{v}_\mathrm{i}/c) \times \mathbf{B}$. Since the ion has no motion in its own rest frame, this centrifugal force cannot be balanced by the Lorentz force associated with the magnetic field $\mathbf{B}' = \mathbf{B}$. Instead, the ion can remain at rest in the rotating frame only because this frame contains an electric field $\mathbf{E}'$ and an associated Lorentz force $Ze\mathbf{E}'$. For the electric force plus the centrifugal force to sum to zero, we require $\mathbf{E}' = (\mathbf{v}_\mathrm{i}/c) \times \mathbf{B}$, which yields a special case of the more general formula, equation (21.15), that we obtained by applying nonrelativistic Lorentz transformations. (Q.E.D.)

THE FIELD EQUATIONS OF MAGNETOHYDRODYNAMICS

We stand in good position now to derive the evolutionary equation for the magnetic field in MHD. We begin by solving equation (21.15) for $\mathbf{E}$,

$$\mathbf{E} = \frac{1}{\sigma}\mathbf{j}_\mathrm{e} - \frac{\mathbf{u}_\mathrm{i}}{c} \times \mathbf{B},$$

where we have used equation (21.12) to eliminate $\mathbf{E}'$ and expressed $\mathbf{j}'_\mathrm{e}$ as $\mathbf{j}_\mathrm{e}$. To eliminate $\mathbf{j}_\mathrm{e}$, we use equation (21.6); thus

$$\mathbf{E} = \frac{c}{4\pi\sigma}\nabla \times \mathbf{B} - \frac{\mathbf{u}_\mathrm{i}}{c} \times \mathbf{B}.$$

If we substitute the above expression into Faraday's law of induction, we obtain the desired relation,

$$\frac{\partial \mathbf{B}}{\partial t} + \nabla \times (\mathbf{B} \times \mathbf{u}_\mathrm{i}) = -\nabla \times (\eta \nabla \times \mathbf{B}), \tag{21.16}$$

where we have denoted the *electrical resistivity* η by

$$\eta \equiv \frac{c^2}{4\pi\sigma}. \tag{21.17}$$

When the conductivity σ has the expression (21.13), η takes the form

$$\eta = \frac{c^2 m_\mathrm{e} \nu_\mathrm{c}}{4\pi n_\mathrm{e} e^2} = \frac{c^2 \nu_\mathrm{c}}{\omega_\mathrm{pe}^2}, \tag{21.18}$$

where $\omega_\mathrm{pe}^2 \equiv 4\pi n_\mathrm{e} e^2/m_\mathrm{e}$ is the square of the electron plasma frequency. Expressed in the form of equation (21.18), we see explicitly that η has the units of a velocity times a length, characteristic for a diffusivity, here, the electrical diffusivity.

Equation (21.16) allows us to find the time evolution of $\mathbf{B}$ given any initial configuration if we specify η and $\mathbf{u}_\mathrm{i}$. The latter are to be found by solving the dynamical equations for the fluid. Henceforth, unless otherwise stated to the contrary, we ignore the difference between the mean ion velocity and the overall fluid velocity $\mathbf{u}$. With these remarks taken care of, we associate with equation (21.16) the initial condition

$$\nabla \cdot \mathbf{B} = 0, \tag{21.19}$$

which will be satisfied for all time if it is satisfied initially [as we can see by taking the divergence of equation (21.16)]. Only for steady-state problems do we need to regard equations (21.16) and (21.19) as two PDEs of equal standing.

THE INFLUENCE ON THE MATTER

The above description concludes formally our derivation of the relevant equations for the magnetic field. We wish now to include the effects of the electromagnetic fields on the matter. The Lorentz force acting on a charge q moving at velocity $\mathbf{v}$ equals

$$\mathbf{F}_\mathrm{L} = q\left(\mathbf{E} + \frac{\mathbf{v}}{c} \times \mathbf{B}\right),$$

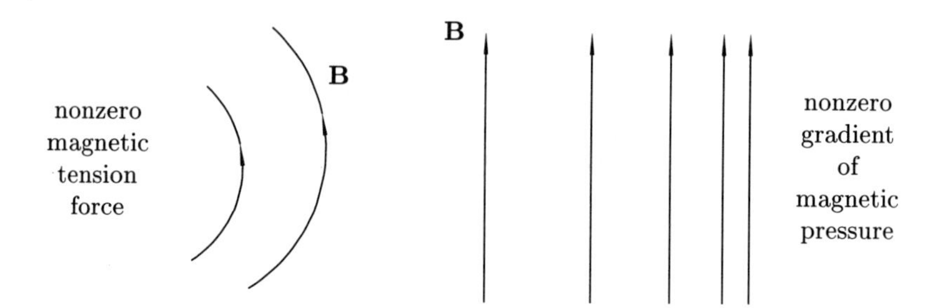

FIGURE 21.2
We can profitably think of the Lorentz force that acts on a collection of charged particles, electrically neutral in bulk, as originating in two forms: magnetic tension if the field lines have curvature, and magnetic pressure if the magnetic field has a gradient in strength.

which demonstrates that the electric force can have a magnitude comparable to that of the magnetic force even if $|\mathbf{E}|$ is only of order u/c compared to $|\mathbf{B}|$, where u equals the bulk speed of the fluid. Per unit volume, the mean Lorentz force that acts on the collection of ions and electrons equals

$$\mathbf{f}_\mathrm{L} = Zen_\mathrm{i}\left(\mathbf{E} + \frac{\mathbf{u}_\mathrm{i}}{c} \times \mathbf{B}\right) - en_\mathrm{e}\left(\mathbf{E} + \frac{\mathbf{u}_\mathrm{e}}{c} \times \mathbf{B}\right).$$

If we make use of the condition of overall charge neutrality, equation (21.7), and the definition of the conduction current, equation (21.8), we obtain

$$\mathbf{f}_\mathrm{L} = \frac{1}{c}\mathbf{j}_\mathrm{e} \times \mathbf{B} = \frac{1}{4\pi}(\boldsymbol{\nabla} \times \mathbf{B}) \times \mathbf{B}, \tag{21.20}$$

where we have made use of equation (21.6) to obtain the final form of equation (21.20). Equation (21.20) expresses the reasonable conclusion that an electrically neutral but conducting medium feels in net only the effects of magnetic forces, not of electric forces.

If we use the formula for the expansion of the triple vector product, we may write equation (21.20) in the alternative form

$$\mathbf{f}_\mathrm{L} = \frac{1}{4\pi}(\mathbf{B} \cdot \boldsymbol{\nabla})\mathbf{B} - \frac{1}{8\pi}\boldsymbol{\nabla}(|\mathbf{B}|^2). \tag{21.21}$$

The second term on the right-hand side represents the contribution to the Lorentz force from the negative gradient of the "magnetic pressure" $P_\mathrm{mag} \equiv |\mathbf{B}|^2/8\pi$; the first term on the right-hand side, the contribution from the "magnetic tension" (see Figure 21.2).

Notice that the force exerted by magnetic tension vanishes for straight field lines, since $\mathbf{B}$ has no variation along the direction of $\mathbf{B}$ if all the field lines are perfectly straight. In contrast, the magnetic pressure can have variations even for a configuration of straight field lines, as long as the field does not have uniform strength (e.g., if it bunches up at some locations relative to others). Hence, magnetic tension generally acts to straighten field lines; magnetic pressure, to spread out such lines of force more uniformly.

We may use dyadic notation to write equation (21.21) in terms of the divergence of a Maxwell stress tensor,

$$\mathbf{f}_\mathrm{L} = \nabla \cdot \overleftrightarrow{T}, \tag{21.22}$$

where $\overleftrightarrow{T}$ has the usual form given in magnetostatics,

$$\overleftrightarrow{T} \equiv \frac{\mathbf{BB}}{4\pi} - \frac{|\mathbf{B}|^2}{8\pi}\mathbf{I}, \tag{21.23}$$

with $\mathbf{I}$ equaling the unit dyadic. Expressed in either form, the vector-invariant equation (21.20) or the tensor-representation equation (21.22), the Lorentz force per unit volume needs to be added to the right-hand side of the usual force equation for the bulk fluid. When we treat the matter in this way, as a single-component conducting fluid with an overall mass density ρ and a well-defined bulk velocity $\mathbf{u}$, we implicitly assume that collisions occur frequently enough to well-couple the individual components: electrons, ions, and neutrals.

In the presence of electrical dissipation, we also need to modify the heat equation. The modification is to add the volumetric rate of work done by the conduction current $\mathbf{j}_\mathrm{c}'$ as it flows down an electric field $\mathbf{E}'$ in the rest frame of the ions,

$$\mathbf{j}_\mathrm{e}' \cdot \mathbf{E}' = \frac{1}{\sigma}|\mathbf{j}_\mathrm{e}'|^2 = \left(\frac{4\pi\eta}{c^2}\right)|\mathbf{j}_\mathrm{e}|^2, \tag{21.24}$$

where we have used equations (21.12) and (21.17). It is conventional to refer to the positive-definite rate of converting electromagnetic energy into heat, equation (21.24), as *Joule or Ohmic dissipation.*

FIELD FREEZING

The left-hand side of equation (21.16) is reminiscent of Kelvin's circulation theorem,

$$\frac{\partial \boldsymbol{\omega}}{\partial t} + \nabla \times (\boldsymbol{\omega} \times \mathbf{u}) = 0,$$

which we interpreted in Chapter 6 as the freezing of the number of vortex lines that threads any element of area that moves with the fluid. If we could drop the diffusive term on the right-hand side of equation (21.16), the analogy between $\mathbf{B}$ and $\boldsymbol{\omega}$ (for a nonmagnetic fluid) would seemingly be complete:

$$\frac{\partial \mathbf{B}}{\partial t} + \boldsymbol{\nabla} \times (\mathbf{B} \times \mathbf{u}) = 0. \tag{21.25}$$

In particular, if we denote by C the circuit forming the circumference of any element of area A that moves with the magnetized fluid, equation (21.25) has the geometric interpretation that the magnetic flux Φ threading A,

$$\Phi \equiv \int_A \mathbf{B} \cdot \hat{\mathbf{n}}\, dA, \tag{21.26}$$

remains constant with time, $d\Phi/dt = 0$. The result that the magnetic flux remains tied to the fluid is called *field freezing*, and applies if the electrical resistivity η can be considered essentially equal to zero ("infinite conductivity").

The approximation of field freezing holds in many applications in astrophysics, not because the conductivity of astrophysical gases has such large values in comparison with normal materials, but because typical systems have enormous dimensions L. As with all diffusive phenomena, we expect the effects of finite electrical resistivity to lead to a gradual spreading of the magnetic field, with a characteristic diffusion time given by

$$t_{\mathrm{D}} \sim L^2/\eta. \tag{21.27}$$

Indeed, if we can treat η as a constant and use the formula for triple vector product to expand the right-hand side of equation (21.16), we get, with $\boldsymbol{\nabla} \cdot \mathbf{B} = 0$,

$$\frac{\partial \mathbf{B}}{\partial t} + \boldsymbol{\nabla} \times (\mathbf{B} \times \mathbf{u}) = \eta \nabla^2 \mathbf{B}.$$

For zero fluid motion, $\mathbf{u} = 0$, this equation differs from the time-dependent heat-conduction equation,

$$\frac{\partial T}{\partial t} = \chi \nabla^2 T,$$

only in the fact that former involves the diffusion of a vector quantity $\mathbf{B}$, while the latter involves a scalar T. In Problem Set 4, you are asked to estimate t_{D} for a variety of magnetized astrophysical objects to justify the approximation that the right-hand side of equation (21.16) can be set equal to zero in many situations.

FURTHER ANALOGIES WITH THE VORTICITY EQUATION

The analogy between the behavior of the magnetic field $\mathbf{B}$ in a conducting medium and the vorticity $\boldsymbol{\omega}$ in a nonmagnetized fluid extends even deeper if the fluid lacks compressibility. In this case, we can manipulate the Navier-Stokes equation as to write the vorticity equation in the form

$$\frac{\partial \boldsymbol{\omega}}{\partial t} + \nabla \times (\boldsymbol{\omega} \times \mathbf{u}) = -\nabla \times (\nu \nabla \times \boldsymbol{\omega}), \tag{21.28}$$

which we may usefully compare with equation (21.16). The formal similarity between these two equations, coupled with the observation that both $\boldsymbol{\omega}$ and $\mathbf{B}$ have zero divergence, has frequently led to the suggestion that the magnetic field in a field of astrophysical turbulence should behave like the vorticity in incompressible fluid turbulence. Kolmogorov's theory describes the latter for homogeneous isotropic turbulence (see Chapter 9). How seriously, then, should we take the analogous concept of a turbulent cascade of magnetic "eddies," ultimately to be dissipated on an inner scale by the effects of magnetic resistivity?

The answer is, probably, not very seriously. A big difference exists between equation (21.16) and (21.28), in that the vorticity relates to the fluid velocity $\mathbf{u}$ appearing in equation (21.28) through $\boldsymbol{\omega} = \nabla \times \mathbf{u}$. This relation ultimately gives the nonlinear coupling that underlies the concept of the turbulent cascade. No such relation exists between $\mathbf{B}$ and $\mathbf{u}$; i.e., $\mathbf{B} = \nabla \times \mathbf{A}$, not $\nabla \times \mathbf{u}$. A better idea appears to be the notion that many chaotic waves make up the phenomenon of MHD turbulence in astrophysics, rather than many chaotic eddies. The subjects of fluid turbulence and magnetohydrodynamics yield so many complications, even when we consider them separately, that attempts to treat them simultaneously via incomplete analogies probably miss many critical points.

SUMMARY

We close this chapter with a brief summary of the two central concepts that underlie field freezing. Ideal MHD belongs to the regime where we consider collisions to occur so infrequently that individual electrons and ions can gyrate many times around field lines before they are knocked off by collisions. Microscopically, then, the charged components of the gas are effectively tied to the field lines. For small collision frequency ν_c, we have a negligible electrical resistivity η [see equation (21.18)], or essentially an "infinite" conductivity σ. Measured in terms of variables that apply to the laboratory frame, then, the conduction current,

$$\mathbf{j}_e = \sigma \left(\mathbf{E} + \frac{\mathbf{u}}{c} \times \mathbf{B} \right), \tag{21.29}$$

will acquire very large values unless the electric and magnetic fields almost exactly satisfy the relationship

$$\mathbf{E} = -\frac{\mathbf{u}}{c} \times \mathbf{B}. \tag{21.30}$$

(This expression explicitly demonstrates, by the way, that E is smaller than B typically by a factor u/c.) If $\mathbf{E}$ and $\mathbf{B}$ did not possess almost precisely this relationship, very large conduction currents would result in accordance with equation (21.29) that would act as a source for changing $\mathbf{B}$ via Ampere's law, equation (21.6), until equation (21.30) did achieve validity. On the other hand, when we substitute $\mathbf{E}$ as given by equation (21.30) into Faraday's law of induction, equation (21.2), we obtain

$$\frac{\partial \mathbf{B}}{\partial t} + \boldsymbol{\nabla} \times (\mathbf{B} \times \mathbf{u}) = 0,$$

which expresses the condition of field freezing. Thus we see that field freezing really encompasses two notions: one arising from the dynamics of the matter (free electrons and ions are well tied to the magnetic lines of force); the other relating to the dynamics of the field (induction always occurs in a highly conducting fluid in such a way as to keep fixed the total number of field lines that thread any circuit moving with the fluid). The matter is tied to the field; the field is tied to the matter: field freezing!

IMPLICATIONS FOR A NONZERO CHARGE DENSITY

Associated with an electric field $\mathbf{E}$ given approximately by equation (21.30) exists an electric-charge density in accordance with Coulomb's law:

$$\rho_{\mathrm{e}} = \frac{1}{4\pi} \boldsymbol{\nabla} \cdot \left(\mathbf{B} \times \frac{\mathbf{u}}{c} \right).$$

For our development in the current chapter to be self-consistent, we require this nonzero value for ρ_{e} to be negligibly small. In order of magnitude, we can estimate

$$\rho_{\mathrm{e}} \sim \frac{Bu}{4\pi c L},$$

where L equals the macroscopic length scale over which $\mathbf{E} = \mathbf{B} \times \mathbf{u}/c$ typically varies. If we identify the electron gyrofrequency as $\omega_{\mathrm{L}} \equiv eB/m_{\mathrm{e}}c$ and the square of the electron plasma frequency as $\omega_{\mathrm{pe}}^2 \equiv 4\pi e^2 n_{\mathrm{e}}/m_{\mathrm{e}}$, we can express the charge density ρ_{e} above as a fraction of a hypothetical fully separated value en_{e}:

$$\rho_{\mathrm{e}}/en_{\mathrm{e}} \sim \frac{(u/L)\omega_{\mathrm{L}}}{\omega_{\mathrm{pe}}^2}.$$

According to the numerical estimates in Chapter 1, the product of the hydrodynamic frequency u/L and the electron gyrofrequency ω_{L} will generally always be very small in comparison with the square of the electron plasma frequency ω_{pe}^2. Only in astrophysical environments where the magnetic field B is very strong and the fluid motion u occurs nearly at the speed of light over very small length scales L (e.g., in the vicinity of pulsars) would we need to worry about any significant charge separation occurring (nonzero ρ_e). Otherwise, within the context of the magnetohydrodynamic approximation, we may safely ignore the formal implications of Coulomb's law.

22

Hydromagnetic Equations and Hydromagnetic Waves

Reference: Spitzer, *Physics of Fully Ionized Gases*, pp. 61–67; Landau, Lifshitz, and Pitaevskii, *Electrodynamics of Continuous Media*, §69.

In this chapter we wish to write down explicitly the equations of magnetohydrodynamics; then we shall apply the nondiffusive version of these equations to examine the number and kinds of MHD waves that can propagate and carry information in a magnetized ideal gas. From this quick survey we can then appreciate how much more complicated magnetohydrodynamics is than ordinary gas dynamics, and why the subject has far fewer solved examples containing levels of realism sufficient for high-precision comparisons with observations of complex astronomical objects. Indeed, theorists in MHD have generally concentrated, with good reason, more on the study of idealized problems that exhibit the gist of important physical processes than on detailed model building. To paraphrase Hamming when speaking about computing: in this field, the goal is insight, not numbers. This realization alerts us both to the pragmatic need for reasonable expectations as well as to the opportunity for substantial future progress.

BASIC EQUATIONS

Within the confines of the magnetohydrodynamics of a single-component conducting fluid, the material effects of electromagnetic fields enter as two terms:

$$\text{Lorentz force per unit volume} = \mathbf{j}_\mathrm{e} \times \mathbf{B}/c = (\boldsymbol{\nabla} \times \mathbf{B}) \times \mathbf{B}/4\pi.$$
$$\text{Heat production by Ohmic dissipation} = |\mathbf{j}_\mathrm{e}|^2/\sigma = \eta|\boldsymbol{\nabla} \times \mathbf{B}|^2/4\pi.$$

Hence the complete set of MHD equations reads

$$\frac{\partial \rho}{\partial t} + \boldsymbol{\nabla} \cdot (\rho \mathbf{u}) = 0, \tag{22.1}$$

$$\rho\left[\frac{\partial \mathbf{u}}{\partial t}+\boldsymbol{\nabla}\left(\frac{|\mathbf{u}|^2}{2}\right)+(\boldsymbol{\nabla}\times\mathbf{u})\times\mathbf{u}\right]$$
$$=-\rho\boldsymbol{\nabla}\mathcal{V}-\boldsymbol{\nabla}P+\boldsymbol{\nabla}\cdot\overleftrightarrow{\pi}+\mathbf{f}_{\text{rad}}+\frac{1}{4\pi}(\boldsymbol{\nabla}\times\mathbf{B})\times\mathbf{B}, \quad (22.2)$$

$$\rho T\left(\frac{\partial s}{\partial t}+\mathbf{u}\cdot\boldsymbol{\nabla}s\right)=-\boldsymbol{\nabla}\cdot\mathbf{F}_{\text{cond}}+\Gamma-\Lambda+\Psi+\frac{\eta}{4\pi}|\boldsymbol{\nabla}\times\mathbf{B}|^2, \quad (22.3)$$

$$\nabla^2\mathcal{V}=4\pi G(\rho+\rho_{\text{ext}}), \quad (22.4)$$

$$\frac{\partial \mathbf{B}}{\partial t}+\boldsymbol{\nabla}\times(\mathbf{B}\times\mathbf{u})=-\boldsymbol{\nabla}\times(\eta\boldsymbol{\nabla}\times\mathbf{B}), \quad (22.5)$$

with

$$\boldsymbol{\nabla}\cdot\mathbf{B}=0 \quad (22.6)$$

to be used as an initial condition on equation (22.5), and with the constitutive relations for the material system as well as the specification of the radiation field to be supplied as additional equations.

In astrophysics, we often content ourselves with the diffusion-free approximations of these equations that we get by dropping the terms proportional to $\overleftrightarrow{\pi}$, $\mathbf{F}_{\text{cond}}$, Ψ, and η. For an ideal gas, we have the constitutive relations

$$P=\frac{\rho}{m}kT, \qquad s=c_v\ln(P\rho^{-\gamma}). \quad (22.7)$$

As an example of the application of the equations of ideal MHD, we consider the problem of hydromagnetic waves.

LINEARIZED HYDROMAGNETIC DISTURBANCES

We ignore the effects of gravitation and radiative interactions by setting $\boldsymbol{\nabla}\mathcal{V}$, $\mathbf{f}_{\text{rad}}$, and $\Gamma-\Lambda$ equal to 0. We may then take the unperturbed state of the medium to be static and homogeneous:

$$\rho_0=\text{constant}, \quad \mathbf{u}_0=0, \quad P_0=\text{constant}, \quad \mathbf{B}_0=\textbf{constant}\equiv B_0\hat{\mathbf{n}}.$$

Consider small perturbations of this system:

$$\rho=\rho_0(1+\alpha_1), \quad \mathbf{u}=\mathbf{u}_1, \quad P=P_0(1+p_1), \quad \mathbf{B}=B_0(\hat{\mathbf{n}}+\mathbf{b}_1). \quad (22.8)$$

We substitute equations (22.8) into equations (22.1)–(22.5), with $\mathcal{V}$, $\mathbf{f}_{\text{rad}}$, $\Gamma-\Lambda$, $\overleftrightarrow{\pi}$, $\mathbf{F}_{\text{cond}}$, Ψ, and η set equal to zero. Upon linearization, the resulting perturbation equations read

$$\frac{\partial\alpha_1}{\partial t}+\boldsymbol{\nabla}\cdot\mathbf{u}_1=0, \quad (22.9)$$

$$\rho_0 \frac{\partial \mathbf{u}_1}{\partial t} = -P_0 \nabla p_1 + \frac{B_0^2}{4\pi} (\nabla \times \mathbf{b}_1) \times \hat{\mathbf{n}}, \tag{22.10}$$

$$\frac{\partial p_1}{\partial t} = \gamma \frac{\partial \alpha_1}{\partial t}, \tag{22.11}$$

$$\frac{\partial \mathbf{b}_1}{\partial t} + \nabla \times (\hat{\mathbf{n}} \times \mathbf{u}_1) = 0. \tag{22.12}$$

To derive the adiabatic perturbation equation (22.11), we have used equation (22.7) to simplify equation (22.3).

Since equations (22.9)–(22.12) contain only constant coefficients, we may look for solutions having the Fourier dependences,

$$e^{i(\omega t - \mathbf{k}\cdot\mathbf{x})},$$

with ω and $\mathbf{k}$ being (real) constants. With this form, operation by $\partial/\partial t$ and ∇ is equivalent to multiplication by $i\omega$ and $-i\mathbf{k}$; thus, equations (22.9)–(22.12) become

$$i\omega\alpha_1 - i\mathbf{k}\cdot\mathbf{u}_1 = 0, \tag{22.13}$$

$$i\omega\mathbf{u}_1 = \frac{P_0}{\rho_0} i\mathbf{k} p_1 - \frac{B_0^2}{4\pi\rho_0} (i\mathbf{k} \times \mathbf{b}_1) \times \hat{\mathbf{n}}, \tag{22.14}$$

$$i\omega p_1 = i\omega\gamma\alpha_1, \tag{22.15}$$

$$i\omega\mathbf{b}_1 - i\mathbf{k} \times (\hat{\mathbf{n}} \times \mathbf{u}_1) = 0. \tag{22.16}$$

Notice that if we dot $\mathbf{k}$ into equation (22.16), we get the condition $\mathbf{k}\cdot\mathbf{b}_1 = 0$, which reproduces the constraint imposed by $\nabla \cdot \mathbf{B} = 0$. In this problem, therefore, equation (22.5) automatically incorporates the information contained in equation (22.6).

Equation (22.15) implies either $\omega = 0$ or $p_1 = \gamma\alpha_1$. The zero-frequency case, $\omega = 0$ for any $\mathbf{k}$, represents the entropy mode, which holds no interest in the current problem (see, however, problem set 3 on thermal instabilities). Hence, we set $p_1 = \gamma\alpha_1$ and eliminate p_1 from equation (22.14). We define the squares of the adiabatic speed of sound and the Alfven speed as

$$a_s^2 = \gamma P_0/\rho_0, \qquad v_{\rm A}^2 = B_0^2/4\pi\rho_0, \tag{22.17}$$

and we use equations (22.13) and (22.16) to eliminate α_1 and $\mathbf{b}_1$ from the rest of the system. This allows us to write equation (22.14) as

$$\omega^2\mathbf{u}_1 + v_{\rm A}^2\{\mathbf{k} \times [\mathbf{k} \times (\hat{\mathbf{n}} \times \mathbf{u}_1)]\} \times \hat{\mathbf{n}} - \mathbf{k} a_s^2 \mathbf{k}\cdot\mathbf{u}_1 = 0.$$

Expanding the vector cross products, we get

$$[\omega^2 - (\hat{\mathbf{n}}\cdot\mathbf{k})^2 v_{\rm A}^2]\mathbf{u}_1 + \mathbf{k}[-(v_{\rm A}^2 + a_s^2)\mathbf{k}\cdot\mathbf{u}_1 + v_{\rm A}^2(\hat{\mathbf{n}}\cdot\mathbf{k})\hat{\mathbf{n}}\cdot\mathbf{u}_1] + \hat{\mathbf{n}} v_{\rm A}^2(\hat{\mathbf{n}}\cdot\mathbf{k})\mathbf{k}\cdot\mathbf{u}_1 = 0. \tag{22.18}$$

To make further progress, we need to introduce a coordinate system. Define the x-y plane so that $\mathbf{k} = k\mathbf{e}_x$ and $\hat{\mathbf{n}} = \mathbf{e}_x \cos\psi + \mathbf{e}_y \sin\psi$, with ψ being the angle between $\mathbf{k}$ and $\hat{\mathbf{n}}$ (angle between $\mathbf{k}$ and $\mathbf{B}_0$). Equation (22.18) now becomes

$$\begin{aligned}&[\omega^2 - k^2 v_\mathrm{A}^2 \cos^2\psi]\mathbf{u}_1\\&+\mathbf{e}_x[-k^2(v_\mathrm{A}^2 + a_s^2)u_{1x} + k^2 v_\mathrm{A}^2 \cos\psi(u_{1x}\cos\psi + u_{1y}\sin\psi) + k^2 v_\mathrm{A}^2 \cos^2\psi u_{1x}]\\&+\mathbf{e}_y[k^2 v_\mathrm{A}^2 \sin\psi\cos\psi u_{1x}] = 0 \end{aligned} \tag{22.19}$$

Clearly, equation (22.19) splits up into modes which have $u_{1z} \neq 0$ and those which have $u_{1z} = 0$, i.e., into modes that do and do not contain motion perpendicular to the plane of the wave propagation, $\hat{\mathbf{k}}$, and the equilibrium magnetic field, $\mathbf{B}_0$.

ALFVEN WAVES

For $u_{1z} \neq 0$, we require $u_{1x} = u_{1y} = 0$ and

$$\omega^2 - k^2 v_\mathrm{A}^2 \cos^2\psi = 0,$$

i.e., for the square of the phase velocity (and group velocity) to equal

$$\frac{\omega^2}{k^2} = v_\mathrm{A}^2 \cos^2\psi. \tag{22.20}$$

If the wavefronts were perpendicular to $\mathbf{B}_0$ ($\mathbf{k} \parallel \hat{\mathbf{n}}$) so that $\psi = 0$, the wave speed would equal the Alfven speed. Consequently, equation (22.20) represents *Alfven waves with inclined wavefronts.* With this interpretation, the $\cos\psi$ factor in the signal speed merely represents a geometric factor arising from the fact that we choose to project the natural speed of propagation v_A along a field line onto a wave vector corresponding to an inclined wave front (Figure 22.1).

Notice also that Alfven waves represent *transverse* waves, since the fluid dispacement $\mathbf{u}_1 = u_{1z}\mathbf{e}_z$ is perpendicular to both $\mathbf{k}$ and $\mathbf{B}_0$. According to equation (22.16), the perturbation magnetic field fraction $\mathbf{b}_1 = -(\mathbf{k} \cdot \hat{\mathbf{n}}/\omega)\mathbf{u}_1$ opposes the fluid displacement for this mode, so a given magnetic field line looks like a plucked string (Figure 22.2).

Indeed, the formula for the Alfven velocity, $v_\mathrm{A} = (B_0^2/4\pi\rho_0)^{1/2}$, reinforces the plucked-string analogy, since $B_0^2/4\pi$ equals the tension of the magnetic field, ρ_0 is its mass density, and the wave speed of a plucked string equals the square root of the tension divided by the mass density (per unit length instead of per unit volume if we approximate the string to have zero cross-sectional area).

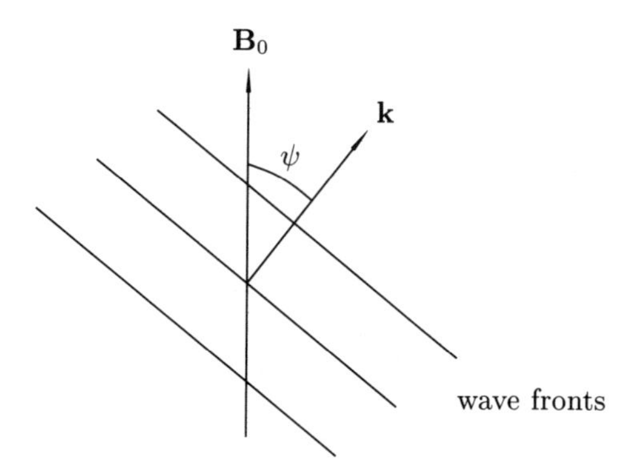

FIGURE 22.1
Geometry for the propagation of a plane Alfven wave with a wavefront inclined at an angle ψ with respect to the undisturbed magnetic field $\mathbf{B}_0$. The resulting phase velocity is $v_A \cos\psi$, where v_A is the usual Alfven velocity for wave vector $\mathbf{k}$ parallel to $\mathbf{B}_0$; i.e., the only component of $\mathbf{B}_0$ that matters is that along $\mathbf{k}$.

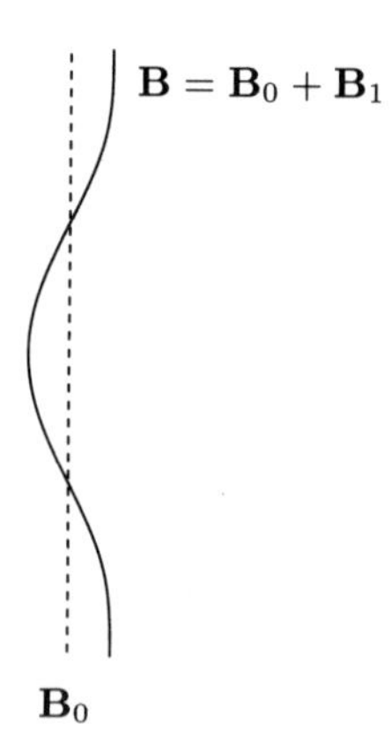

FIGURE 22.2
The total magnetic field, undisturbed plus perturbation, in an Alfven wave looks like a vibrating string for any field line.

FAST AND SLOW MHD WAVES

For the modes which have $u_{1z} = 0$, equation (22.19) with $\mathbf{u}_1 = u_{1x}\mathbf{e}_x + u_{1y}\mathbf{e}_y$ yields a matrix equation:

$$\begin{pmatrix} \omega^2 + k^2 v_A^2 \cos^2\psi - k^2(v_A^2 + a_s^2) & k^2 v_A^2 \sin\psi\cos\psi \\ k^2 v_A^2 \sin\psi\cos\psi & \omega^2 - k^2 v_A^2\cos^2\psi \end{pmatrix} \begin{pmatrix} u_{1x} \\ u_{1y} \end{pmatrix} = \begin{pmatrix} 0 \\ 0 \end{pmatrix}. \tag{22.21}$$

In order for the above equation to have nontrivial solutions, we must set the determinant of the coefficient matrix equal to zero. This condition yields the dispersion relation,

$$\omega^4 - k^2(v_A^2 + a_s^2)\omega^2 + k^4 v_A^2 a_s^2 \cos^2\psi = 0. \tag{22.22}$$

The solution of this quadratic equation for ω^2/k^2 reads

$$\frac{\omega^2}{k^2} = \frac{1}{2}\left\{(v_A^2 + a_s^2) \pm \left[(v_A^2 + a_s^2)^2 - 4v_A^2 a_s^2 \cos^2\psi\right]^{1/2}\right\}. \tag{22.23}$$

The upper sign, which gives a larger wave speed, yields the *fast MHD wave*; the lower sign, the *slow MHD wave*.

To obtain a better feeling for the properties of these waves, we consider the special propagation directions: $\cos^2\psi = 1$ ($\mathbf{k} \parallel \mathbf{B}_0$) and $\cos^2\psi = 0$ ($\mathbf{k} \perp \mathbf{B}_0$).

(a) $\mathbf{k} \parallel \mathbf{B}_0$ ($\cos^2\psi = 1$):

$$\frac{\omega^2}{k^2} = \frac{1}{2}\left[(v_A^2 + a_s^2) \pm |v_A^2 - a_s^2|\right] = v_A^2 \qquad \text{or} \qquad a_s^2.$$

In this case, the fast (slow) mode is the faster (slower) of an acoustic wave or an Alfven wave (with the other polarization from that considered in the previous section).

(b) $\mathbf{k} \perp \mathbf{B}_0$ ($\cos^2\psi = 0$):

$$\frac{\omega^2}{k^2} = v_A^2 + a_s^2 \qquad \text{or} \qquad 0.$$

The slow mode has disappeared (zero phase and group velocities); the fast mode is a magnetosonic wave, which represents alternating compressions and rarefactions of the gas and field (Figure 22.3).

Since the field lines do not get bent in a magnetosonic wave, only the magnetic pressure contributes to the restoring force. We reconcile this interpretation with the formula for the square of the wave-propagation speed,

$$a_s^2 + v_A^2 = \gamma\frac{P_0}{\rho_0} + 2\frac{P_{\text{mag}}}{\rho_0},$$

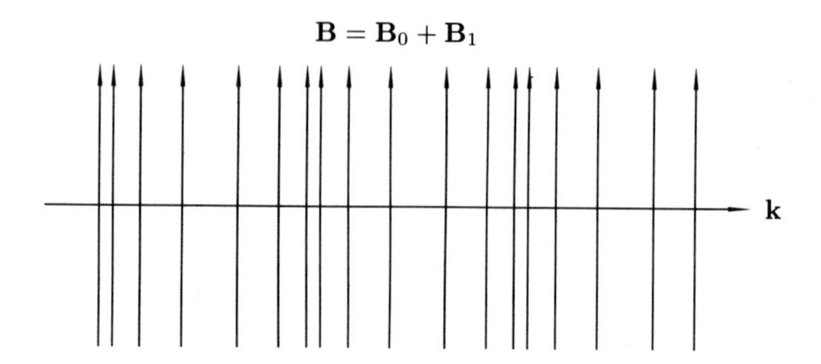

FIGURE 22.3
The distribution of field strengths in a magnetosonic wave shows alternating regions of compression and rarefaction, signatures of the basic acoustic-like nature of the disturbance.

where $P_{\text{mag}} \equiv B_0^2/8\pi$, by identifying an effective $\gamma_{\text{mag}} = 2$ for the one-dimensional compression of a magnetic field. This identification makes sense, since field freezing for such a 1-D compression implies $B \propto \rho$, i.e., $B^2/8\pi \propto \rho^2$. In Chapter 24, we shall see that γ_{mag} effectively has a value equal to 4/3 for compressions in *three* dimensions, with interesting consequences for the amount of mass that can be supported against self-gravitation by frozen-in magnetic fields.

WAVE-PROPAGATION DIAGRAM

Summarizing, we have three kinds of waves: Alfven, with $\mathbf{u}_1 \propto \mathbf{k} \times \hat{\mathbf{n}}$; and fast and slow, with $\mathbf{u}_1 \cdot (\mathbf{k} \times \hat{\mathbf{n}}) = 0$. A polar plot of the phase (and group) velocity for the case of interest when the magnetic forces dominate those of gas pressure, $v_{\text{A}} > a_s$, is shown in Figure 22.4.

This diagram has the following pictorial interpretation. Imagine an emitter of all three kinds of waves to reside at the origin, in a region where an unperturbed magnetic field $\mathbf{B}_0$ points in the vertical direction. At $t = 0$, waves are emitted; after the passage of one unit of time, the fronts of the fast, Alfven, and slow modes will have reached the loci indicated in the graph. Parallel or anti-parallel to the unperturbed magnetic field ($\psi = 0$ or π), the fast mode and the Alfven wave have the same speed of propagation, but orthogonal polarizations. Perpendicular to the unperturbed magnetic field ($\psi = \pi/2$ or $3\pi/2$), the fast mode travels as a magnetosonic speed—the fastest signal possible for any small-amplitude MHD disturbance—while the slow and Alfven waves take infinitely long to propagate away from the

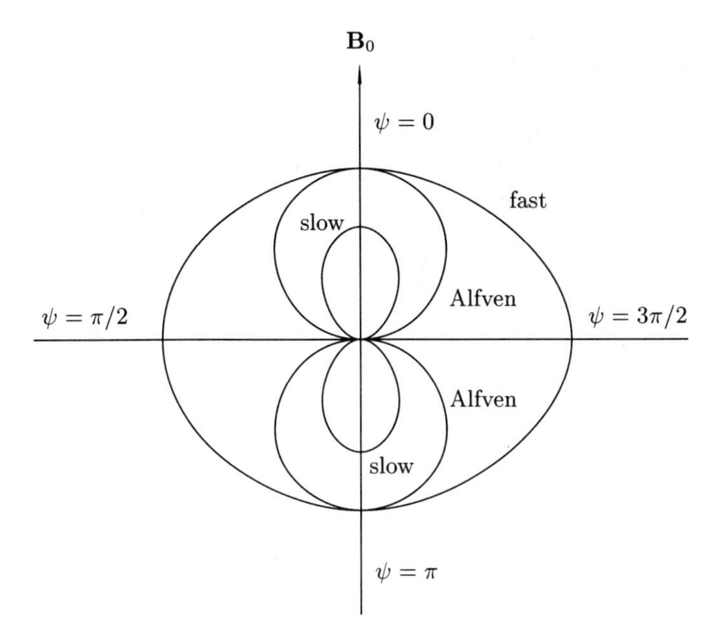

FIGURE 22.4
Wave-propagation diagram displayed in a polar plot for the three waves of classical magnetohydrodynamics: fast, slow, and Alfven. The plot is drawn for the case when the Alfven speed exceeds the sound speed.

origin in these directions. At intermediate angles of wave propagation, the information carried by the fast mode always reaches the observer first; that carried by the slow mode, last.

EIGENVECTORS

Given the satisfaction of the dispersion relation, we can recover the eigenvectors of the disturbance in the usual manner. We have already discussed in some detail the properties of a pure Alfven wave. The following forms a summary for the fast and slow modes. Equation (22.6) shows $\mathbf{b}_1$ to be orthogonal to $\mathbf{k} : \mathbf{k} \cdot \mathbf{b}_1 = 0$. Hence, for the fast and slow modes, we may take $\mathbf{k} = k\mathbf{e}_x$, $\mathbf{b}_1 = b_1\mathbf{e}_y$, $\mathbf{u}_1 = u_{1x}\mathbf{e}_x + u_{1y}\mathbf{e}_y$. The eigenvectors then work out to be

$$\alpha_1 : u_{1x} : u_{1y} : b_1 =$$

$$\cos\psi\sin\psi : \frac{\omega}{k}\cos\psi\sin\psi : \frac{\omega}{k}\left[\sin^2\psi + \frac{a_s^2 - \omega^2/k^2}{v_{\rm A}^2}\right] : -\frac{(a_s^2 - \omega^2/k^2)}{v_{\rm A}^2}\cos\psi.$$

In particular, for $\mathbf{k} \parallel \mathbf{B}_0$, we set $\sin\psi = 0$, $\cos\psi = \pm 1$, and obtain an Alfven wave with $\omega^2/k^2 = v_{\rm A}^2$ in which

$$\alpha_1 : u_{1x} : u_{1y} : b_1 = 0 : 0 : \frac{\omega}{k} : \mp 1;$$

and an acoustic wave with $\omega^2/k^2 = a_s^2$ in which

$$\alpha_1 : u_{1x} : u_{1y} : b_1 = 1 : \frac{\omega}{k} : 0 : 0.$$

The Alfven wave is a transverse disturbance involving no compression of the density, whereas the acoustic wave is a longitudinal one involving no compression of the field.

For $\mathbf{k} \perp \mathbf{B}_0$, we set $\cos\psi = 0$, $\sin\psi = \pm 1$, and obtain a magnetosonic wave with $\omega^2/k^2 = a_s^2 + v_{\rm A}^2$ in which

$$\alpha_1 : u_{1x} : u_{1y} : b_1 = 1 : \frac{\omega}{k} : 0 : \pm 1;$$

and a slow mode with $\omega^2/k^2 = 0$ in which

$$\alpha_1 : u_{1x} : u_{1y} : b_1 = 1 : 0 : \frac{a_s}{v_{\rm A}}(a_s^2 + v_{\rm A}^2)^{1/2} : \mp \frac{a_s^2}{v_{\rm A}^2}.$$

The magnetosonic wave is a longitudinal disturbance that has both compressions of the density and the field, while the (nonpropagating) slow mode is a transverse disturbance that compresses the matter and displaces the field.

At intermediate angles of wave propagation, the fast and slow modes are neither purely transverse nor purely longitudinal, neither purely compressional nor purely distortional. They represent complicated beasts with easily describable properties only for propagation purely along or purely orthogonal to the ambient magnetic field. This complication, arising from the anisotropic nature of magnetic forces, constitutes much of what makes magnetohydrodynamics such a rich subject. The existence of *three* characteristic speeds (in the sense of the solution of hyperbolic PDEs) adds to the technical difficulties.

NONLINEAR ALFVEN WAVES AND THEIR TORSIONAL COUNTERPARTS

Plane Alfven waves of finite amplitude have the peculiar property that they may propagate at constant speed in a homogeneous incompressible medium without any distortion of waveform ("steepening"). On p. 238 of their book, Landau, Lifshitz, and Pitaevskii give the solution for a linearly polarized

Alfven wave. We consider the case of the propagation of a circularly polarized Alfven wave. Astronomers find such *torsional Alfven waves* interesting because they can transport angular momentum. To begin our calculation, let us adopt cylindrical coordinates (ϖ, φ, z) and consider finite-amplitude disturbances that have the following postulated properties:

$$\rho = \text{constant}, \qquad \mathbf{u} = \varpi\Omega(z,t)\mathbf{e}_\varphi, \qquad \mathbf{B} = B_0\mathbf{e}_z + \beta\rho\mathbf{u}, \tag{22.24}$$

where B_0 and β are spatial and temporal constants. With ρ also equal to a constant, the equation of continuity (22.1) reduces to $\nabla \cdot \mathbf{u} = 0$. With $\mathbf{u}$ and $\mathbf{B}$ given by equations (22.24), we easily verify that $\nabla \cdot \mathbf{u} = 0 = \nabla \cdot \mathbf{B}$. The vector equation of field freezing (21.25) reduces to a single nontrivial component in the φ direction, which reads

$$\beta\frac{\partial\Omega}{\partial t} = B_0\frac{\partial\Omega}{\partial z}; \tag{22.25}$$

the ϖ, φ, and z components of the force equation (22.2) (with no gravitational, viscous, or radiation forces) read

$$-\rho\varpi\Omega^2 = -\frac{\partial P}{\partial\varpi} - \frac{\beta^2}{4\pi}2\varpi\Omega^2, \tag{22.26}$$

$$\rho\varpi\frac{\partial\Omega}{\partial t} = \frac{\beta B_0}{4\pi}\varpi\frac{\partial\Omega}{\partial z}, \tag{22.27}$$

$$0 = -\frac{\partial P}{\partial z} - \frac{\beta^2}{4\pi}\varpi^2\Omega\frac{\partial\Omega}{\partial z}. \tag{22.28}$$

Equation (22.28) requires the z-gradient of the pressure P of our incompressible fluid to offset the z-gradient of the magnetic pressure (due to the toroidal field generated by the wrapping action of the fluid's rotation). Integration of equation (22.28) yields that the fluid pressure satisfies

$$P = -\frac{\beta^2}{8\pi}\varpi^2\Omega^2(z,t) + \mathcal{P}(\varpi,t), \tag{22.29}$$

where $\mathcal{P}(\varpi,t)$ is an arbitrary function of ϖ and t. In an incompressible fluid, the satisfaction of equation (22.29) poses no special difficulty, since a liquid's pressure can instantaneously assume any distribution required for consistency with the momentum equation. On the other hand, the kinetic pressure of a gas must satisfy the constraints of an equation of state and the internal energy equation, and cannot respond arbitrarily quickly; thus the buildup of a magnetic pressure gradient in the z direction by the azimuthal wrapping of the magnetic field will generally lead to fluid motions in the z direction, $u_z \neq 0$. We ignore the latter possibility in this discussion.

The substitution of equation (22.29) into equation (22.26), and the requirement that the result hold at all z identifies the function $\mathcal{P}$ as a strict constant. In other words, equation (22.29) requires that the sum of the gas and magnetic pressures, $P + B^2/8\pi$, everywhere equals a constant. The remaining part of equation (22.26) then results in the identification,

$$\beta = \pm(4\pi\rho)^{1/2}. \tag{22.30}$$

With β given by equation (22.30), we see that the solution for the magnetic field $\mathbf{B}$ can be derived from the solution for the angular velocity $\Omega(z,t)$ as

$$\mathbf{B} = B_0 \left[\mathbf{e}_z \pm \frac{\varpi\Omega(z,t)}{v_{A0}} \mathbf{e}_\varphi \right], \tag{22.31}$$

where v_{A0} equals the Alfven speed associated with B_0:

$$v_{A0} \equiv \frac{B_0}{(4\pi\rho)^{1/2}}. \tag{22.32}$$

In other words, the ratio of the magnitude of the perturbational magnetic field to that of the unperturbed field equals the ratio of the transverse fluid speed to the Alfven speed, the same relation as for linear Alfven waves.

When β is given by equation (22.30), equations (22.25) and (22.27) become redundant [thereby justifying our original guesses ("postulates") for the forms of the dependent variables], with both reading

$$\frac{\partial\Omega}{\partial t} = \pm v_{A0} \frac{\partial\Omega}{\partial z}. \tag{22.33}$$

Equation (22.33) has as its most general solution,

$$\Omega(z,t) = F(z \pm v_{A0}t), \tag{22.34}$$

where F is an arbitrary function of its argument. The solution (22.34) represents a torsional Alfven wave of arbitrary and unchanging waveform propagating at constant speed v_{A0} in the $\mp z$ direction. Notice that in equation (22.33) we recover a *linear wave equation* despite the fact that we have made no linearization (small-amplitude) approximation.

The ability of the tension in twisted magnetic fields to change the angular velocity Ω of the surrounding medium underlies the physical mechanism for the transport of angular momentum by torsional Alfven waves (see Figure 22.5). Astronomers believe that this basic process accounts for the spindown of rotating objects as diverse as magnetized stars with outer convective envelopes and interstellar clouds threaded by the Galactic magnetic field. In some fundamental sense, magnetic fields resist being twisted by differential rotation, and in back reaction, try to equalize the angular rotation speeds of the different parts of the system by sending torsional waves back and forth until Ω becomes constant.

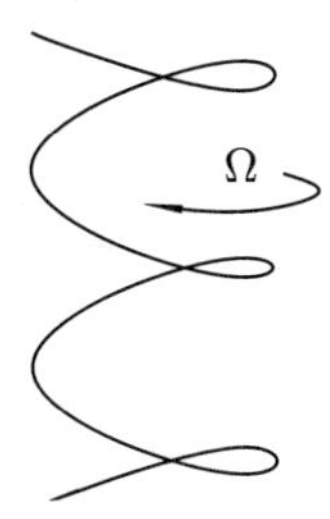

FIGURE 22.5
Torsional Alfven waves in a magnetized conducting medium can transport angular momentum by carrying rotational disturbances from one part of the fluid to another.

APPLICATION TO ACCRETION DISKS

We now stand in a good position to understand the physical basis of the Chandrasekhar instability referred to in Chapters 7 and 8. We follow here the physical description given by S. Balbus and J. Hawley, 1991, *Ap. J.* (**376**, 214). Consider a conducting cylinder of fluid threaded by magnetic field parallel to its axis of rotation in the z direction. Suppose furthermore that this cylinder initially has a distribution of angular velocity $\Omega(\varpi)$ that decreases in magnitude with increasing distance ϖ from the axis of rotation. Imagine now setting up a standing torsional wave, such that at one value of $z = z_1$, we spin up the fluid on an annulus slightly ahead in the direction of rotation, and a half-wavelength up or down the axis, $z = z_2$, we spin down the fluid on another annulus, magnetically connected to the first, slightly behind in the direction against rotation. (See Figure 22.6.)

The same process twists the frozen-in magnetic field in such a way that the field advanced (retarded) in φ at z_1 (z_2) exerts a negative (positive) torque on the attached matter, removing (adding) angular momentum from (to) it. Having its angular momentum decreased (increased), the matter at z_1 (z_2) will want to shrink (expand) in radius ϖ. If the balance of forces due to gravity, inertia, pressure, etc., are such that Ω decreases with ϖ, the matter at z_1 (z_2) with its shrunken (expanded) orbital radius will want spin up (down) even more, thereby accentuating the original φ displacement. If the process runs away, we will have an instability whenever the rotation law satisfies Chandrasekhar's criterion (8.10).

Paradoxically, the process will not run away if the field B_0 is too strong. The generation of a component B_ϖ when the alternate annuli shrink or expand constitutes a stabilizing influence because the associated component

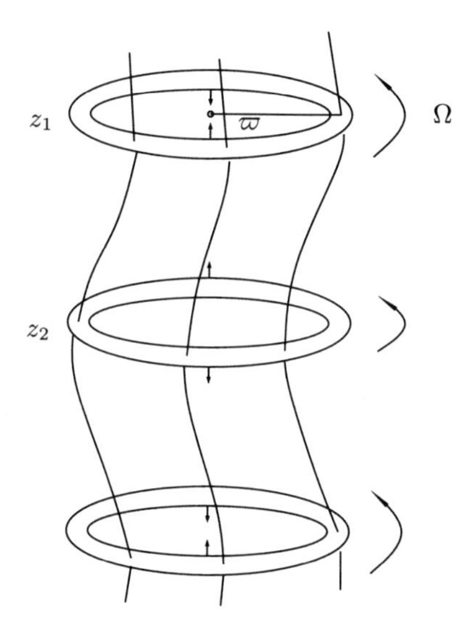

FIGURE 22.6
The physical basis for the Chandrasekhar instability criterion,

$$\frac{d}{d\varpi}(\Omega^2) < 0,$$

for a magnetized, differentially-rotating, medium. (See text for detailed explanation.)

of magnetic tension $B_\varpi B_z/4\pi$ exerts restoring forces in a direction opposed to the radial displacements of the fluid elements (Figure 22.7). The stabilizing influence will overcome the destabilizing tendency of the component of magnetic tension $B_\varphi B_z/4\pi$ acting to spin down or spin up vertically connected fluid elements, (a) if the wavelength $\lambda_z \equiv 2\pi/k_z$ is too large (for a torsional Alfven wave to have time to propagate from oppositely twisted portions of the fluid before the stabilizing tension restores their radial positions), or (b) if the magnetic field B_0 has too much strength to begin with. The actual criterion involves the combination λ_z/B_0, i.e., if the field is weak, we will still get instability by considering disturbances with very small wavelengths (until diffusive effects come into play). For application to a disk geometry, the field cannot get too strong, because disturbances much longer than the vertical scale height of the disk, equal to a/Ω where a is the gaseous sound speed (see Chapter 7), no longer propagate, even approximately, in a region with the effective geometry of an infinite cylinder. In a

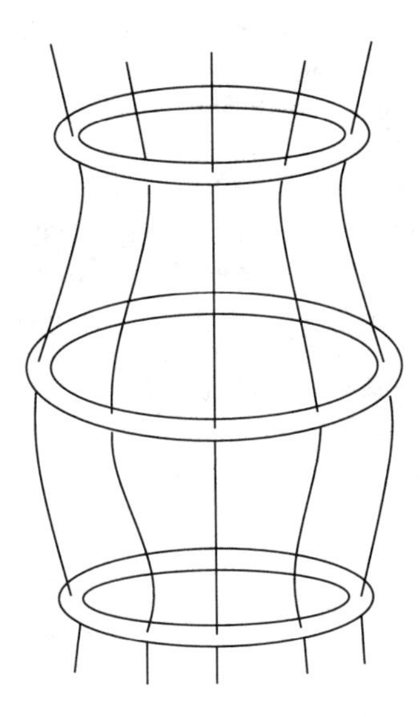

FIGURE 22.7
The alternate pinching and stretching of poloidal magnetic fields as alternate rings of rotating gas contract and expand can provide a stabilizing influence on Chandrasekhar's instability.

vertically stratified disk, other stabilizing influences (e.g., buoyancy forces) might come into play. The restriction on the field strength works out, on dimensional grounds, so that the Alfven speed v_{A0} should be $\lesssim a$ for instability; i.e., the magnetic pressure should be $\lesssim$ the thermal pressure. Notice that this restriction attaches only to the strength of the poloidal field; the instability in its axisymmetric form proceeds independent of the strength of the toroidal component B_φ. To be sure, however, "magnetic buoyancy" effects (see the discussion in Chapter 23 on "Parker's instability") probably limits the strengths of disk toroidal fields to similar values (magnetic pressure $\lesssim$ thermal pressure). In any case, in conducting disks of sufficiently weak magnetic fields, any (realistic) rotation law with $\Omega(\varpi)$ decreasing in magnitude outward should be unstable with respect to Chandrasekhar's instability.

While extremely suggestive, the above discussion leaves unanswered the basic astrophysical question about how this mechanism might contribute to the effective viscosity of an *accretion* disk. After all, the basic trick in accretion disks is to allow material at different *radial* positions ϖ to exchange angular momentum. The formal discussion above involves the transfer of angular momentum between different fluid elements in the *vertical* direction z. Although an interesting result in itself, the process does not address how angular momentum might be *systematically* transported in the outward direction *radially* through the disk, thereby leading to a steady drift of mass inward (accretion, at least, in the inner parts of the disk). Our worry may be excessive, since any disk magnetic field with a nonzero radial component B_ϖ, in addition to a toroidal component B_φ, will give magnetic coupling across ϖ. Of course, radial components of magnetic fields should get sheared by the differential rotation; so their regeneration may be tied to the question of disk *dynamo* action (see Chapter 26). The real utility of the Chandrasekhar instability may then lie in its possible importance as the mechanistic basis for such dynamo operation.

23

Magnetostatics and the Parker Instability

References: J. Dungey, 1953, *M.N.R.A.S.*, **113**, 180;
T. Ch. Mouschovias, 1974, *Ap. J.*, **192**, 37.

We continue our considerations of hydromagnetic effects with the problem of magnetostatics. Studies of static equilibria form necessary preludes to dynamic investigations of oscillations about such states if they are stable, or of steady evolution away from such states if they are unstable. In this chapter we first formulate the equations of magnetostatics in a simplified geometry where spatial variations exist only in two rectilinear dimensions y and z. We then apply the formalism to the issue of the vertical structure of a model of the interstellar gas and magnetic field in the gravitational field provided by the rest of the Galaxy. For this problem, Eugene Parker pointed out that a plane-parallel configuration stratified in z is possible, but is unstable. We then reproduce Mouschovias's demonstration that alternative equilibrium states exist in which occur variations in *both* y and z, with y being the direction of the undisturbed magnetic field. Because the latter states possess the same mass-to-flux distribution as Parker's initial states, they represent possible "final" configurations for the evolution of such initial states (assuming that such "final states" are themselves stable, an issue on which there exists some controversy).

BASIC EQUATIONS

We obtain the conditions of magnetostatics by setting $\mathbf{u}$ and $\partial/\partial t = 0$ in the equations of ideal MHD. With this procedure, the equations of continuity,

$$\frac{\partial \rho}{\partial t} + \nabla \cdot (\rho \mathbf{u}) = 0, \tag{23.1}$$

and field freezing,

$$\frac{\partial \mathbf{B}}{\partial t} + \nabla \times (\mathbf{B} \times \mathbf{u}) = 0, \tag{23.2}$$

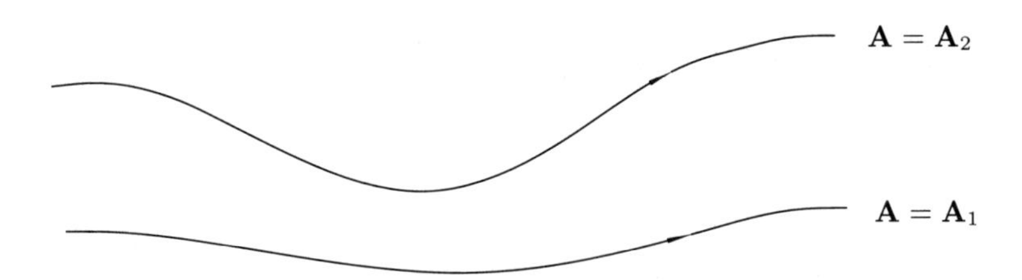

FIGURE 23.1
In a two-dimensional problem, the value of the vector potential A can be used to label field lines.

are satisfied trivially, a feature to which we shall return later. The remaining equations read

$$-\rho\nabla\mathcal{V} - \nabla P + \frac{1}{4\pi}(\nabla \times \mathbf{B}) \times \mathbf{B} = 0, \tag{23.3}$$

$$\Gamma - \Lambda = 0, \tag{23.4}$$

$$\nabla^2\mathcal{V} = 4\pi G(\rho + \rho_{\text{ext}}), \tag{23.5}$$

$$\nabla \cdot \mathbf{B} = 0. \tag{23.6}$$

INTRODUCTION OF VECTOR POTENTIAL

To satisfy $\nabla \cdot \mathbf{B} = 0$, we may introduce a vector potential $\mathbf{A}$ such that

$$\mathbf{B} = \nabla \times \mathbf{A}. \tag{23.7}$$

For a two-dimensional problem, $\mathbf{B} = B_y(y,z)\mathbf{e}_y + B_z(y,z)\mathbf{e}_z$, we may choose $\mathbf{A}$ to have a single component, $\mathbf{A} = A(y,z)\mathbf{e}_x$; thus equation (23.7) becomes

$$B_y = \frac{\partial A}{\partial z}, \qquad B_z = -\frac{\partial A}{\partial y}, \qquad i.e., \qquad \mathbf{B} = -\mathbf{e}_x \times \nabla A. \tag{23.8}$$

Notice that equation (23.8) implies that $\mathbf{B}$ is perpendicular to ∇A. On the other hand, loci of constant A also lie perpendicular to ∇A; thus, lines of constant A in the y-z plane can be used to label magnetic field lines (Figure 23.1).

With $\mathbf{B}$ given by $\nabla \times \mathbf{A}$, we may write

$$\nabla \times \mathbf{B} = -\nabla^2\mathbf{A} + \nabla(\nabla \cdot \mathbf{A}) = -\mathbf{e}_x\nabla^2 A,$$

since $\nabla \cdot \mathbf{A} = \partial A/\partial x = 0$ if $\mathbf{A} = A(y,z)\mathbf{e}_x$. Hence we have for the Lorentz force per unit volume,

$$\frac{1}{4\pi}(\nabla \times \mathbf{B}) \times \mathbf{B} = -\frac{1}{4\pi}(\nabla^2 A)\mathbf{e}_x \times (-\mathbf{e}_x \times \nabla A) = -\frac{1}{4\pi}(\nabla^2 A)\nabla A, \quad (23.9)$$

where

$$\nabla A = \frac{\partial A}{\partial y}\mathbf{e}_y + \frac{\partial A}{\partial z}\mathbf{e}_z.$$

DUNGEY'S FORMULATION FOR MAGNETOSTATICS

To simplify the algebra further, we suppose the balance of heating and cooling, equation (23.4), yields nearly isothermal behavior, so that in the equation of state,

$$P = a^2\rho, \quad (23.10)$$

we may take $a^2 \equiv kT/m$ to be a constant. (For the Galactic problem, we shall assume that we can replace the thermal speed a with an equivalent velocity dispersion of interstellar clouds in the z direction.) With these assumptions, we may write the combination of gravitational and gas pressure forces as

$$-\rho\nabla\mathcal{V} - \nabla P = -\frac{P}{a^2}\nabla\mathcal{V} - \nabla P = -e^{-\mathcal{V}/a^2}\nabla q, \quad (23.11)$$

where we have defined

$$q \equiv Pe^{\mathcal{V}/a^2}. \quad (23.12)$$

Equations (23.9) and (23.11) substituted into the force equation (23.3) now gives

$$-\frac{1}{4\pi}(\nabla^2 A)\nabla A = e^{-\mathcal{V}/a^2}\nabla q, \quad (23.13)$$

which implies that ∇q and ∇A are parallel to one another. This requires surfaces of constant q to correspond to surfaces of constant A; i.e.,

$$q = q(A), \quad (23.14)$$

where the function $q(A)$ remains to be found. With $q = q(A)$, $\nabla q = (dq/dA)\nabla A$; hence we may write equation (23.13) in the scalar form

$$\nabla^2 A = -4\pi e^{-\mathcal{V}/a^2}\frac{dq}{dA}. \quad (23.15)$$

Equation (23.15) forms a quasilinear elliptic PDE for A, given $\mathcal{V}$ as a function of y and z, and q as a function of A. Dungey provided the above formulation for the problem of magnetostatics in two rectilinear dimensions in connection with his theory of solar prominences. In Dungey's formulation, one could freely specify $q(A)$; different choices would give different magnetostatic configurations.

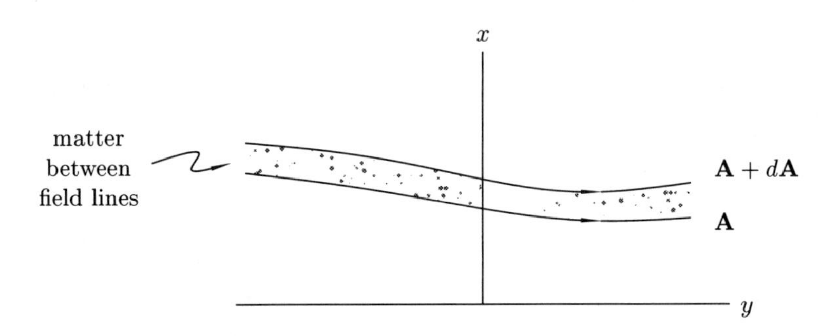

FIGURE 23.2
For a problem in two-dimensionals magnetostatics, we need to replace the usual equation of field freezing by a constraint that the amount of matter between two field lines remain the same as in some reference "initial" state.

THE MASS-TO-FLUX DISTRIBUTION

In 1974 Mouschovias pointed out that an *ad hoc* choice for $q(A)$ would generally lead to an equilibrium state that could not be realized from a continuous deformation of field lines of a *known* initial state. As a matter of principle, the functional choice for $q(A)$ has to be determined from considerations of the constraints of mass conservation and field freezing. In other words, when we set $\partial/\partial t = 0$ and $\mathbf{u} = 0$, we must replace the loss of equations (23.1) and (23.2) by the constraint that *mass-to-flux* distribution of the final state equals that of the initial state.

One might have argued that given no *a priori* information, an arbitrary choice for $q(A)$ for the final state (which corresponds to *some* mass-to-flux distribution) forms as valid a procedure as an arbitrary choice for the initial mass-to-flux distribution. This argument lacks appeal for two reasons. First, there exists the matter of principle, which clearly sides with taking explicit account of a conservation law at the heart of the concept of field freezing. Second, there exists the practical consideration that the mass-to-flux ratio forms a quantity that can be measured observationally, at least in principle, whereas the function $q(A)$ has no such physical prospect.

Consider two field lines, labeled A and $A+dA$, separated by an infinitesimal distance (Figure 23.2). The amount of matter dm per unit length in the x direction between the field lines A and $A + dA$ reads

$$dm = \int dy \rho\, dz,$$

where the y-integration extends from $-Y$ to $+Y$ if we apply periodic boundary conditions, with the configuration repeating after each horizontal length of $2Y$. For a given value of A, let the equation $z = z(y, A)$ describe the locus of a field line. Writing

$$dz = \frac{\partial z}{\partial A}\,dA,$$

where the derivative $\partial z/\partial A$ occurs with y held constant, we now have

$$dm = dA \int_{-Y}^{+Y} \rho \frac{\partial z}{\partial A}\,dy.$$

If we substitute for ρ the expression

$$\rho = \frac{P}{a^2} = \frac{q}{a^2} e^{-\mathcal{V}/a^2}$$

that we may recover from equation (23.12), we get the mass-to-flux ratio

$$\frac{dm}{dA} = \frac{q(A)}{a^2} \int_{-Y}^{+Y} e^{-\mathcal{V}/a^2} \frac{\partial z}{\partial A}\,dy, \tag{23.16}$$

where $\mathcal{V}$ inside the integral is to be evaluated at (y, z) following the field line $z = z(y, A)$. After integration over y, then, the integral on the right-hand side depends only on A; consequently, we may regard equation (23.16) as specifying $q(A)$ once we know dm/dA as a function of A,

$$q(A) = a^2 \frac{dm}{dA} \left(\int_{-Y}^{+Y} e^{-\mathcal{V}/a^2} \frac{\partial z}{\partial A}\,dy \right)^{-1}. \tag{23.17}$$

Equations (23.15) and (23.17) constitute the fundamental set of equations for the problem of (isothermal) magnetostatics with variations in two rectilinear dimensions. A schematic method of solution might read as follows. Given dm/dA, guess a form for $A = A(x, y)$ (e.g., from a previous solution with somewhat different input parameters). Invert this form (numerically) to find $z = z(y, A)$. Substitute the dependences, $z = z(y, A)$ and $\mathcal{V} = \mathcal{V}[y, z(y, A)]$, into equation (23.17) to compute $q(A)$. With this $q(A)$, solve equation (23.15) for $A(y, z)$ (e.g., by using the "relaxation" technique described in Chapter 13 for elliptic PDEs). With the new iterate for $A = A(y, z)$, invert to find a new $z = z(y, A)$. Continue this process (perhaps with under-relaxation or over-relaxation parameters) until satisfactory convergence has been achieved. Notice, in particular, that although the function dm/dA remains fixed from iteration to iteration (if we have field freezing), the function $q(A)$ does not, and the "final" value of $q(A)$ will generally differ from the "initial."

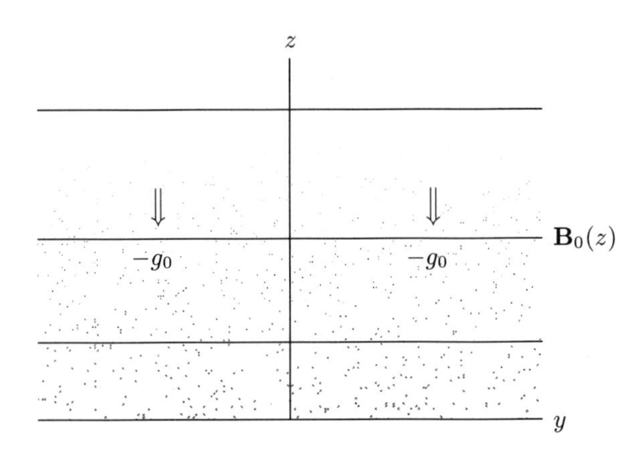

FIGURE 23.3
The simplest picture of the vertical structure of the magnetic field in the disk of the Galaxy and the gas to which it is tied imagines a magnetostatic equilibrium between the gradient of the total pressure, gas plus magnetic, and the vertical component of the Galactic gravitational field. This simple picture produces, however, a vertical stratification that turns out to be unstable with respect to Parker's instability.

THE VERTICAL CONFINEMENT OF THE GALACTIC MAGNETIC FIELD

As an example, we consider the magnetostatic equilibria of the system of interstellar gas and magnetic field in the vertical gravitational field of the Galaxy (E. N. Parker, 1966, *Ap. J.*, **145**, 811). Motivation for the problem comes from the realization that cosmic-ray particles are much too energetic to be bound gravitationally to the Galaxy; on the other hand, neither can they freely stream from the disk of the Galaxy where they are mostly produced (and where they contribute to the synchrotron radiation) or the energy requirements would prove enormous. Consequently, they must be tied to the Galaxy via the interstellar magnetic field.

What then holds the interstellar magnetic field to the disk of the Galaxy? A natural answer might seem to be: the weight of the thermal gas (clouds) that the magnetic fields thread (Figure 23.3).

We calculate the resulting equilibrium without considering the expansive contribution of the cosmic rays (which Parker incorporated with relatively little additional work). For simplicity, we naturally consider as a first attempt, plane-parallel stratified states in the z direction. If we ignore variations in the other two directions (x = radial direction, y = circular direction), we may adopt

$$\mathbf{B}_0 = B_0(z)\mathbf{e}_y, \qquad P_0 = P_0(z) = a^2\rho_0(z), \qquad \mathcal{V} = \mathcal{V}(z). \tag{23.18}$$

To simplify the algebra, we follow Parker in assuming the ratio of magnetic and gas pressures to have a constant value at all z:

$$\frac{B_0^2}{8\pi P_0} = \alpha_0 = \text{constant independent of } z. \tag{23.19}$$

If we substitute equations (23.18), together with equation (23.19), into equation (23.3), we obtain

$$(1+\alpha_0)\frac{dP_0}{dz} = -\frac{P_0}{a^2}\frac{d\mathcal{V}}{dz}, \tag{23.20}$$

which has the solution,

$$P_0(z) = P_0(0)\exp\left[-\frac{\mathcal{V}(z)}{(1+\alpha_0)a^2}\right]. \tag{23.21}$$

Equation (23.19) now implies the desired concentration of the interstellar magnetic field toward the central plane of the Galaxy,

$$B_0(z) = B_0(0)\exp\left[-\frac{\mathcal{V}(z)}{2(1+\alpha_0)a^2}\right], \tag{23.22}$$

with

$$\frac{B_0^2(0)}{8\pi P_0(0)} = \alpha_0.$$

Since

$$\left(\frac{\partial z}{\partial A}\right)_y = \left(\frac{dA}{dz}\right)^{-1} = [B_0(z)]^{-1},$$

equation (23.16), with q given by equation (23.12), gives

$$\frac{dm}{dA} = \frac{P_0 e^{\mathcal{V}/a^2}}{a^2}\int_{-Y}^{+Y}\frac{e^{-\mathcal{V}(z)/a^2}}{B_0(z)}\,dy = \frac{P_0(z)}{a^2 B_0(z)}2Y.$$

Substituting equation (23.22) into the above, we obtain

$$\frac{dm}{dA} = \frac{YB_0(0)}{4\pi a^2\alpha_0}\exp\left[-\frac{\mathcal{V}(z)}{2(1+\alpha_0)a^2}\right]. \tag{23.23}$$

To obtain an explicit expression for dm/dA as a function of A, we must integrate

$$\frac{dA}{dz} = B_0(z) = B_0(0)\exp\left[-\frac{\mathcal{V}(z)}{2(1+\alpha_0)a^2}\right]$$

to obtain $A = A(z)$, and then we would invert for $z = z(A)$. For example, if we adopt the crude approximation that the Galactic gravitational field equals a constant $-g_0$ or $+g_0$ in each half plane, $z > 0$ or $z < 0$, reversing sign across the mid-plane $z = 0$, we have

$$\mathcal{V}(z) = g_0|z|. \tag{23.24}$$

We then obtain

$$A = -\,\mathrm{sgn}(z)\frac{2}{g_0}(1+\alpha_0)a^2 B_0(0)\exp\left[-\frac{g_0|z|}{2(1+\alpha_0)a^2}\right],$$

where $\mathrm{sgn}(z) = \pm 1$ depending on whether $z > 0$ or $z < 0$. Comparison with equation (23.23) now yields

$$\frac{dm}{dA} = \frac{g_0 Y}{8\pi\alpha_0(1+\alpha_0)a^4}|A|, \tag{23.25}$$

which represents the desired mass-to-flux formula for Parker's problem.

THE PARKER INSTABILITY

The equilibrium state considered in the last section has a difficulty: Parker showed it to be unstable with respect to sufficiently long, slow MHD modes, as modified by the presence of the external gravitational field and the vertical stratification. The physical basis for the instability corresponds to the heuristic statement that a light "fluid," represented by the magnetic field, held down by the weight of a "heavy" fluid, represented by the thermal gas, tends to overturn. In other words, magnetic fields tend to be buoyant in a medium not far removed from convective disequilibrium in an external gravitational field. (The buoyancy of magnetic fields in the outer convection zones of cool stars plays an important role in modern theories of magnetic dynamos; see Chapter 26.) The situation does not form the exact analog of the Rayleigh-Taylor instability (when a true light fluid underlies a heavy fluid). Since we regard the conducting interstellar gas to be frozen to the field lines, they can "unload" only by sliding along them, through a process of magnetic "buckling" (Figure 23.4).

When the field lines buckle upward, the gas can drain into the magnetic valleys. A linear stability analysis shows that the unloading of field lines occurs at roughly a free-fall rate if the wavelength $2Y$ has a scale $\sim 2\pi(1+\alpha_0)a^2/g_0$. Disturbances with much shorter wavelengths possess too much magnetic tension to go unstable; disturbances with much longer wavelengths have smaller growth rates.

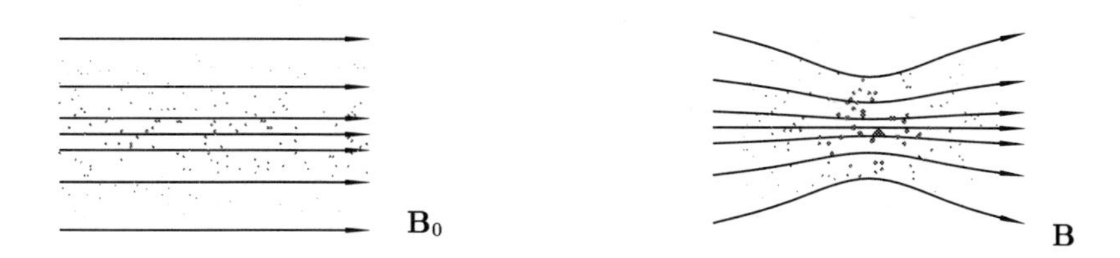

FIGURE 23.4
Parker's instability proceeds by a process of "buckling," wherein the magnetic field lines, initially assumed to lie purely in the horizontal direction, rise in certain portions and sink in others, in such a way that matter loaded onto the field lines slides off the peaks and sinks into the valleys. The energy gained by lowering the heavy gas deeper in the external gravitational potential provided by the rest of the Galaxy more than compensates for the extra energy needed to bend the originally straight field lines, if the wavelength separating alternate peaks exceeds a certain critical value.

FINAL STATES FOR PARKER'S INSTABILITY

We may now ask the following question. Does the nonlinear resolution of Parker's instability produce a steady final state, in which magnetostatic equilibrium again prevails, but with a buckled rather than a plane-parallel stratified state? We may obtain an answer to this question by solving equations (23.15) and (23.17) when $\mathcal{V}$ and dm/dA are given by equations (23.24) and (23.25) and when we apply the boundary conditions,

$$A(y, z = 0^+) = \text{undisturbed value} = -\frac{2}{g_0}(1+\alpha_0)a^2 B_0(0), \qquad (23.26)$$

$$A(y, z = +\infty) = 0, \qquad (23.27)$$

$$\frac{\partial A}{\partial y}(0, z) = \frac{\partial A}{\partial y}(Y, z) = 0. \qquad (23.28)$$

The vertical boundary conditions (23.26) and (23.27) at the midplane and infinity are self-explanatory; the horizontal boundary conditions at $y = 0$ and $y = Y$ reproduce the periodicity requirements under which the linear stability analysis was performed (sinusoidal dependence with wavelength $2Y$ along the direction of the unperturbed field).

The combination of Dirichlet and Neumann conditions, equations (23.26)–(23.28), make the solution of the elliptic PDE (23.15) a well-posed problem. In Mouschovias's actual treatment, he replaced condition (23.27) with the assumption that

$$A(y, Z) = \text{undisturbed value}$$

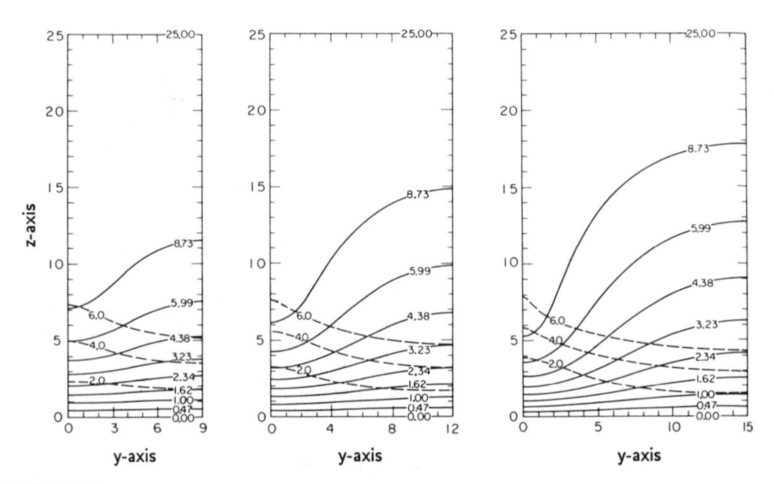

FIGURE 23.5
A sequence of magnetically buckled "final" states. Solid curves represent field lines, with the numbers indicating their vertical position in the initial uniform state; dashed lines represent isodensity contours with the numbers indicating the number of equivalent scales heights off the midplane of the contour level in the initial state. The difference between the panels lies in the multiple of critical wavelengths taken for the periodic boundary conditions in the horizontal direction. (From T. Ch. Mouschovias, 1974, *Ap. J.*, **192**, 37.)

for some large value of $z = Z$. Figure 23.5 is taken from his resulting paper.

Mouschovias claimed that his buckled equilibria represent stable configurations, and thus, they constitute true final states. This claim has been challenged by Catherine Cesarsky, Rene Pellat, and others, who note that even the buckled two-dimensional equilibria would be unstable to sinusoidal deformations in the third dimension (the x or radial direction in the Galaxy). Mouschovias does not dispute this part of their criticism, but he does make the valid point that "crinkling" in the third dimension lies outside of the formal problem that he considered. His "final states" probably do not possess instabilities with respect to additional two-dimensional perturbations in y and z of small amplitude that contain the same periodicity (Figure 23.5).

AN INTERESTING SIDELIGHT

We call attention to an interesting sidelight of the above discussion. We note that *both* the initial plane-parallel stratified state and the final buckled state satisfy the same set of governing equations and boundary conditions. Because of the nonlinearity of the basic PDE, equation (23.15), the solution is not unique. For the same boundary conditions, there can exist in these systems (like a metal plate stressed along its edges) more than one state of equilibrium. In general, the different states will have different energies. In these circumstances, a slow change of the external boundary conditions may lead the system to evolve quasisteadily to a state which, although a local minimum, actually has a higher energy globally than another accessible state. If the system subsequently receives a large enough perturbation, it may suddenly make a transition to the other state, buckling under the external load and releasing a tremendous amount of energy. This facet of the nonlinearity of the underlying equations may explain some of the eruptive properties of solar phenomena such as the collapse of magnetic arches or the onset of solar flares.

24

The Magnetic Virial Theorem

References: Chandrasekhar, *Hydrodynamic and Hydromagnetic Stability*, Chapter 8; L. Mestel, 1985, in *Protostars and Planets II*, ed. D. C. Black and M. S. Matthews (Tucson: University of Arizona Press), p. 320.

The virial theorem forms a theoretical tool used in many areas of astronomy; in this chapter, we consider it in the context of magnetohydrodynamics. From a mathematical point of view, we may regard the technique as an extension of the moment equation method. We started this volume with a derivation of the fluid description by taking various velocity moments of the Boltzmann equation, motivated by the philosophy that many problems do not require a detailed phase-space specification of the state of the system. In a similar manner, observations of many astronomical objects lack sufficient resolution to justify an elaborate dynamical model. In such cases, we might well be satisfied with a contracted description in which we multiply the MHD force equation by various powers of the spatial coordinates x_i and integrate over the entire spatial volume. The resulting equations turn out, however, generally to have limited practical utility since they involve relations between global quantities that have no ready physical interpretation. The equation that corresponds to the linear moment (one factor of x_i) yields a notable exception; this equation, the magnetic virial theorem, yields, as we shall see below, a powerful way to obtain approximate estimates for the conditions of equilibria and stability for self-gravitating configurations of magnetized gases.

SCALAR VIRAL THEOREM FOR SELF-GRAVITATING MAGNETIZED CLOUDS

Consider a mass M of electrically conducting fluid, confined to volume V by a rarefied and hot external medium of pressure P_{ext}. We suppose that the cloud is statically supported against its self-gravitation, by a combination of internal pressure P and magnetic fields $\mathbf{B}$ (Figure 24.1).

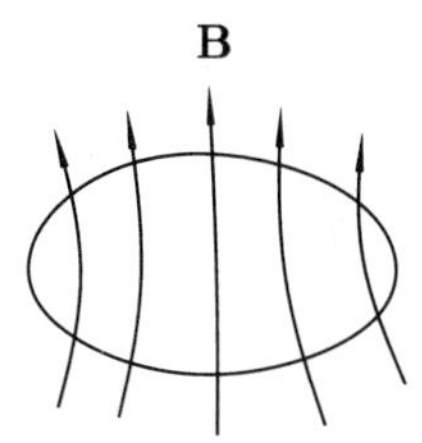

FIGURE 24.1
Schematic depiction of a gas mass supported in part against its self-gravitation by a system of magnetic fields that connect to the ambient medium.

The equation for force balance reads

$$0 = -\rho \frac{\partial \mathcal{V}}{\partial x_i} - \frac{\partial P}{\partial x_i} + \frac{\partial T_{ik}}{\partial x_k}, \tag{24.1}$$

where $-\partial \mathcal{V}/\partial x_i$ equals the gravitational field due to the gas cloud,

$$-\frac{\partial \mathcal{V}}{\partial x_i} = -G \int_V \frac{\rho(\mathbf{x}')(x_i - x_i')}{|\mathbf{x} - \mathbf{x}'|^3}\, d^3x', \tag{24.2}$$

and T_{ik} equals the Maxwell stress tensor associated with the ambient magnetic field,

$$T_{ik} = \frac{B_i B_k}{4\pi} - \frac{|\mathbf{B}|^2}{8\pi}\delta_{ik}. \tag{24.3}$$

In ignoring any contribution of the external medium to the right-hand side of equation (24.2), we have assumed that the external medium has negligible gravity in comparison with that of the mass enclosed within volume V. This assumption is justified if the external gas has a much higher temperature and, consequently, a much lower density than the material within volume V, in such a way that the pressure remains continuous across the surface in the direction along field lines. (Another way in which the assumption could be valid is if the gravitational field of the rest of the system has a negligible value because the surface A corresponds to a tidal boundary, or one of a special symmetry, inside of which the self-gravity of the mass M dominates that of the external medium.)

If we were to multiply equation (24.1) by x_m and integrate over volume V, we would get the *tensor virial theorem*, the off-diagonal elements of which carry information concerning angular-momentum conservation (see Chandrasekhar's book for an exposition). In this chapter, we shall be more interested in the trace of the tensor equation, which we may derive by simply multiplying equation (24.1) by x_i (with an implicit summation over repeated indices) and integrating over V,

$$0 = \int_V \left(-\rho x_i \frac{\partial \mathcal{V}}{\partial x_i} - x_i \frac{\partial P}{\partial x_i} + x_i \frac{\partial T_{ik}}{\partial x_k} \right) d^3x. \tag{24.4}$$

We wish to convert each of the three terms in the integrand of equation (24.4) into a volume contribution plus, perhaps, a surface contribution.

To begin, write the gas pressure term as

$$-x_i \frac{\partial P}{\partial x_i} = -\frac{\partial}{\partial x_i}(x_i P) + \frac{\partial x_i}{\partial x_i} P.$$

Using $\partial x_i / \partial x_i = \delta_{ii} = 3$, we get

$$-\int_V x_i \frac{\partial P}{\partial x_i} d^3x = 2\mathcal{U} - \oint_A P\mathbf{x} \cdot \hat{\mathbf{n}}\, dA, \tag{24.5}$$

where we have used the divergence theorem to convert the volume integral of $\partial(x_i P)/\partial x_i = \boldsymbol{\nabla} \cdot (\mathbf{x}P)$ into the surface integral represented by the second term on the right-hand side, and where $\mathcal{U}$ equals the thermal energy content of the gas in volume V,

$$\mathcal{U} \equiv \frac{3}{2} \int_V P\, d^3x. \tag{24.6}$$

In many applications (e.g. to stars), we can ignore the surface integral on the right-hand side of equation (24.5) in comparison with the volume term $2\mathcal{U}$ on the basis that the surface pressure is small in comparison with typical interior values. On the other hand, such a procedure would obviously be invalid if P has a uniform value throughout the volume V, since the surface and volume terms on the right-hand side must then exactly cancel if the sum is to equal the vanishing left-hand side of equation (24.5).

In a similar manner, write the Maxwell stress term as

$$x_i \frac{\partial T_{ik}}{\partial x_k} = \frac{\partial}{\partial x_k}(x_i T_{ik}) - \frac{\partial x_i}{\partial x_k} T_{ik} = \boldsymbol{\nabla} \cdot (\mathbf{x} \cdot \overleftrightarrow{T}) + \frac{|\mathbf{B}|^2}{8\pi}, \tag{24.7}$$

where we have used $(\partial x_i / \partial x_k) T_{ik} = \delta_{ik} T_{ik} = T_{ii} = -|\mathbf{B}|^2/8\pi$. Hence, we obtain for the magnetic contribution,

$$\int_V x_i \frac{\partial T_{ik}}{\partial x_k} d^3x = \mathcal{M} + \oint_A \mathbf{x} \cdot \overleftrightarrow{T} \cdot \hat{\mathbf{n}}\, dA, \tag{24.8}$$

where $\mathcal{M}$ equals the magnetic energy contained in volume V,

$$\mathcal{M} \equiv \int_V \frac{|\mathbf{B}|^2}{8\pi} d^3x. \tag{24.9}$$

Notice that if the field lines are straight and uniform, $\mathbf{B} = \mathbf{constant}$, the surface and volume integrals on the right-hand side of equation (24.8) must again exactly cancel.

For the gravitational term, substitute the expression for $-\partial\mathcal{V}/\partial x_i$ from equation (24.2) to obtain

$$-\int_V \rho x_i \frac{\partial \mathcal{V}}{\partial x_i}\, d^3x = -G\int_V\int_V \frac{\rho(\mathbf{x})\rho(\mathbf{x}')x_i(x_i - x_i')}{|\mathbf{x}-\mathbf{x}'|^3}\, d^3x'\, d^3x.$$

Except for the term x_i and the minus sign in $(x_i - x_i')$, the integrand is symmetric with respect to the primed and unprimed variables; hence we will recover the same value for the integral if we were to replace x_i by $(x_i - x_i')/2$. With $(x_i - x_i')(x_i - x_i') = |\mathbf{x}-\mathbf{x}'|^2$, we then get

$$-\int_V \rho x_i \frac{\partial \mathcal{V}}{\partial x_i}\, d^3x = \mathcal{W}, \tag{24.10}$$

where $\mathcal{W}$ represents the self-gravitational energy of the gas within volume V,

$$\mathcal{W} \equiv -\frac{1}{2}G\int_V\int_V \frac{\rho(\mathbf{x})\rho(\mathbf{x}')}{|\mathbf{x}-\mathbf{x}'|}\, d^3x'\, d^3x, \tag{24.11}$$

with the factor of 1/2 in front being necessary to account for the unrestricted integration over all pairs of points $\mathbf{x}'$ and $\mathbf{x}$.

Collecting expressions in equations (24.5), (24.8), and (24.10), we obtain the *scalar virial theorem*:

$$\mathcal{W} + 2\mathcal{U} + \mathcal{M} = P_{\text{ext}} \oint \mathbf{x}\cdot\hat{\mathbf{n}}dA - \oint \mathbf{x}\cdot \overleftrightarrow{T}\,\hat{\mathbf{n}}\, dA. \tag{24.12}$$

In the right-hand side of equation (24.12), we have set P on A equal to P_{ext} and assumed for simplicity that the latter has a uniform value, i.e., that the external medium supports no nontrivial gravitational or magnetic stresses. This assumption requires the external medium to differ qualitatively from the material inside volume V and will usually nullify any attempt to apply equation (24.12) to a small piece of a much larger cloud.

Before we proceed to applications, we make one final comment (see Problem Set 4). Had we not assumed quasistatic equilibrium, i.e., had we included the term $\rho Du_i/Dt$ on the left-hand side of equation (24.1), we would have gotten two additional terms in equation (24.12). On the left-hand side, we should add twice the kinetic energy, $2K$, where

$$K \equiv \int_V \frac{1}{2}\rho|\mathbf{u}|^2\, d^3x,$$

is the bulk-motion analog of the random-motion contribution (24.6). On the right-hand side, we should add one-half the second time-derivative of the moment of inertia of the system, $(1/2)d^2I/dt^2$, where

$$I \equiv \int_V |\mathbf{x}|^2 \rho \, d^3x.$$

The scalar virial theorem in this form,

$$\mathcal{W} + 2K + 2\mathcal{U} + \mathcal{M} = \frac{1}{2}\frac{d^2I}{dt^2} + P_{\text{ext}} \oint \mathbf{x} \cdot \hat{\mathbf{n}} dA - \oint \mathbf{x} \cdot \overleftrightarrow{T} \, \hat{\mathbf{n}} \, dA,$$

can be used to examine the gross dynamics of oscillations about equilibrium as well as the equilibrium state itself (see Problem Set 4). Radio astronomers also implicitly adopt the scalar virial theorem in this form when they incorporate the effects of observed nonthermal linewidths as a "turbulent" contribution to the term $2K$.

APPLICATION TO AN ISOTHERMAL GAS CLOUD

To make practical use of equation (24.12), we must evaluate the various integrals without having access to the detailed values of their integrands (which we forfeited any chance of calculating by integrating out the spatial dependence of the detailed force balance). If the volume V were a sphere of radius R, the integral $\oint_A \mathbf{x} \cdot \hat{\mathbf{n}} dA$ would equal $\oint R \, R^2 \, d\Omega$, where the second integral occurs over 4π of solid angle Ω. Consequently, we are motivated to define a mean radius R such that

$$4\pi R^3 \equiv \oint \mathbf{x} \cdot \hat{\mathbf{n}} \, dA, \tag{24.13}$$

which we expect to yield a reasonable representation for the equatorial radius unless the cloud becomes very flattened. [Notice that the divergence theorem always gives $3V$ for the right-hand side of equation (24.13).]

By introducing some nondimensional form factors α and β, we may further write the self-gravitational energy and magnetic terms as

$$\mathcal{W} = -\alpha \frac{GM^2}{R}, \tag{24.14}$$

$$\mathcal{M} + \oint \mathbf{x} \cdot \overleftrightarrow{T} \cdot \hat{\mathbf{n}} \, dA = \beta \frac{\Phi^2}{R}, \tag{24.15}$$

where Φ is the (conserved) magnetic flux that threads through the equatorial cross-sectional area and has approximate magnitude $\pi B R^2$, with B being a typical magnetic field strength inside the cloud. We expect α to have order of magnitude unity, and β order of magnitude $1/6\pi^2$ (unless the field lines are nearly straight and uniform, in which case the surface and volume contributions nearly cancel, and β could be much smaller yet).

For an isothermal cloud composed of an ideal gas, $P = a^2\rho$, with $a^2 = kT/m$ = constant, and we may integrate equation (24.6) to get

$$\mathcal{U} = \frac{3}{2}a^2 M, \tag{24.16}$$

where $M = \int_V \rho\, d^3x$ equals the mass of the cloud. Collecting expressions, we may now solve equation (24.12) for the external pressure needed to contain the cloud within radius R as

$$P_{\text{ext}} = \frac{1}{4\pi}\left(-\alpha\frac{GM^2}{R^4} + \beta\frac{\Phi^2}{R^4} + 3\frac{a^2M}{R^3}\right). \tag{24.17}$$

SPECIAL CASES AND PHYSICAL INTERPRETATION

To interpret equation (24.17) physically, begin with the case when we may ignore magnetic and self-gravitational effects. Setting Φ and G equal to zero, we obtain

$$P_{\text{ext}}V = a^2M = \text{constant}, \tag{24.18}$$

where we have identified $4\pi R^3/3$ as the volume V of the spherical gas cloud. Equation (24.18) represents, of course, Boyle's law, $P_{\text{ext}}V = NkT$ where $N = M/m$. To prevent a non-self-gravitating gas cloud from expanding, the external pressure must equal the internal pressure.

Consider next the case when $\Phi = 0$ but $G \neq 0$. For an isothermal gas cloud confined within a sphere of radius R partially by external pressure and partially by self-gravity, we have

$$P_{\text{ext}} = -\alpha\frac{GM^2}{4\pi R^4} + \frac{3a^2M}{4\pi R^3}. \tag{24.19}$$

At large R, the second term on the right-hand side dominates, and we recover Boyle's law, equation (24.18). For smaller R, the self-gravitational term begins to come into play, and we have a lowering of the amount of external pressure needed to produce a given decrease in cloud size R. For some critical value, the curve of P_{ext} versus R actually reaches a maximum and turns over (Figure 24.2).

Beyond the critical point, self-gravity would decrease R without any additional help from P_{ext}; indeed, formally, we require a reduction of P_{ext} to maintain the cloud in equilibrium. Clearly, equilibria states which require P_{ext} to decrease for decreasing R should be unstable to gravitational collapse, a conclusion verified by more detailed analyses. The classical papers of W. B. Bonnor (1956, *M.N.R.A.S.*, **116**, 351) and R. Ebert (1955, *Z. f. .Ap.*, **37**, 217) give additional information on this problem.

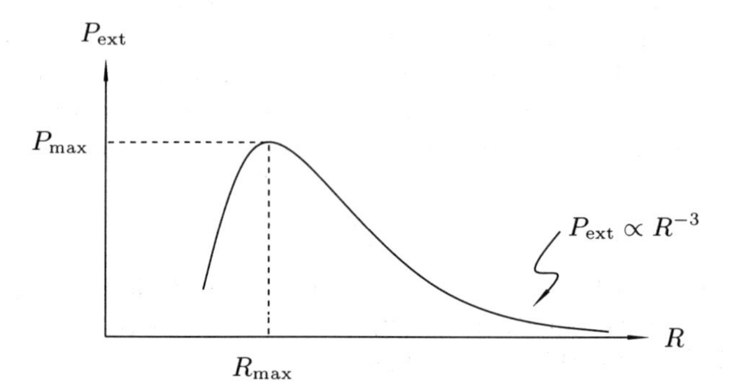

FIGURE 24.2
The pressure-size relation for a spherical gas mass supported against its self-gravitation by a combination of internal (isothermal) pressure and external pressure (assumed to arise from a hot intercloud medium). No equilibrium exists for external pressures beyond a maximum value P_{max}. Clouds of a given mass smaller than the size R_{max}, with $P_{ext} < P_{max}$, are unstable to gravitational collapse (or to expansion to larger sizes).

Consider, finally, the full problem with neither Φ nor G equal to zero. The nonmagnetic case differs fundamentally from the magnetic case, in that the magnetic term on the right-hand side of equation (24.17) contains the *same* R^{-4} dependence as does the gravitational term. Unlike isothermal gas pressure forces, magnetic support scales with the same factor of radius change as does the self-gravitational attraction. Consequently, we may group the two terms in the combination,

$$\frac{1}{4\pi R^4}(-\alpha G M^2 + \beta \Phi^2),$$

which will have a net sign depending on whether the actual cloud mass M exceeds the magnetic critical mass,

$$M_\Phi \equiv \left(\frac{\beta}{\alpha}\right)^{1/2} G^{-1/2}\Phi. \tag{24.20}$$

Numerical MHD calculations of detailed cloud models (cf. the summaries of T. Ch. Mouschovias and Spitzer, 1974, *Ap. J.*, **210**, 326; or of K. Tomisaka, S. Ikeuchi, and T. Nakamura, 1988, *Ap. J.*, **326**, 208) yield $(\beta/\alpha)^{1/2} \approx 0.13$, close to a naive estimate of $(1/6\pi^2)^{1/2}$. For application to magnetized

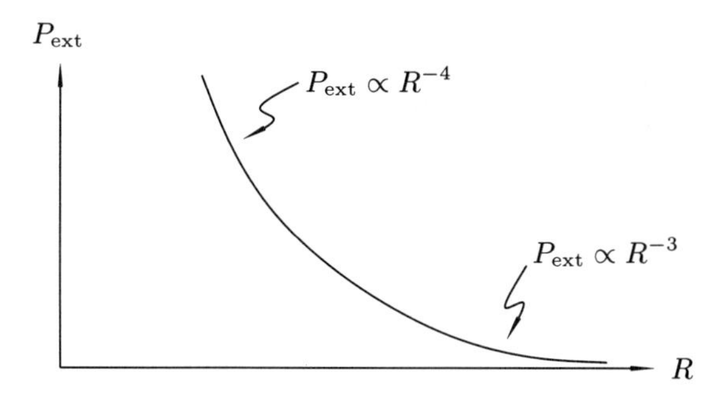

FIGURE 24.3
The pressure-size relation for a gas mass supported against its self-gravitation by a combination of internal (isothermal) pressure, magnetic fields, and external pressure, for a magnetic critical mass M_Φ that exceeds the actual mass M. When $M < M_\Phi$, no amount of external compression can induce gravitational collapse if the magnetic field remains frozen to the matter.

molecular cloud clumps, equation (24.20) with $\Phi = \pi B R^2$ can be expressed as

$$M_\Phi \approx 0.13 G^{-1/2}\Phi \sim 10^3 \ M_\odot \left(\frac{B}{30\ \mu\text{G}}\right)\left(\frac{R}{2\ \text{pc}}\right)^2 .$$

With the definition, equation (24.20), we may write equation (24.12) as

$$P_{ext} = \frac{1}{4\pi}\left[\frac{\alpha G}{R^4}(M_\Phi^2 - M^2) + 3\frac{a^2 M}{R^3}\right]. \tag{24.21}$$

If the actual mass of the cloud $M < M_\Phi$ (see Figure 24.3), the right-hand side is positive definite, and we say that the cloud has a *magnetically subcritical mass* for gravitational collapse because no amount of external compression can induce indefinite contraction of the cloud (for example, to form stars) *if the magnetic flux remains frozen to the matter.*

On the other hand, if the actual mass of the cloud $M > M_\Phi$ (see Figure 24.4), the term $M_\Phi^2 - M^2$ is negative, so that magnetic effects amount merely to an effective *dilution* of the effects of self-gravitation in comparison with the nonmagnetic (Bonnor-Ebert) problem. We say in this case that the cloud has a *magnetically supercritical mass*, because sufficient external compression (or insufficient support by internal thermal pressure) can induce indefinite gravitational collapse, *even if the magnetic flux were to remain perfectly frozen to the matter.*

For application to observed molecular clouds, P_{ext} and $3a^2M/4\pi R^3$ are both usually at least one order of magnitude smaller than $\alpha G M^2/4\pi R^4$; so

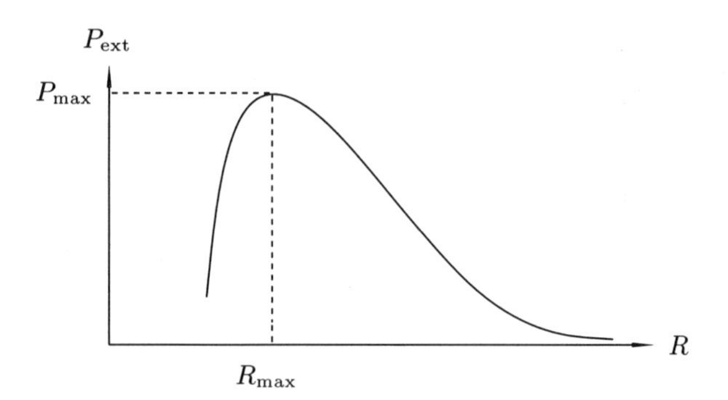

FIGURE 24.4
The same as Figure 24.3 except $M > M_\Phi$. In this case, there exists a maximum external pressure $P_{\rm max}$ that the cloud can withstand before stable equilibrium states become inaccessible to it, and it must undergo gravitational collapse.

M appreciably greater than M_Φ would nominally lead to immediate gravitational collapse. In practice, we must temper this conclusion somewhat by the realization that "interstellar turbulence" (probably many nonlinear MHD waves) may contribute nontrivial amounts of additional support through the term $2K$.

ALTERNATIVE INTERPRETATION

We may think of the existence of a critical mass, equation (24.20), in another way. If we were to compress a magnetized cloud in a more-or-less spherical manner with the conservation of magnetic flux, $\Phi \propto BR^2$, the magnetic field would rise with decreasing R as $B \propto R^{-2}$. In such a compression, the cloud density scales as $\rho \propto M/R^3 \propto R^{-3}$; i.e., the magnetic stresses vary as $B^2 \propto \rho^{4/3}$. Crudely speaking, then, we may say that a magnetic field behaves in three-dimensional compression as a $\gamma_{\rm mag} = 4/3$ "gas," analogous to how the pressure of a relativistically degenerate Fermi-Dirac gas scales with density (see Chapter 5). Self-gravitating configurations of the latter have a limiting mass, the Chandrasekhar mass, which depends only on the constants of the problem. Thus we might intuitively expect that gas clouds supported against their self-gravitation by a frozen-in magnetic field would also have a critical value given by the conserved quantities of the problem. From this point of view, we may regard M_Φ given by equation (24.20) as the *magnetic Chandresekhar limit*, with masses M larger than M_Φ having no other means of support being doomed to gravitational collapse to completely different states of matter. For the original Chandrasekhar problem, this turns out to involve (in stellar evolution theory) a transition from a white-dwarf core to a neutron star (or black hole); for the magnetic "Chandrasekhar" problem, the transition converts interstellar molecular clouds to (clusters of) protostars.

25

Hydromagnetic Shock Waves

Reference: Landau, Lifshitz, and Pitaevskii, *Electrodynamics of Continuous Media*, §§70–73.

Since a magnetized and electrically conducting medium can support three kinds of small amplitude waves (Chapter 23), we may expect the jump conditions associated with hydromagnetic shocks to be more complicated than those for ordinary gas dynamics. In particular, we anticipate three kinds of discontinuities to be possible, one each for the nonlinear counterparts of the linear Alfven, slow MHD, and fast MHD waves. This chapter has the purpose of deriving the jump conditions for these MHD disturbances.

BASIC EQUATIONS IN CONSERVATION FORM

We simplify our problem by ignoring the effects of radiation (see, however, problem set 4). Furthermore, from our previous discussions, we recognize that the diffusive terms in the MHD equations merely spread out the transition layer from the upstream state to the downstream state over a small but finite region of space. If we have no interest in the structure of this region, we may approximate the shock transition as a single discontinuous jump, whose connection relations between the preshock and postshock states may be derived by examining the ideal MHD equations in their conservation forms:

$$\frac{\partial \rho}{\partial t} + \frac{\partial}{\partial x_k}(\rho u_k) = 0; \tag{25.1}$$

$$\frac{\partial}{\partial t}(\rho u_i) + \frac{\partial}{\partial x_k}(\rho u_i u_k + P\delta_{ik} - T_{ik}) = -\rho\frac{\partial \mathcal{V}}{\partial x_i}; \tag{25.2}$$

$$\frac{\partial}{\partial t}\left(\frac{1}{2}\rho|\mathbf{u}|^2 + \rho\mathcal{E} + \rho\mathcal{V} + \frac{|\mathbf{B}|^2}{8\pi}\right) + \nabla\cdot\left[\rho\mathbf{u}\left(h + \frac{1}{2}|\mathbf{u}|^2 + \mathcal{V}\right) + \frac{1}{4\pi}(\mathbf{B}\times\mathbf{u})\times\mathbf{B}\right] = \rho\frac{\partial \mathcal{V}}{\partial t}; \tag{25.3}$$

$$\frac{\partial \mathbf{B}}{\partial t} + \nabla \times (\mathbf{B} \times \mathbf{u}) = 0; \tag{25.4}$$

$$\nabla \cdot \mathbf{B} = 0; \tag{25.5}$$

$$\nabla^2 \mathcal{V} = 4\pi G(\rho + \rho_{\text{ext}}). \tag{25.6}$$

In equation (25.2), T_{ik} is Maxwell's stress tensor

$$T_{ik} = \frac{1}{4\pi}\left(B_i B_k - \frac{1}{2}|\mathbf{B}|^2 \delta_{ik}\right); \tag{25.7}$$

in equation (25.3),

$$\frac{1}{4\pi}(\mathbf{B} \times \mathbf{u}) \times \mathbf{B} = \frac{c}{4\pi}\mathbf{E} \times \mathbf{B}, \tag{25.8}$$

equals the Poynting vector (electromagnetic energy flux) in the field-freezing approximation,

$$\mathbf{E} = \mathbf{B} \times \frac{\mathbf{u}}{c},$$

where we have made use of equation (21.30). To close the set, we assume that the constitutive relations equal those of a perfect gas:

$$\mathcal{E} = \frac{1}{\gamma - 1}\frac{P}{\rho}, \qquad h = \frac{\gamma}{\gamma - 1}\frac{P}{\rho}. \tag{25.9}$$

JUMP CONDITIONS

For hydromagnetic shocks, the incident flow velocity $\mathbf{u}$ and magnetic field $\mathbf{B}$ (in the shock frame) define a plane for the local dynamics. For the local two-dimensional flow (see Figure 25.1), we let n denote the coordinate perpendicular to the shock front (with vector components in the n direction carrying the subscript $\perp$), and s denote the coordinate parallel to the shock front (with vector components parallel to the shock front carrying the subscript $\|$).

Equations (25.1)–(25.5) now read

$$\frac{\partial}{\partial n}(\rho u_\perp) = \ldots, \tag{25.10}$$

$$\frac{\partial}{\partial n}\left[\rho u_\perp u_\perp + P - \frac{1}{8\pi}(B_\perp^2 - B_\|^2)\right] = \ldots, \tag{25.11}$$

$$\frac{\partial}{\partial n}\left[\rho u_\| u_\perp - \frac{1}{4\pi}B_\| B_\perp\right] = \ldots, \tag{25.12}$$

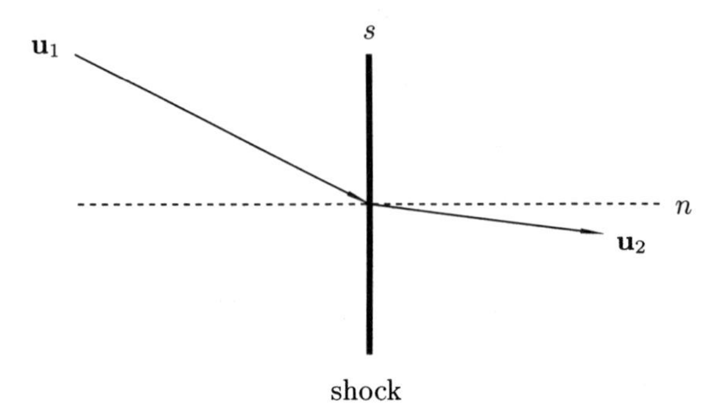

FIGURE 25.1
The geometry of an ideal hydromagnetic shock. The flow is confined to the plane that contains the vectors **u** and **B** of the upstream state. We find it convenient to decompose vectors in this plane to components along n and s, along the shock normal and along the shock front.

$$\frac{\partial}{\partial n}\left\{\rho u_\perp\left[\frac{\gamma}{\gamma-1}\frac{P}{\rho}+\frac{1}{2}(u_\perp^2+u_\parallel^2)\right]-\frac{1}{4\pi}(B_\perp u_\parallel-B_\parallel u_\perp)B_\parallel\right\}=\dots, \tag{25.13}$$

$$\frac{\partial}{\partial n}(B_\perp u_\parallel-B_\parallel u_\perp)=\dots, \tag{25.14}$$

$$\frac{\partial B_\perp}{\partial n}=\dots, \tag{25.15}$$

where the three dots on the right-hand sides denote terms that remain bounded across the shock front. [Among such terms, we include the continuity of the gravitational potential and its first derivatives, as implied by the finiteness of its *second* derivatives in equation (25.6).]

We now integrate equations (25.10)–(25.15) from n_1 to n_2 (in the frame of the shock front) and consider the diffusionless limit $n_1-n_2\to 0$. The jump conditions become

$$[\rho u_\perp]_1^2=0, \tag{25.16}$$

$$\left[\rho u_\perp^2+P+\frac{B_\parallel^2}{8\pi}\right]_1^2=0, \tag{25.17}$$

$$\left[\rho u_\perp u_\parallel-\frac{B_\perp B_\parallel}{4\pi}\right]_1^2=0, \tag{25.18}$$

$$\left[\rho u_\perp\left\{\frac{\gamma}{\gamma-1}\frac{P}{\rho}+\frac{1}{2}(u_\perp^2+u_\parallel^2)\right\}-\frac{1}{4\pi}(B_\perp u_\parallel-B_\parallel u_\perp)B_\parallel\right]_1^2=0, \tag{25.19}$$

$$\left[B_\perp u_\parallel - B_\parallel u_\perp\right]_1^2 = 0, \tag{25.20}$$

$$\left[B_\perp\right]_1^2 = 0. \tag{25.21}$$

Since $\rho u_\perp$ and $B_\perp$ are conserved across the shock in accordance with equations (25.16) and (25.21), the jump condition (25.18) for the conservation of the parallel component of the momentum becomes

$$\left[u_\parallel\right]_1^2 = \frac{B_\perp}{4\pi\rho u_\perp}\left[B_\parallel\right]_1^2. \tag{25.22}$$

Unlike the case of nonmagnetic shocks (Chapter 16), a discontinuity occurs in $u_\parallel$ because a current sheet exists at the shock of strength $(c/4\pi)\left[B_\parallel\right]_1^2$ that can lead to a sudden deflection of the tangential velocity. Except for very special orientations of the upstream flow and magnetic fields, this jump in $u_\parallel$ forms a distinguishing feature of MHD shock waves.

The six equations (25.16)–(25.21) suffice to determine ρ, $u_\perp$, $u_\parallel$, P, $B_\perp$, $B_\parallel$ downstream from the shock, given the values upstream. We wish to derive from them a single equation for $u_\perp(2)$. To do this, first write equations (25.16), (25.18), (25.19), and (25.20) as expressing the continuity of the fluxes of mass, tangential component of the momentum, energy, and the out-of-plane electric field:

$$\rho u_\perp = j_0, \tag{25.23}$$

$$j_0 u_\parallel - \frac{B_\perp B_\parallel}{4\pi} = m_0, \tag{25.24}$$

$$\frac{\gamma}{\gamma-1}P u_\perp + \frac{1}{2}j_0(u_\perp^2 + u_\parallel^2) + f_0 B_\parallel = q_0, \tag{25.25}$$

$$\frac{1}{4\pi}(u_\perp B_\parallel - u_\parallel B_\perp) = f_0, \tag{25.26}$$

where j_0, m_0, q_0, f_0, and $B_\perp$ are the same on both sides of the shock front.

With these definitions, equation (25.17) can have $\rho u_\perp = j_0$ factored from it to read

$$\left[u_\perp + \frac{P}{j_0} + \frac{B_\parallel^2}{8\pi j_0}\right]_1^2 = 0,$$

which we may write, after some algebra, as

$$\left[u_\perp + \frac{c_0^2}{u_\perp} + \left(\frac{2-\gamma}{\gamma+1}\right)\frac{a_0^3}{u_\perp(u_\perp - b_0)} + \frac{1}{\gamma+1}\frac{b_0 a_0^3}{u_\perp(u_\perp - b_0)^2}\right]_1^2 = 0, \tag{25.27}$$

where we have introduced the velocity scales a_0, b_0, and c_0 through the definitions

$$a_0^3 \equiv \frac{(4\pi f_0 + m_0 B_\perp/j_0)^2}{4\pi j_0}, \tag{25.28}$$

$$b_0 \equiv \frac{B_\perp^2}{4\pi j_0}, \tag{25.29}$$

$$c_0^2 \equiv \left(\frac{\gamma-1}{\gamma+1}\right)\left(\frac{2q_0 - m_0^2/j_0}{j_0}\right). \tag{25.30}$$

For nonmagnetic shocks, $\mathbf{B} = 0$, equation (25.27) reduces to the Prandtl-Meyer relation (see Problem Set 3):

$$\left[u_\perp + \frac{c_*^2}{u_\perp}\right]_1^2 = 0, \tag{25.31}$$

where the equation (25.30) for c_0^2 becomes in this case

$$c_*^2 \equiv \left(\frac{\gamma-1}{\gamma+1}\right)\left[h + \frac{1}{2}(u_\perp^2 + u_\parallel^2)\right],$$

with $h = [\gamma/(\gamma-1)]P/\rho$. More generally, equation (25.27) leads to a quartic relation to solve for $u_\perp(2)$. Clearly, however, one root corresponds to the trivial solution $u_\perp(2) = u_\perp(1)$. If we denote $u_\perp(2)$ by u_2 and $u_\perp(1)$ by u_1, and if we factor out $(u_2 - u_1)$, equation (25.27) becomes

$$1 - \frac{c_0^2}{u_1 u_2} - \left(\frac{2-\gamma}{\gamma+1}\right) a_0^3 \frac{[(u_2+u_1) - b_0]}{u_2(u_2-b_0)u_1(u_1-b_0)}$$

$$-\left(\frac{b_0 a_0^3}{\gamma+1}\right)\left[\frac{(u_2-b_0)^2 + +(u_1-b_0)^2 - (u_2u_1 - b_0^2)}{u_2(u_2-b_0)^2 u_1(u_1-b_0)^2}\right] = 0. \tag{25.32}$$

If we were to clear denominators, we could write equation (25.32) as a cubic equation to solve for u_2 (or $u_2 - b_0$). To describe the properties of the solutions, the algebra simplifies considerably if we transform (by sliding tangentially to the front) to a *preferred reference frame* (see Figure 25.2) in which $\mathbf{u}$ and $\mathbf{B}$ are parallel to one another on both sides of the shock:

$$u_\parallel/u_\perp = B_\parallel/B_\perp. \tag{25.33}$$

This assumes that $u_\perp$ does not equal 0, i.e., that we are not talking about *contact discontinuities* (where $B_\perp \neq 0$) or *tangential discontinuities* (where $B_\perp$ also $= 0$).

Not all possible choices of upstream conditions yield three real roots for the cubic equation (25.32). For three roots to exist, u_{n1} must be larger than all three wave speeds (fast MHD, Alfven, and slow MHD) of the upstream medium. In the preferred frame, the strength of the tangential magnetic field $B_\parallel$ increases across a fast shock; it decreases across a slow shock (Figure 25.3).

In particular, it is possible to have $B_\parallel = 0$ behind a slow shock; such a case is called a *switch-off shock*. It is also possible to have $B_\parallel = 0$ ahead

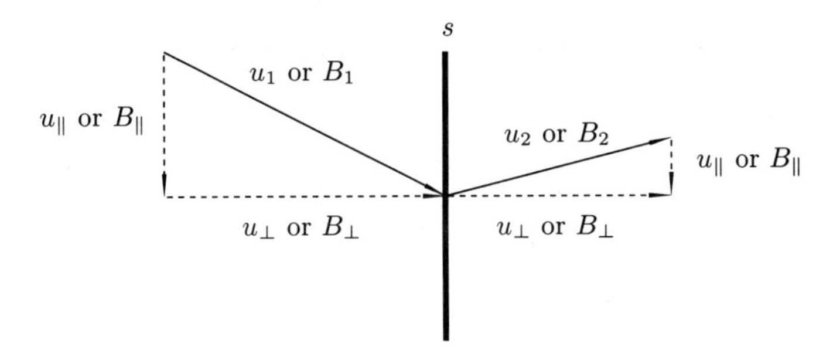

FIGURE 25.2
When $u_{\perp}$ does not equal zero, we can always transform to the preferred reference frame, where **u** and **B** are parallel to one another on both sides of the shock.

of a fast shock; such a case is called a *switch-on shock*. The existence of such cases has the following consequences for shock propagation parallel to the magnetic field. One might think that such a geometry would produce an ordinary gas-dynamic shock in which the magnetic field plays no role. Detailed investigations by Kantrowitz and Petschek (see also Problem 1 on pp. 251–252 of Landau, Lifshitz, and Pitaevskii) showed, however, that a pure gas-dynamic shock propagating exactly along field lines can, under certain conditions, spontaneously disintegrate into a switch-on and

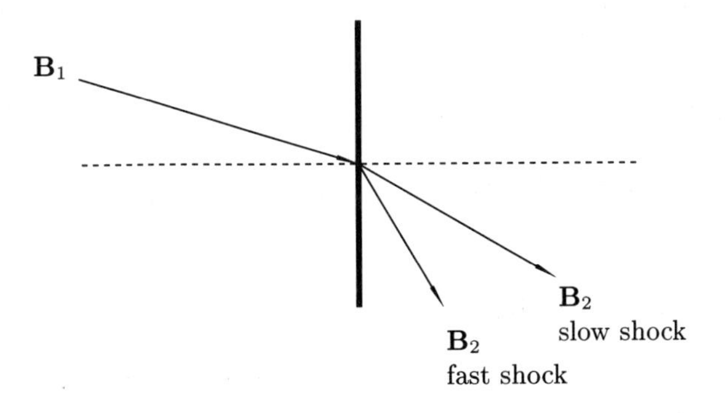

FIGURE 25.3
Schematic diagram of fast and slow shocks in the preferred frame of reference.

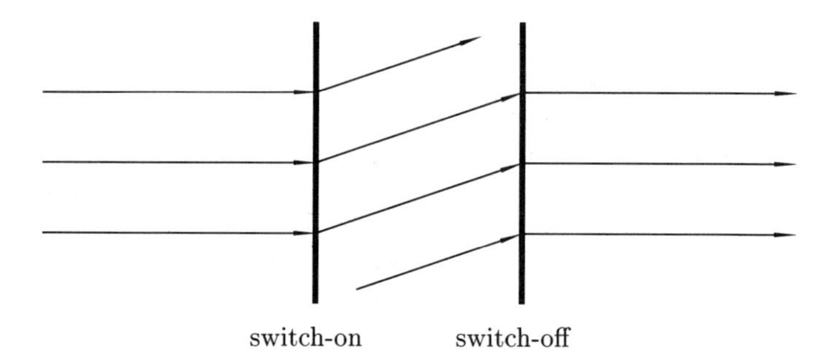

FIGURE 25.4
Shock propagation parallel to the magnetic field in which a single gas-dynamic shock is unstable to disintegration to weaker switch-on and switch-off shocks.

a switch-off shock that leaves the far upstream and far downstream magnetic fields lying in their original directions, but with kinks in the field in a sandwich between the two fronts of the switch-on and switch-off shocks (Figure 25.4).

These considerations generalize also to the case of radiative (or isothermal) shocks. In particular, L. Spitzer (1990, in *Astrophysics: Recent Progress and Future Possibilities*, ed. B. Gustafsson and P. E. Nissen, Matematisk-Fysiske Meddeleslser, **42**:4, pp. 174–176) points out that strong shockwaves that propagate at a small angle with respect to the upstream magnetic field and that produce large density compressions via a single jump (even if it could be established initially) will generally break up into two weaker shockwaves that produce the same net effect via two smaller jumps. Clearly, MHD shock propagation for the parallel or near-parallel orientation contains some surprises. Partially because of such complications, most astrophysical discussions have focused on the perpendicular case, when $u_{\|} = 0$ and $B_{\perp} = 0$, and when only the fast MHD shock (magnetosonic shock) remains nontrivial and okay ("evolutionary" in the language of Landau, Lifshiz, and Pitaevskii). Problem Set 4 asks you to derive the properties of such shocks for both radiative and nonradiative situations.

ROTATIONAL DISCONTINUITIES

A pure Alfven wave does not correspond to a compressional mode; hence we should not be surprised to find a kind of MHD discontinuity which involves no compression even in the nonlinear regime (see Figure 25.5). We follow Landau, Lifshitz, and Pitaevskii in calling a disturbance in which

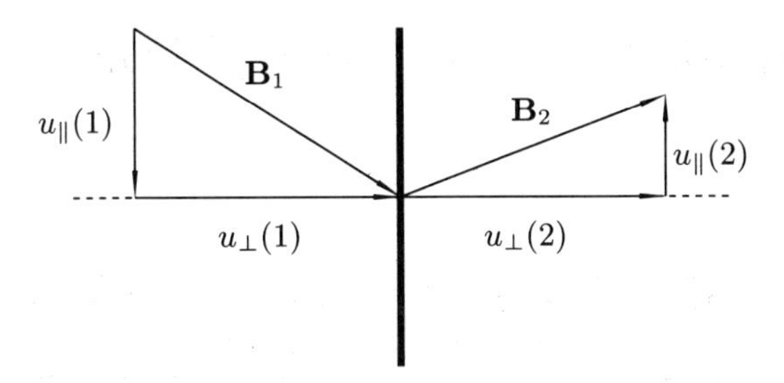

FIGURE 25.5
Magnetic field structure in a rotational discontinuity.

$[\rho]_1^2 = 0 = [u_\perp]_1^2$, with $\left[u_\parallel\right]_1^2$ and $\left[B_\parallel\right]_1^2 \neq 0$, not a "shock," but a "rotational discontinuity" (because the discontinuity ends up only rotating the velocity and magnetic field vectors). Alternatively, we refer to it as an "Alfven discontinuity." (Some books call this type of discontinuity a "shear shock," but this nomenclature seems misleading.)

Equation (25.32) demonstrates that the only nontrivial solution with $u_\perp$ and $B_\perp$ continuous across the front has

$$u_\perp = b_0 = \frac{B_\perp^2}{4\pi\rho u_\perp} \qquad \Rightarrow \qquad u_\perp = \frac{B_\perp}{(4\pi\rho)^{1/2}}, \tag{25.34}$$

which corresponds to the Alfven speed associated with the component of $\mathbf{B}$ perpendicular to the wave front. (Review the discussion of Chapter 22 for Alfven waves with inclined wave fronts.) For such a pure nonlinear Alfven wave, equation (25.22) becomes

$$B_\perp \left[u_\parallel\right]_1^2 = u_\perp \left[B_\parallel\right]_1^2, \tag{25.35}$$

which re-expresses the constancy of mass-to-flux, equation (25.20), across the discontinuity.

Rotational discontinuities of the type represented by equation (25.35) are usually smoothed out by dissipative effects. But because dissipative effects occur slowly in normal astrophysical circumstances, sudden changes of the magnetic field direction, without corresponding changes in the fluid density, can arise under the proper configuration for the external conditions of the flow. In such special regions, the important process of magnetic reconnection can take place, a problem that we shall discuss in Chapter 26.

26

Magnetic Reconnection and Dynamos

Reference: Parker, *Cosmical Magnetic Fields: Their Origin and Their Activity*, Chapters 15 and 18.

In this chapter, we discuss the basic ideas in two complex problems at the forefront of MHD research: magnetic reconnection and dynamos. We begin with the observation that astronomical objects as diverse as planets, stars, and galaxies all have magnetic fields. Where do such fields come from? For a few, such as the Galactic magnetic field, one might argue that the Ohmic dissipation time $t_{\rm D} = L^2/\eta$ is longer than the age of the universe, so perhaps we can blame the existence of the fields on an unknown primordial origin. This answer does not possess much philosophical appeal, nor does it work for the Earth, where $t_{\rm D} \ll$ the age of the Earth (Problem Set 4). Furthermore, the magnetic field of the Sun reverses directions every 11 years, and other stars have been observed to possess similar sorts of magnetic cycles.

When Gilbert first discussed the magnetic properties of the Earth, he compared them with those of a bar magnet. Indeed, he even constructed a scale model of the Earth from a collection of small bar magnets (Figure 26.1). Geophysical research has shown that, although the core of the Earth probably does consist of iron, a ferromagnetic material, this iron exists at a temperature considerably in excess of its Curie point. Consequently, the magnetic field of the Earth cannot correspond to that of a permanent magnet; i.e., unlike a true bar magnet, it does not owe its origin to the orderly alignment of many atomic magnetic moments. Instead, the magnetic field of the Earth must arise from large-scale electric currents that flow in its electrically conducting fluid core. Indeed, the pattern of magnetic stripes on successive deposits of volcanic rock and on spreading seafloors indicates that the magnetic field of the Earth reverses directions at irregular intervals of 10^5-10^6 years. Magnetic *dynamo theory* attempts to understand what patterns of magnetohydrodynamic flow (most popularly, the interaction of differential rotation and convection) are needed to produce spatially coherent (but, perhaps, temporally chaotic) magnetic

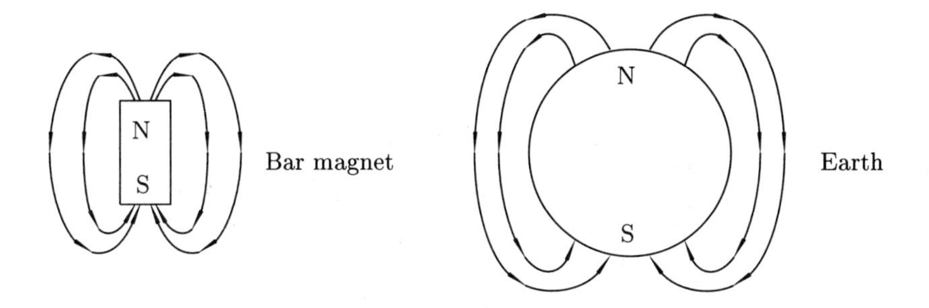

FIGURE 26.1
The dipole magnetic fields of a bar magnet and the Earth (not drawn to scale) bear a superficial similarity to one another.

fields of large scale in cosmic bodies. And in the twentieth century, a virtual dogma has grown that almost *all* astrophysical magnetic fields have an origin in dynamo activity.

To construct a theory of a magnetic dynamos, one might naturally start with the simplest geometry and temporal behavior, say, an axisymmetric time-independent system. Such an attempt would be doomed to failure, because an extremely important result by T.G. Cowling proves that it is generally *impossible to construct an axisymmetric dynamo*, time-independent or not. We shall give the proof of Cowling's theorem shortly, but for now we comment briefly on its consequence: a complete theory of dynamo action must necessarily be quite complicated, involving probably a fully three-dimensional and time-dependent treatment of the equations of magnetohydrodynamics, including the effects of finite resistivity. Thus, we should not be surprised that geophysicists and astrophysicists have achieved only partial progress on this fundamental problem. We shall not attempt here to record all the milestones in this large subject, but only to describe some of the central concepts and qualitative highlights.

COWLING'S THEOREM

We prove Cowling's theorem by assuming its opposite and arriving at a contradiction. Let us suppose it possible to construct an axisymmetric dynamo, where some pattern of axisymmetric fluid motions $\mathbf{u}$ would allow a solution of the magnetic induction equation,

$$\frac{\partial \mathbf{B}}{\partial t} + \nabla \times (\mathbf{B} \times \mathbf{u}) = -\nabla \times (\eta \times \nabla \times \mathbf{B}), \tag{26.1}$$

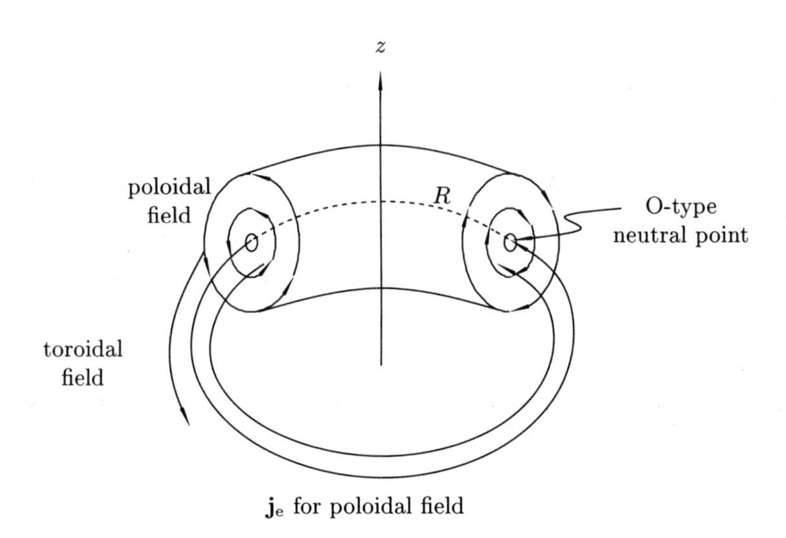

FIGURE 26.2
The geometry near an O-type neutral point for a MHD system with perfect axial symmetry.

that yields a non-decaying, axisymmetric, configuration for $\mathbf{B}$. If our hypothesis holds, and if the system of electric currents that support the magnetic field have a finite spatial extent, so that the magnetic configuration forms closed field lines, the toroidal and poloidal fields—the B_φ and (B_ϖ, B_z) components in cylindrical coordinates (ϖ, φ, z)—must look generically as depicted in Figure 26.2).

In particular, because the poloidal fields close on themselves, there must exist in the meridional plane (the ϖ-z plane) one or more neutral points of the O-type. At an O-type neutral point, the poloidal field $(B_\varpi^2 + B_z^2)^{1/2}$ must go to zero if the current density

$$\mathbf{j}_e = \frac{c}{4\pi}\nabla \times \mathbf{B},$$

does not become infinite there (which it cannot with a finite resistivity). Because of the assumption of axial symmetry, the O-type neutral point actually circles completely around the z-axis; if we computed the integral of $\mathbf{j}_e$ about a complete circuit C, we would get a nonzero result. For sake of definiteness, suppose the poloidal field to have the sense of winding depicted in Figure 26.2, so that the indicated line integral has positive value:

$$\oint_C \mathbf{j}_e \cdot \mathbf{d}\boldsymbol{\ell} > 0. \tag{26.2}$$

From Ohm's law, we have

$$\mathbf{j}_e = \sigma\left(\mathbf{E} + \frac{\mathbf{u}}{c} \times \mathbf{B}\right), \tag{26.3}$$

where $\mathbf{E}$ equals the (induced) electric field, $\boldsymbol{\ell} = \mathbf{e}_\varphi R\,d\varphi$, with R being the radius of the circuit. For simplicity of discussion, we shall assume that we may take the conductivity σ to be a constant without loss of physical generality. Since the poloidal component of $\mathbf{B}$ equals zero at the O-point, and since $\mathbf{u} \times \mathbf{B}$ has no φ-component if $\mathbf{B}$ lies in the direction of $\mathbf{e}_\varphi$, the substitution of equation (26.3) into equation (26.2) now implies

$$\sigma \oint_C \mathbf{E} \cdot \mathbf{d}\boldsymbol{\ell} = \sigma \int_0^R (\boldsymbol{\nabla} \times \mathbf{E}) \cdot \mathbf{e}_z 2\pi\varpi\, d\varpi > 0, \tag{26.4}$$

where we have used Stokes' theorem to convert the line integral to an integral over the area enclosed by the circuit C.

Faraday's law of induction allows us to write $\boldsymbol{\nabla} \times \mathbf{E}$ as $-c^{-1}\partial\mathbf{B}/\partial t$; consequently, we are motivated to introduce the magnetic flux threading through the area enclosed by the circuit C as

$$\Phi \equiv \int_0^R \mathbf{B} \cdot \mathbf{e}_z 2\pi\varpi\, d\varpi. \tag{26.5}$$

Notice that only the poloidal magnetic field makes a contribution to Φ, and that for the configuration drawn in Figure 26.2, Φ is positive. Equation (26.4) now becomes

$$-\frac{\sigma}{c}\frac{d\Phi}{dt} > 0 \qquad \Rightarrow \qquad \frac{d\Phi}{dt} < 0.$$

Hence the positive poloidal flux interior to the O-type neutral circuit decreases with time, and the axisymmetric flow field can not support a non-decaying poloidal magnetic field. This result provides the first part of our proof.

Suppose, then, that the configuration has no poloidal field. Can we prevent the decay, at least, of the toroidal field? With $\boldsymbol{\nabla} \cdot \mathbf{B} = 0$, expand the cross products in equation (26.1) to obtain

$$\frac{\partial \mathbf{B}}{\partial t} + (\mathbf{u} \cdot \boldsymbol{\nabla})\mathbf{B} + \mathbf{B}(\boldsymbol{\nabla} \cdot \mathbf{u}) - -(\mathbf{B} \cdot \boldsymbol{\nabla})\mathbf{u} = \eta\nabla^2\mathbf{B}, \tag{26.6}$$

where we have used $\eta = c^2/4\pi\sigma =$ constant to simplify the right-hand side. When $\mathbf{B} = B_\varphi \mathbf{e}_\varphi$, the identities $\partial \mathbf{e}_\varpi/\partial\varphi = \mathbf{e}_\varphi$, $\partial \mathbf{e}_\varphi/\partial\varphi = -\mathbf{e}_\varpi$, and $\partial \mathbf{e}_z/\partial\varphi = 0$ allow us to write

$$-(\mathbf{B}\cdot\boldsymbol{\nabla})\mathbf{u} = -\varpi^{-1}B_\varphi \frac{\partial \mathbf{u}}{\partial\varphi} = -\varpi^{-1}B_\varphi\left(u_\varpi \frac{\partial \mathbf{e}_\varpi}{\partial\varphi} + u_\varphi \frac{\partial \mathbf{e}_\varphi}{\partial\varphi}\right)$$
$$= \varpi^{-1}B_\varphi(-u_\varpi \mathbf{e}_\varphi + u_\varphi \mathbf{e}_\varpi),$$

since u_ϖ, u_φ, and u_z depend only on ϖ and z. With the help of the equation of continuity,

$$\boldsymbol{\nabla}\cdot\mathbf{u} = \rho\frac{D}{Dt}\left(\rho^{-1}\right),$$

where D/Dt equals the substantial derivative,

$$\frac{D}{Dt} \equiv \frac{\partial}{\partial t} + \mathbf{u}\cdot\boldsymbol{\nabla},$$

we may now express equation (26.6) as

$$\rho\frac{D}{Dt}\left(\rho^{-1}B_\varphi \mathbf{e}_\varphi\right) + \varpi^{-1}B_\varphi(-u_\varpi \mathbf{e}_\varphi + u_\varphi \mathbf{e}_\varpi) = \eta\nabla^2(B_\varphi \mathbf{e}_\varphi). \tag{26.7}$$

We note that

$$\frac{D\mathbf{e}_\varphi}{Dt} = \left(u_\varpi \frac{\partial}{\partial\varpi} + \frac{u_\varphi}{\varpi}\frac{\partial}{\partial\varphi}\right)\mathbf{e}_\varphi = -\varpi^{-1}u_\varphi \mathbf{e}_\varpi,$$

since $\partial \mathbf{e}_\varphi/\partial\varpi = 0$. Hence we may write the left-hand side of equation (26.7) as

$$\mathbf{e}_\varphi \varpi\rho\frac{D}{Dt}\left(\frac{B_\varphi}{\varpi\rho}\right),$$

if we make the identification

$$\varpi\frac{D\varpi^{-1}}{Dt} = -\varpi^{-1}\frac{D\varpi}{Dt} = -\varpi^{-1}u_\varpi.$$

Since $\nabla^2 \mathbf{e}_\varphi = -\varpi^{-2}\mathbf{e}_\varphi$, on the other hand, the right-hand side of equation (26.7) may be expressed as

$$\eta\left\{\frac{\partial}{\partial\varpi}\left[\frac{1}{\varpi}\frac{\partial}{\partial\varpi}(\varpi B_\varphi)\right] + \frac{\partial^2 B_\varphi}{\partial z^2}\right\}\mathbf{e}_\varphi.$$

Collecting expressions, we may now write equation (26.7) in the single-component form,

$$\frac{D}{Dt}\left(\frac{B_\varphi}{\varpi\rho}\right) = \frac{\eta}{\varpi\rho}\left\{\frac{\partial}{\partial\varpi}\left[\frac{1}{\varpi}\frac{\partial}{\partial\varpi}(\varpi B_\varphi)\right] + \frac{\partial^2 B_\varphi}{\partial z^2}\right\}. \tag{26.8}$$

Except for a passive advection by the fluid velocity, equation (26.8) represents a homogeneous diffusion equation for B_φ. It contains no source terms (coming from the poloidal component of the magnetic field); thus all its solutions will eventually decay in time. In other words, no axisymmetric pattern of differential motion—no matter how complicated or time-dependent—can indefinitely maintain either a poloidal or a toroidal magnetic field against Ohmic dissipation. This provides the complete statement of Cowling's powerful theorem: there can exist in nature no self-sustained axisymmetric dynamos. (Q.E.D.)

ELEMENTS NEEDED FOR A DYNAMO

The above proof of Cowling's theorem suggests a few ingredients needed for successful operation of a MHD dynamo. For the sake of definiteness, consider the case of the Sun. First, notice no pattern of fluid motions $\mathbf{u}$ can lead to the spontaneous appearance of a magnetic field from equation (26.1) if $\mathbf{B} = 0$ everywhere to begin with. Dynamo action can only amplify existing fields; it cannot create any where none existed before. Thus, dynamo theories always need seed fields, either brought in from elsewhere (e.g., from the interstellar medium for stars) or generated from currents that arise from other effects (e.g., thermoelectric "batteries").

Second, once we have a poloidal field, we still need to offset its Ohmic decay, e.g., some process to convert toroidal fields to poloidal ones. Radial convection would provide such a mechanism, and it would be helped by magnetic buoyancy effects (e.g., the Parker instability) associated with toroidal fields held down by the weight of thermal gas. (Recall from Chapters 8 and 10 that gas slightly less dense than average in a stellar convection zone tends to rise toward the surface. Gas threaded with extra magnetic flux would be less dense than average, because the extra magnetic pressure partially replaces some of the gas pressure. Thus such flux ropes would naturally tend to rise to the top of the Sun's outer convection zone.)

Third, we probably need some process to regenerate the toroidal fields from the poloidal fields. Differential rotation would provide such a mechanism, since it would take any poloidal fields and stretch them in the azimuthal direction, thereby amplifying the strength of the toroidal magnetic fields. The toroidal fields, in turn, get advected radially or buckle upwards to become poloidal fields, as in the first part of our discussion.

Finally, we need some process that takes the tangle of field lines produced by the above mechanisms and untangles them to produce the large-scale coherence observed in most astrophysical environments. Magnetic *reconnection* to destroy opposed field lines on a small scale while preserving the large-scale order (or somehow periodically reversing it) probably provides this mechanism. We shall discuss the prevailing ideas concerning rapid

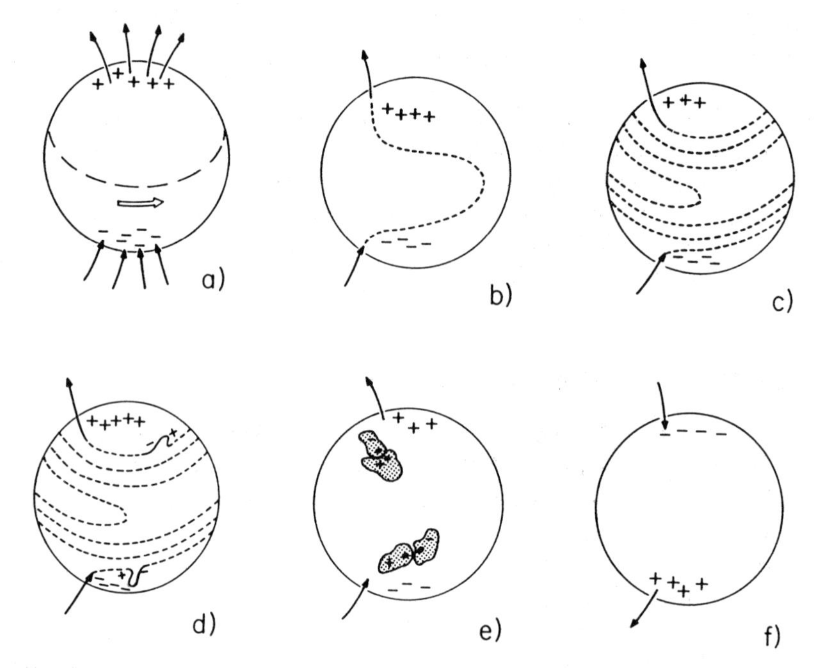

FIGURE 26.3
Babcock's model of the solar cycle. (From Noyes, *The Sun, Our Star.*)

magnetic reconnection after we survey the observed phenomenology of the solar cycle.

THE OBSERVED SOLAR CYCLE

Observations of magnetic activity cycle on the surface of the Sun justifies the general viewpoints expressed in the previous section. In particular, Figure 26.3 illustrates Babcock's model of the sunspot cycle.

At the start of the cycle, (a) magnetic fields run primarily, say, from south to north. The Sun rotates differentially, with the equator taking only $\sim$ 25 days to go once around, while the polar regions take $\sim$ 35 days. As a consequence, (b) differential rotation pulls the fields at the equator ahead of the polar regions, stretching and amplifying the toroidal field. After many rotation periods, (c) the field lines are wrapped several times around the Sun. The strong toroidal fields are buoyant, (d) and they begin to buckle upward out of the Sun in localized "hernias." The magnetic field lines become especially concentrated (for incompletely understood reasons)

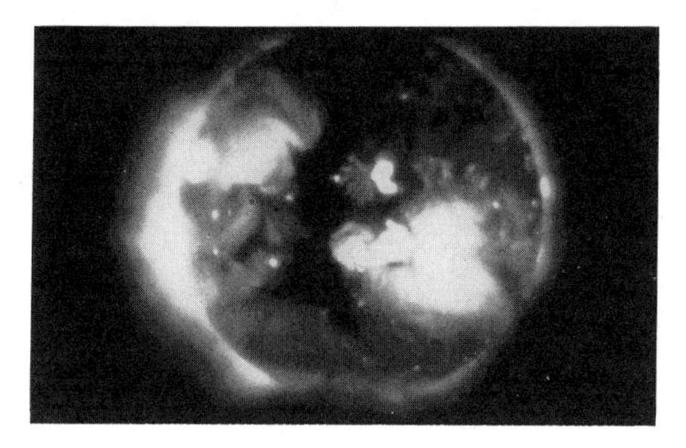

FIGURE 26.4
Skylab photograph of x-rays from the Sun show bright coronal loops and dark coronal holes. (NASA: Harvard-Smithsonian Center for Astrophysics.)

at the ends of the "hernias" to give (e) sunspots. The sunspot groups tend to come in pairs (one at each footpoint) that satisfy *Hale's law of polarity*, namely, in this phase of the cycle and for the northern hemisphere, the field lines protrude out of the Sun in the leading sunspot group (leading in the sense of rotation) while they plunge into the Sun in the trailing group. In the southern hemisphere of the Sun, they do just the reverse. Reconnection of oppositely directed field lines occurs during this phase, occasionally producing violent solar flares (to be discussed below). After much of the disordered field of small scale has been destroyed, the field lines of the trailing polarity somehow drift to the poles where they cancel and replace the existing fields, so that (f) the large-scale poloidal field acquires at the end of 11 years the polarity opposite to that which it possessed at the beginning. The situation then begins anew in another 11-year (half)cycle, except that all polarities reverse.

SOLAR FLARES

Observations carried out from Skylab showed that X-rays from the Sun come from coronal loops; in between the loops occur dark "holes" of much less emission. The coronal loops correspond to closed magnetic field structures that trap the hot gas of the solar corona, making the emissive power rise to readily observable levels. In contrast, the coronal holes correspond to open structures, where the magnetic field has insufficient strength to prevent the hot gases from expanding into a rarefied solar wind (Figure 26.4).

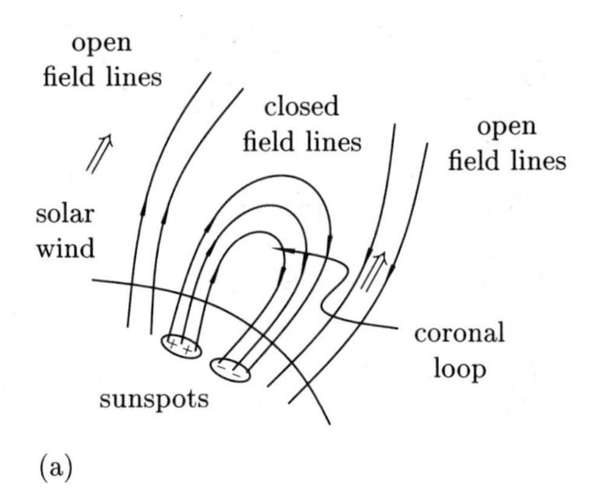

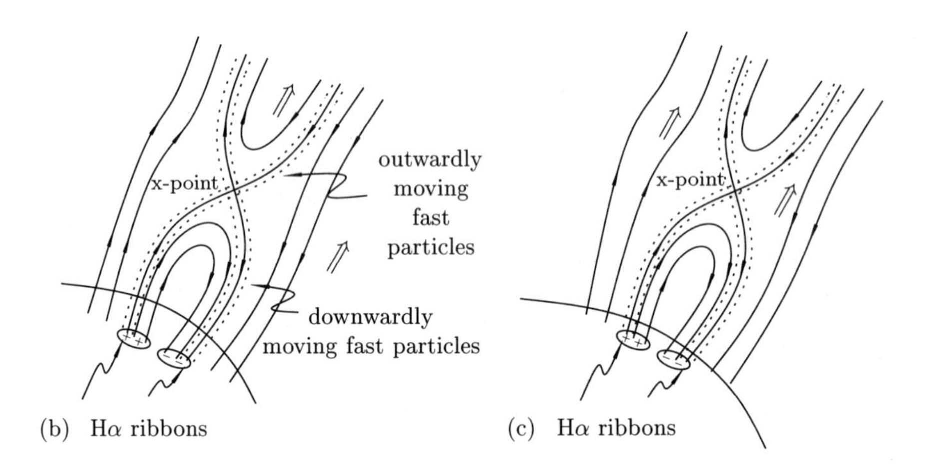

FIGURE 26.5
Schematic depiction of the sequence of events in a solar flare. (Adapted from Noyes, *The Sun, Our Star.*)

The base of the coronal loops often yields sunspots, and the arch, solar prominences. Occasionally, an arch will erupt into a giant solar flare. Thus solar astronomers have come to regard coronal loops as the basic building blocks of stellar activity. The mechanism of a solar flare is quite complex, but a fundamental element seems to be rapid magnetic reconnection between a closed field geometry and an open one (Figure 26.5).

In the preflare configuration, (a) a magnetic loop arches from a sunspot group of one polarity to another. A neutral line separates the reversal of fields near the base, and an X-point does the same at the top of the arch, where the magnetic field changes from a closed to an open configuration. In a steady state, a slow process of reconnection may occur near the X-point as material blowing off the Sun from the neighboring regions erode away the arch. However, subphotospheric convective motions also continually bring in new magnetic pores that add to the strength of the sunspot groups, and under certain circumstances, the opposing fields at the top of the arch may be pressed together sufficiently rapidly as to undergo a violent flare. (b) The onset of the flare evidently initiates a rapid phase of reconnection near the top of the loop. Energy stored in the magnetic field is released suddenly to accelerate charged particles. Some of the high-speed electrons and ions travel downward to heat the chromosphere at the footpoints of the loop, yielding bright ribbons of Hα emission. Others spew outward, giving rise to detectable radio emission, and perhaps race toward artificial satellites that pick them up as a burst of high-energy particles. (c) As the flare progresses, the site of rapid reconnection moves upward, and a tongue of activity shoots away from the surface of the Sun. The downward shower of high-speed particles then occurs along more widely spaced field lines, causing the Hα ribbons to spread apart with time.

MAGNETIC RECONNECTION

How does magnetic reconnection work? In 1957 Eugene Parker gave the following description, which elucidates some of the principal ideas, and reveals some of the difficulties of the theory. Imagine a region near the origin of an x-y coordinate system in which the magnetic field, pointing mostly parallel or antiparallel to the x-direction, reverses directions as we move up the y direction (Figure 26.6).

Let the characteristic dimension of the field reversal be $\pm L_y$. Field annihilation occurs via Ohmic dissipation in the neighborhood of the null sheet at $y = 0$ because

$$\mathbf{j}_e = (c/4\pi)\nabla \times \mathbf{B} \tag{26.9}$$

is large even if $\mathbf{B} \to 0$, as long as L_y is very small. Matter and new field must press upward and downward along the y axis with some speed u_y to replace the annihilated magnetic field. The vertical flow into the current sheet must, by the equation of continuity, cause gas to squirt out horizontally from the reconnection region with some speed u_x. We suppose the characteristic dimension for the horizontal flow to equal L_x. We now wish to find order-of-magnitude relationships between the various variables that have been introduced so far.

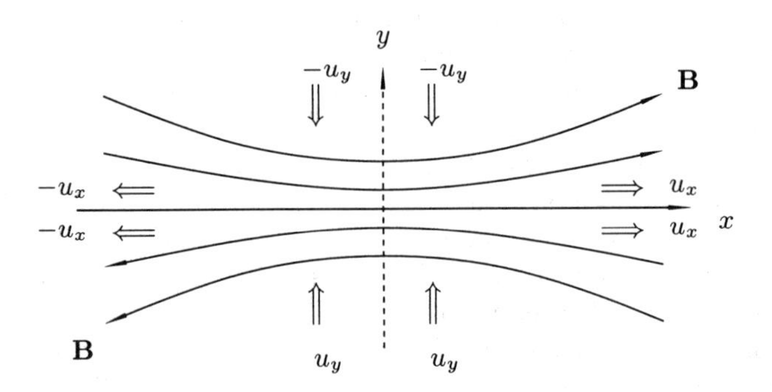

FIGURE 26.6
The reconnection geometry near a null surface where the pattern of the magnetic field reverses directions across a plane where field annihilation takes place.

We assume that the dimensions L_x and L_y are sufficiently small that we may ignore any variations of the gravitational potential across the region of interest. We also suppose (as we can verify *a posteriori*) that the motions occur sufficiently slowly that we may regard the fluid to be effectively incompressible (see Chapter 6) and the y direction to satisfy approximate force balance, i.e., the constancy of total pressure,

$$P + \frac{B^2}{8\pi} = \text{constant}. \tag{26.10}$$

The gas pressure P must then be highest in the central plane, where $B = 0$. The high pressure there will eject the fluid along the x axis if we assume that after a distance $\sim L_x$, there is an overall pressure drop,

$$\Delta P = \frac{B^2}{8\pi}, \tag{26.11}$$

because the gas comes into mechanical equilibrium again with the general surroundings. In the x-direction, for flow along field lines, we have Bernoulli's theorem (with $\rho =$ constant),

$$\frac{1}{2}u_x^2 + \frac{P}{\rho} = \text{constant}, \tag{26.12}$$

which implies that if the beginning speed equals zero at $x = 0$, the end speed at $x = \pm L_x$ equals

$$u_x = (2\Delta P/\rho)^{1/2} = v_\text{A}, \tag{26.13}$$

where we have used equation (26.11) and defined the Alfven speed as

$$v_{\rm A} \equiv B/(4\pi\rho)^{1/2}.$$

Steady-state mass conservation requires that the outward horizontal mass flow (in x) for every length Δz in the third dimension be offset by the inward vertical mass flow (in y) for the same length Δz:

$$2\rho u_x(2L_y\Delta z) = 2\rho u_y(2L_x\Delta z).$$

If we cancel common factors, we get the relation

$$u_y = (L_y/L_x)v_{\rm A}, \tag{26.14}$$

where we have made use of equation (26.13) to eliminate u_x. Equation (26.14) shows that the reconnection speed, the speed at which the opposing magnetic fields are being pushed together, occurs at a fraction of the local Alfven speed, with the fraction given by the aspect ratio L_y/L_x.

To get another formula for the reconnection speed, let us consider the rate at which we can annihilate magnetic energy. We have a Joule dissipation rate per unit volume given by

$$\frac{|\mathbf{j}_{\rm e}|^2}{\sigma} = \frac{\eta}{4\pi}|\nabla\times\mathbf{B}|^2 = \frac{\eta}{4\pi}\frac{B^2}{L_y^2}.$$

The time rate of producing heat in the volume $L_xL_y\Delta z$ must therefore equal

$$\frac{\eta B^2}{4\pi L_y}L_x\Delta z, \tag{26.15}$$

which we must set equal to the rate of annihilation of magnetic field energy pressed vertically into the region:

$$\frac{B^2}{8\pi}u_yL_x\Delta z. \tag{26.16}$$

If we equate equations (26.15) and (26.16), we obtain

$$u_y = 2\eta/L_y. \tag{26.17}$$

Finally, we eliminate L_y from equations (26.14) and (26.17) to derive the desired result for Parker's reconnection speed,

$$u_y = 2\,{\rm Re}_{\rm mag}^{-1/2}\,v_{\rm A}, \tag{26.18}$$

where ${\rm Re}_{\rm mag}$ is the magnetic Reynolds number associated with the horizontal flow,

$${\rm Re}_{\rm mag} \equiv (2L_x)u_x/\eta = 2L_xv_{\rm A}/\eta. \tag{26.19}$$

For the Sun, $\mathrm{Re}_{\mathrm{mag}}$ is a big number, on the order of 10^5 for the smallest pores to 10^9 for the largest sunspot groups. Hence equation (26.18) would imply that magnetic reconnection always occurs at a very small fraction of the dynamical speed v_{A}. Nevertheless, notice that Parker's formula already gives a considerable increase $\sim \mathrm{Re}_{\mathrm{mag}}^{1/2}$ over the speed $v_{\mathrm{A}}/\mathrm{Re}_{\mathrm{mag}}$ that we might have naively expected for magnetic diffusion had we not accounted for the dynamics of the field-reversal geometry.

BOHM DIFFUSION AND PETSCHEK'S MECHANISM

Be that as it may, if we were to compute $\mathrm{Re}_{\mathrm{mag}}$ by using standard kinetic-theory values for the electrical resistivity η, the resulting reconnection speeds come out vastly too slow to account for the quick rise times of observed solar flares. This has suggested to some workers that actual magnetic diffusivity values must be considerably amplified from their normal microscopic values by some anomalous process—generically referred to as *Bohm diffusion.* The idea gains some support from plasma fusion experiments, which show that plasma can often escape from confining magnetic fields much faster than simple kinetic estimates would predict. For Bohm diffusion, η supposedly acquires values comparable to $v_T^2/\omega_{\mathrm{L}}$, where v_T and ω_{L} are the thermal velocity and gyrofrequency, respectively, of the diffusing particles. Perhaps subtle instabilities play an equally important role in the reconnection of astrophysical magnetic fields. If so, the subject rests on an equally uncertain base as many current theories of viscous accretion disks (which require an anomalous source of viscosity; review Chapter 7).

A.G. Petschek [1964, *The Physics of Solar Flares, AAS-NASA Symposium*, NASA SP-50, ed. W.N. Hess (Greenbelt, Maryland), p. 425; see also the discussion in Parker's book] advocates a more optimistic point of view. According to Petschek, we get the low estimate for u_y given by equation (26.18) only because we assumed in the previous section that magnetic reconnection occurs over a finite *plane.* If we assume instead that reconnection occurs through an X-point (see Figure 26.7), as seems true for solar flares, then u_y can approach values much closer to the dynamical speed v_{A}.

Indeed, Petschek's analysis yields (and subsequent work confirms) that the reconnection speed in such a geometry approaches

$$u_y \sim v_{\mathrm{A}}/\ln(\mathrm{Re}_{\mathrm{mag}}). \tag{26.20}$$

The replacement of $\mathrm{Re}_{\mathrm{mag}}^{-1/2}$ by $\ln(\mathrm{Re}_{\mathrm{mag}})$ makes an enormous difference, of as much as $\sim 10^4$ for the Sun. Nevertheless, even Petschek's speed (26.20)

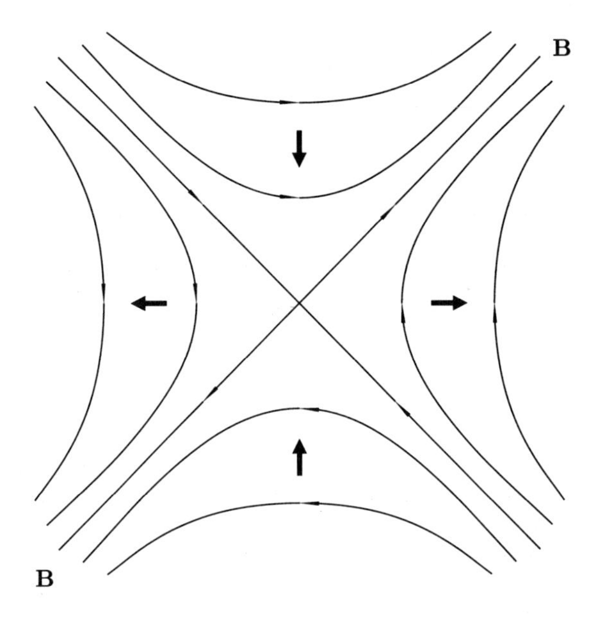

FIGURE 26.7
Magnetic reconnection in the neighborhood of an X-point. Conducting matter from the top and bottom presses together field lines of opposing polarity, causing them to annihilate at a point of zero **B**. Continuity results in the matter leaving the annihilation region in the horizontal directions, carrying with them field lines of a different magnetic topology.

leaves the reconnection rate more than an order of magnitude smaller than the dynamical value $v_{\rm A}$ that some workers feel is necessary to explain the phenomenon of solar flares. This problem, then, continues to be central for space-plasma physicists.

27

Ambipolar Diffusion

References: L. Mestel and L. Spitzer, 1956, *M.N.R.A.S.*, **116**, 503; T. Nakano, 1984, *Fund. Cosmic Physics*, **9**, 139.

In Chapter 26 we discussed how magnetic reconnection can change the topology of how the matter of a conducting fluid is frozen to field lines. In this chapter we wish to consider another method by which a system—now lightly ionized—might be able to alter its mass-to-flux distribution: *ambipolar diffusion*. The process arises from the fundamental need for a relative drift of the charged and neutral species of the gas because the first experiences electromagnetic forces directly, whereas the second experience them only through collisions with the first. Indeed, were it not for these collisions and the frictional forces set up because of the relative drift, the neutral gas, which constitutes the vast bulk of the medium by assumption, would not be affected at all by the presence of the embedded magnetic fields (as the air in a wind can blow without any appreciable interaction with the Earth's magnetic field). Thus, for a lightly ionized gas, the process of ambipolar diffusion is intrinsically linked to the issue of mechanical support by magnetic fields, and does not form a separate topic invented by theorists to complicate the dynamics.

DRAG FORCE BETWEEN NEUTRALS AND IONS

Ambipolar diffusion arises in a lightly ionized gas (e.g., a molecular cloud or a "neutral" wind from the cool surfaces of young and evolved stars), because the mean velocity $\mathbf{u}_\mathrm{n}$ of the neutral atoms or molecules will not generally equal the mean velocity $\mathbf{u}_\mathrm{i} \approx \mathbf{u}_\mathrm{e}$ of the ions and electrons. For example, in a magnetized molecular cloud, the charged species feel a Lorentz force per unit volume,

$$\mathbf{f}_\mathrm{L} = \frac{1}{4\pi}(\boldsymbol{\nabla} \times \mathbf{B}) \times \mathbf{B}, \tag{27.1}$$

that the neutrals do not feel directly. This Lorentz force, acting in addition to all the other forces in the system (gravity, pressure, etc.), will generally cause the charged species to move at a slightly different mean velocity than the neutrals.

Resisting the relative drift will be a frictional (drag) force (per unit volume) that arises because of mutual collisions between the ions and neutrals (we focus on the ions rather than the electrons because for a given slip relative to the neutrals, the ions carry much more difference momenta),

$$\mathbf{f}_{\rm d} = \gamma\rho_{\rm n}\rho_{\rm i}(\mathbf{u}_{\rm i} - \mathbf{u}_{\rm n}), \tag{27.2}$$

where $\rho_{\rm n}$ and $\rho_{\rm i}$ equal, respectively, the mass density of the neutrals and ions, and γ represents a drag coefficient (not to be confused with the ratio of specific heats). In equation (27.2), we have defined $\mathbf{f}_{\rm d}$ to be the drag force exerted per unit volume on the neutrals by the ions; the drag force exerted on the ions by the neutrals equals $-\mathbf{f}_{\rm d}$.

FRICTIONAL DRAG COEFFICIENT

We give below a brief derivation of equation (27.2). The rate of collisions of ions with any neutral equals

$$n_{\rm i}\langle w\sigma_{\rm in}\rangle,$$

where $\sigma_{\rm in}$ is the elastic scattering cross section for ion-neutral encounters, w equals the relative velocity of the ion as seen from the rest frame of the neutral, $n_{\rm i}$ is the number density of ions, and the angular brackets denote an average over the distribution function of the ions. If the masses of the neutrals and ions are, respectively, $m_{\rm n}$ and $m_{\rm i}$, each collision yields a momentum transfer from ion to neutral equal to $m_{\rm i}(\mathbf{u}_{\rm i} - \mathbf{u}_{\rm n})$ times the fraction of the mass of the collision pair contained by the neutral: $m_{\rm n}/(m_{\rm n}+m_{\rm i})$. (If $m_{\rm n}$ were much smaller than $m_{\rm i}$, little of the momentum of the ion would be transferred to the neutral; on the other hand, if $m_{\rm n} = m_{\rm i}$, on average half the difference will be transferred per three-dimensional collision. In actual molecular clouds, the mean molecular mass of neutrals and ions have the approximate values, $m_{\rm n} \sim 2.3\ m_{\rm H}$ and $m_{\rm i} \sim 29\ m_{\rm H}$, where $m_{\rm H}$ equals the mass of the hydrogen atom.)

In any case, if $n_{\rm n}$ denotes the number density of neutrals, the rate of momentum transfer, per unit volume, from ions to neutrals equals

$$n_{\rm n}\frac{m_n m_{\rm i}}{(m_{\rm n} + m_{\rm i})}(\mathbf{u}_{\rm i} - \mathbf{u}_{\rm n})n_{\rm i}\langle w\sigma_{\rm in}\rangle, \tag{27.3}$$

where we recognize the combination $m_{\rm n}m_{\rm i}/(m_{\rm n}+m_{\rm i})$ as the reduced mass of the collision. Denoting $m_{\rm n}n_{\rm n}$ as $\rho_{\rm n}$ and $m_{\rm i}n_{\rm i}$ as $\rho_{\rm i}$, and setting expression

(27.3) equal to $\mathbf{f}_\mathrm{d}$, we recover equation (27.2), with the following expression for the drag coefficient γ:

$$\gamma = \frac{\langle w\sigma_\mathrm{in}\rangle}{(m_\mathrm{n} + m_\mathrm{i})}. \tag{27.4}$$

Osterbrock pointed out in the 1960s that the ion-neutral cross section σ_in has a much larger value than one might infer from the essentially geometric values that apply to neutral-neutral collisions. At low relative velocities w, a passing ion can induce a temporary dipole moment in the neutral atom or molecule that considerably enhances the effective cross-section. Indeed, in the so-called Langevin approximation, $\sigma_\mathrm{in} \propto w^{-1}$, so that the expression $\langle w\sigma_\mathrm{in}\rangle$ effectively equals a constant independent of temperature T or $|\mathbf{u}_\mathrm{i} - \mathbf{u}_\mathrm{n}|$. The resulting value for γ from equation (27.4) for conditions that apply to molecular clouds has been computed by B. Draine, W. Roberge, and A. Dalgarno (1983, *Ap. J.*, **270**, 519) as

$$\gamma = 3.5 \times 10^{13}\,\mathrm{cm}^3\ \mathrm{g}^{-1}\ \mathrm{s}^{-1}. \tag{27.5}$$

If the slip speed $|\mathbf{u}_\mathrm{i} - \mathbf{u}_\mathrm{n}|$ exceeds $\sim 10\,\mathrm{km\,s}^{-1}$, which in turn much exceeds the thermal velocity (for molecular clouds), T. Ch. Mouschovias and E. V. Paleologou (1981, *Ap. J.*, **246**, 48) point out that the Langevin cross section formally becomes smaller than the geometric cross section, $4\pi(r_\mathrm{n} + r_\mathrm{i})^2$, where r_n and r_i are the effective atomic radii of the neutrals and ions, respectively. We should then use the latter to compute $\langle w\sigma_\mathrm{in}\rangle$. In this case,

$$\gamma \approx \frac{4\pi(r_\mathrm{n} + r_\mathrm{i})^2}{(m_\mathrm{n} + m_\mathrm{i})}|\mathbf{u}_\mathrm{i} - \mathbf{u}_\mathrm{n}|,$$

and the drag force (27.2) would depend quadratically on the slip speed $|\mathbf{u}_\mathrm{i} - \mathbf{u}_\mathrm{n}|$ instead of linearly. However, slip speeds in excess of $10\,\mathrm{km\,s}^{-1}$ rarely arise in the interstellar medium; so we shall assume henceforth that γ can be taken to be a constant.

FRACTIONAL IONIZATION

Cosmic rays keep molecular clouds partially ionized by introducing a mean rate of ionization per neutral atom or molecule, $\zeta \sim 10^{-17}\,\mathrm{s}^{-1}$. If the volumetric rate of recombinations of electrons and ions took place in the gas phase, it would be proportional to $n_\mathrm{e}n_\mathrm{i} \propto n_\mathrm{i}^2$. In a steady state, the volumetric rate of recombinations of electrons and ions $\propto n_\mathrm{i}^2$ would equal the volumetric rate of ionizations of neutrals by cosmic rays, ζn_n, so that we have $n_\mathrm{i} \propto n_\mathrm{n}^{1/2}$, or equivalently,

$$\rho_\mathrm{i} = C\rho_\mathrm{n}^{1/2}, \tag{27.6}$$

where C equals a constant. In point of fact, the actual ionization balance in molecular clouds is more complicated, involving recombinations of electrons and ions on charged grains as well as those taking place in the gas phase. Nevertheless, the theoretical calculations of B. G. Elmegreen (1979, *Ap. J.*, **232**, 729) and of T. Umebayashi and T. Nakano (1980, *Publ. Astr. Soc. Japan*, **32**, 405) show that equation (27.6) is not bad as an approximation for a fairly wide range of molecular cloud densities. In what follows, we shall suppose for fiducial purposes that

$$C = 3 \times 10^{-16}\,\text{cm}^{-3/2}\,\text{g}^{1/2}. \tag{27.7}$$

Such a choice implies a fractional ionization $n_\text{i}/n_\text{n} \sim 10^{-7}$ for $n_\text{n} \sim 10^4$ cm^{-3}, within the limits 10^{-6}–10^{-8} set by observations of molecular cloud cores.

TYPICAL DRIFT SPEEDS

Because of the very small fractional ionization that exists in molecular clouds, the Lorentz force $\mathbf{f}_\text{L}$ and the drag force $-\mathbf{f}_\text{d}$ exerted by the neutrals on the ions must vastly dominate every other force that acts on the charged species (e.g., gravity or gradients of the partial pressures), and must sum to zero. Setting $\mathbf{f}_\text{d}$ in equation (27.1) equal to $\mathbf{f}_\text{L}$ in equation (27.2) allows us to solve for the drift velocity,

$$\mathbf{v}_\text{d} \equiv \mathbf{u}_\text{i} - \mathbf{u}_\text{n} = \frac{1}{4\pi\gamma\rho_\text{n}\rho_\text{i}}(\boldsymbol{\nabla} \times \mathbf{B}) \times \mathbf{B}. \tag{27.8}$$

In order of magnitude, if $\mathbf{B}$ changes on a typical scale of L,

$$v_\text{d} \sim \frac{B^2}{4\pi\gamma\rho_\text{n}\rho_\text{i}L} \approx \frac{v_\text{A}^2}{V} \qquad \text{where} \qquad V \equiv \gamma\rho_\text{i}L = \gamma C L \rho^{1/2}, \tag{27.9}$$

and v_A^2 is the square of the Alfven speed of the combined medium,

$$v_\text{A}^2 = B^2/4\pi\rho. \tag{27.10}$$

In writing equation (27.9), we have used equation (27.6) and assumed that $\rho = \rho_\text{n} + \rho_\text{i} \approx \rho_\text{n}$ when the medium is lightly ionized, $\rho_\text{i} \ll \rho_\text{n}$. Substituting in equations (27.5) and (27.7), we get $V = 6\,\text{km}\,\text{s}^{-1}$ when $L = 0.1\,\text{pc}$ and $n_\text{n} = \rho/m_\text{n} = 10^4\,\text{cm}^{-4}$. Since $V \gg v_\text{A} \approx 0.4\,\text{km s}^{-1}$ (for $B \sim 30\ \mu\text{G}$), the drift speed as given by equation (27.9) typically satisfies $v_\text{d} \ll v_\text{A}$, and ambipolar diffusion normally occurs fairly slowly even in the cores of dense molecular clouds. [Notice that V retains the same value if ρ scales as L^{-2}, as in a singular isothermal sphere. On larger scales, molecular radio astronomers have claimed that ρ scales as L^{-1} (constant column density).

In this case, $V \propto L^{1/2}$, whereas, $v_{\rm A}$ remains roughly constant if B scales as $\rho^{1/2}$ (as roughly indicated by theory and observations), and ambipolar diffusion would be less important for molecular cloud envelopes than for their cores.]

AMBIPOLAR DIFFUSION

The resulting equation for the slow-time evolution of $\mathbf{B}$ may be obtained from the approximation that the magnetic field is frozen in the plasma of ions and electrons:

$$\frac{\partial \mathbf{B}}{\partial t} + \nabla \times (\mathbf{B} \times \mathbf{u}_{\rm i}) = 0. \tag{27.11}$$

To obtain the drift of the field relative to the neutrals, we use equation (27.8) to eliminate $\mathbf{u}_{\rm i}$ from equation (27.11):

$$\frac{\partial \mathbf{B}}{\partial t} + \nabla \times (\mathbf{B} \times \mathbf{u}_{\rm n}) = \nabla \times \left\{ \frac{\mathbf{B}}{4\pi\gamma\rho_{\rm n}\rho_{\rm i}} \times [\mathbf{B} \times (\nabla \times \mathbf{B})] \right\}. \tag{27.12}$$

If the right-hand side were zero (e.g., if the collisional coupling constant γ could be regarded as infinite), equation (27.12) would imply that the magnetic field would be well-tied to the motion of the neutrals. As it is, equation (27.12) constitutes a nonlinear diffusion equation for $\mathbf{B}$, with an effective diffusion coefficient $\mathcal{D}$ given by

$$\mathcal{D} \sim v_{\rm A}^2 t_{\rm ni},$$

where $v_{\rm A}^2$ equals the square of the Alfven speed of the combined medium [cf. equation (27.10)], and $t_{\rm ni} \equiv (\gamma\rho_{\rm i})^{-1}$ equals the mean collision time of a neutral molecule in a sea of ions. The nonlinearity enters in that the diffusion coefficient itself depends on the quantity $\mathbf{B}$ which is diffusing (relative to the neutrals). If L represents the characteristic dimension over which the magnetic field varies, we may estimate the time scale for ambipolar diffusion as

$$t_{\rm AD} \sim L^2/\mathcal{D} \sim L/v_{\rm d}, \tag{27.13}$$

where $v_{\rm d}$ is the typical drift speed, equation (27.9).

DYNAMICAL EQUATIONS FOR NEUTRALS

Equation (27.12), together with the initial condition $\nabla \cdot \mathbf{B} = 0$, provides the basic evolutionary equation for the magnetic field $\mathbf{B}$. To obtain a closed problem, we need the (ideal) fluid equations for the neutrals:

$$\frac{\partial \rho_{\rm n}}{\partial t} + \nabla \cdot (\rho_{\rm n}\mathbf{u}_{\rm n}) = 0, \tag{27.14}$$

$$\rho_{\rm n}\left[\frac{\partial \mathbf{u}_{\rm n}}{\partial t}+\nabla\left(\frac{1}{2}|\mathbf{u}_{\rm n}|^2\right)+(\nabla\times\mathbf{u}_{\rm n})\times\mathbf{u}_{\rm n}\right]=-\rho_{\rm n}\nabla\mathcal{V}-\nabla P_{\rm n}+\mathbf{f}_{\rm d}, \tag{27.15}$$

$$\rho_{\rm n}T_{\rm n}\left(\frac{\partial s_{\rm n}}{\partial t}+\mathbf{u}_{\rm n}\cdot\nabla s_{\rm n}\right)=\Gamma_{\rm n}-\Lambda_{\rm n}+\Gamma_{\rm AD}, \tag{27.16}$$

$$\nabla^2\mathcal{V}=4\pi G(\rho_{\rm n}+\rho_{\rm ext}). \tag{27.17}$$

In writing equation (27.17), we have assumed that the mass density of ions (and electrons) can be ignored in comparison with that of the neutrals. To the same order of approximation, we can approximate the frictional heat input into the medium by ambipolar diffusion, which equals the net rate of work done per unit volume on ions and neutrals,

$$\Gamma_{\rm AD}=\mathbf{f}_{\rm d}\cdot(\mathbf{u}_{\rm i}-\mathbf{u}_{\rm n})=\gamma\rho_{\rm n}\rho_{\rm i}|\mathbf{u}_{\rm i}-\mathbf{u}_{\rm n}|^2, \tag{27.18}$$

as entirely deposited into the neutral medium. In equation (27.16), $\Gamma_{\rm n}$ represents all other forms of heat input into the neutral medium (e.g., cosmic-ray heating), and $\Lambda_{\rm n}$ represents all forms of heat losses (e.g., radiative cooling by the CO molecule).

The substitution of equation (27.8) into equation (27.18) allows us to write the rate of ambipolar diffusion heating in terms of the spatial variations of the magnetic field:

$$\Gamma_{\rm AD}=\frac{1}{16\pi^2\gamma\rho_{\rm n}\rho_{\rm i}}|(\nabla\times\mathbf{B})\times\mathbf{B}|^2. \tag{27.19}$$

In turn, if we use equation (27.8) to eliminate $\mathbf{u}_{\rm i}-\mathbf{u}_{\rm n}$ in equation (27.2), we get for the magnitude and direction of the ion-neutral drag force:

$$\mathbf{f}_{\rm d}=\frac{1}{4\pi}(\nabla\times\mathbf{B})\times\mathbf{B}. \tag{27.20}$$

Substituting equation (27.20) into the right-hand side of equation (27.15), we get the illusion that the Lorentz force acts on the neutrals (as we have implicitly assumed in previous chapters when we adopted the single-component fluid approximation). The above derivation shows that the illusion applies only because we can usually assume that the true Lorentz force acting on the charged species comes into rapid balance with the their frictional drag with respect to the neutrals. The net difference in explicitly accounting for this drift velocity shows up, then, not in the force equation for the combined medium (which we regard henceforth as essentially the neutrals by dropping the subscript n everywhere), but in the magnetic field equation (27.12).

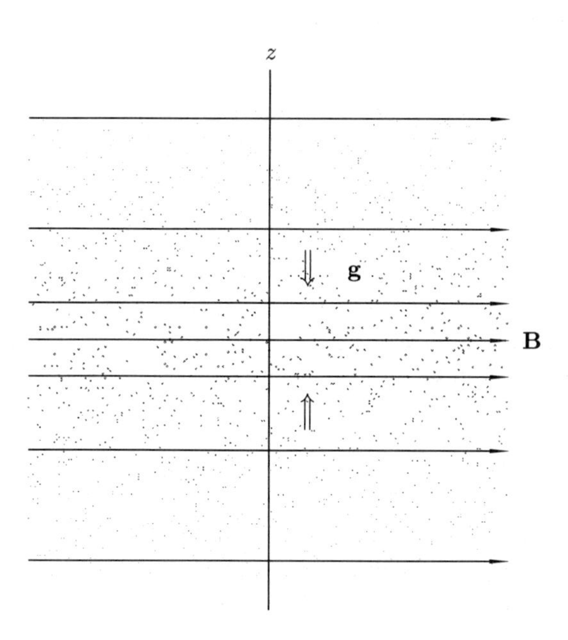

FIGURE 27.1
The geometry for ambipolar diffusion in a magnetized, self-gravitating, slab of lightly ionized gas.

A SIMPLE EXAMPLE

We illustrate the principles of ambipolar diffusion with a simple example: the quasistatic evolution of a plane-parallel self-gravitating slab of isothermal gas that is lightly ionized (Figure 27.1).

The basic configuration is unstable with respect to the development of Jeans's and Parker's instabilities if we allow perturbations in the horizontal (x and y directions), but we ignore this complication for the purpose of calculating the drift of magnetic field relative to the neutrals in z. The basic equations now read

$$\frac{\partial \rho}{\partial t} + \frac{\partial}{\partial z}(\rho u) = 0, \tag{27.21}$$

$$\rho\left(\frac{\partial u}{\partial t} + u\frac{\partial u}{\partial z}\right) = g - \frac{\partial}{\partial z}\left(P + \frac{B^2}{8\pi}\right), \tag{27.22}$$

$$P = a^2\rho, \qquad \text{with} \qquad a^2 = \text{constant}, \tag{27.23}$$

$$\frac{\partial g}{\partial z} = -4\pi G\rho. \tag{27.24}$$

$$\frac{\partial B}{\partial t} + \frac{\partial}{\partial z}(Bu) = \frac{\partial}{\partial z}\left(\frac{B^2}{4\pi\gamma\rho\rho_\mathrm{i}}\frac{\partial B}{\partial z}\right). \tag{27.25}$$

If we introduce the substantial derivative,

$$\frac{D}{Dt} = \frac{\partial}{\partial t} + u\frac{\partial}{\partial z},$$

we may use the equation of continuity (27.21) to rewrite equation (27.25) as

$$\frac{D}{Dt}\left(\frac{B}{\rho}\right) = \frac{1}{\rho}\frac{\partial}{\partial z}\left(\frac{B^2}{4\pi\gamma\rho\rho_\mathrm{i}}\frac{\partial B}{\partial z}\right). \tag{27.26}$$

With ρ_i given by equation (27.6), we now have a complete set of (Eulerian) equations to solve for ρ, u, g and B.

LAGRANGIAN DESCRIPTION

Physically, if the magnetic field plays a role initially in supporting the gas against its vertical self-gravity, we expect the gas to pull itself slowly toward the midplane as it loses magnetic flux to ambipolar diffusion, until the gas eventually becomes supported by its thermal pressure alone. To follow this drift, we find it advantageous to adopt a Lagrangian description. Introduce the Lagrangian coordinate, the surface density of neutrals between the midplane $z = 0$ and the height z:

$$\sigma(z,t) \equiv \int_0^z \rho(z',t)\,dz'. \tag{27.27}$$

If we integrate in z, equation (27.24) becomes Gauss's law,

$$g = -4\pi G\sigma, \tag{27.28}$$

where the gravitational field (and σ) reverse signs as we cross the mid-plane $z = 0$ ($\sigma = 0$). Henceforth, we consider only solutions that are even in z and σ; so we restrict both to positive values.

Transform now from (z,t) to (σ,t); in this transformation,

$$\frac{D}{dt} \equiv \frac{\partial}{\partial t} + u\frac{\partial}{\partial z} \to \left(\frac{\partial}{\partial t}\right)_\sigma,$$

$$\frac{\partial}{\partial z} \to \rho\left(\frac{\partial}{\partial \sigma}\right)_t,$$

$$u \to \left(\frac{\partial z}{\partial t}\right)_\sigma.$$

Henceforth all our derivatives occur with σ and t as the independent variables, and we drop the subscript σ and t notations.

The equation of continuity (27.21) gets replaced by the differential form of equation (27.27),

$$\frac{\partial z}{\partial \sigma} = \frac{1}{\rho}, \tag{27.29}$$

while the equation of motion (27.22) becomes

$$\frac{\partial^2 z}{\partial t^2} = -4\pi G\sigma - a^2\frac{\partial \rho}{\partial \sigma} - \frac{\partial}{\partial \sigma}\left(\frac{B^2}{8\pi}\right), \tag{27.30}$$

and the nonlinear equation (27.26) for the ambipolar diffusion of the magnetic field takes the form

$$\frac{\partial}{\partial t}\left(\frac{B}{\rho}\right) = \frac{\partial}{\partial \sigma}\left(\frac{B^2}{4\pi\rho_{\rm i}}\frac{\partial B}{\partial \sigma}\right). \tag{27.31}$$

QUASISTATIC APPROXIMATION

For ambipolar diffusion timescales $t_{\rm AD}$ much longer than dynamical times scales $t_{\rm dyn}$, the acceleration term $\partial^2 z/\partial t^2$ on the right-hand side of equation (27.30) is typically smaller by a factor $\sim (t_{\rm dyn}/t_{\rm AD})^2$ than any of the terms on the right-hand side. If we drop the acceleration term, we may integrate the resulting force balance to obtain

$$\frac{B^2}{8\pi} + a^2\rho = 2\pi G(\sigma_\infty^2 - \sigma^2), \tag{27.32}$$

where σ_∞ equals the half-sided surface density at infinity:

$$\sigma_\infty \equiv \int_0^\infty \rho\, dz, \tag{27.33}$$

and is a strictly conserved quantity of the problem.

In writing down equation (27.32), we have assumed that B and ρ both go to zero as $\sigma \to \sigma_\infty$. Under these circumstances, equation (27.32) has the simple interpretation that the total pressure, magnetic plus gas, at any level σ equals the weight of the material above it. With B and ρ satisfying the algebraic relation (27.32) and with $\rho_{\rm i}$ given by equation (27.6), the solution of equation (27.31) decouples from that of equation (27.29); i.e., we may solve for the dependence in (σ, t)-space without worrying how the system looks in configuration space z. Once B and ρ have been obtained from equations (27.31) and (27.32) as functions of σ and t, we may then worry about finding $z(\sigma, t)$, and transforming, if we wish, back to an Eulerian description in (z, t).

To have a well-posed problem, we impose on the nonlinear diffusion equation (27.31), the boundary conditions:

$$\frac{\partial B}{\partial \sigma} = 0 \quad \text{at} \quad \sigma = 0, \quad \text{and} \quad B = 0 \quad \text{at} \quad \sigma = \sigma_\infty. \tag{27.34}$$

The first boundary condition expresses the condition of reflection symmetry about the midplane $z = 0$; the second, that B vanishes at $z = \infty$. Given appropriate initial conditions, we now easily find numerical solutions for equations (27.31) and (27.32). (See Figure 27.2.)

ASYMPTOTIC SOLUTION

The numerical solutions all have the interesting property of converging on a unique asymptotic state (an "attractor" in the language of nonlinear mechanics). The properties of this state may be found as follows. As $t \to \infty$, we expect the magnetic field B to decay (slip relative to the neutrals) to negligibly small values. In this limit, equation (27.32) yields

$$\rho = \frac{2\pi G}{a^2}(\sigma_\infty^2 - \sigma^2), \tag{27.35}$$

which, upon substitution into equation (27.29), gives the ODE

$$\frac{dz}{d\sigma} = \frac{a^2}{2\pi G(\sigma_\infty^2 - \sigma^2)}. \tag{27.36}$$

Defining the scale length

$$z_0 \equiv a^2/2\pi G\sigma_\infty, \tag{27.37}$$

and the nondimensional surface density

$$\mu \equiv \sigma/\sigma_\infty, \tag{27.38}$$

we may integrate equation (27.36) to obtain

$$z/z_0 = \int_0^\mu \frac{d\mu}{1-\mu^2} = \operatorname{arctanh} \mu;$$

consequently, we get the Eulerian relations

$$\sigma = \sigma_\infty \tanh(z/z_0), \qquad \rho = \frac{\sigma_\infty}{z_0}\operatorname{sech}^2(z/z_0), \tag{27.39}$$

which correspond to the classic solution for the self-gravitating isothermal slab found by L. Spitzer (1942, *Ap. J.*, **95**, 329).

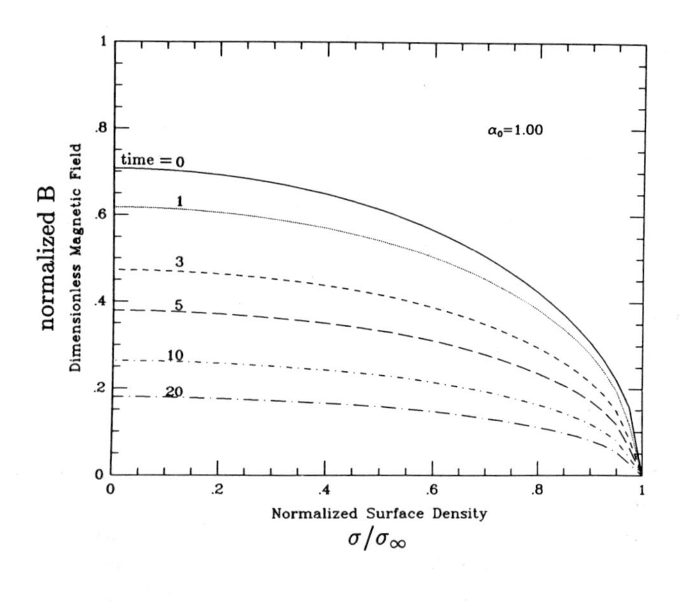

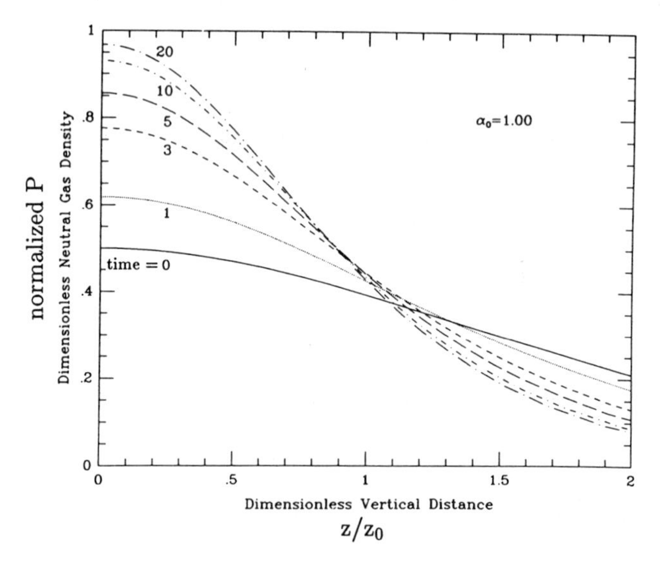

FIGURE 27.2
Numerical solutions for the problem of ambipolar diffusion in magnetized, isothermal, lightly ionized, self-gravitating layers. As a function of (dimensionless) time, the normalized magnetic field gradually drifts from each Lagrangian layer of gas with surface density σ to the midplane. In the process, the gas density ρ transforms from a state with relatively little central concentration to one that resembles a dense molecular cloud core embedded in a more diffuse envelope. (From F. H. Shu, 1983, *Ap. J.*, **273**, 202.)

For us, equations (27.39) are reached asymptotically in time only when the magnetic field has completely diffused out of the neutral slab. A detailed analysis shows that the late-time behavior of B satisfies (see F. H. Shu, 1983, *Ap. J.*, **273**, 202, for details; see also S. Lizano and F. H. Shu, 1989, *Ap. J.*, **342**, 834, for more-realistic simulations of the process of molecular-cloud core formation by the process of ambipolar diffusion)

$$B \to 4\pi G^{1/2}\sigma_\infty \frac{Z(\mu)}{[2(\tau - \tau_0)]^{1/2}}, \tag{27.40}$$

where Z is a universal function of equation (27.38) that we may obtain by substituting equations (27.35) and (27.6) into equation (27.31), and then assuming that B has separable functional dependence in the dimensionless surface density μ and dimensionless time τ, with the latter defined through

$$t \equiv \left[\frac{\gamma C}{2(2\pi G)^{1/2}}\right]\left[\frac{a}{2\pi G\sigma_\infty}\right]\tau. \tag{27.41}$$

In equation (27.40), the quantity τ_0 is an integration constant of order unity, whose precise value cannot be determined within the context of the asymptotic analysis. Instead it represents a characteristic dimensionless time that must be much exceeded before the asymptotic solution (27.40) becomes a good representation for the actual solution. The numerical determination of τ_0 can be obtained by comparing the numerical solutions of equation (27.31) for given initial states to the asymptotic values given by equation (27.40). Barenblatt in his book *Similarity, Self-Similarity, and Intermediate Asymptotics* claims that this appearance of a constant whose value cannot be obtained except by numerical solution of the governing nonlinear PDE is characteristic of solutions that contain *intermediate asymptotics.* Notice that τ_0 introduces a characteristic time scale for ambipolar diffusion via equation (27.41) only for intermediate times. For $\tau \gg \tau_0$, the dependence of B on time in equation (27.40) enters as a power law, $B \propto t^{-1/2}$, and power-laws have no characteristic scale. The nonlinearity of the present diffusion process forces the magnetic field to stack with universal profile $Z(\mu)$ in such a way that the field everywhere decays eventually at the same rate, as the inverse square root of time.

28

Plasma Physics

Reference: Tidman and Krall, *Shock Waves in Collisionless Plasmas.*

In this volume so far, we have considered the dynamics of gases that are collisionally dominated. When electromagnetic forces act on (partially) ionized gases, this corresponds to the magnetohydrodynamic regime. The MHD approximation involves two crucial assumptions: first, ω, the dynamic frequency in the problem, $\ll \omega_{\rm pe}$, the electron plasma frequency, so that we may ignore the effects of charge separation; second, the collision frequency $\nu_{\rm c}$ is sufficiently large that it keeps the distribution functions nearly Maxwellian and prevents large slip velocities between the different components of the system. Plasma physics concerns, in the main, the regime when one or more of these assumptions become violated.

In the remaining chapters, we briefly consider the subject of plasma physics. Much of the theory in plasma physics has developed in support of the fusion program, and as a consequence, there is a large literature about the instabilities that develop when one tries to confine a hot plasma with magnetic fields. We cannot here investigate this vast and interesting area; instead, we shall focus on a few basic principles, and indicate possible applications to astronomy.

A GENERIC ASTROPHYSICAL PROBLEM

As an example of the kinds of plasma-physics problems that interest astronomers, consider the interaction of the solar wind with the magnetosphere of a planet like the Earth (see Figure 28.1). A rarefied solar wind, containing on average several particles per cm^3 at 1 AU from the Sun, sweeps past the Earth at several hundred km s^{-1}. The charged particles are incident supersonically on a magnetic obstructing body, so we might expect the development of a bow shock, as is indeed observed by artificial satellites.

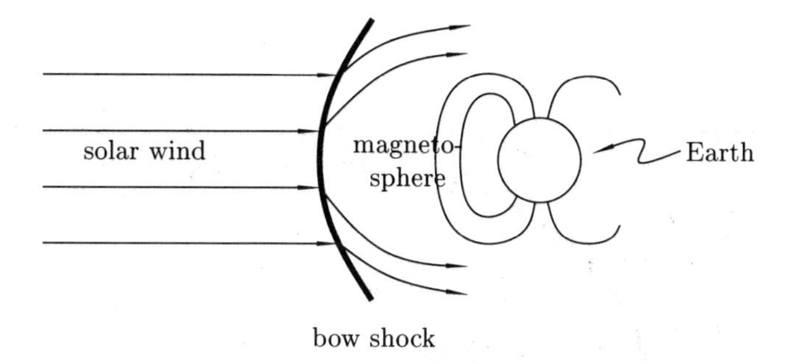

FIGURE 28.1
Structure of the bow shock of the solar wind's interaction with the magnetosphere of the Earth.

On the other hand, if we calculate the collisional time for momentum transfer between proton-proton scatterings (see Chapter 1), we obtain

$$t_c(\mathrm{p-p}) = 11.7\,T^{3/2}/n_\mathrm{p} \ln\Lambda \text{ s},$$

if T and n are expressed in cgs units, with the Coulomb logarithm having a value ~ 40 in the solar wind. With $T \sim 10^6$ K and $n_\mathrm{p} \sim 5\,\mathrm{cm}^{-3}$, we then obtain $t_c(\mathrm{p-p}) \sim 6 \times 10^7$ s. Traveling at $\sim 5 \times 10^7\,\mathrm{cm\,s^{-1}}$, protons should go $\sim 3 \times 10^{15}$ cm, greatly in excess of the dimensions of the Earth or its magnetosphere (10^9–10^{10} cm).

Clearly, particle collisions play little role in the solar wind plasma. How, then, does the bow shock of the Earth develop? A clue to this problem of a "collisionless shock" can be obtained by conducting the following thought experiment. The "ram pressure" ρu^2 of a solar wind with the properties described above incident on the Earth roughly equals $2 \times 10^{-8}\,\mathrm{dyne\,cm^{-2}}$. This equals a magnetic pressure, $B^2/8\pi$, for a field strength $B \sim 7 \times 10^{-4}$ G, a value reached by the Earth's magnetic field at a distance of ten or more Earth radii. Consequently, the solar wind must "squash" the windward magnetic field inside a dimension of this order, with all nonzero values of the Earth's magnetic field confined to within its *magnetopause* (Figure 28.2).

Imagine now a freely streaming solar-wind proton incident on a idealized magnetic field structure, which has zero strength to one side of some boundary, and rises to $\sim 10^{-3}$ G on the other side. A proton with speed $v = 5 \times 10^7\,\mathrm{cm\,s^{-1}}$ and gyrofrequency $\omega_\mathrm{L} = eB/m_\mathrm{p}c \sim 10\,\mathrm{s^{-1}}$ in a 10^{-3} G field has a gyroradius $v/\omega_\mathrm{L} \sim 5 \times 10^6$ cm that is much smaller than the standoff distance of the magnetopause. Consequently, this proton must en-

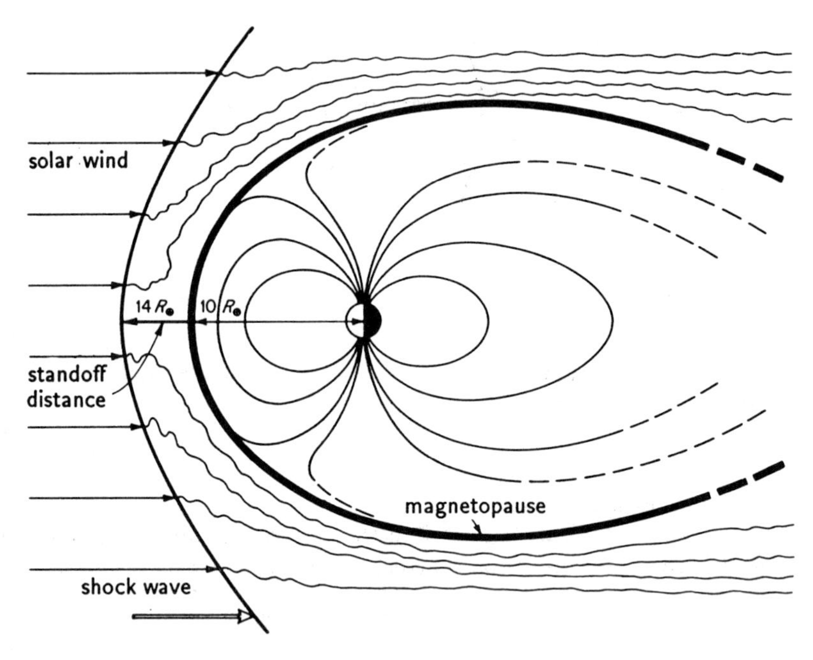

FIGURE 28.2
In a vacuum the Earth's dipole magnetic field would extend to infinity, dropping off in strength with distance as r^{-3}. Exposed to the ram pressure of the solar wind, the Earth's field is squashed inside a finite magnetopause. (From Brandt and Hodge, *Solar System Astrophysics.*)

ter the Earth's magnetic field, make about half a gyroturn, and exit back out into the field-free region (Figure 28.3).

In effect, then, the Earth's magnetosphere tries to "reflect" incident solar protons and to send them back as a counterstream to the oncoming wind. The electrons similarly reflect off the magnetic "wall," but their radius of gyration is considerably smaller. Thus, in the idealized situation envisaged above, there exists a region of charge separation after the electrons have already turned around, but before the protons have, where a substantial electric field tries to develop. Such a potential drop cannot be sustained in the presence of copious numbers of free charges. Moreover, the protons and electrons can not really "reflect" as a counterstream into the oncoming solar wind, because this solar wind usually carries its own magnetic field that would "re-reflect" the counterstream. As a consequence, the interface proves to be unstable to the generation of a variety of plasma instabilities,

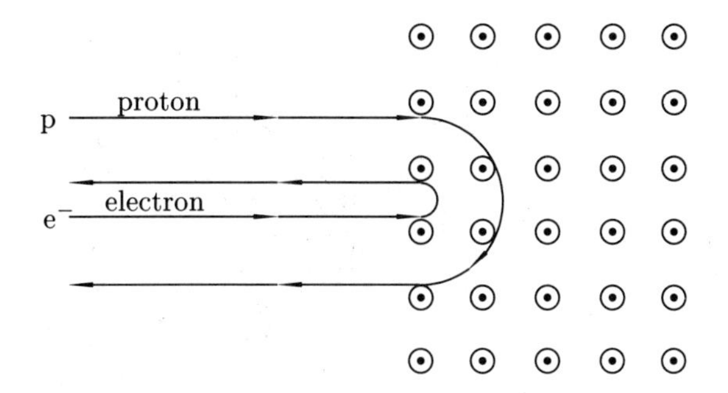

FIGURE 28.3
If the solar-wind plasma contained no embedded solar fields and had no collective behavior, the protons and electrons in it incident on the Earth's magnetosphere in the immediate neighborhood of the magnetopause would be "reflected" into a counterstream. In fact, collective effects and an embedded solar field prevent the formation of such a counterstream. Instead, the system develops a collisionless bow shock in which the postshock plasma is swept downstream by the general flow around the Earth's magnetosphere (see Figure 28.2).

which produce fluctuating collective electromagnetic fields that violently scatter the ordered stream of the incident solar-wind particles. The net result is that the solar wind shocks, with the directed bulk momentum and kinetic energy of the wind transformed to the pressure and heat of random motions, according to the laws of conservation of mass, momentum, and energy. The resultant jump conditions do not differ appreciably from their collisional counterparts, but notice that the randomizing agent—many collective waves rather than many two-particle scatterings—has a completely distinct source. In particular, experiments show that a fraction of the particles in the distribution gets accelerated to very high (i.e., non-Maxwellian) energies by the collisionless shock phenomenon. A central goal of current astrophysical theory attempts to understand whether such a process (in a more general context) can explain the origin of cosmic-ray particles (see the discussion in the last section of this chapter).

Other interesting phenomena associated with the problem of the Earth's bow shock concerns the process of magnetic reconnection at the head and tail of the magnetosheath. This process, which opens and recloses the Earth's magnetic field to the matter and field of the solar wind, plays an

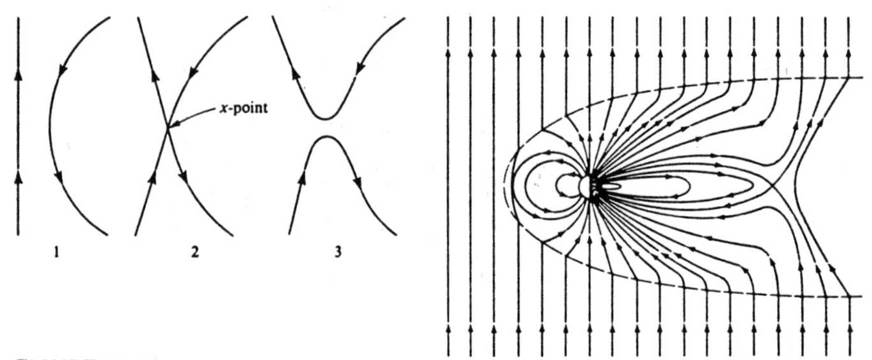

FIGURE 28.4
The presence of misaligned solar-wind fields gives rise to reconnection possibilities at the head and tail of the Earth's magnetopause. The first diagram depicts a reconnection sequence as a function of time; the second, in a quasisteady flow.

important role in the production of high-energy particles and their entry into the Van Allen radiation belts (Figure 28.4). Although these represent specialized topics beyond the scope of this volume, we hope that this brief discussion indicates the motivation for the types of investigations carried out by space-plasma physicists.

THE BASIC KINETIC EQUATION OF COLLISIONLESS PLASMA PHYSICS

The most complete statistical treatment of a collisionless plasma uses a kinetic description. For each charged species of mass m and charge q, we introduce a distribution function $f(\mathbf{x}, \mathbf{v}, t)$ such that $f(\mathbf{x}, \mathbf{v}, t)\, d^3x d^3v$ equals the number of particles at time t in volume $d^3x d^3v$ centered about the phase-space point $(\mathbf{x}, \mathbf{v})$. If we ignore completely the effects of particle collisions, the (nonrelativistic) equation which governs the evolution of f reads

$$\frac{\partial f}{\partial t} + \mathbf{v} \cdot \frac{\partial f}{\partial \mathbf{x}} + \frac{\mathbf{F}}{m} \cdot \frac{\partial f}{\partial \mathbf{v}} = 0, \tag{28.1}$$

where $\mathbf{F}$ is the force felt by a particle at $(\mathbf{x}, \mathbf{v}, t)$,

$$\mathbf{F} = q\left(\mathbf{E} + \frac{\mathbf{v}}{c} \times \mathbf{B}\right) + m\mathbf{g}. \tag{28.2}$$

Equation (28.1) is often called the *Vlasov equation*, but we shall reserve this terminology for the entire collection of equations that govern the dynamics (including those that specify the fields). After all, Vlasov's essential contribution does not entail so much the writing down of the kinetic equation (28.1), which had been done earlier in other contexts by Liouville (statistical mechanics) and Jeans (stellar dynamics), as the emphasizing of the need for *self-consistency* within the laws of electrodynamics in the computation of $\mathbf{F}$ (see below). Equation (28.1) also goes by the name of "the collisionless Boltzmann equation," but this terminology is an even worse misnomer, since the whole point of the Boltzmann equation is the collisional term, which we have explicitly set equal to zero on the right-hand side of equation (28.1). Thus we shall refer to equation (28.1) as "Vlasov's kinetic equation."

SELF-CONSISTENT FIELDS

Since the number density n of the particular species under consideration equals the integral of f over $\mathbf{v}$,

$$n(\mathbf{x},t) \equiv \int f(\mathbf{x},\mathbf{v},t)\,d^3v, \tag{28.3}$$

we may obtain the total mass density ρ and charge density ρ_e through a summation over all charge species,

$$\rho(\mathbf{x},t) \equiv \sum mn(\mathbf{x},t), \tag{28.4}$$

$$\rho_e(\mathbf{x},t) \equiv \sum qn(\mathbf{x},t), \tag{28.5}$$

where, for notational simplicity, we do not indicate the charge species by attaching bothersome subscripts. In a similar fashion, we can compute the electric current density as

$$\mathbf{j}_e \equiv \sum \int q\mathbf{v}f\,d^3v. \tag{28.6}$$

We suppose that the electric, magnetic, and gravitational fields, $\mathbf{E}$, $\mathbf{B}$, and $\mathbf{g}$ in equation (28.2), have a decomposition into external and internal contributions

$$\mathbf{E} = \mathbf{E}_{\text{ext}} + \mathbf{E}_{\text{int}}, \qquad \mathbf{B} = \mathbf{B}_{\text{ext}} + \mathbf{B}_{\text{int}}, \qquad \mathbf{g} = \mathbf{g}_{\text{ext}} + \mathbf{g}_{\text{int}}. \tag{28.7}$$

We regard the external fields to be specified *a priori*—e.g., given as vacuum fields—but we require the internal fields to satisfy the self-consistent field equations:

$$\nabla \cdot \mathbf{E}_{\text{int}} = 4\pi\rho_e, \tag{28.8}$$

$$\nabla \cdot \mathbf{B}_{\text{int}} = 0, \tag{28.9}$$

$$\nabla \times \mathbf{E}_{\text{int}} = -\frac{1}{c}\frac{\partial \mathbf{B}_{\text{int}}}{\partial t}, \tag{28.10}$$

$$\nabla \times \mathbf{B}_{\text{int}} = \frac{4\pi}{c}\mathbf{j}_{\text{e}} + \frac{1}{c}\frac{\partial \mathbf{E}_{\text{int}}}{\partial t}, \tag{28.11}$$

$$\nabla \cdot \mathbf{g}_{\text{int}} = -4\pi G\rho, \tag{28.12}$$

where ρ, ρ_{e}, $\mathbf{j}_{\text{e}}$ are given as the summed integrals over f given by equations (28.4), (28.5), and (28.6). Notice that equation (28.1) is a linear equation if $\mathbf{F}$ is regarded as known, whereas equations (28.8)–(28.12) are also linear if the f's (and therefore, ρ, ρ_{e}, and $\mathbf{j}_{\text{e}}$) are known. The combined situation (Vlasov's equations), when both the distribution function and the fields have to be solved *self-consistently*, makes the coupled problem highly nonlinear [through the term $m^{-1}\mathbf{F} \cdot \partial f/\partial \mathbf{v}$ in equation (28.1)]. Real life yields much more complicated problems than the idealized ones a conventional physics training teaches one to solve.

RELATIONSHIP TO PARTICLE ORBITS

Despite the need eventually to solve the whole set of Vlasov's equations simultaneously, we can obtain considerable insight by attacking the individual pieces as subproblems. An important subproblem is the solution of the collisionless kinetic equation (28.1). If we regard $\mathbf{F}$ as given, equation (28.1) has the standard form of a (quasi)linear PDE for f in the variables t, $\mathbf{x}$, and $\mathbf{v}$. The characteristic equations associated with this PDE read (review Chapter 13):

$$\frac{dt}{1} = \frac{d\mathbf{x}}{\mathbf{v}} = \frac{d\mathbf{v}}{\mathbf{F}/m} = \frac{df}{0},$$

i.e.,

$$\frac{df}{dt} = 0 \qquad \text{on} \qquad \frac{d\mathbf{x}}{dt} = \mathbf{v}, \qquad \frac{d\mathbf{v}}{dt} = \frac{\mathbf{F}}{m}. \tag{28.13}$$

Equation (28.13) states that f equals a constant following a particle trajectory in phase space when the latter is subjected to the force $\mathbf{F}$. Thus, the most general solution for f reads

$$f(\mathbf{x}, \mathbf{v}, t) = f(I_1, I_2, \ldots, I_6), \tag{28.14}$$

where I_1, I_2, ..., I_6 are the six integrals (constants) of motion associated with the particle orbit.

As a straightforward example for which we can analytically obtain six integrals, consider the problem of the nonrelativistic motion of a charged particle in a steady and uniform magnetic field of strength B directed in the z-direction. If we introduce Cartesian coordinates (x, y, z), we easily derive that the velocity components perpendicular and parallel to the magnetic field,

$$(v_x^2 + v_y^2)^{1/2} = \text{constant} \equiv V,$$

$$v_z = \text{constant} \equiv W,$$

are integrals of the motion. We also know from elementary mechanics that the orbit in space has the form of a helix,

$$x = X + \frac{V}{\omega_{\mathrm{L}}} \sin\left[\omega_{\mathrm{L}}(t - T)\right], \tag{28.15}$$

$$y = Y + \frac{V}{\omega_{\mathrm{L}}} \cos\left[\omega_{\mathrm{L}}(t - T)\right], \tag{28.16}$$

$$z = Z + W(t - T), \tag{28.17}$$

where

$$\omega_{\mathrm{L}} \equiv qB/mc, \tag{28.18}$$

and X, Y, Z, and T are integration constants. If we solve for X, Y, Z, and T in terms of the present phase-space variables $(x, y, z, v_x, v_y, v_z, t)$, where

$$\begin{aligned} v_x &= \dot{x} = V \cos\left[\omega_{\mathrm{L}}(t - T)\right], \\ v_y &= \dot{y} = -V \sin\left[\omega_{\mathrm{L}}(t - T)\right), \end{aligned}$$

we will have derived four more integrals of the motion in addition to V and W. Thus we have the six integrals of motion:

$$V = (v_x^2 + v_y^2)^{1/2}, \tag{28.19}$$

$$W = v_z, \tag{28.20}$$

$$T = t + \frac{1}{\omega_{\mathrm{L}}} \arctan\left(\frac{v_y}{v_x}\right), \tag{28.21}$$

$$X = x + \frac{v_y}{\omega_{\mathrm{L}}}, \tag{28.22}$$

$$Y = y - \frac{v_x}{\omega_{\mathrm{L}}}, \tag{28.23}$$

$$Z = z + \frac{v_z}{\omega_{\mathrm{L}}} \arctan\left(\frac{v_y}{v_x}\right). \tag{28.24}$$

The most general solution to Vlasov's kinetic equation in this case reads

$$f(x, y, z, v_x, v_y, v_z, t) = F(T, V, W, X, Y, Z), \tag{28.25}$$

where F has an arbitrary dependence on the functions of $(x, y, z, v_x, v_y, v_z, t)$ defined by equations (28.19)–(28.24).

PHASE MIXING AND ISOLATING INTEGRALS

Not all of the integrals of motion are useful for the realistic statistical problem. As we will show later, we can effectively ignore integrals that do not offer any real constraints on the motion of a particle in phase space. We call such integrals *nonisolating* or *ignorable*. The necessity of including only *isolating integrals* in the dependence of $f = F(I_1, I_2, \ldots)$ was first recognized in the context of collisionless stellar dynamics, where it goes by the name of the *strong Jeans theorem* (see Chapter 4 of Binney and Tremaine, *Galactic Dynamics*).

To fix ideas, let us consider the six formal integrals (28.19)–(28.24). The constants of motion V and W clearly represent isolating integrals, since specifying their values implies that the corresponding particle cannot visit those parts of velocity space where $(v_x^2+v_y^2)^{1/2} \neq V$ and $v_z \neq W$. Similarly, the x and y positions of the guiding center, X and Y, represent isolating integrals, since specifying their values, along with that of V, bounds the x and y locations of the particle for all time to lie within $X \pm V/\omega_{\rm L}$ and $Y \pm V/\omega_{\rm L}$. On the other hand, one might reasonably question whether the temporal phase T of the y-oscillation relative to the x-oscillation contains useful statistical information (and similarly for the fiducial coordinate marker Z for the translation in z).

To be sure, two neighboring particles oscillating at exactly the same radian frequency $\omega_{\rm L}$ would always maintain synchronous phase relative to one another, but this isolating property of T represents a feature peculiar to pure harmonic motion. It would break down if we had any departures from the idealized state. For example, the gyromotion does not occur synchronously if the magnetic field $\mathbf{B}$ is not uniform, or if relativistic effects are important. To fix ideas pedagogically, let us discuss the latter case in greater detail.

For the relativistic motion of a charged particle in a homogeneous magnetic field, $V = (v_x^2 + v_y^2)^{1/2}$ and $W = v_z$ still represent good integrals of the motion. More familiarly, we may replace V and W by the Lorentz factor of the particle,

$$\gamma \equiv \left(1 - \frac{v^2}{c^2}\right)^{-1/2}, \qquad \text{with} \qquad v^2 \equiv v_x^2 + v_y^2 + v_z^2, \tag{28.26}$$

and the parallel component of the momentum,

$$p_{\parallel} \equiv \gamma m v_z. \tag{28.27}$$

For relativistic particles, the gyromotion occurs at the frequency

$$\omega_B = \frac{qB}{\gamma mc}, \tag{28.28}$$

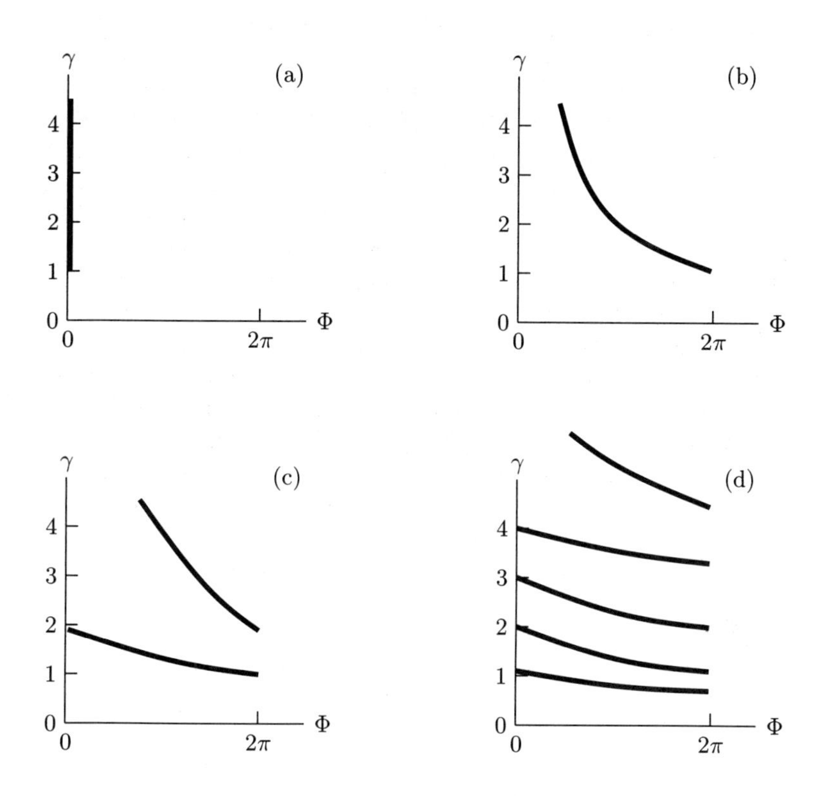

FIGURE 28.5
The relative angular phase Φ as a function of time t between the x- and y-oscillations of relativistic charges moving in a uniform magnetic field oriented in the z-direction. Even if we were to artificially arrange the angular phases of all particles to lie initially at $\Phi = 0$ (diagram a), the inverse dependence of the gyroperiod on the Lorentz factor γ would cause the particles to lose synchronism after several gyro-orbits (diagrams b-d). This phenomenon of phase mixing implies that Φ is associated with an ignorable integral of the motion (denoted T in the text) for any practical, coarse-grained, statistical description of the plasma system.

and differs from the nonrelativistic formula, equation (28.18), by an extra factor of $1/\gamma$; otherwise, the formulae of the previous section continue to hold if we merely replace ω_{L} by ω_B.

The gyroperiod $2\pi/\omega_B$ now depends on the particle energy $\epsilon = \gamma mc^2$; as a consequence, the temporal phase (28.21) between the x and y oscillations no longer represents an isolating integral. Figure 28.5 gives a pictorial explanation for this modification.

Let $\Phi = \arctan[(x - X)/(y - Y)]$ between 0 and 2π equal the relative angular phase of the oscillations in x and y. Imagine arranging the gyro-orbits of particles of different Lorentz factors so that initially they all have nearly the same value of Φ, say, $\Phi = 0$ at $t = 0$. The distribution function plotted as a contour diagram in Φ-γ space at time $t = 0$ then looks like a narrow vertical strip, as show in Figure 28.5a. At some later time $t = t_1$, the low-energy particles ($\gamma = 1$) will have completed one gyrocircuit and moved to phase $\Phi = 2\pi$, but the higher-energy particles, having longer periods by a factor of $\gamma > 1$, will be in phases of oscillation $\Phi = 2\pi/\gamma < 2\pi$. In other words, the initially vertical contour gets distorted into the curve shown in Figure 28.5b. At some still later time $t = t_2$, the particles with $\gamma = 1$ will have completed 2 circuits; those with $\gamma = 2$ will have completed 1 circuit, etc.; and the contour diagram for F will look as depicted in Figure 28.5c. After the passage of a very long time, the stretching and folding of the contour where F has a nonzero (fixed) value (a process known generically as *phase mixing*) will have occurred so often that even particles with only slightly different Lorentz factors γ will have very different phases Φ [Figure 28.5d]. One might conclude in these circumstances that the distribution of particles in the Φ-coordinate carries no useful information in the asymptotic limit $t \gg \omega_B^{-1}$, and that we can therefore treat the effective statistics as if the particles were uniformly distributed in gyrophase Φ. Mathematically, this means that we may ignore any dependence of F on the (nonisolating) integral of motion T.

To prove this result analytically, let us note that for the same (v_x, v_y, v_z), the quantity T in equation (28.21) acquires multiple numerical values depending on how we interpret the arctan function. In order for $f = F(T, \ldots)$ to remain a single-valued function of v_x, v_y, and v_z, we insist that $F(T, \ldots)$ must have a Fourier series expansion in T with periodicity $2\pi/\omega_B$, i.e.,

$$F(T, \ldots) = \sum_{m=-\infty}^{+\infty} F_m(\ldots) e^{im\omega_B T}, \tag{28.29}$$

where the dots in the argument of $F_m(\ldots)$ denote that the Fourier coefficients of F may depend on any of the other integrals of the problem. For F to be real, we require $F_{-m}(\ldots) = F_m^*(\ldots)$. Consider now any (velocity) moment of the distribution function F, obtained by multiplying arbitrary powers of velocity components $Q = v_x^p v_y^r v_z^s$ into F and integrating over all velocities,

$$n\langle Q\rangle \equiv \int\int\int QF\, dv_x dv_y dv_z.$$

Substituting in the Fourier expansion (28.29), we get

$$n\langle Q\rangle = \sum_{m=-\infty}^{+\infty} \int\int\int \left[QF_m e^{im \arctan(v_y/v_x)}\right] e^{imqBt/mc\gamma}\, dv_x dv_y dv_z.$$

The quantity in the brackets on the right-hand side varies smoothly with (v_x, v_y, v_z), but as $t \to \infty$, the term $e^{imqBt/mc\gamma}$, for all m except $m = 0$, oscillates wildly as we integrate in (v_x, v_y, v_z) because of the changes in γ. Thus, for $t \gg \omega_B^{-1}$, the contributions to $n\langle Q\rangle$ integrate to zero for all m except $m = 0$, i.e.,

$$n\langle Q\rangle \to \int\int\int QF_0 \, dv_x dv_y dv_z.$$

Since this result holds for *any* moment of the distribution function, and since the *coarse-grained description* of a (velocity) distribution function is equivalent to a low-resolution specification of all its (velocity) moments, we might have effectively used only F_0 in the first place, i.e.,

$$n\langle Q\rangle = \int Qf \, d^3v, \qquad \text{with} \qquad f = F_0(\ldots), \tag{28.30}$$

if we ignore a role for the nonisolating integral T from the start. (Q.E.D.)

Similar comments apply to all nonisolating integrals; they are all ignorable. We have here a special example of the problem in statistical mechanics known as the *ergodic hypothesis* (see Chapter 6 of Volume I). Finally, we should state that integrals which are isolating under some circumstances can become nonisolating when those circumstances alter sufficiently. The subject of nonlinear dynamics phrases this question in terms of *regular orbits* becoming *chaotic.* We shall not address the important issue of chaotic orbits in this volume; instead, we focus our attention in the next section on the opposite regime, when the modifications are small, and there exist in the problem *adiabatic invariants.*

ADIABATIC INVARIANTS

The discussion of the previous section illustrates the importance of being able to give analytical expressions for the isolating integrals of the problem. However, when the electromagnetic fields change with time or space, we cannot usually derive exact expressions for the true constants of the motion. Fortunately, if the changes occur slowly in comparison with the gyrofrequency or gyroradius, we can frequently find quantities that approximate true constants of motion to high order; such quantities are called *adiabatic invariants.*

We refer to a quantity as an adiabatic invariant if it satisfies the following property. Suppose the external changes in the fields occur slowly, with the slow change characterized by some small parameter ε. If no changes occurred at all, i.e., if $\varepsilon = 0$, then elementary considerations may suffice to recover the relevant conserved quantities of the problem, e.g., the energy of the particle, ϵ. In the presence of slight variations, however, $\varepsilon \neq 0$, and these quantities, e.g., ϵ, would generally also vary on the order of ε. These former constants of motion then represent nothing special; they depart from being integrals of motion for the new problem by an amount that is just what one expects intuitively. The special quantities are those entities that vary, in fact, with a higher power of ε than the first, say, as ε^2 or better. These special entities we name adiabatic invariants.

The word "adiabatic" has its origin in thermodynamics. The total entropy S of a thermally isolated body equals an adiabatic invariant when subjected only to slow mechanical changes.

Proof: Suppose $\lambda(t)$ represents some mechanical characteristic of the body that changes slowly. The time rate of change of the entropy, dS/dt, cannot be proportional to $\dot{\lambda}$, because $\dot{\lambda}$ can have either sign, whereas the second law of thermodynamics requires $dS/dt \geq 0$ (see, e.g., Problem Set 1). Thus, a Taylor-series expansion of dS/dt starts with a term proportional to $\dot{\lambda}^2$, or a yet higher power, and therefore

$$\frac{dS}{d\lambda} = \frac{1}{\dot{\lambda}}\frac{dS}{dt} \propto \dot{\lambda}.$$

The above result implies that we can effect a finite modification in λ without inducing any variation in S, provided we carry out the rate of change of λ slowly enough, i.e., if we take the limit $\dot{\lambda} \to 0$. (Q.E.D.)

Such slow processes are called adiabatic variations ("impassable to heat") because they keep the entropy constant, and this nomenclature has carried over to fields other than thermodynamics.

There exists a rule of wide applicability for finding adiabatic invariants in classical mechanics. If a Hamiltonian formulation of the problem contains a nearly periodic motion associated with some generalized coordinate q and conjugate momentum p, then the action integral,

$$J \equiv \frac{1}{2\pi}\oint p\,dq, \tag{28.31}$$

where the integration occurs over one (quasi)period of the oscillation, represents an adiabatic invariant when the external conditions (or force fields) of the problem change slowly. For a proof of this claim, see *Classical Mechanics* by Landau and Lifshitz. For here, we merely note that action integrals have the dimension of energy divided by frequency, or angular momentum, the same units as Planck's constant. Indeed, in the transformation from classical mechanics to quantum mechanics, the old atomic theory identifies the adiabatic invariants of the problem as those entities that become quantized in integer units of $\hbar$. Going backward, we see that quantized entities in quantum mechanics must become adiabatic invariants in classical mechanics, because slow changes can not transform one integer to another, even if the integers are fairly large. (The same advantage applies to digital technology over analog technology.)

TRANSVERSE ADIABATIC INVARIANT

As an application of the above rule, note that the relativistically correct Lagrangian for a point particle of charge q and mass m in an electromagnetic field characterized by the vector potential $\mathbf{A}$ and the scalar potential ϕ is given by (see Chapter 21 in Volume I)

$$L = mc^2\left(1 - \frac{v^2}{c^2}\right)^{1/2} + \frac{q}{c}\mathbf{v}\cdot\mathbf{A} - q\phi. \tag{28.32}$$

The Euler-Lagrange equation from the principle of least action,

$$\frac{d}{dt}\left(\frac{\partial L}{\partial \mathbf{v}}\right) - \frac{\partial L}{\partial \mathbf{x}} = 0,$$

yields the relativistically correct equation of motion,

$$\frac{d\mathbf{p}}{dt} = q\left(\mathbf{E} + \frac{\mathbf{v}}{c}\times\mathbf{B}\right), \tag{28.33}$$

where $\mathbf{p} \equiv \gamma m\mathbf{v}$ represents the particle momentum, and the electric and magnetic fields, $\mathbf{E}$ and $\mathbf{B}$, are given by the usual expressions,

$$\mathbf{E} = -\nabla\phi - \frac{1}{c}\frac{\partial \mathbf{A}}{\partial t}, \qquad \mathbf{B} = \nabla\times\mathbf{A}. \tag{28.34}$$

In a Hamiltonian formulation, the generalized momentum conjugate to $\mathbf{x}$ reads

$$\mathbf{P} \equiv \frac{\partial L}{\partial \mathbf{v}} = \mathbf{p} + \frac{q}{c}\mathbf{A}, \tag{28.35}$$

which differs from the particle momentum $\mathbf{p}$ by the electromagnetic contribution $q\mathbf{A}/c$. If the variations of $\mathbf{E}$ and $\mathbf{B}$ are slow, and if $|\mathbf{E}| \ll |\mathbf{B}|$, we expect the motion of the charge to be nearly periodic, with period $2\pi/\omega_B$, where ω_B is given by equation (28.28). In such a situation, the discussion of the previous section implies that the action integral,

$$J \equiv \frac{1}{2\pi}\oint \mathbf{P}\cdot d\mathbf{x} = \frac{1}{2\pi}\oint \mathbf{p}\cdot d\mathbf{x} + \frac{1}{2\pi}\frac{q}{c}\oint \mathbf{A}\cdot d\mathbf{x},$$

should be an adiabatic invariant. Taking the circuit in the above integration paths about one gyro-orbit of radius a, and applying Stokes' theorem to the second term, we obtain

$$J = \frac{1}{2\pi}\oint p_\perp a\, d\varphi + \frac{q\Phi}{2\pi c},$$

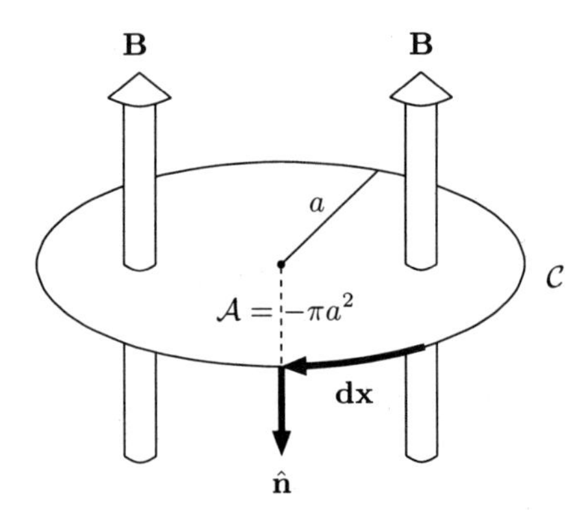

FIGURE 28.6
The part of the motion of a particle with positive charge q associated with the component of momentum $p_\perp$ perpendicular to a quasi-uniform magnetic field $\mathbf{B}$ oriented in the $+z$ direction yields a circuit C that rotates in the clockwise sense in the x-y plane. Such a circuit has a gyroradius $a = cp_\perp/qB$ and an associated area $\mathcal{A} = \pi a^2$, whose unit normal $\hat{\mathbf{n}} = -\hat{\mathbf{z}}$. For a charge with a negative charge q, the sense of the circuit reverses, and we have to introduce a minus sign in the definition of a. However, the net result is the same, namely, $J = cp_\perp^2/2|q|B$ constitutes an adiabatic invariant of the motion.

where

$$p_\perp = m\gamma v_\perp, \tag{28.36}$$

$$a = \frac{v_\perp}{\omega_B} = \frac{cp_\perp}{qB}, \tag{28.37}$$

and Φ equals the magnetic flux threading the circular orbit of area $\mathcal{A} = \pi a^2$ (see Figure 28.6):

$$\Phi \equiv \int (\nabla \times \mathbf{A}) \cdot \hat{\mathbf{n}}\, d\mathcal{A} = \int \mathbf{B} \cdot \hat{\mathbf{n}}\, d\mathcal{A} \approx -\pi a^2 B.$$

[We previously denoted $v_\perp$ and $v_\parallel$ by $V = (v_x^2 + v_y^2)^{1/2}$ and $W = v_z$, but we avoid the latter notation here because we do not require $\mathbf{B}$ to lie in a fixed direction z.] If $\mathbf{B}$ is slowly varying, we obtain

$$J = p_\perp a - \frac{qBa^2}{2c} \approx \text{constant}.$$

Substituting in the expression (28.37) for a, we obtain the identification

$$J = \frac{c}{2q}\frac{p_\perp^2}{B}, \qquad \Rightarrow \qquad \frac{p_\perp^2}{B} \approx \text{constant}. \tag{28.38}$$

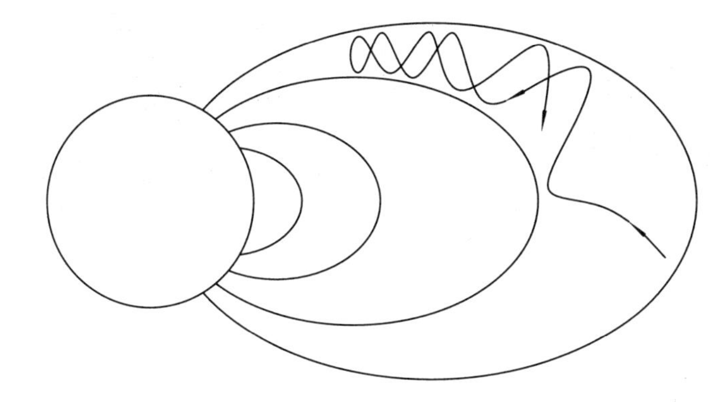

FIGURE 28.7
The converging field lines at the magnetic poles of the Earth act as "magnetic mirrors" to trap charged particles in the Van Allen radiation belts. (See Brandt and Hodge, *Solar System Astrophysics.*)

In this approximation, therefore, the enclosed magnetic flux $\Phi = -\pi a^2 B$, and the magnetic moment associated with the gyromotion, $|\boldsymbol{\mu}| = (q/2c)p_\perp a$ (see Chapter 30), both represent (relativistically correct) adiabatic invariants associated with the components of motion transverse to the local magnetic field $\mathbf{B}$. Equation (28.38) implies that $p_\perp^2$ will increase if B gets stronger, either spatially or temporally. In terrestrial application, this effect yields the *Bevatron acceleration mechanism.*

Another important application involves the *magnetic mirror effect.* Suppose we have only a static but inhomogenous magnetic field $\mathbf{B}$, e.g., the motion of a charged particle in the radiation belt of the Earth (see Figure 28.7). When we dot equation (28.33) with $\mathbf{p} = m\gamma\mathbf{v}$, we recover the result that the square of the particle's momentum, $|\mathbf{p}|^2$, remains a strict constant of the motion when acted upon by $\mathbf{B}(\mathbf{x})$ alone. Thus,

$$p_\perp^2 + p_\parallel^2 = \text{constant}, \tag{28.39}$$

where $p_\parallel \equiv m\gamma v_\parallel$. Since $p_\perp^2 \propto B$ according to equation (28.38), $p_\perp^2$ increases as the particle approaches the magnetic poles, where B gets stronger if we follow a given line of force. The increase of $p_\perp^2$ must lead to a decrease of $p_\parallel^2$ according to equation (28.39), and if $p_\parallel^2$ decreases enough (which will happen if the particle's helical orbit has a sufficiently large initial pitch angle), $p_\parallel$ will eventually reach zero and switch sign. The particle will then reverse its motion, so that it ends up shuttling back and forth between the two poles

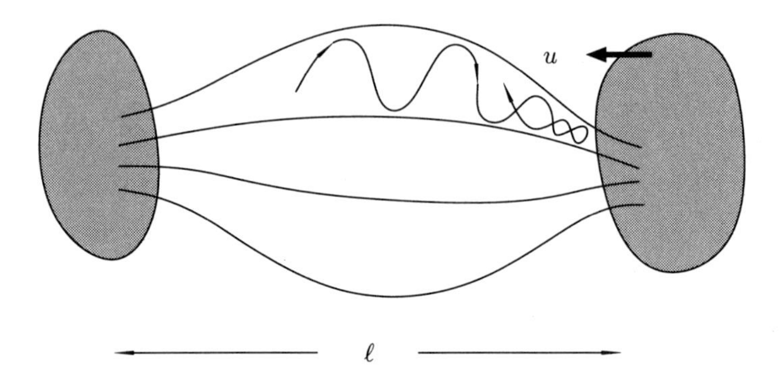

FIGURE 28.8
A "magnetic bottle" formed by two magnetized regions separated by a distance ℓ but converging upon one another at relative speed u can act to accelerate charged particles trapped within the bottle. In practical application, magnetic inhomogeneities that scatter charged particles back and forth between the upstream and downstream sides of a collisionless shock can serve equally effectively as an acceleration mechanism for cosmic rays.

of the Earth's dipole magnetic field. This magnetic mirror effect explains the essentials of how charged particles are trapped in the Earth's radiation belts. It also underlies "magnetic-bottle" schemes for plasma containment in the thermonuclear fusion program.

LONGITUDINAL ADIABATIC INVARIANT

Suppose two material regions threaded by the (same) magnetic fields converge upon one another at speed u (see Figure 28.8). Charged particles (cosmic rays), magnetically trapped between two points separated by a macroscopic length ℓ, shuttle back and forth between the two converging regions. If the particle travels much faster than u (easily the case if the particle's motion is relativistic), the action integral associated with the quasiperiodic longitudinal motion should be an adiabatic invariant,

$$J \equiv \frac{1}{2\pi} \oint p_{\|} d\ell \propto p_{\|} \ell \approx \text{constant}. \tag{28.40}$$

If we take the logarithmic derivative of the above, we find that the fractional time rate of change of $p_{\|}$ (averaged over one cycle)

$$\frac{1}{p_{\|}} \frac{dp_{\|}}{dt} = -\frac{1}{\ell} \frac{d\ell}{dt} = \frac{u}{\ell},$$

where we have used $d\ell/dt = -u$. Each reflection therefore increases the magnitude of $p_{\|}$, yielding a possible explanation for how the particle acquired relativistic energies in the first place. If we imagine that the converging magnetized regions represent the two sides of a collisionless shock (see Figure 28.1), and that the two turning points of the longitudinal motion represent particle scattering off magnetic irregularities in the upstream and downstream sides of the shock front, then we have a schematic representation of the basics of cosmic-ray acceleration. The shuttling of a cosmic-ray particle between two regions of converging flow (first-order *Fermi acceleration*) serves to increase the energy of that particle as effectively as if a tennis player were to use a moving racket to continually whack an idealized tennis ball (having unit coefficient of restitution and zero frictional coupling to the air) against a stationary wall.

From a statistical-mechanical point of view, we may say that the cosmic-ray particle tries to come into equipartition with the motions of the regions upstream and downstream from the shock front. Since the energy contained in the bulk flow (many particles) always much exceeds the energy of a single particle (the cosmic ray), the latter can, in principle, gain an almost indefinite amount of energy from the macroscopic flow. In actual practice, the cosmic-ray particle would statistically wander away from the region of the collisionless shock as it suffers repeated "whacks," so that it would eventually escape from the acceleration zone after having its energy increased by a large but finite amount. Detailed theory suggests a reason why the emergent distribution of energies of the totality of cosmic-ray particles from this acceleration process should take the form of a (declining) power law, but exposition of this theory lies beyond the scope of this volume. (For a review, see R. Blandford and D. Eichler, 1987, *Physics Reports*, vol. **154**, No. 1, 1–75.)

29

Longitudinal Plasma Oscillations and Landau Damping

Reference: Stix, *The Theory of Plasma Waves*, pp. 118–124, 131–148

In Chapter 28 we showed that the formal solution to Vlasov's kinetic equation reduces to finding the relevant integrals of particle motion if the force fields are regarded as known. Much of the complication of plasma physics stems, however, from the dominance of *internal* electromagnetic fields in affecting the motions of charged particles. These fields have a dynamics self-consistent with the dynamics of the particles. Finding the integrals of motion under these conditions may not be the most convenient way to solve the problem; a direct attack on the governing partial differential equations may offer a better approach.

In this chapter we discuss an example of such a fully dynamic phenomenon, the high-frequency oscillations that can take place if charge separation occurs with a periodic structure. In particular, we wish to reconsider the problem of longitudinal plasma oscillations when the electrons possess nonzero random motions. For a cold plasma in which we regard only the electrons to be mobile, spatial displacement of the electrons from the ions introduces restoring electric forces that set up oscillations at a frequency (see Chapter 1 and Volume I)

$$\omega_{\rm pe} = (4\pi e^2 n_{\rm e0}/m_{\rm e})^{1/2}. \tag{29.1}$$

How does a finite temperature for the electrons modify the above result?

PERTURBATIONAL VLASOV TREATMENT

To simplify the considerations, we continue to adopt the model of a two-component plasma in which the ions provide only a positively charged background of uniform density. The electrons have a phase space description given by a distribution function $f(\mathbf{x}, \mathbf{v}, t)$. In the absence of magnetic fields or collisions, f satisfies Vlasov's kinetic equation,

$$\frac{\partial f}{\partial t} + \mathbf{v} \cdot \frac{\partial f}{\partial \mathbf{x}} - \frac{e}{m_\mathrm{e}} \mathbf{E} \cdot \frac{\partial f}{\partial \mathbf{v}} = 0, \tag{29.2}$$

where the self-consistent electric field arises from the charge distribution of electrons and ions:

$$\boldsymbol{\nabla} \cdot \mathbf{E} = 4\pi e(Zn_\mathrm{i} - n_\mathrm{e}) \qquad \text{with} \qquad n_\mathrm{e} = \int f \, d^3v. \tag{29.3}$$

We assume the equilibrium state to be uniform and electrically neutral:

$$f_0 = f_0(\mathbf{v}), \qquad Zn_\mathrm{i} = n_\mathrm{e0} = \int f_0 \, d^3v. \tag{29.4}$$

With the ions immobile, the linearized equations governing small-amplitude perturbations read

$$\frac{\partial f_1}{\partial t} + \mathbf{v} \cdot \frac{\partial f_1}{\partial \mathbf{x}} = \frac{e}{m_\mathrm{e}} \mathbf{E}_1 \cdot \frac{\partial f_0}{\partial \mathbf{v}}, \tag{29.5}$$

$$\boldsymbol{\nabla} \cdot \mathbf{E}_1 = -4\pi e \int f_1 \, d^3v. \tag{29.6}$$

We look for solutions with temporal and spatial variations given by

$$\exp[i(\omega t - \mathbf{k} \cdot \mathbf{x})].$$

Equations (29.5) and (29.6) then become

$$f_1 = -i \frac{e}{m_\mathrm{e}} \frac{\mathbf{E}_1 \cdot \partial f_0 / \partial \mathbf{v}}{(\omega - \mathbf{k} \cdot \mathbf{v})}, \tag{29.7}$$

$$-i\mathbf{k} \cdot \mathbf{E}_1 = -4\pi e \int f_1 \, d^3v. \tag{29.8}$$

Faraday's law of induction,

$$\boldsymbol{\nabla} \times \mathbf{E} = -\frac{1}{c} \frac{\partial \mathbf{B}}{\partial t},$$

with $\mathbf{B} = 0$ if we consider disturbances that generate no magnetic fields, requires

$$\mathbf{k} \times \mathbf{E}_1 = 0.$$

This result implies that the oscillations are purely *longitudinal*; i.e., $\mathbf{E}_1$ is parallel to $\mathbf{k}$. Set up a rectangular coordinate system such that $\mathbf{k} = k\mathbf{e}_x$,

$\mathbf{E}_1 = E_1\mathbf{e}_x$. Without loss of generality, we suppose that ω_R, the real part of ω, is positive, but we allow k (restricted to real values) to be either positive or negative. We now define the one-dimensional distribution functions:

$$F_0(v_x) \equiv \iint_{-\infty}^{+\infty} f_0(v_x, v_y, v_z)\, dv_y\, dv_z,$$

$$F_1(x, v_x, t) \equiv \iint_{-\infty}^{+\infty} f_1(x, v_x, v_y, v_z, t)\, dv_y\, dv_z.$$

Upon integration over v_y and v_z, equations (29.7) and (29.8) become (if we let v_x be denoted by v)

$$F_1(x, v, t) = -i\frac{e}{m_e}E_1(x, t)\frac{F_0'(v)}{(\omega - kv)}, \tag{29.9}$$

$$-ikE_1 + 4\pi e \int_{-\infty}^{+\infty} F_1\, dv = 0. \tag{29.10}$$

Substituting equation (29.9) into equation (29.10), we now obtain

$$-ik\epsilon E_1 = 0, \tag{29.11}$$

where ϵ is the plasma dielectric for longitudinal oscillations,

$$\epsilon \equiv 1 + \frac{4\pi e^2}{m_e k}\int_{-\infty}^{+\infty} \frac{F_0'(v)}{\omega - kv}\, dv. \tag{29.12}$$

We can interpret equation (29.11) by introducing the idea of a displacement vector $\mathbf{D}_1 \equiv \epsilon\mathbf{E}_1$ that accounts for the polarization of the plasma by the electric field $\mathbf{E}_1$, so that the macroscopic divergence equation reads

$$\nabla \cdot \mathbf{D}_1 = 0. \tag{29.13}$$

If we are to have nontrivial solutions for E_1 in equation (29.11), we require $\epsilon = 0$. Equation (29.12) with $\epsilon = 0$ represents the *dispersion relationship* between the wave frequency ω and the wavenumber k of longitudinal plasma oscillations. Notice that the functional form of this dispersion relationship depends on the derivative of the equilibrium velocity distribution $F_0'(v)$.

LANDAU'S ANALYSIS

Before we embark on a detailed analysis of the dispersion relation $\epsilon = 0$, we introduce some simplifying notation. We first define a normalized equilibrium distribution function $\Psi_0(v)$ by

$$F_0(v) \equiv n_{e0}\Psi_0(v),$$

so that

$$\int_{-\infty}^{+\infty} \Psi_0(v)\, dv = 1.$$

We can then rewrite equation (29.12) as

$$\epsilon = 1 - \frac{\omega_{\mathrm{pe}}^2}{k^2} W(\omega/k) = 0, \tag{29.14}$$

where we have defined $W(\omega/k)$ as the integral function,

$$W(\omega/k) \equiv \int_{-\infty}^{+\infty} \frac{\Psi_0'(v)\, dv}{v - \omega/k}. \tag{29.15}$$

In Vlasov's analysis of equation (29.14), he naturally but erroneously assumed that longitudinal oscillations set up initially in a plasma with a nonpathological electron distribution function should be able to persist forever in the absence of dissipative collisions. In other words, it should be possible to consider real values for both ω and k. For ω/k real and finite (equal to the phase velocity of the longitudinal wave), however, the mathematical meaning of equation (29.15) presents a substantive difficulty. How should we interpret the integral for W if the pole at $v = \omega/k$ lies along the path of integration (along the real v axis)? In other words, what should we do about electrons that travel at a material speed exactly equal to the wave speed?

Vlasov proposed that one should interpret the integral as a principal-value integral, i.e., integrate from $-\infty$ to a little less than ω/k; jump over the singularity at $v = \omega/k$; and continue to integrate from a little greater than ω/k to $+\infty$. The procedure yields a finite value for the integral W since the change in signs of the two sides of the singularity cancel their individual, logarithmically divergent, contributions as $v \to \omega/k$. Nevertheless, in a classic paper, Landau showed that Vlasov's proposal overlooks an important physical phenomenon.

By directly analyzing the initial-value problem, Landau demonstrated that if modes with real ω exist, they should be obtained by first considering slightly growing modes, $\omega = \omega_{\mathrm{R}} + i\omega_{\mathrm{I}}$ with $\omega_{\mathrm{I}} < 0$ (so that $e^{-\omega_{\mathrm{I}} t}$ yields a growing exponential), and then taking the limit of vanishing growth rate,

$$\omega_{\mathrm{I}} \to 0^-.$$

We now briefly describe why this procedure is the correct one mathematically.

When we formulate the problem as an initial-value problem rather than as a normal-mode problem, we take Laplace transforms (in time) rather than Fourier transforms. The Laplace coefficient $Q(s)$ associated with any quantity $q(t)$ is given by

$$Q(s) \equiv \int_0^{\infty} q(t) e^{-st}\, dt,$$

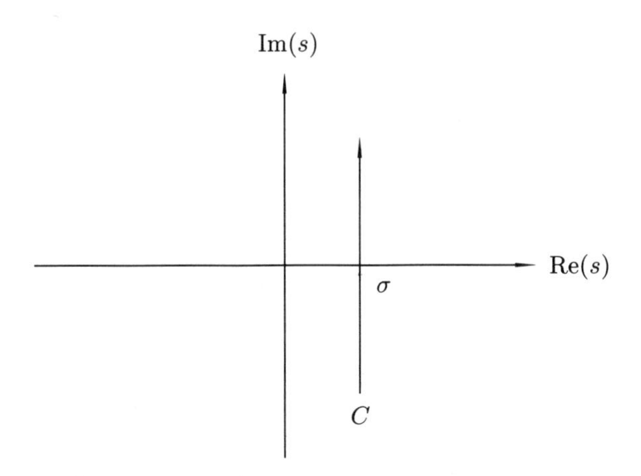

FIGURE 29.1
The starting contour in taking inverse Laplace transforms.

and the inverse transform recovers the original function,

$$q(t) = \frac{1}{2\pi i} \int_{\sigma - i\infty}^{\sigma + i\infty} Q(s) e^{st}\, ds.$$

The real value of σ in the contour integration over s must be chosen to be large and positive enough to lie to the right of all singularities of the function $Q(s)$ (see Figure 29.1). We can deform the path of integration by pushing it to the left, but analytical continuation requires that we keep all singularities to the left of the resulting contour (see Figure 29.2).

Now, let us apply the above lesson to the current problem. The combination $Q(s)e^{st}$ is analogous to the combination $W(\omega/k)e^{i\omega t}$ if we identify $s = i\omega$. The singularities of $Q(s)$ in s-space have an analogy in the pole of the integrand of W, which occurs in ω-space at $\omega = kv$, or in v-space at $v = \omega/k$. The requirement that $\sigma = \text{Re}(s)$ start off positive and to the right of all singularities of $Q(s)$ then has its counterpart in the requirement that $\text{Im}(\omega) \equiv \omega_\text{I}$ start off negative, so that the pole $v = \omega/k$ lies *below* the real line (for k real and positive) in complex-v space. The integral (29.15) is easy to interpret in this case, since the complex pole does not lie along the path of integration, and we get analytic functions by simply performing the integration straightforwardly along the real-v axis (see Figure 29.3):

$$W = \int_{-\infty}^{+\infty} \frac{\Psi_0'(v)}{v - \omega/k}\, dv \qquad \text{for} \qquad \omega_\text{I} < 0. \tag{29.16}$$

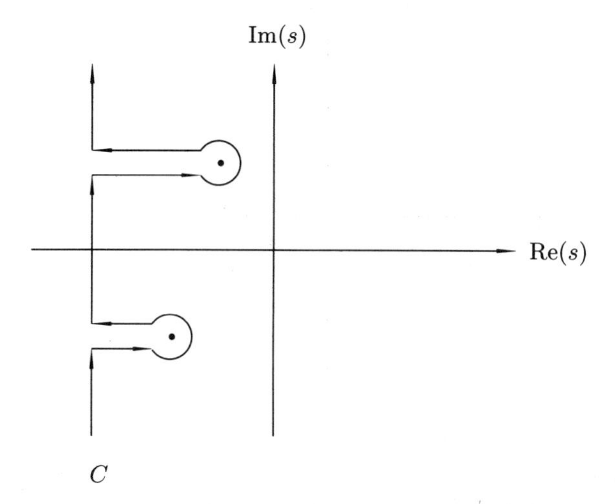

FIGURE 29.2
Analytic continuation of the inverse Laplace-transform integral if we choose to deform the contour in Figure 29.1 and the integrand contains isolated singularities (poles) denoted schematically by the dots.

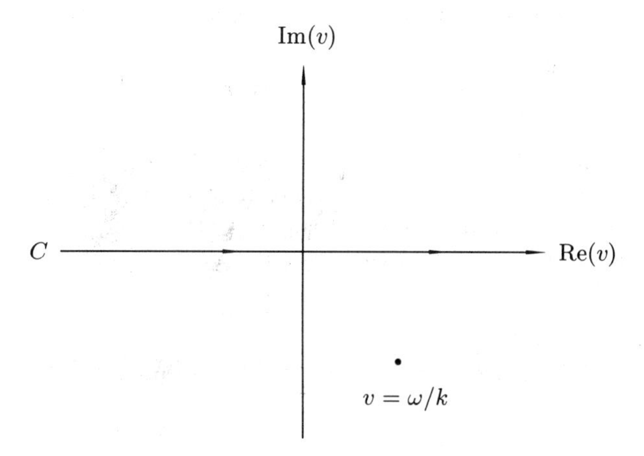

FIGURE 29.3
Interpretation of the path of integration for W in complex-v space when ω contains a negative imaginary part (growing disturbances).

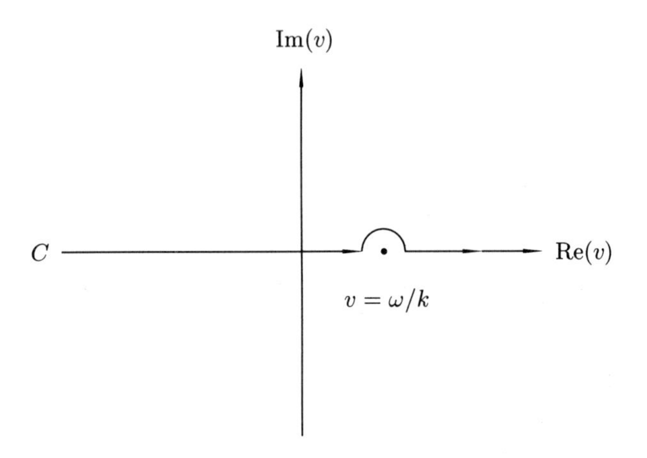

FIGURE 29.4
Analytic deformation of the path of integration for W in Figure 29.3 as the imaginary part of ω goes to zero.

As we allow ω_I to go to zero from the negative side, and the pole at $v = \omega/k$ approaches the real line (from below for k real and positive), however, we need to deform the contour to keep $W(\omega/k)$ analytic. In addition to the principal-value integral, this picks up half a residue contribution [negative or positive depending on whether $\mathrm{sgn}(k) = \pm 1$; see Figure 29.4].

$$W = \wp \int_{-\infty}^{+\infty} \frac{\Psi_0'(v)}{v - \omega/k}\, dv - \pi i \Psi_0'(\omega/k)\, \mathrm{sgn}(k) \qquad \text{for} \qquad \omega_\mathrm{I} = 0. \quad (29.17)$$

In equation (29.17), the symbol $\wp$ stands for "principal value." If ω_I were to continue and become positive (damped disturbance), then analytical continuation yields, in addition to the integral along the real line (which again presents no difficulty of interpretation), a full residue contribution (see Figure 29.5).

$$W = \int_{-\infty}^{+\infty} \frac{\Psi_0'(v)}{v - \omega/k}\, dv - 2\pi i \Psi_0'(\omega/k)\, \mathrm{sgn}(k) \qquad \text{for} \qquad \omega_\mathrm{I} > 0. \quad (29.18)$$

ABSENCE OF GROWING DISTURBANCES FOR MONOTONICALLY DECREASING DISTRIBUTION FUNCTIONS

We now wish to prove that there are no growing disturbances if $\Psi_0(v)$ is an even, positive definite, and monotonically decreasing function of v for

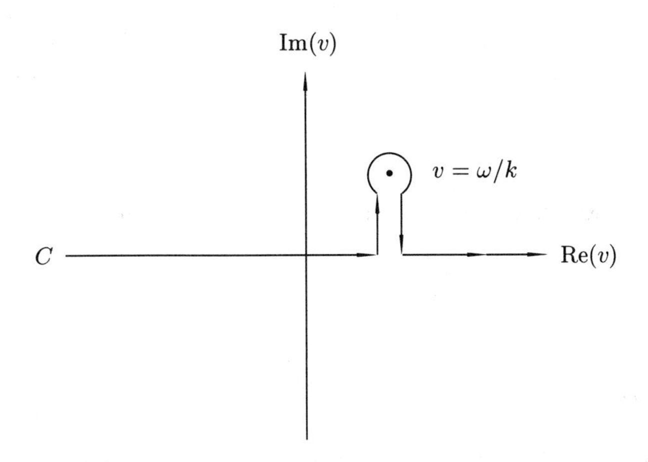

FIGURE 29.5
Continuation of the process began in Figure 29.4 as the imaginary part of ω takes on positive values (damped disturbances).

$v > 0$, as it is, for example, for a Maxwellian. For growing modes, we can multiply top and bottom by the complex conjugate of the denominator in the integrand and write equation (29.16) as

$$W = \int_{-\infty}^{+\infty} \frac{(v - \omega_\text{R}/k)\Psi_0'(v)}{(v - \omega_\text{R}/k)^2 + \omega_\text{I}^2/k^2}\, dv + i\frac{\omega_\text{I}}{k}\int_{-\infty}^{+\infty} \frac{\Psi_0'(v)\, dv}{(v - \omega_\text{R}/k)^2 + \omega_\text{I}^2/k^2}.$$

Since the dispersion relationship requires that a complex function $\epsilon = 1 - \omega_\text{pe}^2 W/k^2$ vanishes, the imaginary part of W must equal 0. With $\Psi_0'(v)$ an odd function of v, we obtain

$$\frac{\omega_\text{I}}{k}\left[\int_0^\infty \frac{\Psi_0'(v)\, dv}{(v - \omega_\text{R}/k)^2 + \omega_\text{I}^2/k^2} - \int_0^\infty \frac{\Psi_0'(v)\, dv}{(v + \omega_\text{R}/k)^2 + \omega_\text{I}^2/k^2}\right] = 0.$$

Collecting terms, we get

$$4\frac{\omega_\text{R}\omega_\text{I}}{k^2}\int_0^\infty \frac{v\Psi_0'(v)\, dv}{[(v - \omega_\text{R}/k)^2 + \omega_\text{I}^2/k^2][(v + \omega_\text{R}/k)^2 + \omega_\text{I}^2/k^2]} = 0.$$

The integral above is strictly negative if $\Psi'(v) \le 0$ for $v \ge 0$. As a consequence, we require either ω_I or ω_R to equal 0. But $\omega_\text{I} < 0$ by assumption that we have a growing disturbance, and $\omega_\text{R} = 0$ would generally contradict the requirement of the real part of the dispersion relationship, $\epsilon = 0$. Therefore, we conclude that no growing modes exist in the plasma for the assumed case. All longitudinal plasma oscillations must damp in time if $\Psi_0(v)$ is a monotonically decreasing function of $|v|$. The novel method by which oscillations in a collisionless plasma damp forms the subject of the next section.

LANDAU DAMPING IN THE LONG-WAVELENGTH LIMIT

We specialize our discussion of *Landau damping* to disturbances of long wavelength. For small k, we may expand

$$\frac{1}{(v-\omega/k)} = -\frac{k}{\omega}\left[1+\left(\frac{v}{\omega/k}\right)+\left(\frac{v}{\omega/k}\right)^2+\left(\frac{v}{\omega/k}\right)^3+\cdots\right]. \tag{29.19}$$

We assume again that $\Psi_0(v)$ is a monotonically decreasing function of $|v|$, disappearing at infinity faster than any power of v (e.g., a Maxwellian). We may then safely carry out the expansion (29.19) for the integrand and perform the integrals term-by-term:

$$\int_{-\infty}^{+\infty}\frac{\Psi_0'(v)\,dv}{(v-\omega/k)}$$
$$= -\frac{k}{\omega}\int_{-\infty}^{+\infty}\left[1+\left(\frac{v}{\omega/k}\right)+\left(\frac{v}{\omega/k}\right)^2+\left(\frac{v}{\omega/k}\right)^3+\cdots\right]\Psi_0'(v)\,dv,$$

and we need not be careful whether we write a $\wp$ in front of the integral on the right-hand side. All the even powers in v vanish if $\Psi_0'(v)$ is odd in v; hence, if we factor out a factor of kv/ω from the remaining terms, we have

$$\int_{-\infty}^{+\infty}\frac{\Psi_0'(v)\,dv}{(v-\omega/k)} = -\frac{k^2}{\omega^2}\int_{-\infty}^{+\infty} v\left[1+\left(\frac{v}{\omega/k}\right)^2+\cdots\right]\Psi_0'(v)\,dv.$$

Integration by parts yields

$$\int_{-\infty}^{+\infty}\frac{\Psi_0'(v)\,dv}{(v-\omega/k)} = \frac{k^2}{\omega^2}\left[1+3\frac{\langle v^2\rangle}{\omega^2/k^2}+\cdots\right].$$

Collecting results, we may now express the dispersion relationship as

$$\epsilon = 1-\frac{\omega_{\rm pe}^2}{\omega^2}\left[1+3\frac{k^2\langle v\rangle^2}{\omega^2}+\cdots\right]+i\frac{\omega_{\rm pe}^2}{k^2}p = 0, \tag{29.20}$$

where p represents the pole contribution:

$$p = \pi\Psi_0'(\omega/k)\,\mathrm{sgn}(k) \qquad \text{for} \qquad \omega_{\rm I}=0, \tag{29.21}$$

$$p = 2\pi\Psi_0'(\omega/k)\,\mathrm{sgn}(k) \qquad \text{for} \qquad \omega_\mathrm{I} > 0. \tag{29.22}$$

For small k, $\Psi_0'(\omega/k)$ is to be evaluated out in the tail of the distribution function; hence the self-consistent value for ω_I will turn out to be exponentially small in comparison with ω_R. We therefore choose to solve for ω_R and ω_I by successive approximation. If we assume that $\omega_\mathrm{I} \ll \omega_\mathrm{R}$, the real and imaginary parts of equation (29.20) have the expressions

$$1 - \frac{\omega_\mathrm{pe}^2}{\omega_\mathrm{R}^2}\left[1 + 3\frac{k^2\langle v^2\rangle}{\omega_\mathrm{R}^2} + \cdots\right] = 0, \tag{29.23}$$

$$2\frac{\omega_\mathrm{I}\omega_\mathrm{pe}^2}{\omega_\mathrm{R}^3} = -\pi\frac{\omega_\mathrm{pe}^2}{k^2}\Psi_0'(\omega_\mathrm{R}/k)\ \mathrm{sgn}(k), \tag{29.24}$$

where we have used the value (29.21) for p when we approximate ω/k by ω_R/k in the pole contribution.

For small k, equation (29.23) can be solved by iteration,

$$\omega_\mathrm{R}^2 \approx \omega_\mathrm{pe}^2 + 3k^2\langle v^2\rangle, \tag{29.25}$$

which demonstrates that only waves with frequency higher than the plasma frequency can propagate (see Chapter 20 of Volume I for an analogous phenomenon involving *transverse* waves, i.e., electromagnetic radiation). The additional term $3k^2\langle v^2\rangle$ in equation (29.25) represents the effect of finite velocity dispersion, with the factor 3 first obtained correctly for a collisionless plasma by Bohm and Gross (see discussion below). In a fluid analogy the effect arises from finite "pressure," and we would substitute $\langle v^2\rangle = \mathcal{R}T$ for a Maxwellian distribution, where $\mathcal{R}$ equals Boltzmann's constant divided by the mean molecular mass and T equals the electron temperature.

Equation (29.25) therefore provides the answer to the question that motivated our introductory discussion. Notice, in particular, the similarities and differences with the dispersion relation for Jeans' problem of gravitational stability in an infinite medium (review Problem Set 3):

$$\omega^2 = -4\pi G\rho_0 + k^2 a_s^2.$$

The term $3\langle v^2\rangle$ replaces $a_s^2 = \gamma\mathcal{R}T$ because $c_v = \mathcal{R}/2$ and $c_P = c_v + \mathcal{R} = 3\mathcal{R}/2$ for the specific heats associated with the one-dimensional variation of a plasma in which the effects of compressional heating do not get shared by the other two directions through randomizing collisions. Thus, the square of the adiabatic sound speed for longitudinal plasma oscillations is larger than the square of the velocity dispersion in one dimension, $\langle v^2\rangle$, by the ratio $\gamma_\mathrm{eff} = c_P/c_v = 3$. The term $\omega_\mathrm{pe}^2 = +4\pi e^2 n_\mathrm{e0}/m_\mathrm{e}$ replaces $-4\pi G\rho_0$ because we deal here with electric forces rather than gravitational ones. The plus sign instead of a minus sign reminds us that electricity (between like particles) is repulsive, but gravity is attractive. Thus, mutual Coulomb repulsion among electrons crowded together makes the compressed regions bounce back faster than velocity dispersion or pressure can accomplish alone.

Even more interesting are the implications of the imaginary part of the dispersion relationship, equation (29.24). For $k^2 \ll \omega_{\rm pe}^2/\langle v^2\rangle$, so that $\omega_{\rm R}^2 \approx \omega_{\rm pe}^2$ in equation (29.23), we may approximate equation (29.24) by

$$\omega_{\rm I} \approx -\frac{\pi}{2}\frac{\omega_{\rm pe}^3}{k^2}\Psi_0'(\omega_{\rm R}/k)\ \mathrm{sgn}(k). \tag{29.26}$$

Notice that we follow J. D. Jackson rather than L. D. Landau in not replacing $\omega_{\rm R}$ by $\omega_{\rm pe}$ in the argument of Ψ_0'; this is because we are working out in the tail of the distribution function, where small changes in the velocity can make substantial changes in the value of Ψ_0 and its derivatives.

For a one-dimensional Maxwellian distribution,

$$\Psi_0(v) = \left(2\pi\langle v^2\rangle\right)^{-1/2}\exp\left(-\frac{v^2}{2\langle v^2\rangle}\right), \tag{29.27}$$

and equation (29.26) becomes

$$\omega_{\rm I} \approx \left(\frac{\pi}{8}\right)^{1/2}\langle v^2\rangle^{-3/2}\frac{\omega_{\rm pe}^3\omega_{\rm R}}{|k|^3}\exp\left(-\frac{\omega_{\rm R}^2}{2k^2\langle v^2\rangle}\right). \tag{29.28}$$

Notice that $\omega_{\rm I} > 0$ independent of the sign of k; all waves are damped whether they travel to the right or to the left.

If we now substitute the expression (29.25) for $\omega_{\rm R}^2$, we obtain

$$\frac{\omega_{\rm I}}{\omega_{\rm R}} = \left(\frac{\pi}{8}\right)^{1/2}\left(\frac{k_{\rm D}}{|k|}\right)^3\exp\left[-\frac{3}{2}-\frac{1}{2}\left(\frac{k_{\rm D}}{|k|}\right)^2\right], \tag{29.29}$$

where $k_{\rm D}$ is the Debye wavenumber (see Chapter 1),

$$k_{\rm D} \equiv \frac{\omega_{\rm pe}}{\langle v^2\rangle^{1/2}}. \tag{29.30}$$

Equation (29.29) represents Jackson's correction (the term 3/2 in the exponent) to Landau's damping decrement for longitudinal plasma oscillations. Notice that the imaginary part of ω is (exponentially) small in comparison to the real part to the extent that $|k| \ll k_{\rm D}$. The damping decrement formally becomes of order unity for $|k| \sim k_{\rm D}$; thus only disturbances much longer than the Debye length do not suffer heavy Landau damping. At scales much larger than the Debye length, the plasma has a collective behavior; below the Debye length, collective oscillations are much less important than the graininess manifest in individual charges (see Chapter 1).

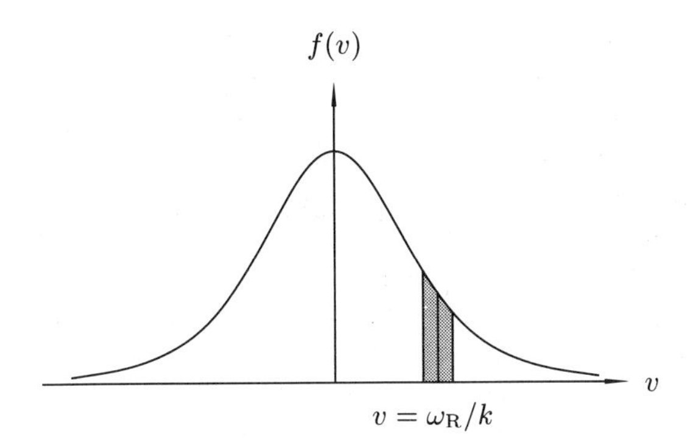

FIGURE 29.6
For a longitudinal plasma wave of infinitesimal amplitude, resonant interactions of electrons with the wave are possible only for electrons with velocities v (component along $\mathbf{k}$) within an infinitesimal range centered on $v = \omega_R/k$. Electrons going slightly faster than the wave have an opposite effect to electrons going slightly slower than the wave; thus the Landau damping decrement depends only on the derivative of the electron distribution function evaluated at the exactly resonant value $v = \omega_R/k$.

PHYSICAL INTERPRETATION FOR LANDAU DAMPING

Resonant wave-particle interaction underlies the physical mechanism behind Landau damping (see Figure 29.6 and the discussion cited in Stix's book at the beginning of the chapter). Electrons with random velocities v substantially different from the phase speed of the wave, ω_R/k, drift in and out of the crests and troughs of the wave. Sometimes they are accelerated by the collective electric field E_1, sometimes decelerated; but integrated over time, no net interaction results (in the linear approximation) because the time spent in crests and troughs averages out for a sinusoidal disturbance.

Consider, however, the behavior of particles with velocities v very nearly equal to the phase speed $v_p \equiv \omega_R/k$. Such particles maintain the same phase relationship to the wave for practically all time; they can "ride" the electrostatic field (as seen in their frame) as effectively as a surfer can ride a water wave. Such resonant particles have secular interactions with the wave that do not average out with time. This description lends insight into the necessity of referring to an *initial-value* treatment to sort out the ambiguities in the integration path. When we adopt an initial-value treatment (instead of a normal-mode or steady-state treatment), we are forced to specify an arrow of time—namely, that the secular buildup of resonant interactions began in the *past*. This specification makes possible damped modes in the absence of growing ones, when the time-reversibility of the formal equations of motion might have led us naively to expect that eigenfrequencies must come in complex-conjugate pairs.

In its purest form, Landau damping represents a phase-space behavior peculiar to collisionless systems. Analogs to Landau damping exist, for example, in the interactions of stars in a galaxy at the Lindblad resonances of a spiral density wave. Such resonances in an inhomogeneous medium can produce wave absorption (in space rather than in time), which does not usually happen in fluid systems in the absence of dissipative forces (an exception is the behavior of corotation resonances for density waves in a gaseous medium). For our current phase-space problem, electrons in the distribution function moving at velocities v slightly faster than the wave "push" on it and thereby give energy to the wave, whereas electrons slightly slower the wave "drag" on it and thereby remove energy from the wave. Since longitudinal plasma oscillations (unlike spiral density waves) have only positive energies, the former tend to amplify the wave; the latter, to damp it. Because a Maxwellian has more slow particles than fast ones, $\Psi'(v_{\rm p})\,\mathrm{sgn}(v_{\rm p}) < 0$; and longitudinal oscillations in a thermal plasma will always suffer a net amount of damping. Landau amplification can take place only for charged-particle distribution functions that display some sort of anomalous behavior in phase space. Strange distribution functions can result, for example, in the vicinity of collisionless shocks, and phase-space instabilities can be important in these and other plasma contexts.

30

Orbit Theory and Drift Currents

Reference: Chandrasekhar, *Plasma Physics*, pp. 14–36, 71–74.

Electric currents constitute the material sources for astronomical magnetic fields. In the collisionally dominated regime (MHD), we obtain the current density via Ohm's law,

$$\mathbf{j}_{\mathrm{e}} = \sigma \left(\mathbf{E} + \frac{\mathbf{u}}{c} \times \mathbf{B} \right), \tag{30.1}$$

where σ is the electrical conductivity,

$$\sigma = \frac{\omega_{\mathrm{pe}}^2}{4\pi \nu_{\mathrm{c}}}, \tag{30.2}$$

with ω_{pe} being the electron plasma frequency and ν_{c} being the collision frequency of electrons with ions (review Chapter 21). In the MHD approximation we assume that the mean difference in velocity between electrons and ions reaches a terminal value in which frictional forces due to collisions balance the macroscopic electromagnetic forces. Clearly, this approximation must break down in the collisionless regime—the usual assumption of plasma physics. In a collisionless plasma, the electromagnetic forces can be balanced only by inertial effects. This realization motivates us to study how free electric charges move in the presence of large-scale electromagnetic (and gravitational) fields.

SOME BASIC CONSIDERATIONS

A charge q of mass m spiraling nonrelativistically in a static uniform magnetic field $\mathbf{B}$ possesses a gyrofrequency given by Larmor's formula,

$$\boldsymbol{\omega}_{\mathrm{L}} = -\frac{q\mathbf{B}}{mc}, \tag{30.3}$$

where the minus sign indicates that the angular velocity is pointed in a direction such that the magnetic moment (see Volume I) arising from the particle's gyromotion,

$$\boldsymbol{\mu} = \frac{q}{2mc} m a^2 \boldsymbol{\omega}_{\rm L} = -\frac{q^2}{2mc^2} a^2 \mathbf{B}, \tag{30.4}$$

opposes $\mathbf{B}$, independent of the sign of q. In equation (30.4), $ma^2\boldsymbol{\omega}_{\rm L}$ represents the angular momentum of the particle, with a equaling its gyroradius,

$$a \equiv \left| \frac{v_\perp}{\omega_{\rm L}} \right|, \tag{30.5}$$

where $v_\perp$ is the component of the particle's velocity perpendicular to $\mathbf{B}$. The aligning of the magnetic moments of charged particles antiparallel to $\mathbf{B}$ implies, as we shall see, that a plasma is *diamagnetic*.

To obtain an order-of-magnitude feel for the scale of the problem, we note that the gyrofrequency eB/mc equals $10^{3.98}$ (B/gauss) s^{-1} for protons, but equals $10^{7.24}$ (B/gauss) s^{-1} for electrons. For comparison, the relaxation time for electron-electron collisions (see Chapter 1) reads

$$t_{\rm c}({\rm e-e}) = 0.27 \frac{(T)^{3/2}}{n_{\rm e} \ln \Lambda} \ \text{s},$$

if the electron temperature T and density $n_{\rm e}$ are measured, respectively, in degrees Kelvin and cm^{-3}. In many regimes of density, temperature, and magnetic field strength outside of stars, the collision time much exceeds the inverse of the gyrofrequency, and the plasma may be considered effectively collisionless in terms of deflections of individual particle orbits. Moreover, the gyrofrequency $\omega_{\rm L}$ is usually much larger than any macroscopic frequency ω at which external fields vary; so we can ignore the explicit time-dependence of $\mathbf{B}$.

The thermal gyroradius, defined from equation (30.5) with $v_\perp = (2kT/m)^{1/2}$,

$$a_T = \frac{mc}{qB}(2kT/m)^{1/2},$$

equals $1.34\,(T/\text{K})^{1/2}(B/\text{gauss})^{-1}$ cm for protons and $0.0313\,(T/\text{K})^{1/2}$ $(B/\text{gauss})^{-1}$ cm for electrons. For almost any reasonable astrophysical values of T and B, these values are small in comparison with the macroscopic dimensions of astronomical systems. Thus, except in special regions such as collisionless shocks, spatial variations of the force fields usually occur slowly on dimensions of the order of the gyroradii of the particles making up the thermal plasma. In what follows we shall develop a general theory for charged-particle orbits that exploits these facts.

ORBITS OF CHARGED PARTICLES IN CROSSED ELECTROMAGNETIC AND GRAVITATIONAL FIELDS

Consider the nonrelativistic equation of motion of a charged particle in a region containing an electric field $\mathbf{E}$, a magnetic field $\mathbf{B}$, and a gravitational field $\mathbf{g}$:

$$m\dot{\mathbf{v}} = q\left(\mathbf{E} + \frac{\mathbf{v}}{c} \times \mathbf{B}\right) + m\mathbf{g}. \tag{30.6}$$

For simplicity we suppose that $\mathbf{E}$ and $\mathbf{g}$ are both perpendicular to $\mathbf{B}$ (but not necessarily parallel to each other). If $\mathbf{E}$ or $\mathbf{g}$ had a component parallel to $\mathbf{B}$, the unopposed acceleration of the charged mass would eventually have to be balanced by frictional forces due to collisions, a possibility whose further elaboration we wish to avoid here.

We motivate an approximate solution of equation (30.6), under the assumption that $\mathbf{g}$ and $\mathbf{E}$ are $\perp$ to $\mathbf{B}$, by noting that we know how to solve the case

$$\dot{\mathbf{v}}_1 = \frac{q}{mc}\mathbf{v}_1 \times \mathbf{B} \equiv \boldsymbol{\omega}_{\mathrm{L}} \times \mathbf{v}_1 \tag{30.7}$$

if we ignore any time dependence or spatial inhomogeniety in $\mathbf{B}$. If $\mathbf{B}$ is constant in direction and strength (on spacetime scales of the gyroradius and gyroperiod), the (helical-motion) solution to equation (30.7) corresponds, of course, to a precession of $\mathbf{v}_1$ about $\mathbf{B}$ at the gyrofrequency (30.3). Our strategy is to try to reduce equation (30.6) to equation (30.7).

We begin by writing the equation of motion as

$$\dot{\mathbf{v}} = \frac{q}{mc}\left(\mathbf{v} - c\frac{\mathbf{E} \times \mathbf{B}}{B^2} - \frac{mc}{q}\frac{\mathbf{g} \times \mathbf{B}}{B^2}\right) \times \mathbf{B}, \tag{30.8}$$

which is completely equivalent to equation (30.6) when $\mathbf{g}$ and $\mathbf{E}$ are $\perp$ to $\mathbf{B}$. Next, we define

$$\mathbf{v}_2 \equiv \mathbf{v} - c\frac{\mathbf{E} \times \mathbf{B}}{B^2} - \frac{mc}{q}\frac{\mathbf{g} \times \mathbf{B}}{B^2}. \tag{30.9}$$

We assume that $\mathbf{E}$ can vary in time as we follow the particle motion, but that $\mathbf{g}$ and $\mathbf{B}$ remain constant. We allow $\mathbf{E}$ to have more general properties than $\mathbf{g}$, because when an electric field is present, it will generally be more important than the gravitational field. Moreover, we shall find that a time-independent electric field yields bulk motion (the so-called $\mathbf{E} \times \mathbf{B}$ drift), but no current; so inclusion of the time dependence represents an important consideration for the latter effect. Similarly, we shall find that a spatially homogeneous magnetic field has limited capacity for inducing current flow (resulting in a negative magnetization of the plasma); so we shall later generalize the discussion to situations when $\mathbf{B}$ has spatial curvature and/or gradients in strength.

With the definition (30.9), we may now write equation (30.8) as

$$\dot{\mathbf{v}}_2 + c\frac{\dot{\mathbf{E}} \times \mathbf{B}}{B^2} = \frac{q}{mc}\mathbf{v}_2 \times \mathbf{B}. \tag{30.10}$$

By introducing the definition

$$\mathbf{v}_\mathrm{p} \equiv \frac{mc^2\dot{\mathbf{E}}}{qB^2}, \tag{30.11}$$

we may rewrite equation (30.10) as

$$\dot{\mathbf{v}}_2 = \frac{q}{mc}(\mathbf{v}_2 - \mathbf{v}_\mathrm{p}) \times \mathbf{B}. \tag{30.12}$$

Equation (30.12) is not yet in the desired form, so we define

$$\mathbf{v}_1 \equiv \mathbf{v}_2 - \mathbf{v}_\mathrm{p}, \tag{30.13}$$

and express equation (30.12) as

$$\dot{\mathbf{v}}_1 + \dot{\mathbf{v}}_\mathrm{p} = \frac{q}{mc}\mathbf{v}_1 \times \mathbf{B}. \tag{30.14}$$

Now, compare the magnitude of the second term on the left-hand side of equation (30.14) with the term on the right-hand side:

$$\frac{mc|\dot{\mathbf{v}}_\mathrm{p}|}{q|\mathbf{v}_1 \times \mathbf{B}|} \sim \frac{1}{{\omega_\mathrm{L}}^2}\frac{c\ddot{E}}{v_1 B} \sim \left(\frac{\omega}{\omega_\mathrm{L}}\right)^2 \frac{cE}{v_1 B} \ll 1,$$

if $\omega \ll \omega_\mathrm{L}$ and $E \sim v_1 B/c$ or smaller (as is likely to be the case). Thus, to a high order of approximation, we may replace equation (30.14) with equation (30.7). (Q.E.D.)

POLARIZATION, ELECTRIC, AND GRAVITY DRIFTS

The results of the previous section show that the solution for the velocity of the charged particle in crossed electromagnetic and gravitational fields reads

$$\mathbf{v} = \mathbf{v}_1 + \mathbf{v}_\mathrm{p} + c\frac{\mathbf{E} \times \mathbf{B}}{B^2} + \frac{mc}{q}\frac{\mathbf{g} \times \mathbf{B}}{B^2}. \tag{30.15}$$

where $\mathbf{v}_1$ represents ordinary gyromotion about the magnetic field $\mathbf{B}$ and $\mathbf{v}_\mathrm{p}$ is given by equation (30.11). Thus, equation (30.15) states that superimposed on top of the ordinary gyromotion are three other motions: the polarization, electric, and gravitational drifts. The latter two are also called $\mathbf{E} \times \mathbf{B}$ and $\mathbf{g} \times \mathbf{B}$ drifts because they are perpendicular to both $\mathbf{E}$

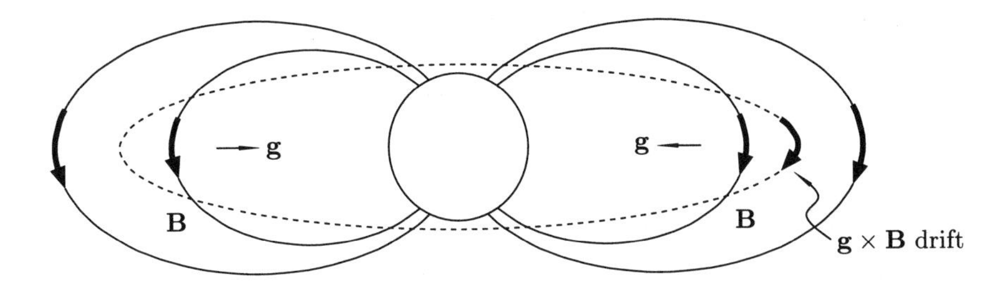

FIGURE 30.1
The $\mathbf{g} \times \mathbf{B}$ drift causes charged particles in the radiation belts of the Earth to circulate in the azimuthal direction as well as to shuttle back and forth between the poles (see Figure 28.7).

and $\mathbf{B}$ in the first case, and to $\mathbf{g}$ and $\mathbf{B}$ in the second. Anyone who has played with a gyroscope should not find the behaviors of the electric and gravitational drifts surprising. When one exerts a force perpendicular to a gyroscope's spin, the gyroscope responds, not by moving in the direction of the applied force, but by moving in a direction perpendicular to both the spin and the applied force. The same holds true for electric and gravitational fields applied perpendicularly to the spinning motion of charged particles about magnetic fields. The example most familiar to astrophysicists of the gravitational drift of charged particles in crossed magnetic and gravitational fields involves the gradual azimuthal spreading of solar-wind particles introduced inhomogeneously into the Earth's radiation belt (see Figure 30.1).

The polarization drift acquires its name for a reason that will be made clear in a following section. For now, we simplify its expression by noting that $\dot{\mathbf{E}}$ in equation (30.11) refers to taking the time derivative of $\mathbf{E}$ following the motion of the particle. But the fastest motion perpendicular to $\mathbf{B}$ corresponds to rotation with a gyroradius a that is much smaller than the scale over which $\mathbf{E}$ is likely to vary. Thus, we may approximate $\dot{\mathbf{E}}$ by $\partial\mathbf{E}/\partial t$.

To summarize these deliberations, we write

$$\mathbf{v} = \mathbf{v}_{\text{mag}} + \mathbf{v}_{\text{pol}} + \mathbf{v}_{\text{elec}} + \mathbf{v}_{\text{grav}}, \tag{30.16}$$

where $\mathbf{v}_{\text{mag}}$ satisfies

$$\dot{\mathbf{v}}_{\text{mag}} = \boldsymbol{\omega}_{\text{L}} \times \mathbf{v}_{\text{mag}}, \tag{30.17}$$

and $\mathbf{v}_{\rm pol}$, $\mathbf{v}_{\rm elec}$, and $\mathbf{v}_{\rm grav}$ are the polarization, electric, and gravitational drifts,

$$\mathbf{v}_{\rm pol} = \frac{mc^2}{qB^2}\frac{\partial \mathbf{E}}{\partial t}, \tag{30.18}$$

$$\mathbf{v}_{\rm elec} = c\frac{\mathbf{E}\times\mathbf{B}}{B^2} \ll c \qquad \text{if} \qquad E \ll B, \tag{30.19}$$

$$\mathbf{v}_{\rm grav} = \frac{mc}{q}\frac{\mathbf{g}\times\mathbf{B}}{B^2}. \tag{30.20}$$

Notice that the electric drift is the same for all species independent of charge or mass; hence we expect it to give rise to *bulk* motions of a plasma and not to the flow of any currents. On the other hand, the small-scale gyromotions arising from the random velocities of individual electrons and ions clearly contribute nothing to the mean bulk motion, although they can contribute to a flow of electric current. If the mean and random velocities of plasma particles, $\mathbf{u}$ and $\mathbf{w}$, have similar values, then $\mathbf{v}_{\rm elec}$ and $\mathbf{v}_{\rm mag}$ will have comparable magnitudes. Nevertheless, because most of the action in $\mathbf{v}_{\rm mag}$ involves going around in circles, the net current generated by this term is much smaller than might be measured simply by taking the magnitude of $\mathbf{v}_{\rm mag}$ (see the next section).

In contrast, the drift speeds induced by polarization and gravitational effects start off small in comparison with either the gyromotion or the electric drift. This is clearly true for $|\mathbf{v}_{\rm grav}|$ compared to $|\mathbf{v}_{\rm elec}| \approx u$ if $|m\mathbf{g}| \ll |q\mathbf{E}|$. For $|\mathbf{v}_{\rm pol}|$, equation (30.18) allows us to estimate

$$|\mathbf{v}_{\rm pol}| \sim \left(\frac{\omega}{\omega_{\rm L}}\right)\left(\frac{cE}{B}\right) \sim \left(\frac{\omega}{\omega_{\rm L}}\right) u \ll u$$

if $\omega \ll \omega_{\rm L}$ and $cE/B \sim u$ (see below). The above argument represents a variation on the one given in Chapter 21, where we showed that the drift speed of electrons relative to ions providing the currents needed to maintain any astrophysical magnetic field usually turns out to be an almost absurdly small number. Thus, $\mathbf{v}_{\rm grav}$ and $\mathbf{v}_{\rm pol}$ usually make negligible direct contributions to the overall bulk velocity (see, however, the later discussion of the *indirect* response of a magnetized plasma blob to the action of an external gravitational field). Nevertheless, because negatively and positively charged particles drift in opposite directions, these charge-dependent terms can contribute appreciably to the electric current (see below).

If we now identify the local mean ("fluid") velocity $\mathbf{u}$ of the plasma with $\mathbf{v}_{\rm elec}$, we can cross equation (30.19) with $\mathbf{B}$, rearrange terms, and get (with $\mathbf{E}\perp\mathbf{B}$):

$$\mathbf{E} = -\frac{\mathbf{u}}{c}\times\mathbf{B}. \tag{30.21}$$

We recognize equation (30.21) as the relationship between the electric and magnetic fields in the laboratory frame provided the plasma can be considered a fluid with infinite conductance (review Chapter 21). In particular, if we substitute equation (30.21) into Faraday's law of induction,

$$\nabla \times \mathbf{E} = -\frac{1}{c}\frac{\partial \mathbf{B}}{\partial t}, \tag{30.22}$$

we recover the equation of field freezing,

$$\frac{\partial \mathbf{B}}{\partial t} + \nabla \times (\mathbf{B} \times \mathbf{u}) = 0. \tag{30.23}$$

The derivation of this result from the point of view of the $\mathbf{E} \times \mathbf{B}$ drift provides additional insight into the statement in Chapter 21 that field freezing occurs because the matter is tied to the field and the field is tied to the matter. We see here that an electric field $\mathbf{E}$ perpendicular to $\mathbf{B}$ causes the entire plasma to try to drift across lines of $\mathbf{B}$. But the spatial curl of the same electric field causes, by Faraday's law of inductance, the magnetic field lines also to drift at the same rate. (Of course, if $\mathbf{E}$ and $\mathbf{B}$ are both spatially uniform, all lines of $\mathbf{B}$ are equivalent, and we gain nothing by trying to label any given field line.)

DRIFT CURRENTS IN THE LOW-FREQUENCY PLASMA APPROXIMATION

We compute the current density internal to the plasma by integrating $q\mathbf{v}$ over the distribution function of each charge species and summing over all species. If we use equation (30.16) to express $\mathbf{v}$, we obtain

$$\mathbf{j}_{\text{int}} = \sum \int q\mathbf{v} f \, d^3v = \mathbf{j}_{\text{mag}} + \mathbf{j}_{\text{pol}} + c\frac{\mathbf{E} \times \mathbf{B}}{B^2} \sum qn + \mathbf{j}_{\text{grav}}, \tag{30.24}$$

where n is the number density of each species (unlabeled to simplify the notation):

$$n = \int f \, d^3v,$$

and

$$\mathbf{j}_{\text{mag}} = \sum \int q\mathbf{v}_{\text{mag}} f \, d^3v, \tag{30.25}$$

$$\mathbf{j}_{\text{pol}} = \frac{c^2}{B^2}\frac{\partial \mathbf{E}}{\partial t} \sum mn = \frac{\rho c^2}{B^2}\frac{\partial \mathbf{E}}{\partial t}, \tag{30.26}$$

$$\mathbf{j}_{\text{grav}} = c\frac{\mathbf{g} \times \mathbf{B}}{B^2} \sum mn = \rho c \frac{\mathbf{g} \times \mathbf{B}}{B^2}. \tag{30.27}$$

In equations (30.26) and (30.27), $\rho = \sum mn$ represents the mass density of the plasma.

For low-frequency phenomena, $\omega \ll \omega_{\text{pe}}$, we do not expect much charge separation; hence we shall approximate $\sum qn = 0$, and the $\mathbf{E} \times \mathbf{B}$ term contributes nothing to $\mathbf{j}_{\text{int}}$ in equation (30.24), as advertised. For later reference, we pause here briefly to note the nature of this low-frequency *plasma approximation.* We take $\sum qn$ to equal zero, not for computing $\mathbf{E}$ via Coulomb's law, $\nabla \cdot \mathbf{E} = 4\pi\rho_{\text{e}}$, but for calculating electric currents. These currents may then imply a *small* amount of space-charge build-up via the equation of continuity,

$$\frac{\partial \rho_{\text{e}}}{\partial t} + \nabla \cdot \mathbf{j}_{\text{e}} = 0,$$

and it is this resulting charge density that we plug into Coulmb's law to make sure we have an (approximately) consistent set of equations. Indeed, except for the problem of longitudinal plasma oscillations (Chapter 29), we usually avoid altogether using Coulomb's law, in the sense that a *given* charge density ρ_{e} acts as a source for computing $\nabla \cdot \mathbf{E}$. The plasma particles prove too mobile to allow us to guess *a priori* how they will distribute themselves (apart from the general statement that where the ions go, the electrons will follow). If we have to use Coulomb's law at all, experience show it much safer generally to compute the electric field $\mathbf{E}$ from the other equations, and take its divergence to estimate the charge density ρ_{e}.

With the adoption of the low-frequency plasma approximation, we expect the component of velocity $v_{\|}$ parallel to $\mathbf{B}$ in $\mathbf{v}_{\text{mag}}$ not to give rise to any currents either (unless there is a component of $\mathbf{E}$ parallel to $\mathbf{B}$, whose effect would then have to be balanced by collisional friction, yielding the usual conduction current as a consequence). At first sight, it might appear that the component of velocity $v_{\perp}$ perpendicular to $\mathbf{B}$ in $\mathbf{v}_{\text{mag}}$ should also give nothing, since it corresponds merely to circular motion. We speak somewhat loosely, since equation (30.25), as it stands, contains mixed notation if we think of $\mathbf{v}_{\text{mag}}$ in a Lagrangian description (following a particle's motion), when the integration over $f\, d^3v$ involves an Eulerian description (in velocity space). In what follows, we solve this problem by considering the current carried by the gyromotion past a fixed area of the plasma. In any case, the argument given below reveals the error in thinking that $\mathbf{v}_{\text{mag}}$ contributes no current. Basically, $\mathbf{v}_{\text{mag}}$ intrinsically involves such a large term, as far as current-carrying capacity goes, that even slight incomplete cancellations due to spatial gradients allow non-negligible contributions to $\mathbf{j}_{\text{mag}}$.

Consider an area A bounded by a closed circuit C within the plasma (see Figure 30.2). Orbits that penetrate A twice, such as the orbit labeled 2, clearly make no net contribution to the current I passing through A. Only orbits near the edge, such as the orbit labeled 1, which intersects A

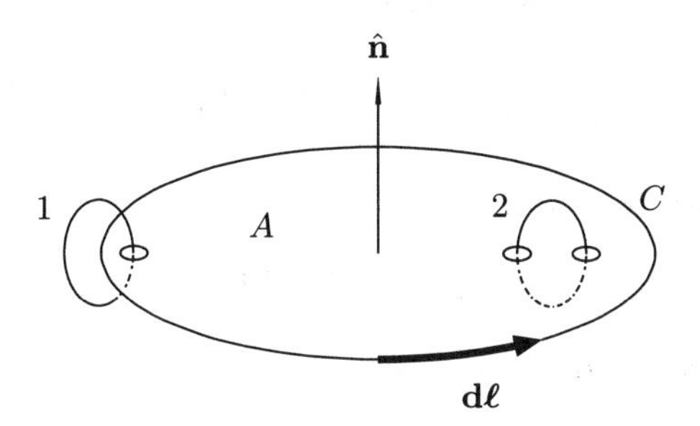

FIGURE 30.2
The geometry relevant for a discussion of the magnetization associated with the gyromotions of the charged particles of a plasma placed in a magnetic field **B**.

once, make a contribution to I. Single intersections occur for orbits that lie within radial distance from the edge less than the gyroradius a. For each line segment $\mathbf{d\ell}$ on the circuit C, a volume element $\pi a^2 d\ell$ contains singly intersecting orbits, each case carrying a charge q passing once through A per gyroperiod $2\pi/\omega_L$. Thus the current contribution from particles of each species with gyroradius a inside the relevant volume element adjacent to $\mathbf{d\ell}$ equals $q(\boldsymbol{\omega}_L/2\pi)\cdot(\pi a^2 \mathbf{d\ell})$. To get the total current I_{mag} due to this effect, we need to integrate over the distribution function f and the circuit C as well as sum over all species:

$$I_{\text{mag}} = \sum \oint_C \frac{q\boldsymbol{\omega}_L}{2\pi}\cdot \mathbf{d\ell} \int \pi a^2 f\, d^3v.$$

Substituting in the expressions of equations (30.3) and (30.5), we obtain

$$I_{\text{mag}} = -c\oint_C \frac{\mathbf{B}}{B^2}\cdot \mathbf{d\ell} \sum \int \frac{1}{2} m v_\perp^2 f\, d^3v = -c\oint_C \rho\mathcal{E}_\perp \frac{\mathbf{B}}{B^2}\cdot \mathbf{d\ell}, \qquad (30.28)$$

where we have defined $\rho\mathcal{E}_\perp$ as the thermal energy per unit volume associated with the gyromotion about field lines,

$$\rho\mathcal{E}_\perp \equiv \sum \int \frac{1}{2} m v_\perp^2 f\, d^3v. \qquad (30.29)$$

Stokes' theorem applied to equation (30.28) now yields

$$I_{\text{mag}} = -c\int_A \nabla \times \left(\frac{\rho\mathcal{E}_\perp}{B^2}\mathbf{B}\right)\cdot \hat{\mathbf{n}} dA. \qquad (30.30)$$

If we compare equation (30.30) with the requirement

$$I_{\rm mag} \equiv \int_A \mathbf{j}_{\rm mag} \cdot \hat{\mathbf{n}} dA,$$

we obtain the identification,

$$\mathbf{j}_{\rm mag} = c\boldsymbol{\nabla} \times \mathbf{M}, \tag{30.31}$$

where $\mathbf{M}$ is the *magnetization* of the plasma,

$$\mathbf{M} = -\frac{\rho \mathcal{E}_\perp}{B^2}\mathbf{B}. \tag{30.32}$$

Since $\boldsymbol{\nabla} \cdot \mathbf{j}_{\rm mag} = 0$ if $\mathbf{j}_{\rm mag}$ is given by equation (30.31), no charge buildup $\rho_{\rm e,mag}$ occurs associated with this contribution to the current flow.

A PLASMA AS A DIAMAGNETIC AND DIELECTRIC MATERIAL

Let us now consider the effect of these considerations on Maxwell's equation. We begin with Ampere's law, including Maxwell's inclusion of the displacement current (one of the differences in treatment between magnetohydrodynamics and plasma physics),

$$\boldsymbol{\nabla} \times \mathbf{B} = \frac{4\pi}{c}(\mathbf{j}_{\rm ext} + \mathbf{j}_{\rm int}) + \frac{1}{c}\frac{\partial \mathbf{E}}{\partial t}, \tag{30.33}$$

where $\mathbf{j}_{\rm int}$ is given by

$$\mathbf{j}_{\rm int} = \mathbf{j}_{\rm mag} + \mathbf{j}_{\rm pol} + \mathbf{j}_{\rm grav}.$$

We find it convenient to incorporate $\mathbf{j}_{\rm mag}$, as given by equation (30.31), into the $\boldsymbol{\nabla} \times \mathbf{B}$ term on the left-hand side of equation (30.33), and $\mathbf{j}_{\rm pol}$, as given by equation (30.26), into the $\partial \mathbf{E}/\partial t$ term on the right-hand side. Thus,

$$\boldsymbol{\nabla} \times \mathbf{B} - \frac{4\pi}{c}\mathbf{j}_{\rm mag} \equiv \boldsymbol{\nabla} \times \mathbf{H}, \qquad \text{where} \qquad \mathbf{H} = \mathbf{B} - 4\pi\mathbf{M},$$

and

$$\frac{1}{c}\frac{\partial \mathbf{E}}{\partial t} + \frac{4\pi}{c}\mathbf{j}_{\rm pol} \equiv \frac{1}{c}\frac{\partial \mathbf{D}}{\partial t},$$

where

$$\mathbf{D} = \epsilon \mathbf{E}, \tag{30.34}$$

with

$$\epsilon = 1 + \frac{4\pi\rho c^2}{B^2}. \tag{30.35}$$

To derive the expression (30.35) for the *dielectric constant* of the plasma, we have neglected any time variations of ρ and $|\mathbf{B}|$. With the magnetization $\mathbf{M}$ given by equation (30.32), we can write a similar (but inverted) expression for the relation between $\mathbf{B}$ and $\mathbf{H}$,

$$\mathbf{B} = \mu \mathbf{H}, \tag{30.36}$$

where μ is now the *magnetic permeability* of the plasma (and not the magnetic moment of a charged particle):

$$\mu = \left(1 + \frac{4\pi \rho \mathcal{E}_\perp}{B^2}\right)^{-1}. \tag{30.37}$$

Since $\mu < 1$, and $\epsilon > 1$ (in many circumstances, $\epsilon \gg 1$, since the magnetic energy density is usually small in comparison with the rest energy density of the plasma), the plasma acts as both a diamagnetic and a dielectric material.

With these definitions, equation (30.33) now becomes

$$\nabla \times \mathbf{H} = \frac{4\pi}{c}(\mathbf{j}_{\text{ext}} + \mathbf{j}_{\text{grav}}) + \frac{1}{c}\frac{\partial \mathbf{D}}{\partial t}. \tag{30.38}$$

Equation (30.38) follows Maxwell's program in regarding the nonelectromagnetic contributions to the current density as the basic source terms in Ampere's law. This program yields the fundamental motivation for introducing the subsidiary fields, $\mathbf{H}$ and $\mathbf{D}$.

We still have to deal with Coulomb's law,

$$\nabla \cdot \mathbf{E} = 4\pi(\rho_{\text{e,ext}} + \rho_{\text{e,int}}),$$

where $\rho_{\text{e,int}} = \rho_{\text{e,pol}} + \rho_{\text{e,grav}}$ since $\rho_{\text{e,mag}} = 0$. The equation of (polarization) charge continuity gives the relationship between $\rho_{\text{e,pol}}$ and $\mathbf{j}_{\text{pol}}$,

$$\frac{\partial \rho_{\text{e,pol}}}{\partial t} + \nabla \cdot \mathbf{j}_{\text{pol}} = 0. \tag{30.39}$$

With $\mathbf{j}_{\text{pol}}$ given by equation (30.26), we can integrate equation (30.39) to obtain

$$\rho_{\text{e,pol}} = -\frac{\rho c^2}{B^2}\nabla \cdot \mathbf{E}, \tag{30.40}$$

where we have again assumed that the spacetime variations of ρ and B are negligible in comparison with those of $\mathbf{E}$. The substitution of equation (30.40) into Coulomb's law now yields

$$\nabla \cdot \mathbf{D} = 4\pi(\rho_{\text{e,ext}} + \rho_{\text{e,grav}}), \tag{30.41}$$

where $\mathbf{D}$ is again given by equation (30.34).

To summarize, for a plasma with crossed electric, magnetic, and gravitational fields, Maxwell's equations can be written in the form,

$$\nabla \cdot \mathbf{B} = 0, \tag{30.42}$$

plus equations (30.41), (30.38), and (30.22). In this set, **H** and **D** are derived from **B** and **E** through equations (30.36) and (30.34). We stress, however, that in more complicated situations (e.g., more rapidly varying fields), **H** and **D** need not be related to **B** and **E** via scalar proportionality factors. In such cases, it is always safest to go back to the form of Maxwell's equations that use only the fundamental fields, **B** and **E**, and take into account *all* the material sources for charge and current densities.

MAGNETIZED PLASMA BLOB FALLING UNDER GRAVITY

We now comment on the counterintuitive notion that only the $\mathbf{E} \times \mathbf{B}$ drift gives rise to bulk motions. How does a magnetized plasma fall under the action of gravity if the gravitational field does not directly yield motion in the direction of **g**, but only to relative drifts perpendicular to both **g** and **B**?

Let us therefore consider the problem of the motion of a blob of magnetized plasma under the action of a uniform gravitational field **g** (see Figure 30.3). We assume that **g** points downward in the plane of the page, and that we try to support the plasma with an internal magnetic field **B** that points out of the page. The combination produces a gravitational drift that sends current flowing in the direction of $\mathbf{g} \times \mathbf{B}$, i.e., from right to left in the plane of the page. This must lead to a charge buildup that gives a compensating electric field **E** that points from left to right. The resulting $\mathbf{E} \times \mathbf{B}$ drift then causes the whole plasma blob to move downward, i.e., in the direction of **g**, as we might have expected intuitively. What is the actual acceleration so produced?

For simplicity, we assume that $\rho_{e,\text{ext}}$ and $\mathbf{j}_{\text{ext}}$ are zero, but that there exist nonzero values of $\rho_{e,\text{grav}}$ and $\mathbf{j}_{\text{grav}}$ satisfying the equation of (gravitational) charge continuity,

$$\frac{\partial \rho_{e,\text{grav}}}{\partial t} + \nabla \cdot \mathbf{j}_{\text{grav}} = 0 \qquad \text{with} \qquad \mathbf{j}_{\text{grav}} = \rho c \frac{\mathbf{g} \times \mathbf{B}}{B^2}. \tag{30.43}$$

Equation (30.41) now becomes

$$\nabla \cdot (\epsilon \mathbf{E}) = 4\pi \rho_{e,\text{grav}}.$$

If we differentiate the above with respect to time and use equation (30.43), we obtain

$$\nabla \cdot \left(\epsilon \frac{\partial \mathbf{E}}{\partial t} \right) = -4\pi \nabla \cdot \mathbf{j}_{\text{grav}},$$

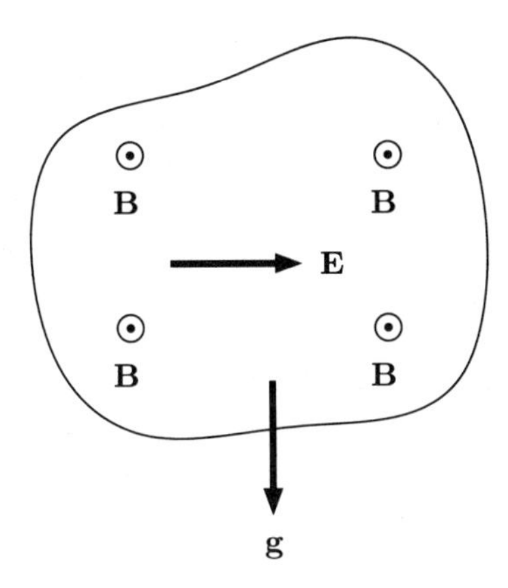

FIGURE 30.3
A blob of magnetized plasma under the action of a vertical gravitational field **g**. The $\mathbf{g} \times \mathbf{B}$ drift points from right to left if **B** points out of the page. The resulting current leads to a space charge distribution that yields an electric field **E** that points from left to right. The $\mathbf{E} \times \mathbf{B}$ drift then points down the page, the same direction as **g**.

which we may integrate to get

$$\frac{\partial \mathbf{E}}{\partial t} = -\frac{4\pi}{\epsilon}\dot{\mathbf{j}}_{\text{grav}},$$

providing we make the reasonable assumption that no electric fields arise except those that result from the slight charge separation due to the $\mathbf{g} \times \mathbf{B}$ drift. Since $\mathbf{j}_{\text{grav}}$, as given by equation (30.27), is perpendicular to **B**, so will be the resultant **E**. The presence of such a crossed electric field gives rise to an $\mathbf{E} \times \mathbf{B}$ drift that affects the plasma as a whole [see equation (30.19)]:

$$\mathbf{u} = c\frac{\mathbf{E} \times \mathbf{B}}{B^2}.$$

Assuming that the internal magnetic field **B** merely gets advected by the fluid motion (through field freezing), we may compute the acceleration as

$$\frac{D\mathbf{u}}{Dt} \approx c\frac{\partial \mathbf{E}/\partial t \times \mathbf{B}}{B^2} = -\frac{4\pi c}{\epsilon}\frac{\dot{\mathbf{j}}_{\text{grav}} \times \mathbf{B}}{B^2}.$$

With $\mathbf{j}_{\text{grav}}$ given by equation (30.27), and with **g** assumed perpendicular to **B**, we may expand the resultant triple vector product and obtain

$$\frac{D\mathbf{u}}{Dt} = \frac{4\pi\rho c^2}{\epsilon B^2}\mathbf{g} = \frac{\mathbf{g}}{1 + B^2/4\pi\rho c^2}. \tag{30.44}$$

Notice that the bulk acceleration is not quite **g** because the magnetic field has an associated inertial density $B^2/4\pi c^2$. We have not included the general relativistic correction that the presence of the magnetic field might also affect the "effective" gravitational attraction for the magnetized plasma. Presumably if we did, the bulk acceleration *would* equal **g**, or else the principle of equivalence would not hold. In any case, the fractional correction for the plasma inertia, $B^2/4\pi\rho c^2$, will be very small for nonrelativistically strong fields.

The real point that we wish to make in this section is that a plasma blob with only interior magnetic fields *does* indeed ultimately respond as a whole to the presence of an external gravitational field in the intuitively expected manner. However, the route to this answer involves counterintuitive reasoning, because the electromagnetic forces usually so dominate the immediate plasma interactions. Although the answer satisfies intuition, the reasoning does not.

CURVATURE AND GRADIENT DRIFTS

If **B** is nonuniform, there will be additional drifts. The two most important additional ones are *curvature drift* and *gradient drift*. Curvature drift has the mathematical expression

$$\mathbf{v}_{\text{curv}} = \left(\frac{cmv_\parallel^2}{qB^4}\right)\mathbf{B}\times\left[\boldsymbol{\nabla}\left(\frac{1}{2}B^2\right) + (\boldsymbol{\nabla}\times\mathbf{B})\times\mathbf{B}\right], \tag{30.45}$$

where the term in the bracket on the right-hand side represents the vector-invariant way to write $(\mathbf{B}\cdot\boldsymbol{\nabla})\mathbf{B}$. Gradient drift has the mathematical expression

$$\mathbf{v}_{\text{grad}} = \frac{cmv_\perp^2}{2qB^3}(\mathbf{B}\times\boldsymbol{\nabla}B). \tag{30.46}$$

We notice that gyroscopic action cause these drifts to occur in directions perpendicular to **B**, just like the other drifts discussed in this chapter. Moreover, because the curvature and gradient drifts depend inversely on the charge-to-mass ratio, they can also give rise to drift currents. We give below the derivations of the formulae (30.45) and (30.46).

Curvature drift arises essentially because forcing the guiding center of a spiraling charge to follow a curved path introduces, in the frame of motion of the guiding center, a centrifugal force (see Figure 30.4). This centrifugal force behaves as a fake gravitational field $\mathbf{g}_{\text{eff}}$ acting perpendicularly to **B**, and therefore introduces a "$\mathbf{g}_{\text{eff}}\times\mathbf{B}$" drift. From Figure 30.4 we compute

$$\mathbf{g}_{\text{eff}} = \frac{v_\parallel^2}{R}\hat{\mathbf{n}},$$

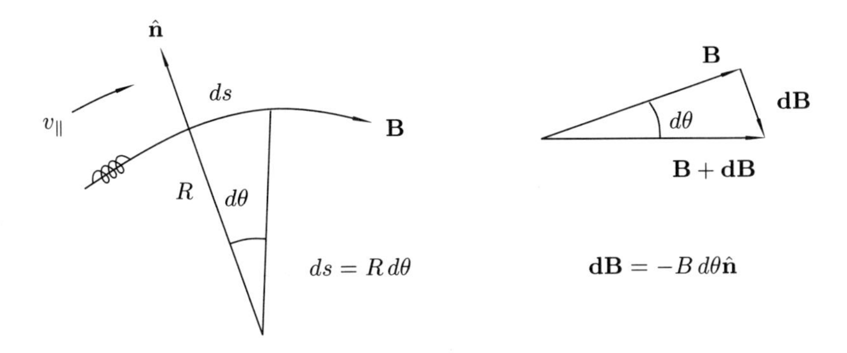

FIGURE 30.4
The geometry of curvature drift.

where R is the radius of curvature of the magnetic line of force about which the particle spirals. On the other hand, the assumed geometry shows that

$$d\mathrm{B} = -B\,d\theta\,\hat{\mathbf{n}} = -B\frac{ds}{R}\hat{\mathbf{n}},$$

if we ignore variations in the scalar strength of the field (to be considered below). To be able to compute $\mathbf{g}_{\text{eff}}$, we want the combination $\hat{\mathbf{n}}/R$,

$$\frac{\hat{\mathbf{n}}}{R} = -\frac{1}{B}\frac{d\mathbf{B}}{ds} \equiv -\frac{1}{B^2}(\mathbf{B}\cdot\boldsymbol{\nabla})\mathbf{B},$$

which yields

$$\mathbf{g}_{\text{eff}} = -\frac{v_\parallel^2}{B^2}\left[\boldsymbol{\nabla}\left(\frac{1}{2}B^2\right) + (\boldsymbol{\nabla}\times\mathbf{B})\times\mathbf{B}\right], \qquad (30.47)$$

if we use the usual vector identity to transform $(\mathbf{B}\cdot\boldsymbol{\nabla})\mathbf{B}$. Substitution of equation (30.47) for the effective gravity into equation (30.20) yields equation (30.45). (Q.E.D.)

Gradient drift arises essentially because the gyroradius is slightly larger on one side of the orbit than the other if the magnetic field $\mathbf{B}$ has a gradient in strength (see Figure 30.5). The larger gyroradius on the side that has the weaker field leads, with each passage, to a steady drift directed perpendicular to both $\mathbf{B}$ and its spatial gradient (assumed itself to lie in a direction perpendicular to $\mathbf{B}$, because we do not wish to duplicate the calculation of the curvature drift).

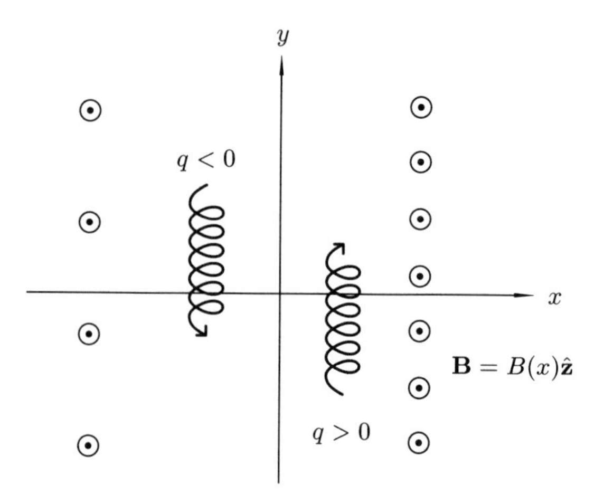

FIGURE 30.5
Schematic orbits for positively charged and negatively charged particles placed in a region with magnetic fields pointing in the $+z$ direction (out of the page) and with a gradient in strength from left to right in the x direction. The resulting drift rate in the y direction is computed in the text via perturbation theory.

Let us adopt rectangular coordinates to make a detailed calculation of the effect. We suppose the magnetic field $\mathbf{B}$ to lie in the z-direction, and that it increase in strength toward the x-direction. We anticipate therefore that a steady drift will occur in the y-direction, and be oppositely directed for positive and negative charges (drift toward positive and negative y, respectively).

In the adopted coordinate system, the motion in the x-y plane reads

$$\ddot{x} = \frac{q}{mc}\dot{y}B, \qquad \ddot{y} = -\frac{q}{mc}\dot{x}B. \tag{30.48}$$

For inhomogeneities in B that vary slowly on the scale of the instantaneous gyroradius, we may Taylor-expand $B(x)$ about the equilibrium (guiding center) position, taken to be $x = 0$,

$$B(x) = B(0) + xB'(0) + \cdots. \tag{30.49}$$

Since we expect the resulting motion to be a small departure superimposed onto a basic circular rotation, we write

$$x = a\sin\omega_{\rm L}t + \xi(t), \qquad y = a\cos\omega_{\rm L}t + \eta(t), \tag{30.50}$$

with $\omega_{\rm L}$ being the magnitude of the unperturbed gyrofrequency,

$$\omega_{\rm L} \equiv \frac{qB(0)}{mc},$$

and a being the unperturbed gyroradius

$$a = \frac{v_\perp}{\omega_{\rm L}}.$$

The substitution of equations (30.49) and (30.50) into equation (30.48) and standard perturbation-theory techniques yield

$$\ddot{\xi} = -\omega_{\rm L}^2 a^2 \frac{B'(0)}{B(0)} \sin^2 \omega_{\rm L} t + \omega_{\rm L} \dot{\eta}, \qquad \ddot{\eta} = -\omega^2 a^2 \frac{B'(0)}{B(0)} \sin \omega_{\rm L} t \cos \omega_{\rm L} t - \omega_{\rm L} \dot{\xi}.$$

We average over one cycle of the fast gyromotion, taking $\langle \ddot{\xi} \rangle = 0 = \langle \ddot{\eta} \rangle$. We then obtain

$$\langle \dot{\eta} \rangle = \frac{1}{2} \omega_{\rm L} a^2 \frac{B'(0)}{B(0)}, \qquad \langle \dot{\xi} \rangle = 0.$$

This result demonstrates that, to lowest perturbation order, the secular drift takes place entirely in the y-direction, i.e.,

$$\mathbf{v}_{\rm grad} = \frac{1}{2} \omega_{\rm L} a^2 \frac{B'(0)}{B(0)} \mathbf{e}_y.$$

Writing out the expressions for a and $\omega_{\rm L}$, we get

$$\mathbf{v}_{\rm grad} = \frac{cmv_\perp^2}{2qB^2} \frac{dB}{dx} \mathbf{e}_y,$$

which is equation (30.46) expressed in coordinate-dependent form. (Q.E.D.)

FORCE ASSOCIATED WITH GRADIENT DRIFT

From equations (30.19) and (30.20), we see that every drift velocity $\mathbf{v}_{\rm drift}$ perpendicular to $\mathbf{B}$ may be associated with a force $\mathbf{F}_{\rm drift}$ such that

$$\mathbf{v}_{\rm drift} = \frac{c}{q} \frac{\mathbf{F}_{\rm drift} \times \mathbf{B}}{B^2}. \tag{30.51}$$

If we apply formula (30.51) to equation (30.46), we may identify

$$\mathbf{F}_{\rm grad} = -\frac{mv_\perp^2}{2B} \nabla B = -|\boldsymbol{\mu}| \nabla B, \tag{30.52}$$

where $\boldsymbol{\mu}$ is the magnetic moment associated with the charge's gyromotion [see equation (30.4)]:

$$\boldsymbol{\mu} = -\frac{mv_\perp^2}{2B^2}\mathbf{B}. \tag{30.53}$$

We may give an alternative, more revealing, derivation for equation (30.52). The potential energy of a magnetic dipole $\boldsymbol{\mu}$ placed in an external magnetic field $\mathbf{B}$ reads (see, e.g., Chapter 5 of Jackson, *Classical Electrodynamics*)

$$W = -\boldsymbol{\mu} \cdot \mathbf{B}. \tag{30.54}$$

If $\boldsymbol{\mu}$ is a constant or an adiabatic invariant (see Chapter 28) and lies antiparallel to $\mathbf{B}$, $W = |\boldsymbol{\mu}|B$ will be higher in regions of higher field strength, leading to a corresponding force $-\nabla W = -|\boldsymbol{\mu}|\nabla B$ acting on the dipole that directs the dipole to slide toward regions of lower B.

These comments give a different perspective on the magnetic mirror effect of the Earth's radiation belt discussed in Chapter 28. We now see from another vantage point how the kinetic energy of a charge's motion perpendicular to $\mathbf{B}$, $mv_\perp^2/2 = -\boldsymbol{\mu} \cdot \mathbf{B}$ according to equations (30.53), can act as a potential energy reservoir, W according to equation (30.54), for the motion parallel to $\mathbf{B}$, trapping the latter between two regions of high magnetic field strength B. The little magnetic dipole represented by the gyrating charge always approaches a pole of the big magnetic dipole represented by the Earth with the "opposing ends" of the two dipoles oriented toward to each other. As a consequence, the big dipole repels the little dipole, shuttling it back and forth between the magnetic north and south poles of the Earth.

The most famous application of these ideas lies historically not in classical mechanics, but in quantum mechanics—in the famous experiments carried out by Stern and Gerlach that led to the discovery of electron spin as a physical quantity more concrete than the double-valued theoretical entity that Pauli had constructed to help explain the periodic table. According to the Bohr-Sommerfeld model, silver atoms may carry uncompensated circulating electronic charge and therefore may act as miniature magnetic dipoles. Associated with the electronic angular momentum $\mathbf{L}$ will be a magnetic moment $\mathbf{m} = -(e/2m_e c)\mathbf{L}$ (see Volume 1). The vector model of orbital angular momentum allows $2\ell + 1$ discrete orientations of $\mathbf{L}$ with respect to $\mathbf{B}$, where ℓ equals the integer associated with the quantization of $\mathbf{L}$. Thus, there should be $2\ell + 1$ possible values for $W = -\mathbf{m} \cdot \mathbf{B}$ for a given external magnetic field. If one passes a beam of silver atoms through a region with an intensely inhomogeneous field, the beam should split into $2\ell + 1$ beams, since each orientation of the atomic dipoles should feel a different force $-\nabla W$. The experiments showed the actual beam to split into only *two* beams (Figure 30.6). But $2\ell + 1$ can not equal 2 unless ℓ equals 1/2, which is impossible, since orbital angular momentum comes in

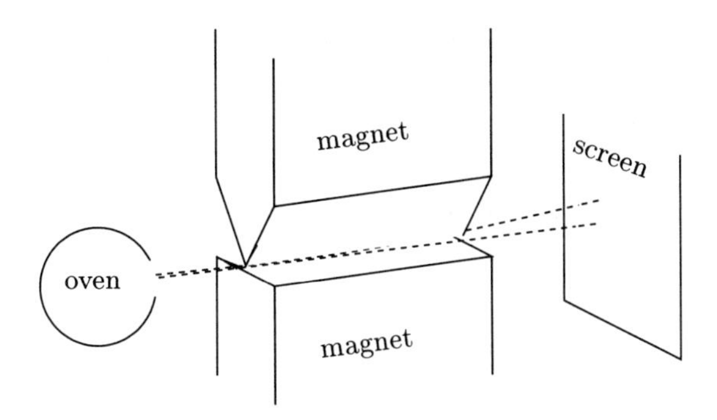

FIGURE 30.6
Schematic representation of the Stern-Gerlach experiment. An oven issues a beam of silver atoms that passes between the poles of a magnet. The knife-edge shape of one of the pole-pieces yields an intensely inhomogeneous magnetic field, and splits the original beam into two beams that impinge on a screen.

whole integer units of $\hbar$, $\ell = 0, 1, 2, \ldots$. Evidently, something in silver atoms is capable of carrying a half-unit of angular momentum (on top of, say, $\ell = 0$). That something we know today as electron spin, with $s = 1/2$ and $2s + 1 = 2$ possible orientations ("parallel" and "antiparallel") with respect to an external field $\mathbf{B}$.

A SIMPLE FUSION APPLICATION

As a last application of the formulas derived in the previous sections, let us consider a problem in magnetic-confinement fusion. Suppose we try to confine a hot plasma inside a simple magnetic torus, one that has no material ends for the plasma to contact (see Figure 30.7). We suppose the externally generated magnetic fields are very strong (to better confine the plasma), so that currents inside the plasma make a negligible contribution to the total field. In this case, we may take the field to be current free, i.e.,

$$\nabla \times \mathbf{B} = 0.$$

Since $\mathbf{B} = B(\varpi, z)\mathbf{e}_\varphi$ for an axisymmetric toroidal field, we have

$$(\nabla B) \times \mathbf{e}_\varphi = -B\nabla \times \mathbf{e}_\varphi = -B\frac{\mathbf{e}_z}{\varpi}.$$

Crossing the above with $\mathbf{e}_\varphi$, we may rewrite the result as

$$\nabla B = -B\frac{\mathbf{e}_\varpi}{\varpi}. \tag{30.55}$$

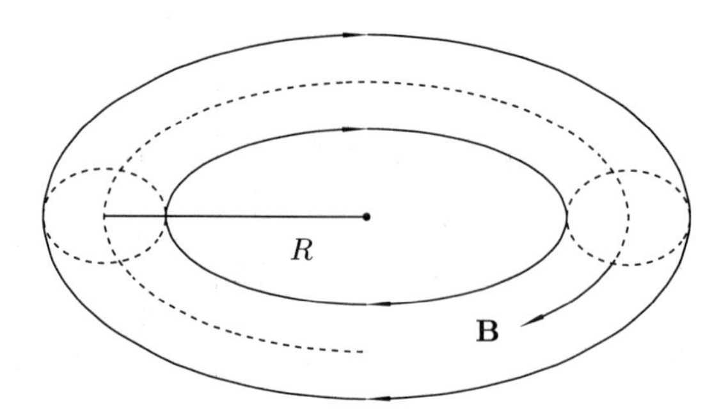

FIGURE 30.7
A naive scheme to confine a hot plasma within a simple toroidal field will not work, because the plasma will escape from the configuration for the reasons described in the text.

Substituting equation (30.55) into equation (30.52), we get

$$\mathbf{F}_{\text{grad}} = \frac{mv_\perp^2}{2} \frac{\mathbf{e}_\varpi}{\varpi}. \tag{30.56}$$

Since the toroidal magnetic-field configuration has both curvature and a gradient in strength, an effective gravitational field acts on a charged particle in the plasma that arises partially from the centrifugal acceleration from the curvature of the field lines and partially from the acceleration associated with the gradient force:

$$\mathbf{g}_{\text{eff}} = \frac{v_\parallel^2}{\varpi}\mathbf{e}_\varpi + \frac{1}{m}\mathbf{F}_{\text{grad}}.$$

Equation (30.56) now implies

$$\mathbf{g}_{\text{eff}} = \left(v_\parallel^2 + \frac{1}{2}v_\perp^2\right)\frac{\mathbf{e}_\varpi}{\varpi}. \tag{30.57}$$

In a thermal plasma, $v_\perp^2/2$ has the same average value as $v_\parallel^2$, since $v_\perp$ contains random motions in two spatial dimensions, and $v_\parallel$, in one. In any case, equation (30.57) shows that the effective gravitational field acting on charged particles of either sign in a plasma that is contained by a simple toroidal magnetic configuration always points radially outward from the central axis. Without any restoring forces, our previous discussion demonstrates that the plasma will fall in the direction of $\mathbf{g}_{\text{eff}}$; i.e., the plasma will fall out of the torus and will not remain confined. Working Tokamaks must have a more complex magnetic structure (an additional poloidal component that wraps around the ring of the torus).

A

Useful Formulae in Curvilinear Coordinates

VECTOR INVARIANT FORMS

$$\frac{D\mathbf{u}}{Dt} = \frac{\partial \mathbf{u}}{\partial t} + \frac{1}{2}\nabla(|\mathbf{u}|^2) + (\nabla \times \mathbf{u}) \times \mathbf{u}.$$

$$\nabla \cdot \overleftrightarrow{\pi} = \nabla(\mathbf{u} \cdot \nabla\mu) - \nabla \times [\nabla \times (\mu\mathbf{u})] - \mathbf{u}\nabla^2\mu + \nabla\mu \times (\nabla \times \mathbf{u})$$

$$+ \frac{4}{3}\nabla(\mu\nabla \cdot \mathbf{u}) - (\nabla \cdot \mathbf{u})\nabla\mu + \nabla(\beta\nabla \cdot \mathbf{u}).$$

$$\Psi = \frac{\mu}{2}|| \overleftrightarrow{D} ||^2 + \beta(\nabla \cdot \mathbf{u})^2$$

$$= \mu\left\{\nabla^2(|\mathbf{u}|^2) + 2\nabla \cdot [(\nabla \times \mathbf{u}) \times \mathbf{u}] - 2\mathbf{u} \cdot \nabla(\nabla \cdot \mathbf{u}) + |\nabla \times \mathbf{u}|^2 - \frac{2}{3}(\nabla \cdot \mathbf{u})^2\right\} + \beta(\nabla \cdot \mathbf{u})^2.$$

COORDINATES (q_1, q_2, q_3) AND METRIC COEFFICIENTS (h_1, h_2, h_3)

$$d\mathbf{s} = h_1\, dq_1\hat{\mathbf{e}}_1 + h_2\, dq_2\hat{\mathbf{e}}_2 + h_3\, dq_3\hat{\mathbf{e}}_3.$$

$$\nabla\mathcal{V} = \frac{1}{h_1}\frac{\partial\mathcal{V}}{\partial q_1}\hat{\mathbf{e}}_1 + \frac{1}{h_2}\frac{\partial\mathcal{V}}{\partial q_2}\hat{\mathbf{e}}_2 + \frac{1}{h_3}\frac{\partial\mathcal{V}}{\partial q_3}\hat{\mathbf{e}}_3.$$

$$\nabla \cdot \mathbf{B} = \frac{1}{h_1h_2h_3}\left[\frac{\partial}{\partial q_1}(h_2h_3B_1) + \frac{\partial}{\partial q_2}(h_3h_1B_2) + \frac{\partial}{\partial q_3}(h_1h_2B_3)\right].$$

$$\nabla \times \mathbf{B} = \frac{1}{h_2h_3}\left[\frac{\partial}{\partial q_2}(h_3B_3) - \frac{\partial}{\partial q_3}(h_2B_2)\right]\hat{\mathbf{e}}_1$$

$$+ \frac{1}{h_3h_1}\left[\frac{\partial}{\partial q_3}(h_1B_1) - \frac{\partial}{\partial q_1}(h_3B_3)\right]\hat{\mathbf{e}}_2$$

$$+ \frac{1}{h_1h_2}\left[\frac{\partial}{\partial q_1}(h_2B_2) - \frac{\partial}{\partial q_2}(h_1B_1)\right]\hat{\mathbf{e}}_3.$$

CYLINDRICAL COORDINATES (ϖ, φ, z): $h_\varpi = 1$, $h_\varphi = \varpi$, $h_z = 1$.

$$\nabla P = \frac{\partial P}{\partial \varpi}\hat{\mathbf{e}}_\varpi + \frac{1}{\varpi}\frac{\partial P}{\partial \varphi}\hat{\mathbf{e}}_\varphi + \frac{\partial P}{\partial z}\hat{\mathbf{e}}_z.$$

$$\nabla \cdot \mathbf{u} = \frac{1}{\varpi}\frac{\partial}{\partial \varpi}(\varpi u_\varpi) + \frac{1}{\varpi}\frac{\partial u_\varphi}{\partial \varphi} + \frac{\partial u_z}{\partial z}.$$

$$\nabla \times \mathbf{u} = \left(\frac{1}{\varpi}\frac{\partial u_z}{\partial \varphi} - \frac{\partial u_\varphi}{\partial z}\right)\hat{\mathbf{e}}_\varpi + \left(\frac{\partial u_\varpi}{\partial z} - \frac{\partial u_z}{\partial \varpi}\right)\hat{\mathbf{e}}_\varphi + \frac{1}{\varpi}\left[\frac{\partial}{\partial \varpi}(\varpi u_\varphi) - \frac{\partial u_\varpi}{\partial \varphi}\right]\hat{\mathbf{e}}_z.$$

$$\frac{D\mathbf{u}}{Dt} = \left(\frac{\partial u_\varpi}{\partial t} + u_\varpi\frac{\partial u_\varpi}{\partial \varpi} + \frac{u_\varphi}{\varpi}\frac{\partial u_\varpi}{\partial \varphi} + u_z\frac{\partial u_\varpi}{\partial z} - \frac{u_\varphi^2}{\varpi}\right)\hat{\mathbf{e}}_\varpi + \left(\frac{\partial u_\varphi}{\partial t} + u_\varpi\frac{\partial u_\varphi}{\partial \varpi} + \frac{u_\varphi}{\varpi}\frac{\partial u_\varphi}{\partial \varphi} + u_z\frac{\partial u_\varphi}{\partial z} + \frac{1}{\varpi}u_\varpi u_\varphi\right)\hat{\mathbf{e}}_\varphi + \left(\frac{\partial u_z}{\partial t} + u_\varpi\frac{\partial u_z}{\partial \varpi} + \frac{u_\varphi}{\varpi}\frac{\partial u_z}{\partial \varphi} + u_z\frac{\partial u_z}{\partial z}\right)\hat{\mathbf{e}}_z.$$

$$\nabla\cdot \overleftrightarrow{\pi} = \left[\frac{1}{\varpi}\frac{\partial}{\partial \varpi}(\varpi \pi_{\varpi\varpi}) + \frac{1}{\varpi}\frac{\partial \pi_{\varpi\varphi}}{\partial \varphi} + \frac{\partial \pi_{z\varpi}}{\partial z} - \frac{1}{\varpi}\pi_{\varphi\varphi}\right]\hat{\mathbf{e}}_\varpi + \left[\frac{1}{\varpi}\frac{\partial}{\partial \varpi}(\varpi \pi_{\varpi\varphi}) + \frac{1}{\varpi}\frac{\partial \pi_{\varphi\varphi}}{\partial \varphi} + \frac{\partial \pi_{\varphi z}}{\partial z} + \frac{1}{\varpi}\pi_{\varpi\varphi}\right]\hat{\mathbf{e}}_\varphi + \left[\frac{1}{\varpi}\frac{\partial}{\partial \varpi}(\varpi \pi_{z\varpi}) + \frac{1}{\varpi}\frac{\partial \pi_{\varphi z}}{\partial \varphi} + \frac{\partial \pi_{zz}}{\partial z}\right]\hat{\mathbf{e}}_z.$$

$$D_{\varpi\varpi} = 2\frac{\partial u_\varpi}{\partial \varpi}.$$

$$D_{\varphi\varphi} = 2\left(\frac{1}{\varpi}\frac{\partial u_\varphi}{\partial \varphi} + \frac{u_\varpi}{\varpi}\right).$$

$$D_{zz} = 2\frac{\partial u_z}{\partial z}.$$

$$D_{\varpi\varphi} = \frac{\partial u_\varphi}{\partial \varpi} + \frac{1}{\varpi}\frac{\partial u_\varpi}{\partial \varphi} - \frac{u_\varphi}{\varpi}.$$

$$D_{z\varpi} = \frac{\partial u_z}{\partial \varpi} + \frac{\partial u_\varpi}{\partial z}.$$

$$D_{\varphi z} = \frac{1}{\varpi}\frac{\partial u_z}{\partial \varphi} + \frac{\partial u_\varphi}{\partial z}.$$

SPHERICAL COORDINATES (r, θ, φ): $h_r = 1$, $h_\theta = r$, $h_\varphi = r \sin\theta$.

$$\nabla P = \frac{\partial P}{\partial r}\hat{\mathbf{e}}_r + \frac{1}{r}\frac{\partial P}{\partial \theta}\hat{\mathbf{e}}_\theta + \frac{1}{r\sin\theta}\frac{\partial P}{\partial \varphi}\hat{\mathbf{e}}_\varphi.$$

$$\nabla \cdot \mathbf{u} = \frac{1}{r^2}\frac{\partial}{\partial r}(r^2 u_r) + \frac{1}{r\sin\theta}\frac{\partial}{\partial \theta}(u_\theta \sin\theta) + \frac{1}{r\sin\theta}\frac{\partial u_\varphi}{\partial \varphi}.$$

$$\nabla \times \mathbf{u} = \frac{1}{r\sin\theta}\left[\frac{\partial}{\partial \theta}(u_\varphi \sin\theta) - \frac{\partial u_\theta}{\partial \varphi}\right]\hat{\mathbf{e}}_r + \left[\frac{1}{r\sin\theta}\frac{\partial u_r}{\partial \varphi} - \frac{1}{r}\frac{\partial}{\partial r}(r u_\varphi)\right]\hat{\mathbf{e}}_\theta$$
$$+ \frac{1}{r}\left[\frac{\partial}{\partial r}(r u_\theta) - \frac{\partial u_r}{\partial \theta}\right]\hat{\mathbf{e}}_\varphi.$$

$$\frac{D\mathbf{u}}{Dt} = \left[\frac{\partial u_r}{\partial t} + u_r\frac{\partial u_r}{\partial r} + \frac{u_\theta}{r}\frac{\partial u_r}{\partial \theta} + \frac{u_\varphi}{r\sin\theta}\frac{\partial u_r}{\partial \varphi} - \frac{1}{r}(u_\theta^2 + u_\varphi^2)\right]\hat{\mathbf{e}}_r$$
$$+ \left[\frac{\partial u_\theta}{\partial t} + u_r\frac{\partial u_\theta}{\partial r} + \frac{u_\theta}{r}\frac{\partial u_\theta}{\partial \theta} + \frac{u_\varphi}{r\sin\theta}\frac{\partial u_\theta}{\partial \varphi}\right.$$
$$\left. + \frac{1}{r}u_r u_\theta - \frac{u_\varphi^2}{r}\cot\theta\right]\hat{\mathbf{e}}_\theta$$
$$+ \left[\frac{\partial u_\varphi}{\partial t} + u_r\frac{\partial u_\varphi}{\partial r} + \frac{u_\theta}{r}\frac{\partial u_\varphi}{\partial \theta} + \frac{u_\varphi}{r\sin\theta}\frac{\partial u_\varphi}{\partial \varphi}\right.$$
$$\left. + \frac{1}{r}(u_r u_\varphi + u_\theta u_\varphi \cot\theta)\right]\hat{\mathbf{e}}_\varphi.$$

$$\nabla \cdot \overleftrightarrow{\pi} = \left[\frac{1}{r^2}\frac{\partial}{\partial r}(r^2 \pi_{rr}) + \frac{1}{r\sin\theta}\frac{\partial}{\partial \theta}(\pi_{r\theta}\sin\theta)\right.$$
$$\left. + \frac{1}{r\sin\theta}\frac{\partial \pi_{\varphi r}}{\partial \varphi} - \frac{1}{r}(\pi_{\theta\theta} + \pi_{\varphi\varphi})\right]\hat{\mathbf{e}}_r$$
$$+ \left[\frac{1}{r^2}\frac{\partial}{\partial r}(r^2 \pi_{r\theta}) + \frac{1}{r\sin\theta}\frac{\partial}{\partial \theta}(\pi_{\theta\theta}\sin\theta) + \frac{1}{r\sin\theta}\frac{\partial \pi_{\theta\varphi}}{\partial \varphi} + \frac{1}{r}(\pi_{r\theta}\right.$$
$$\left. - \pi_{\varphi\varphi}\cot\theta)\right]\hat{\mathbf{e}}_\theta$$
$$+ \left[\frac{1}{r^2}\frac{\partial}{\partial r}(r^2 \pi_{\varphi r}) + \frac{1}{r}\frac{\partial}{\partial \theta}(\pi_{\theta\varphi}\sin\theta)\right.$$
$$\left. + \frac{1}{r\sin\theta}\frac{\partial \pi_{\varphi\varphi}}{\partial \varphi} + \frac{1}{r}(\pi_{\varphi r} + \pi_{\theta\varphi})\right]\hat{\mathbf{e}}_\varphi.$$

$$D_{rr} = 2\frac{\partial u_r}{\partial r}.$$

$$D_{\theta\theta} = 2\left(\frac{1}{r}\frac{\partial u_\theta}{\partial \theta} + \frac{u_r}{r}\right).$$

$$D_{\varphi\varphi} = 2\left(\frac{1}{r\sin\theta}\frac{\partial u_\varphi}{\partial \varphi} + + \frac{u_r}{r} + \frac{u_\theta}{r}\cot\theta\right).$$

$$D_{r\theta} = \frac{\partial u_\theta}{\partial r} + \frac{1}{r}\frac{\partial u_r}{\partial \theta} - \frac{u_\theta}{r}.$$

$$D_{\varphi r} = \frac{\partial u_\varphi}{\partial r} + \frac{1}{r\sin\theta}\frac{\partial u_r}{\partial \varphi} - \frac{u_\varphi}{r}.$$

$$D_{\theta\varphi} = \frac{1}{r}\frac{\partial u_\varphi}{\partial \theta} + \frac{1}{r\sin\theta}\frac{\partial u_\theta}{\partial \varphi} - \frac{u_\varphi}{r}\cot\theta.$$

PROBLEM SETS

Problem Set 1

1. Define the total matter entropy S as the integral over the entire spatial volume V of the system

$$S = \int_V \rho s \, d^3x,$$

where ρs is the Boltzmann entropy per unit volume,

$$\rho s \equiv -k \int f \ln f \, d^3v.$$

(a) Suppose f to satisfy the Boltzmann equation and that $f(\mathbf{x}, -\mathbf{w}, t) = f(\mathbf{x}, \mathbf{w}, t)$ on the surface A (if any) that bounds V [i.e., particles bounce elastically off any walls (possibly moving) so that the walls neither add nor subtract heat from the gas]. Prove the second law of thermodynamics, i.e., that $dS/dt \geq 0$, with equality reached only when f is an exact Maxwellian.

Hint: Consult any book on kinetic theory or statistical mechanics; however, note that most discussions avoid the (interesting) consideration of possibly moving walls. Show that when walls with area $A(t)$ enclosing volume $V(t)$ move with local velocity $\mathbf{u}_{\text{wall}}$, the time rate of change of entropy can be calculated as

$$\frac{dS}{dt} = \int_{V(t)} \frac{\partial}{\partial t}(\rho s) \, d^3x + \oint_{A(t)} \rho s \mathbf{u}_{\text{wall}} \cdot \hat{\mathbf{n}} dA.$$

Assume that the gas in the neighborhood of the wall moves with a *mean velocity* equal to the wall velocity (although individual molecules may possess *random velocities* $\mathbf{w}$ with respect to the wall), so that we may make the substitution $\mathbf{u} = \mathbf{u}_{\text{wall}}$. For later application in part (c) below, apply now the divergence theorem to obtain

$$\frac{dS}{dt} = \int_V \left[\frac{\partial}{\partial t}(\rho s) + \nabla \cdot (\rho s \mathbf{u}) \right] d^3x.$$

(b) For f equal to a Maxwellian, show that, up to an additive constant, s defined above in terms of the integral of $f \ln f$ equals the conventional expression given by thermodynamics or statistical mechanics for the specific entropy of a classical perfect gas,

$$s = c_v \ln(P\rho^{-\gamma}),$$

where $c_v = 3k/2m$ and $\gamma = 5/3$.

(c) Suppose s to satisfy equation (3.26) of the text. Divide by T and integrate this expression over V, and use the equation of continuity to show

$$\frac{dS}{dt} = \int_V \left(-\frac{1}{T} \nabla \cdot \mathbf{F}_{\text{cond}} + \frac{\Psi}{T} \right) d^3x,$$

where S is the integral of ρs over V and Ψ is the viscous dissipation rate: $\Psi = (\mu/2) || \overleftrightarrow{D} ||^2$. With $\mathbf{F}_{\text{cond}}$ given by Fourier's law, $\mathbf{F}_{\text{cond}} = -\mathcal{K}\nabla T$, perform an integration by parts of the heat conduction term, assuming $\mathbf{F}_{\text{cond}}$ to vanish on the bounding surface A [explain the relationship to the boundary condition of part (a)], and show that

$$\frac{dS}{dt} = \int_V \left(\frac{\mathcal{K}}{T^2} |\nabla T|^2 + \frac{\mu}{2T} || \overleftrightarrow{D} ||^2 \right) d^3x,$$

which is positive definite if $\mathcal{K}$ and $\mu > 0$. In order for dS/dt to equal 0, ∇T and $\overleftrightarrow{D}$ must both vanish. Notice that these requirements hold independently of the nature of any external force fields. What do these conditions mean physically for the state of complete thermodynamic equilibrium? In particular, does the uniform expansion of the universe change the matter entropy content?

2. Adopt the Krook equation as governing the evolution of f:

$$\mathcal{L}f = -\nu_{\text{c}}(f - f_0),$$

where f_0 is the *local* Maxwell-Boltzmann distribution [cf. equation (3.10) of the text]. Assume a series expansion of the form given by equation (3.6) of the text and show that f_1 has the solution

$$f_1 = -\nu_{\text{c}}^{-1} \mathcal{L} f_0.$$

(a) Show, to this order of approximation,

$$\begin{pmatrix} \pi_{ik} \\ F_i \end{pmatrix} = -\nu_{\text{c}}^{-1} \int \begin{bmatrix} m \left(\frac{1}{3} |\mathbf{w}|^2 \delta_{ik} - w_i w_k \right) \\ \frac{m}{2} |\mathbf{w}|^2 w_i \end{bmatrix} \mathcal{L} f_0 \, d^3w.$$

In the transformation $(\mathbf{x}, \mathbf{v}, t) \to (\mathbf{x}, \mathbf{w}, t)$, where $\mathbf{w} \equiv \mathbf{v} - \mathbf{u}(\mathbf{x}, t)$, show that the partial derivatives transform as

$$\left(\frac{\partial}{\partial t} \right)_{\mathbf{x},\mathbf{v}} = \left(\frac{\partial}{\partial t} \right)_{\mathbf{x},\mathbf{w}} - \frac{\partial u_p}{\partial t} \left(\frac{\partial}{\partial w_p} \right)_{\mathbf{x},t},$$

$$\left(\frac{\partial}{\partial x_s} \right)_{\mathbf{v},t} = \left(\frac{\partial}{\partial x_s} \right)_{\mathbf{w},t} - \frac{\partial u_p}{\partial x_s} \left(\frac{\partial}{\partial w_p} \right)_{\mathbf{x},t},$$

$$\left(\frac{\partial}{\partial v_s}\right)_{\mathbf{x},t} = \left(\frac{\partial}{\partial w_s}\right)_{\mathbf{x},t}.$$

Thus, demonstrate that $\mathcal{L}f_0$ becomes

$$\mathcal{L}f_0 = \frac{\partial f_0}{\partial t} + (u_s + w_s)\frac{\partial f_0}{\partial x_s} - \left[\frac{\partial \mathcal{V}}{\partial x_p} + \frac{\partial u_p}{\partial t} + (u_s + w_s)\frac{\partial u_p}{\partial x_s}\right]\frac{\partial f_0}{\partial w_p}.$$

(b) Since f_0 is an isotropic function of $|\mathbf{w}|^2$, show that the only nonzero contribution in $\mathcal{L}f_0$ to π_{ik} comes from the term

$$-w_s \frac{\partial u_p}{\partial x_s}\frac{\partial f_0}{\partial w_p}.$$

Perform now an integration by parts to show

$$\pi_{ik} = m\nu_c^{-1}\frac{\partial u_p}{\partial x_s}\int\left[\delta_{ip}w_k w_s + \delta_{kp}w_i w_s + \delta_{sp}w_i w_k - \frac{\delta_{ik}}{3}(\delta_{sp}|\mathbf{w}|^2 + 2w_p w_s)\right] f_0\, d^3w.$$

Argue for the property,

$$\int m w_i w_k f_0 d^3w = \delta_{ik}\int \frac{m}{3}|\mathbf{w}|^2 f_0\, d^3w = nkT\delta_{ik},$$

to derive

$$\pi_{ik} = \nu_c^{-1}nkT\left(\frac{\partial u_i}{\partial x_k} + \frac{\partial u_k}{\partial x_i} - \frac{2}{3}\frac{\partial u_p}{\partial x_p}\delta_{ik}\right),$$

which is the first relation (3.19) of the text if we identify

$$\mu = \frac{nkT}{\nu_c}.$$

Notice that $\mu \sim mv_T/\sigma$ if we use $\nu_c \sim n\sigma v_T$ with $v_T \equiv (kT/m)^{1/2}$.

(b) In the integration for F_i, argue that, to the order we are working, it suffices to use the force equation in the zeroth-order form to eliminate

$$\frac{\partial u_p}{\partial t} + u_s\frac{\partial u_p}{\partial x_s} + \frac{\partial \mathcal{V}}{\partial x_p}$$

in $\mathcal{L}f_0$ as $-\rho^{-1}\partial P/\partial x_p$. With this replacement, show that the only terms in $\mathcal{L}f_0$ that contribute to f_0 read

$$w_s\frac{\partial f_0}{\partial x_s} + \rho^{-1}\frac{\partial P}{\partial x_p}\frac{\partial f_0}{\partial w_p}.$$

Use the considerations of part (b) to demonstrate

$$F_i = \nu_c^{-1}\left[\frac{5kT}{2m}\frac{\partial P}{\partial x_i} - \frac{\partial}{\partial x_i}\left(n\frac{m}{6}\langle|\mathbf{w}|^4\rangle\right)\right].$$

For a Maxwellian, $\langle|\mathbf{w}|^4\rangle = 15(kT/m)^2$, but $P = nkT$; so

$$F_i = -\left(\frac{nkT}{\nu_c}\right)\left(\frac{5k}{2m}\right)\frac{\partial T}{\partial x_i},$$

which is the second relation of equation (3.19) of the text if we identify

$$\mathcal{K} = \left(\frac{nkT}{\nu_c}\right)\left(\frac{5k}{2m}\right).$$

Notice that this procedure produces a factor $5c_v/3$ for Eucken's constant instead of the correct (Boltzmann) value $5c_v/2$. The discrepancy arises because a Boltzmann treatment yields unequal values for the effective collision frequencies associated with given amounts of momentum and energy transfer through collisions, while the Krook treatment assumes a unique value for ν_c.

Problem Set 2

1. For a completely degenerate gas of spin $s = 1/2$, the number density n is given by

$$n = \frac{(2s+1)}{h^3} \int_0^{p_0} 4\pi\, p^2\, dp = \frac{(2s+1)}{h^3} \frac{4\pi}{3} p_0^3,$$

where p_0 is the momentum corresponding to the Fermi energy $\epsilon = \mu$, with μ being the chemical potential. In other words, μ and p_0 satisfy

$$\mu = (p_0^2 c^2 + m^2 c^4)^{1/2} - mc^2 \qquad \text{where} \qquad p_0 = h \left[\frac{3n}{(2s+1)4\pi} \right]^{1/3},$$

with m being the rest mass of the gas particle. Using whatever statistical mechanics you know, demonstrate in the same notation that the pressure of this gas equals

$$\begin{aligned} P &= \frac{(2s+1)}{h^3} \frac{4\pi c^2}{3} \int_0^{p_0} \frac{p^4\, dp}{(p^2c^2 + m^2c^4)^{1/2}} \\ &= \frac{(2s+1)}{h^3} \frac{4\pi c}{3} (mc)^4 \int_0^{\theta_0} \sinh^4\theta\, d\theta, \end{aligned}$$

where we have defined θ so that $\sinh\theta \equiv p/mc$. Show that we may write

$$\begin{aligned} \sinh^4\theta &= \sinh^2\theta(\cosh^2\theta - 1) = \frac{1}{4}\sinh^2 2\theta - \frac{1}{2}(\cosh 2\theta - 1) \\ &= \frac{1}{8}(\cosh 4\theta - 1) - \frac{1}{2}(\cosh 2\theta - 1). \end{aligned}$$

Perform now the integration for P; define $x \equiv p_0/mc$; and show that a completely degenerate gas has the parametric representation

$$n = (2s+1)\frac{4\pi}{3}\left(\frac{mc}{h}\right)^3 x^3, \qquad P = (2s+1)\frac{\pi}{6}\left(\frac{mc}{h}\right)^3 mc^2 f(x),$$

where

$$f(x) \equiv x(x^2+1)^{1/2}(2x^2 - 3) + 3\,\text{arcsinh}\, x.$$

Demonstrate that $f(x)$ has the limiting (nonrelativistic and ultrarelativistic) behaviors

$$f(x) \approx 8x^5/5 \quad \text{for} \quad x \ll 1, \qquad f(x) \approx 2x^4 \quad \text{for} \quad x \gg 1.$$

2. Apply the result of problem 1 to a completely ionized gas in which the electron component is completely degenerate. For a completely ionized gas, each ion of charge Z and atomic weight A_Z contributes Z free electrons and a mass of $A_Z m_H$, where m_H is the mass of the hydrogen atom.

(a) If X_Z represents the mass fraction of the species, show that the electron number density n_e is related to the mass density ρ of the fluid by

$$n_e = \frac{\rho}{m_H} \sum_Z \frac{X_Z Z}{A_Z} \equiv \frac{\rho}{\mu_e m_H},$$

where the last relation defines μ_e, the mean molecular weight per electron (in atomic mass units), and is not to be confused with the electron chemical potential. Show that μ_e equals 2 if $Z/A_Z = 1/2$ (most of the elements in the periodic table likely to make up a white dwarf) for all nonzero X_Z whose sum equals 1 (since the X_Z's are mass *fractions*):

$$\sum_Z X_Z = 1.$$

With $s = 1/2$, write now the P-ρ relation in the parametric form:

$$\rho = \rho_0 x^3, \qquad P = P_0 f(x),$$

where $f(x)$ is given by problem 1 and

$$\rho_0 \equiv \frac{8\pi}{3}\left(\frac{m_e c}{h}\right)^3 \mu_e m_H \qquad \text{and} \qquad P_0 \equiv \frac{\pi}{3}\left(\frac{m_e c}{h}\right)^3 m_e c^2$$

are constants.

(b) The condition of hydrostatic equilibrium reads

$$\frac{dP}{dM} = -\frac{GM}{4\pi r^4} \qquad \text{with} \qquad \frac{dr}{dM} = \frac{1}{4\pi r^2 \rho}.$$

Nondimensionalize by defining

$$r \equiv r_0 \xi \qquad \text{and} \qquad M = M_0 \mathcal{M},$$

where

$$r_0 \equiv (P_0/\rho_0)^{1/2}(G\rho_0)^{-1/2} \qquad \text{and} \qquad M_0 \equiv \rho_0 r_0^3.$$

Compute r_0 and M_0 numerically for $\mu_e = 2$. Show that the nondimensional equations read

$$\frac{df}{d\mathcal{M}} = -\frac{\mathcal{M}}{4\pi\xi^4} \qquad \text{with} \qquad \frac{d\xi}{d\mathcal{M}} = \frac{1}{4\pi\xi^2 x^3},$$

where $f(x)$ is the dimensionless pressure and x^3 is the dimensionless density. But $df/d\mathcal{M} = 8x^4(x^2+1)^{-1/2}dx/d\mathcal{M}$; thus obtain the pair of ODEs

$$\frac{dx}{d\mathcal{M}} = -\frac{\mathcal{M}(x^2+1)^{1/2}}{32\pi x^4\xi^4}, \qquad \frac{d\xi}{d\mathcal{M}} = \frac{1}{4\pi\xi^2 x^3}.$$

(c) Integrate the above equations for $dx/d\mathcal{M}$ and $d\xi/d\mathcal{M}$ subject to the two-point boundary conditions,

$$\xi = 0 \quad \text{at} \quad \mathcal{M} = 0 \quad \text{and} \quad x = 0 \quad \text{at} \quad \mathcal{M} = \mathcal{M}_*.$$

The easiest way to do this in the present context, when we have complete freedom to choose the nondimensional mass $\mathcal{M}_*$, is to guess a starting value for x at $\mathcal{M} = 0$ (i.e., guess the central density), and integrate outward (toward larger values of $\mathcal{M}$) by a Runge-Kutta technique (or some other) until x equals 0, which then defines $\mathcal{M}_*$. Continue to do this until you have spanned the range from very small to very large central values for x. In particular, extrapolate your results to central values of $1/x \to 0$ to obtain the Chandrasekhar limit.

Hint: The governing ODEs as written have singularities as one approaches the surface, where $x \to 0$. On the principle that it is better to multiply by zero rather than divide by it, you may wish to integrate the equations in the form:

$$\frac{d\mathcal{M}}{dx} = -\frac{32\pi x^4\xi^4}{\mathcal{M}(x^2+1)^{1/2}}, \qquad \frac{d\xi}{dx} = -\frac{8x\xi^2}{\mathcal{M}(x^2+1)^{1/2}}.$$

Notice that this transformation does not eliminate the need to begin the integrations by stepping slightly off the stellar center, where $\mathcal{M} \approx 4\pi x^3\xi^3/3 \to 0$ as $\xi \to 0$ for a finite central value of the dimensionless density x^3. Also notice that it is not a good idea to integrate with a fixed step size when one has a large dynamic range. Make sure each step does not change $\mathcal{M}$ and ξ by more than a small fraction. (Graphs look smooth when the errors are less than ~ 1 percent.)

Plot up your results for the resulting $R_* \equiv r_0\xi_*$ and $M_* \equiv M_0\mathcal{M}_*$. Use solar units ($R_\odot$ and $M_\odot$) for a log-log plot, and cover the range from $10^{-3}\,M_\odot$ to $1.44\,M_\odot$. Indicate the central values of the Fermi energy (in eV) for various fiducial stellar masses, noting especially the expected transition values for giant planets and neutron stars.

3. Convince yourself of the correctness of the dimensionless equations (6.38) –(6.41) in the text for the Bondi problem:

$$x^2\alpha v = \lambda,$$

$$\frac{v^2}{2} + H(\alpha) - \frac{1}{x} = 0,$$

where

$$H(\alpha) = \ln \alpha \qquad \text{for } \gamma = 1, \qquad H(\alpha) = \frac{\gamma}{\gamma - 1}(\alpha^{\gamma-1} - 1) \qquad \text{for } \gamma \neq 1.$$

(a) To start, consider the isothermal case, $\gamma = 1$. To obtain the condition for a sonic transition, show that the differentials of the relevant equations satisfy

$$2\frac{dx}{x} + \frac{d\alpha}{\alpha} + \frac{dv}{v} = 0,$$
$$v\,dv + \frac{d\alpha}{\alpha} + \frac{dx}{x^2} = 0.$$

If we subtract the above two equations to eliminate $d\alpha/\alpha$, we obtain

$$\left(v - \frac{1}{v}\right) dv = \left(\frac{2}{x} - \frac{1}{x^2}\right) dx.$$

A sonic transition occurs when $v = 1$; for this to occur without dv/dx becoming infinite, we want $x = 1/2$. Show that this combination requires λ to have the critical value

$$\lambda_{\rm c} = \frac{1}{4}e^{3/2} = 1.1204223.$$

Demonstrate that curves with $\lambda > \lambda_{\rm c}$ have $dv/dx = \infty$ at $v = 1$, whereas curves with $\lambda < \lambda_{\rm c}$ have $dv/dx = 0$ at $x = 1/2$. Plot up a few cases to show that they have the forms drawn schematically in Figure 6.10.

(b) Consider the general case. Show that

$$[v^2 - \alpha H'(\alpha)]\frac{dv}{v} = \left[2\alpha H'(\alpha) - \frac{1}{x}\right]\frac{dx}{x},$$

and hence that the sonic transition

$$v^2 = \alpha H'(\alpha)$$

occurs at $1/x = 2\alpha H'(\alpha)$. Thus, at the sonic point of the critical solution, show that α must satisfy

$$H(\alpha) - \frac{3}{2}\alpha H'(\alpha) = 0,$$

i.e.,

$$\alpha = e^{3/2} \quad \text{for} \quad \gamma = 1, \qquad \text{and} \qquad \alpha^{\gamma-1} = \frac{2}{5 - 3\gamma} \quad \text{for} \quad 1 \neq \gamma \leq 5/3.$$

(c) Show that the sonic transition is reached at the origin for $\gamma = 5/3$. Explain physically the impossibility of steady supersonic accretion when $\gamma > 5/3$, i.e., when we have too "stiff" a pressure response.

Hint: For steady supersonic flow, inertia must be capable of dominating the pressure buildup; i.e., we want $v^2/2 \gg H(\alpha)$ so that $v \to (2/x)^{1/2}$ (free-fall) as $x \to 0$. Is such a situation possible when $\gamma > 5/3$?

How do the pressure and gravitational forces scale? Can you think of a scenario in which $\gamma > 5/3$ might simulate a realistic situation?

Hint: What about accretion onto an energetic X-ray source that heats the matter faster than it can radiatively cool? Would such a (monatomic) gas compress more stiffly than adiabatically?

(d) Compute the numerical values of the dimensionless accretion rate for the curves with a sonic transition needed to get Table 2.1: In particular,

Table 2.1

γ	1	1.2	7/5	1.5	5/3
$\gamma^{3/2}\lambda_c$	1.12	0.871	0.625	0.500	0.250
λ_c	1.12	0.663	0.377	0.272	0.116

show that λ_c can be computed for $\gamma = 5/3$ despite the fact that the sonic point is formally reached at $x = 0$.

(e) Suppose a star has speed V with respect to the interstellar medium. Argue dimensionally for the Hoyle-Lyttleton formula,

$$\dot{M} = \lambda_* 4\pi\rho_\infty \frac{(GM)^2}{(V^2 + c_\infty^2)^{3/2}},$$

where λ_* is a numerical coefficient of order unity. A drawing to indicate your physical ideas would help. The mean density in the interstellar medium (say, in H I regions, where $T \sim 100\,\mathrm{K}$) equals $\rho_* \sim 2 \times 10^{-24}\,\mathrm{g\,cm^{-3}}$, but the typical r.m.s. stellar velocity dispersion equals $V \sim 30\,\mathrm{km\,s^{-1}}$. Estimate $\dot{M}$ for $M = 1\ M_\odot$. Convert your answer to $M_\odot\,\mathrm{yr}^{-1}$. Is accretion from H I clouds by stars an important process? In a molecular cloud core (mostly H_2), the density ρ_∞ might be 10^5 times larger; the temperature T, 10 times cooler ($\sim 10\,\mathrm{K}$). Is accretion inside a molecular cloud core an important process if the star has a random velocity $V \sim 30\,\mathrm{km\ s^{-1}}$ with respect to this core?

Hint: How long does it take for the star to traverse the core if the core has a diameter of 0.1 pc [for a small core] or 1 pc [for a large core]?

What is $\dot{M}$ if $V = 0$ (as in the original Bondi problem)? Under what circumstances might a star have *zero* random velocity with respect to a molecular cloud core?

Hint: Consider the process of star formation from a molecular cloud core. In this case, do you think it is correct to ignore the self-gravity of the gas? What about time dependence?

(f) In the isothermal case, discuss the relevance to Parker's wind problem of the branch of the solutions which starts off with small velocities at small x, accelerates through a sonic point at $x = 1/2$, and acquires large velocities as $x \to \infty$. In particular, given that conditions at infinity are not specified *a priori* for an *outflow* problem, comment on the need to interpret equation (6.37) of the text as defining the density at the sonic point $\rho_s = e^{3/2}\rho_\infty$ in terms of a given $\dot{M}$ (when $\lambda = \lambda_c$), where $\dot{M}$ now corresponds to an observationally measured *mass-loss* rate. (Physically, of course, the Sun has a certain mass-loss rate $\dot{M}$ because heating mechanisms are able to maintain coronal temperatures in regions that contain a sonic point with density ρ_s.) With this renormalization, the symbols c_∞ and $e^{3/2}\rho_\infty$ no longer mean the sound speed and ($e^{3/2}$ times) the density at infinity, but relate to their values at the sonic point. Compute numerically the sonic radius $r_s = GM/2c_\infty^2$ and the sonic density if the Sun possesses a completely ionized hydrogen corona at a temperature of 2×10^6 K and an average mass-loss rate $\dot{M} = 2 \times 10^{-14}\ M_\odot\ \mathrm{yr}^{-1}$. Do the answers appear reasonable to you?

Hint: What is the visual appearance of the static corona (roughly, the part inside the sonic point, not counting coronal loops, where the densities, emissivities, and number of scattering electrons are relatively high)? Moreover, can you extrapolate the mean proton density of $\sim 5\ \mathrm{cm}^{-3}$ measured for the solar wind at the Earth back to your computed sonic point?

Argue also that the dimensionless density $\alpha \to 0$ as $x \to \infty$, so that the dimensionless wind speed $v \to [\ln(\alpha^{-2})]^{1/2}$ diverges as $x \to \infty$. The terminal velocity of the wind becoming infinite rather than reaching a finite value represents an artifact of the isothermal equation of state, which provides an infinite thermal reservoir for accelerating an expansional flow. Parker (1963) discusses this peculiarity, along with others, that arises when one does not take into account the detailed physical mechanisms that heat and cool the solar wind.

Problem Set 3

1. In cylindrical coordinates (ϖ, φ, z), where ϖ is the distance from the rotation axis in the z direction, we suppose a self-gravitating cylinder of ideal gas to rotate with uniform angular velocity $\Omega \hat{\mathbf{e}}_z$. The equations of motion in the rotating frame of reference need to include the centrifugal and Coriolis forces per unit mass, $\varpi \Omega^2 \hat{\mathbf{e}}_\varpi$ and $-2\Omega \hat{\mathbf{e}}_z \times \mathbf{u}$; thus, Euler's equations become

$$\frac{\partial \rho}{\partial t} + \nabla \cdot (\rho \mathbf{u}) = 0, \tag{1}$$

$$\frac{\partial \mathbf{u}}{\partial t} + \frac{1}{2}\nabla(|\mathbf{u}|^2) + (\nabla \times \mathbf{u}) \times \mathbf{u} = -\frac{1}{\rho}\nabla P - \nabla \mathcal{V} + \varpi \Omega^2 \hat{\mathbf{e}}_\varpi - 2\Omega \hat{\mathbf{e}}_z \times \mathbf{u}, \tag{2}$$

$$\frac{\partial P}{\partial t} + \mathbf{u} \cdot \nabla P = \frac{\gamma P}{\rho}\left(\frac{\partial \rho}{\partial t} + \mathbf{u} \cdot \nabla \rho\right), \tag{3}$$

where $\mathcal{V}$ satisfies Poisson's equation,

$$\nabla^2 \mathcal{V} = 4\pi G \rho. \tag{4}$$

(a) Assume $\rho = \rho_0 =$ constant, $P = P_0 =$ constant, $\mathbf{u} = 0$, represent the state of equilibrium. Show that this state satisfies

$$\mathcal{V}_0 = \frac{1}{2}\Omega^2 \varpi^2 + \text{constant}, \qquad \text{where} \qquad \Omega^2 = 2\pi G \rho_0. \tag{5}$$

Consider now spatially and temporally dependent small perturbations about this uniformly rotating homogeneous state,

$$\rho = \rho_0 + \rho_1, \qquad \mathbf{u} = \mathbf{0} + \mathbf{u}_1, \qquad P = P_0 + P_1, \qquad \mathcal{V} = \mathcal{V}_0 + \mathcal{V}_1, \tag{6}$$

and show that the substitution of equations (6) into equations (1)–(4) yields the linearized perturbation set,

$$\frac{\partial \rho_1}{\partial t} + \rho_0 \nabla \cdot \mathbf{u}_1 = 0, \tag{7}$$

$$\frac{\partial \mathbf{u}_1}{\partial t} = -\frac{1}{\rho_0}\nabla P_1 - \nabla \mathcal{V}_1 - 2\Omega \hat{\mathbf{e}}_z \times \mathbf{u}_1, \tag{8}$$

$$\frac{\partial P_1}{\partial t} = a_s^2 \frac{\partial \rho_1}{\partial t}, \tag{9}$$

$$\nabla^2 \mathcal{V}_1 = 4\pi G \rho_1, \tag{10}$$

where $a_s^2 \equiv \gamma P_0/\rho_0$ is the square of the unperturbed adiabatic speed of sound. Notice that, apart from the Coriolis term in equation (8), the perturbation equations (7)–(10) have identically the same expressions as in the original Jeans problem, where a "swindle" is needed to cancel out the zeroth-order state. The present problem differs therefore from Jeans's case, in that (i) rotation (which can balance the zeroth-order gravity) removes the need to invoke a "swindle," and (ii) fluid elements moving in directions other than z tend to conserve their specific angular momentum, which leads to a deflectional Coriolis force.

(b) Since the coefficients in equations (7)–(10) are constants, we can adopt Cartesian coordinates (x, y, z) and look for a solution with the Fourier decomposition,

$$\left(\frac{\rho_1}{\rho_0}, \frac{\mathbf{u}_1}{a_s}, \frac{P_1}{\rho_0 a_s^2}, \frac{\mathcal{V}_1}{a_s^2}\right) = (\alpha, \mathbf{U}, \Pi, V)e^{i(\omega t - \mathbf{k}\cdot\mathbf{x})}, \tag{11}$$

where ω, $\mathbf{k}$, α, $\mathbf{U}$, Π, V are constants (possibly complex), with the last four being dimensionless. Substitute equation (11) into equations (7)–(10); eliminate α, Π, V; and show that equation (8) becomes

$$\omega \mathbf{U} = \left(\frac{a_s^2}{\omega} - \frac{4\pi G \rho_0}{\omega k^2}\right) \mathbf{k}(\mathbf{k}\cdot\mathbf{U}) + 2i\Omega \hat{\mathbf{e}}_z \times \mathbf{U}. \tag{12}$$

(c) Equation (12) can be solved as a simultaneous set of linear homogeneous equations for (U_x, U_y, U_z), leading to a eigenvalue determination (dispersion relationship) for ω in terms of (k_x, k_y, k_z). Here, however, we remain content to study the following special cases.

Wave propagation parallel to the rotation axis: $\mathbf{k} = k\hat{\mathbf{e}}_z$. Show that this case yields either transverse (circularly polarized) inertial waves where $U_z = 0$, $U_y = \pm i U_x \neq 0$, with

$$\omega = \pm 2\Omega$$

(notice 2Ω is the epicyclic frequency when Ω is constant); or longitudinal disturbances with $\mathbf{U} \parallel \mathbf{k}$ and the dispersion relation,

$$\omega^2 = k^2 a_s^2 - 4\pi G \rho_0, \tag{13}$$

which gives the solution examined originally by Jeans. Interpret this result.

Wave propagation perpendicular to the rotation axis: Since we can choose the x axis arbitrarily, we may pick $\mathbf{k} = k\hat{\mathbf{e}}_x$ without loss of generality. Show that the dispersion relationship now becomes

$$\omega^2 = k^2 a_s^2 - 4\pi G \rho_0 + 4\Omega^2. \tag{14}$$

Substitute in the equilibrium relation (5) between Ω^2 and ρ_0 and demonstrate that no gravitational instability can arise in this case. Explain this result in terms of the (horizontal) stabilization provided by rotation.

(d) For extra credit: In a thin disk rather than an infinite cylinder, disturbances longer than the critical wavelength in the vertical direction [$k_z = [4\pi G\rho_0]^{1/2}/a_s$ according to equation (13)] may not fit within the finite thickness of the disk, while disturbances with arbitrary horizontal variations may be stabilized by a combination of rotation and acoustic effects [cf. equation (14)]. Will such a disk be completely stable to all gravitational disturbances? Hint: This is a very open-ended question, on which there remains much active research (see the references in Chapters 11 and 12).

2. This problem asks you to reconstruct a simplified version of the two-phase model of the interstellar medium.

(a) In diffuse interstellar clouds, gaseous carbon exists primarily in the form of C^+. The ground electronic state of this species has two fine-structure levels, separated by an energy difference measured in temperature units of $\epsilon_{21}/k = h\nu_{21}/k = 91\,\text{K}$. A radiative transition from the upper level $^2P_{3/2}$ to the lower level $^2P_{1/2}$ releases far-infrared photons of wavelength $\lambda_{21} = 158\,\mu\text{m}$, an important coolant for the general interstellar medium. (For a review of the spectroscopic notation associated with LS coupling, see Chapter 27 of Volume I.) The volumetric rate for spontaneous de-excitation equals $n_2 A_{21}$, where Einstein's A for this particular transition has the value $A_{21} = 2.4 \times 10^{-6}\,\text{s}^{-1}$.

Superelastic collisions with thermal electrons will depopulate C^+ in the upper state at a volumetric rate equal to $n_e n_2 \gamma_{21}$, with the collisional rate-coefficient γ_{21} given by

$$\gamma_{21} = \int_0^\infty v\sigma_{21}(\epsilon) f_e(v) 4\pi v^2\, dv,$$

where $\epsilon \equiv m_e v^2/2$ and $\sigma_{21}(\epsilon)$ is the energy-dependent superelastic cross section. The principle of detailed balance allows us to relate excitation and de-excitation cross sections (see Chapter 8 of Volume I)

$$(\epsilon + \epsilon_{21}) g_1 \sigma_{12}(\epsilon + \epsilon_{21}) = \epsilon g_2 \sigma_{21}(\epsilon), \tag{15}$$

where g_1 and g_2 are the statistical weights of the lower and upper states. Notice that if the excitation cross section σ_{12} possesses a nonzero value at threshold ϵ_{21}, equation (15) implies that the de-excitation cross section varies with energy for small values of ϵ (the most important part of a thermal distribution) as

$$\sigma_{21}(\epsilon) \approx \text{constant} \times \epsilon^{-1}. \tag{16}$$

Use a normalized Maxwellian distribution function,

$$f_{\rm e} = \left(\frac{m_{\rm e}}{2\pi kT}\right)^{3/2} \exp\left(-\frac{m_{\rm e}v^2}{2kT}\right), \tag{17}$$

and the standard approximation (16) for electronic de-excitation cross sections, and show that $\gamma_{21} \propto T^{-1/2}$. In cgs units of $\text{cm}^{-3}\,\text{s}^{-1}$, γ_{21} has a value $\sim 10^{-5}\ T^{-1/2}$ if the temperature T is given in degrees K. The corresponding rate coefficient for collisions with H atoms, $\gamma_{21}^{(\rm H)} \sim 10^{-11}\ T^{-1/2}$, so collisional de-excitation by electrons is more important than that by hydrogen atoms as long as the fractional ionization $n_{\rm e}/n_{\rm H}$ exceeds $\sim 10^{-6}$. Assume the latter to be true, and show that both collisional processes fail to compete with radiative de-excitation if $n_{\rm e} \ll 1\,\text{cm}^{-3}$ and $T \gtrsim 10^2\,\text{K}$, as is indicated for the general interstellar medium by measurements of pulsar dispersion measure and line absorption in the 21-cm line.

If we assume that the medium is very optically thin, we may ignore both the absorption and the stimulated emission of the 158-μm line photons. In this case, upward transitions occur almost exclusively by inelastic collisions of C^+ with thermal electrons, and has a volumetric rate,

$$n_{\rm e}n_1\gamma_{12},$$

where

$$\gamma_{12} = \int_{m_{\rm e}v^2/2=\epsilon_{21}}^{\infty} v\sigma_{12}(\epsilon) f_{\rm e} 4\pi v^2\, dv.$$

Now use equations (15) and (17) to demonstrate

$$\gamma_{12} = \frac{g_2}{g_1}\gamma_{21}\exp(-\epsilon_{21}/kT).$$

Argue that the last result could have been anticipated on general principles, e.g., by the recovery of the Boltzmann distribution if LTE applied and upward and downward collisional transitions balanced, which doesn't actually happen here. Argue instead that, in steady state, every upward collisional transition is followed by a downward *radiative* transition; so the rate of cooling by the C^+ 158-μm line equals

$$\Lambda(158\,\mu\text{m}) = n_2 A_{21} h\nu_{21} = n_{\rm e}n_1\gamma_{12}\epsilon_{21}.$$

For realistic interstellar densities, show that almost all of the C^+ will exist in the lower state. Let n equal the number density of atomic nuclei (hereafter assumed to be mostly hydrogen) and denote the fractional (number) abundance of electrons n_e/n by x_e. We further assume that the carbon not resident in interstellar grains exists entirely in its singly ionized form (all in the ground state), so that the fractional abundance of gaseous C^+, n_1/n, is a constant. We now get

$$\Lambda(158\,\mu\text{m}) = n^2 x_e L_C(T),$$

with

$$L_C(T) \equiv 2.4 \times 10^{-24} T^{-1/2} e^{-91/T},$$

where the numerical value for the coefficient comes from A. Dalgarno and R. McCray, 1972, *Ann. Rev. Astr. Ap.*, **10**, 375, if we take a depleted carbon abundance $n_1/n = 3 \times 10^{-5}$.

(b) If T becomes sufficiently large, the collisional excitation by electronic impact of atomic hydrogen, from the ground state to the first excited state, becomes possible. The subsequent radiative de-excitation yields Lyman-alpha radiation, and a corresponding cooling rate

$$\Lambda(\text{Ly}\alpha) = n_e n_1 \gamma_{12} \epsilon_{21}$$

where n_1, γ_{12}, and ϵ_{21} now refer to the ground-level population, collisional excitation rate, and energy associated with the Lyman-alpha transition, of atomic hydrogen. Almost all H exists in the ground electronic state (see, e.g., Chapter 10 of Volume I), $n_1 \approx n_H$. We may also assume (see below) that most of the free electrons in the general interstellar medium come from the ionization of H, with $n_H + n_e = n$, i.e.,

$$n_e \approx (n - n_1).$$

With these approximations, we may now write

$$\Lambda(\text{Ly}\alpha) = n^2 x_e (1 - x_e) L_H(T),$$

with

$$L_H(T) \equiv 9.2 \times 10^{-17} T^{1/2} \left(1 + \frac{17,500}{T}\right) e^{-118,000/T},$$

where the numerical coefficients come from using the cross-sections cited in D. M. Peterson and S. E. Strom, 1969, *Ap. J.*, **157**, 1341. These cross-sections do not follow the standard law (16), because σ_{12} vanishes at threshold ϵ_{21} instead of having a finite value, and this explains the differences in the T dependences of the nonexponential factors in $L_C(T)$ and $L_H(T)$.

Ignore the other coolants in the practical problem (of which there are many), and show that the total volumetric cooling function becomes

$$\Lambda = \Lambda(158\,\mu\text{m}) + \Lambda(\text{Ly}\alpha) = n^2 L(x_e, T),$$

where $L(x_e, T)$ has the functional form

$$L(x_e, T) = x_e \left[L_C(T) + (1 - x_e)L_H(T)\right]. \tag{18}$$

(c) The controversial part of the two-phase model comes from its assumption concerning the source of heating: the ionization of hydrogen by low-energy cosmic rays at a rate per hydrogen atom $\zeta \approx 10^{-15}\ s^{-1}$. Today, we know that this rate overestimates the true rate (principally by high-energy cosmic rays) by about two orders of magnitude. Nevertheless, the original proposal had interest because each ionization kicks out an electron with average excess energy $\epsilon_h \approx 50\,\mathrm{eV}$. This excess is large in comparison with that possessed per particle by the thermal reservoir; so on subsequent impacts with other particles, the ejected electron can act as a source of heat for the medium. In what follows, we shall ignore the secondary ionizations that can occur via hydrogen collisions with the primary electrons.

Adopt the approximations of the previous section, and show that ionization balance requires

$$\zeta(1 - x_e) = x_e^2 n\alpha(T), \tag{19}$$

where $\alpha(T)$ is the "case B" recombination coefficient for atomic hydrogen. [For the temperature dependence of $\alpha(T)$, see Chapter 10 of Volume I.] Argue furthermore that the volumetric heating rate equals

$$\Gamma = \zeta(1 - x_e)n\epsilon_h.$$

For thermal balance, set $\Lambda = \Gamma$, and show

$$\frac{n}{\zeta} = \frac{\epsilon_h(1 - x_e)}{L(x_e, T)}. \tag{20}$$

Substituting the expression (20) for n/ζ into equation (19), we obtain

$$x_e^2 = \frac{L(x_e, T)}{\epsilon_h \alpha(T)}. \tag{21}$$

(d) Given $\epsilon_h = 50\,\mathrm{eV}$, as well as the functional form (18) for $L(x_e, T)$, the relation (21) represents a quadratic equation for the ionization fraction x_e when we know the temperature T. Solve x_e numerically for T between 10 K and a few times 10^4 K. Substitute the resulting value for x_e into equation (20) and obtain n/ζ also as a function of T. Since n scales simply as ζ, the actual choice for ζ can be made after the fact to best match observations. Plot $\log(n/\zeta)$ versus $\log(T)$ and explain your results.

Hint: At high n/ζ what mechanism of cooling dominates? At low n/ζ? What happens at intermediate T, say, $T \sim 10^3$ K? See the discussion of Chapter 8.

Compute also the total gas pressure,

$$P = (n + n_e)kT = n(1 + x_e)kT,$$

and plot $\log(P/\zeta k)$ versus n/ζ. Notice that in the temperature range between a few hundred K and a few thousand K, the temperature drops so fast for increasing density n that P is a decreasing function of n. For an intermediate range of pressures, notice that, in principle, three phases can exist in pressure equilibrium with each other [the so-called "FGH theory"; see G. B. Field, D. W. Goldsmith, and H. J. Habing, 1969, *Ap. J. (Letters)*, **155**, 49; see also L. Spitzer and E. H. Scott, 1969, *Ap. J.*, **158**, 161, and S. B. Pikelner, 1968, *Soviet Astronomy*, **11**, 737]. Phase F has a low density and a high temperature; it can be identified with an intercloud medium. Phase H has a high density and a low temperature; it can be identified with diffuse interstellar clouds. Phase G has intermediate densities and temperatures; it turns out to be thermally unstable (see Problem 3), so it cannot physically exist, except possibly as a transient.

What range of values of ζ, consistent with the simultaneous existence of phases F and H, yields an intercloud electron density $n_e = x_e n$ equal to the measured pulsar dispersion measure ($n_e \approx 0.03\,\text{cm}^{-3}$)? What sort of densities and temperatures does this imply for diffuse interstellar clouds? Do you now see why the two-phase model, with its simplicity and economy of assumptions, proved so seductive to theorists and observers during the early 1970s as a description for the general interstellar medium (outside of H II regions and molecular clouds)?

3. Consider the thermal stability of an initially static homogeneous gas. If we ignore the effects of self-gravitation, the inviscid equations, including the effects of optically thin heating and cooling, read

$$\frac{\partial \rho}{\partial t} + \nabla \cdot (\rho \mathbf{u}) = 0, \tag{22}$$

$$\frac{\partial \mathbf{u}}{\partial t} + \frac{1}{2}\nabla(|\mathbf{u}|^2) + (\nabla \times \mathbf{u}) \times \mathbf{u} = -\frac{1}{\rho}\nabla P, \tag{23}$$

$$T\left(\frac{\partial s}{\partial t} + \mathbf{u} \cdot \nabla s\right) = -\mathcal{L}(\rho, T), \tag{24}$$

where

$$\rho\mathcal{L} \equiv \Lambda - \Gamma, \tag{25}$$

and

$$s = c_v \ln(P\rho^{-\gamma}) + \text{constant}, \qquad P = \mathcal{R}\rho T, \tag{26}$$

with $\mathcal{R} \equiv k/m = c_P - c_v =$ constant, and $\gamma = c_P/c_v$.

In writing the net (volumetric) cooling function $\rho\mathcal{L}$ as a function only of (mass) density ρ and temperature T, we are glossing over the fact that $\Lambda - \Gamma$, as computed, for example, in Problem 2, usually depends also on the electron fraction x_e. The implicit elimination of x_e in favor of $n = \rho/m$ and T implies we assume either that the medium remains wholly neutral or completely ionized, or that ionization balance of a partially ionized medium, equation (19), occurs on a time scale short in comparison with the rest of the problem (mechanical and thermal balance). In point of fact, recombination often occurs at a rate comparable to or slower than thermal relaxation; so the last assumption usually cannot stand rigorous scrutiny. The inclusion of a time-dependent ionization equation (see, e.g., Chapter 20) represents the right thing to do, but for pedagogical reasons, we avoid this complication here and continue to express $\mathcal{L}$ as dependent only on the instantaneous local values of ρ and T.

(a) Given the preceding prefacing remarks, suppose the equilibrium state corresponds to $\rho = \rho_0 =$ constant, $T = T_0 =$ constant, $\mathbf{u} = \mathbf{0}$, and $\mathcal{L}(\rho_0, T_0) = 0$. Consider spatially and temporally dependent small perturbations about this state of equilibrium of the form

$$\rho = \rho_0 + \rho_1, \qquad T = T_0 + T_1, \qquad \mathbf{u} = \mathbf{0} + \mathbf{u}_1, \tag{27}$$

and show that the linearized fluid equations become

$$\frac{\partial \rho_1}{\partial t} + \rho_0 \nabla \cdot \mathbf{u}_1 = 0, \tag{28}$$

$$\frac{\partial \mathbf{u}_1}{\partial t} = -\frac{1}{\rho_0}\nabla P_1, \tag{29}$$

$$T_0 \frac{\partial s_1}{\partial t} = -\left[\left(\frac{\partial \mathcal{L}}{\partial \rho}\right)_T \rho_1 + \left(\frac{\partial \mathcal{L}}{\partial T}\right)_\rho T_1\right], \tag{30}$$

$$s_1 = c_v \frac{P_1}{P_0} - c_P \frac{\rho_1}{\rho_0}, \qquad \frac{P_1}{P_0} = \frac{\rho_1}{\rho_0} + \frac{T_1}{T_0}, \tag{31}$$

with $P_0 = \mathcal{R}\rho_0 T_0$. In equation (30), the partial derivatives of the net cooling function $\mathcal{L}$ are to be evaluated at (ρ_0, T_0). Eliminate $\mathbf{u}_1$, T_1, and s_1 by differentiating equation (28) once in time, using equations (29) and (31), to obtain

$$\frac{\partial^2 \rho_1}{\partial t^2} - \nabla^2 P_1 = 0, \tag{32}$$

$$\frac{c_v}{P_0}\frac{\partial P_1}{\partial t} - \frac{c_P}{\rho_0}\frac{\partial \rho_1}{\partial t} = \left(\frac{\partial \mathcal{L}}{\partial T}\right)_P \left(\frac{\rho_1}{\rho_0}\right) - \left(\frac{\partial \mathcal{L}}{\partial T}\right)_\rho \left(\frac{P_1}{P_0}\right), \tag{33}$$

where we have written (and you should use multivariate calculus to demonstrate)

$$\left(\frac{\partial \mathcal{L}}{\partial T}\right)_P = \left(\frac{\partial \mathcal{L}}{\partial T}\right)_\rho - \frac{\rho_0}{T_0}\left(\frac{\partial \mathcal{L}}{\partial \rho}\right)_T .$$

Notice that

$$\left(\frac{\partial \mathcal{L}}{\partial T}\right)_P < \left(\frac{\partial \mathcal{L}}{\partial T}\right)_\rho , \tag{34}$$

if $(\partial \mathcal{L}/\partial \rho)_T > 0$, which is true for any gas where cooling gains on heating when we increase the density at constant temperature.

(b) Take the Laplacian of equation (33), and use equation (32) to eliminate $\nabla^2 P_1$,

$$\frac{\partial}{\partial t}\left(\frac{\partial^2 \rho_1}{\partial t^2} - a_s^2 \nabla^2 \rho_1\right) = N_P a_s^2 \nabla^2 \rho_1 - N_v \frac{\partial^2 \rho_1}{\partial t^2}, \tag{35}$$

where we have denoted

$$N_P \equiv \frac{1}{c_P}\left(\frac{\partial \mathcal{L}}{\partial T}\right)_P , \qquad N_v \equiv \frac{1}{c_v}\left(\frac{\partial \mathcal{L}}{\partial T}\right)_\rho . \tag{36}$$

Since the differential increases of heat (per unit mass) at constant volume and constant pressure equal $c_v\, dT$ and $c_P\, dT$, respectively, and the differential of net heat loss equals $d\mathcal{L}$, we see that, up to a sign, N_v and N_P represent the inverse of the characteristic cooling times at constant volume and constant pressure, respectively. Notice also that $N_v > N_P$ if $c_P > c_v$ and the inequality (34) holds.

Interpret equation (35) if the timescales for radiative cooling are very long compared to acoustic timescales, i.e., if we can ignore the terms proportional to N_v and N_P. In particular, explain why the relevant propagation speed in this case is the *adiabatic* speed of sound.

Make no assumption about the signs or magnitudes of N_v and N_P, and try a solution of the form,

$$\rho_1 = A\, e^{nt + i\mathbf{k}\cdot\mathbf{x}},$$

where n is the growth rate and $\mathbf{k}$ is the wavenumber. Demonstrate that equation (35) now gives the cubic dispersion relationship (cf. G.B. Field, 1965, *Ap. J.*, **142**, 531):

$$n^3 + N_v n^2 + k^2 a_s^2 n + N_P k^2 a_s^2 = 0. \tag{37}$$

(c) Although cubic equations have known closed-form solutions, the formulae are messy, and we prefer to examine the solutions in various interesting limits.

In the limit of very short wavelengths (large k), recover the three roots

$$\text{isobaric thermal mode,} \qquad n = -N_P, \tag{38}$$

$$\text{nearly adiabatic acoustic waves,} \qquad n = \pm ika_s + \frac{1}{2}(N_P - N_v). \tag{39}$$

Notice that the isobaric ("at constant pressure") thermal mode gives instability if $N_P < 0$, i.e., if $(\partial\mathcal{L}/\partial T)_P < 0$. This is *Field's criterion.* The criterion involves constant P because small blobs tend to maintain pressure equilibrium with their surroundings when heating and cooling get out of balance. Notice also that short-wavelength acoustic waves are damped by radiative effects if $N_v > N_P$, the usual case as we have discussed earlier. To interpret equation (39) further, show that

$$N_P - N_v = \left(\frac{1}{c_P} - \frac{1}{c_v}\right)\left(\frac{\partial\mathcal{L}}{\partial T}\right)_s,$$

where the last partial derivative is taken at constant s (the lowest order of approximation for short-wavelength sound waves propagating nearly adiabatically). Alternatively, write

$$\left(\frac{\partial\mathcal{L}}{\partial T}\right)_s = \left(\frac{\partial\mathcal{L}}{\partial P}\right)_s\left(\frac{\partial P}{\partial T}\right)_s;$$

and use $P \propto T^{\gamma/(\gamma-1)}$ for adiabatic variations of an ideal gas to evaluate $(\partial P/\partial T)_s = \gamma P/(\gamma-1)T$. With $P = \mathcal{R}\rho T$ and $c_v = \mathcal{R}/(\gamma-1)$ and $c_P = \gamma\mathcal{R}/(\gamma-1)$, show now that we have

$$N_P - N_v = -(\gamma-1)\rho\left(\frac{\partial\mathcal{L}}{\partial P}\right)_s,$$

and interpret equation (39) accordingly.

In the limit of very long wavelengths (small k), recover the three roots

$$\text{isochoric thermal mode,} \qquad n = -N_v, \tag{40}$$

$$\text{effective acoustic waves,} \qquad n = \pm ika_{\text{eff}}, \tag{41}$$

where we have defined the effective speed of sound as

$$a_{\text{eff}} = (N_P/N_v)^{1/2}a_s. \tag{42}$$

Notice that the isochoric ("at constant volume") thermal mode gives instability if $N_v < 0$, i.e., if $(\partial\mathcal{L}/\partial T)_\rho < 0$. This is *Parker's criterion.* The criterion involves constant ρ because large blobs cannot have their volumes (and therefore their densities) changed appreciably by the pressure of the ambient medium when cooling and heating get out of balance. To interpret equation (42) in the thermally stable case, N_P and N_v both greater than zero, use multivariate calculus to show that

$$a_{\text{eff}}^2 = \left(\frac{\partial P}{\partial \rho}\right)_{\mathcal{L}=0}, \tag{43}$$

where the last partial derivative is evaluated as the slope of the P-ρ relation following the equilibrium curve $\mathcal{L} = 0$. Explain the result (43) physically; in particular, explain why for sound waves to be stable in an environment where heating and cooling processes come to equilibrium quickly in comparison with the acoustic propagation time, the pressure had better increase with increasing density.

(d) Which of the two thermal instability criteria, isobaric, $N_P < 0$, or isochoric, $N_v < 0$, is the more relevant (i.e., easier to achieve)? Explain your answer clearly. Are the phases F, G, H in Problem 2 thermally stable or unstable by Field's criterion? How about Parker's criterion? Are they stable or unstable with respect to the propagation of large-scale sound waves?

4. (a) Given the jump conditions for normal shocks,

$$\rho_2 u_2 = \rho_1 u_1, \qquad P_2 + \rho_2 u_2^2 = P_1 + \rho_1 u_1^2, \qquad h_2 + u_2^2/2 = h_1 + u_1^2/2,$$

where $h = \mathcal{E} + P/\rho = \gamma P/(\gamma - 1)\rho$ is the specific enthalpy of a perfect gas, derive the elegant Prandtl-Meyer relation,

$$u_1 u_2 = c_*^2,$$

where c_*^2 is given by

$$\left(\frac{\gamma+1}{\gamma-1}\right)\frac{c_*^2}{2} = h + \frac{u^2}{2}$$

and equals a conserved quantity across the shock.

Hint: Manipulate the jump conditions into the form

$$(u_1 - u_2)\left(\frac{c_*^2}{u_1 u_2} - 1\right) = 0,$$

i.e., for $u_1 \neq u_2$, $u_1 u_2 = c_*^2$.

Prove, moreover, that

$$a_1^2 \equiv \gamma P_1/\rho_1 < c_*^2 < a_2^2 \equiv \gamma P_2/\rho_2,$$

and that the Prandtl-Meyer relation requires, for a (compressive) shock, the upstream flow to be supersonic, $u_1 > a_1$, and the downstream flow to be subsonic, $u_2 < a_2$.

(b) We now stand ready to consider the structure of a viscous shock layer. The one-dimensional steady fluid equations, including the effects of viscosity and conductivity, read

$$\rho u = \text{constant} \equiv j_0, \tag{44}$$

$$\rho u^2 + P - \frac{4}{3}\mu\frac{du}{dx} = \text{constant}, \tag{45}$$

$$\rho u\left(h + \frac{u^2}{2}\right) - \mathcal{K}\frac{dT}{dx} - \frac{4}{3}\mu u\frac{du}{dx} = \text{constant} \equiv f_0. \tag{46}$$

We may analytically perform a qualitatively correct integration of equation (46) by taking advantage of a fortuitous circumstance. Eucken's formula for a neutral gas implies

$$\frac{3}{4}\frac{\mathcal{K}}{\mu c_P} = \frac{3(9\gamma - 5)}{16\gamma},$$

which equals 9/8 for $\gamma = 5/3$ (monatomic gas) and 57/56 for $\gamma = 7/5$ (diatomic gas with only rotational degrees of freedom excited). If we approximate 9/8 or 57/56 by unity, demonstrate that equation (46), with $h = c_P T$, becomes

$$j_0\left(h + \frac{u^2}{2}\right) - \frac{4}{3}\mu\frac{d}{dx}\left(h + \frac{u^2}{2}\right) = f_0.$$

Define $Q \equiv h + u^2/2 - f_0/j_0$, and integrate the above equation to obtain

$$Q = Q_0 \exp\left\{\int_{x_0}^{x} \frac{3j_0}{4\mu}\,dx\right\},$$

where Q_0 is an integration constant. Show that the exponent has order of magnitude $(x - x_0)/\ell$, where ℓ equals the particle m.f.p.; argue thereby that Q reaches unbounded values as $x - x_0$ becomes much larger than ℓ unless Q_0 identically equals 0. For $Q_0 = 0$, we have, *even inside the shock layer*,

$$h + \frac{u^2}{2} = \frac{f_0}{j_0} = \text{constant} = \left(\frac{\gamma + 1}{\gamma - 1}\right)\frac{c_*^2}{2}. \tag{47}$$

State in words what our trick has accomplished. In particular, how do the rate of viscous dissipation and heat conduction offset one another here?

(c) With the help of the Prandtl-Meyer relation, show now that equation (45) can be written as

$$\frac{c_*^2}{u} + u - \lambda \frac{du}{dx} = u_1 + u_2, \tag{48}$$

where u_1 and u_2 are the far upstream and far downstream fluid velocities, with $\lambda \equiv 8\gamma\mu/3j_0(\gamma + 1)$. Show that λ, which varies as μ varies, has an order of magnitude comparable to the collision m.f.p., ℓ, inside the shock layer. Define then the stretched dimensionless coordinate ξ by

$$d\xi \equiv \frac{dx}{\lambda}, \tag{49}$$

which serves an analogous purpose as the optical depth variable in radiative transfer problems. With the definition (49), solve equation (48) under the (arbitrary) reference-point condition

$$u = \frac{1}{2}(u_1 + u_2) \qquad \text{at} \qquad \xi = 0,$$

to obtain the solution (using the method of partial fractions):

$$\xi = \left(\frac{u_1}{u_1 - u_2}\right) \ln\left[\frac{2(u_1 - u)}{(u_1 - u_2)}\right] - \left(\frac{u_2}{u_1 - u_2}\right) \ln\left[\frac{2(u - u_2)}{(u_1 - u_2)}\right].$$

Clearly, $u \to u_1$ as $\xi \to -\infty$ and $u \to u_2$ as $\xi \to +\infty$. Plot u/u_2 versus ξ for the case $u_1/u_2 = 3$. Comment on the important features of your graph, e.g., the effective thickness of the shock transition layer.

Problem Set 4

1. Suppose the ions of a plasma move with a mean velocity $\mathbf{u}_\mathrm{i}$, and let the mean velocity of an electron with respect to the ions equal

$$\mathbf{v}'_\mathrm{e} \equiv \mathbf{u}_\mathrm{e} - \mathbf{u}_\mathrm{i}.$$

The equation of motion for electrons in the rest frame of the ions then reads

$$m_\mathrm{e}\frac{d\mathbf{v}_\mathrm{e}}{dt} = -e\left(\mathbf{E}' + \frac{\mathbf{v}'_\mathrm{e}}{c} \times \mathbf{B}'\right) - m_\mathrm{e}\nu_\mathrm{c}\mathbf{v}'_\mathrm{e} + \text{gravitational and inertial terms,}$$

where ν_c is the mean electron-collision frequency for transferring momentum to the ions.

In the pure MHD limit, we suppose the electron inertia to be negligible; i.e., we discard the left-hand side of of the equation of motion, as well as any inertial or gravitational forces. The approximation amounts to neglecting the effects of a finite electron gyrofrequency

$$\omega_\mathrm{L} \equiv eB/m_\mathrm{e}c.$$

Astrophysical situations often do not satisfy this stringent condition; so we need to examine the consequences of finite electron gyromotion when we discuss plasma physics. In the pure MHD limit, when electrodynamic forces otherwise dominate the scene, electrons quickly reach a terminal velocity,

$$\mathbf{v}'_\mathrm{e} = -\frac{e}{m_\mathrm{e}\nu_\mathrm{c}}\left(\mathbf{E}' + \frac{\mathbf{v}'_\mathrm{e}}{c} \times \mathbf{B}'\right),$$

where, as we shall see, inclusion of the term involving the unknown $\mathbf{v}'_\mathrm{e}$ does not substantially complicate a practical solution.

(a) Show that the electric current density reads

$$\mathbf{j}'_\mathrm{e} = -n_\mathrm{e}e\mathbf{v}'_\mathrm{e} = \sigma\left(\mathbf{E}' + \frac{\mathbf{v}'_\mathrm{e}}{c} \times \mathbf{B}'\right),$$

where the electrical conductivity is given by

$$\sigma = n_\mathrm{e}e^2/m_\mathrm{e}\nu_\mathrm{c}.$$

Solve for $\mathbf{E}'$ and obtain

$$\mathbf{E}' = \frac{1}{\sigma}\mathbf{j}_\mathrm{e} - \frac{1}{c}(\mathbf{u}_\mathrm{e} - \mathbf{u}_\mathrm{i}) \times \mathbf{B},$$

where we have used $\mathbf{j}_\mathrm{e}' = \mathbf{j}_\mathrm{e}$ and $\mathbf{B}' = \mathbf{B}$ for nonrelativistic frame transformations. To order u/c, the electric field in the ion and laboratory frame are related by

$$\mathbf{E}' = \mathbf{E} + \frac{\mathbf{u}_\mathrm{i}}{c} \times \mathbf{B};$$

thus show that

$$\mathbf{E} = \frac{1}{\sigma}\mathbf{j}_\mathrm{e} - \frac{\mathbf{u}_\mathrm{e}}{c} \times \mathbf{B},$$

which differs from the derivation given in the text only in that $\mathbf{u}_\mathrm{e}$ replaces $\mathbf{u}_\mathrm{i}$. Thus we have that the replacement of $\mathbf{u}_\mathrm{i}$ by $\mathbf{u}_\mathrm{e}$ in the evolutionary equation for the magnetic field, which now reads

$$\frac{\partial \mathbf{B}}{\partial t} + \nabla \times (\mathbf{B} \times \mathbf{u}_\mathrm{e}) = \nabla \times (\eta \nabla \times \mathbf{B}),$$

where η equals the electrical resistivity,

$$\eta \equiv \frac{c^2}{4\pi\sigma} = \frac{c^2 m_\mathrm{e} \nu_\mathrm{c}}{4\pi n_\mathrm{e} e^2}.$$

To the extent that the drift speed of electrons relative to ions $|\mathbf{u}_\mathrm{e} - \mathbf{u}_\mathrm{i}|$ needed to maintain the electromagnetic fields in the problem remains negligibly small in comparison with macroscopic speeds, we may ignore the difference between the discussion here and that given in the text for the derivation of the fundamental equations for MHD.

(b) For a completely ionized hydrogen plasma,

$$\eta \sim 10^{13} T^{-3/2} \mathrm{cm}^2\ \mathrm{s}^{-1},$$

when T is expressed in degrees Kelvins. (Our η differs by a factor of $1/4\pi$ from Spitzer's definition in *Physics of Fully Ionized Gases*, p. 139.) Estimate numerically the resistive diffusion time,

$$t_\mathrm{D} = L^2/\eta,$$

for the interior of the Sun. Do the same for an H II region. Comment on the validity of the assumption of field freezing to the plasma in these cases. Estimate η for the corona of the Sun. How does this value compare with the Bohm diffusivity?

(c) Provided we can use Ohm's law, $\mathbf{j}'_\mathrm{e} = \sigma \mathbf{E}'$, our derivation of the evolutionary equation for $\mathbf{B}$ applies to conducting liquids as well as to plasmas; we only need a different expression for η. Molten iron has an electrical resistivity $\eta = 1.1 \times 10^4\ \mathrm{cm^2\,s^{-1}}$. Estimate the resistive diffusive time t_D for the core of the Earth. Does the magnetic field of the Earth require regeneration against Ohmic decay? Compare your estimate of t_D with the average reversal time for the magnetic field observed empirically for the Earth of 10^5–10^6 yr. Comment qualitatively on the relevance of this comparison for dynamo theories of the Earth's magnetic field. In particular, do we nominally require anomalous diffusion to reconnect opposing small-scale fields generated by differential motions in the Earth's core?

2. (a) Show that the scalar virial theorem has the general form

$$\frac{1}{2}\frac{d^2 I}{dt^2} = \mathcal{W} + 2K + 2\mathcal{U} + \mathcal{M} - P_\mathrm{ext}\oint_A \mathbf{x}\cdot\hat{\mathbf{n}}\,dA + \oint_A \mathbf{x}\cdot \overleftrightarrow{T} \cdot\hat{\mathbf{n}}\,dA,$$

where

$$I \equiv \int_V \rho|\mathbf{x}|^2\,d^3x,$$

$$K \equiv \int_V \frac{1}{2}\rho|\mathbf{u}|^2\,d^3x,$$

and the other symbols have the definitions given in Chapter 24. In other words, demonstrate that the inclusion of the inertial term,

$$\rho\frac{Du_i}{Dt},$$

on the left-hand side of the force equation introduces the extra terms, $(1/2)d^2I/dt^2 - 2K$.

Hint: Note $u_i = Dx_i/Dt$ is a mathematical identity if $D/Dt \equiv \partial/\partial t + u_k\partial/\partial x_k$. Use the equation of continuity to prove also the identity

$$\frac{d}{dt}\int_V \rho\chi\,d^3x = \int_V \rho\frac{D\chi}{Dt}\,d^3x,$$

for any fluid quantity χ, when V is a volume across whose surface A no fluid passes.

(b) Use similar approximations to those of Chapter 24, and argue that the scalar virial theorem applied to an isothermal gas cloud yields an equation that governs the (linear or nonlinear) motion of the radius R,

$$\frac{d^2}{dt^2}\left(\frac{\gamma}{2}MR^2\right) - \epsilon M\left(\frac{dR}{dt}\right)^2 = 4\pi R^3\left[P_0(R) - P_\mathrm{ext}\right],$$

where we have defined the sum of gravitational, magnetic, and thermal "pressures" as

$$P_0(R) \equiv \frac{1}{4\pi R^3}\left(-\alpha\frac{GM^2}{R} + \beta\frac{\Phi^2}{R} + 3a^2M\right),$$

with α, γ, and ϵ being constants of order unity, and $\beta \sim 1/6\pi^2$. Note that we expect the approximation for $2K = \epsilon M(dR/dt)^2$ to hold, with $\epsilon \sim 1$, only if the interior velocity increases monotonically from 0 at the center to dR/dt at the surface. The assumption that the amplitude of the motion is largest at the surface of the cloud and decreases inward may hold for the fundamental mode of oscillation (one that has no radial nodes) of a safely stable cloud—see part (c)—but it fails badly for unstable configurations that are strongly centrally concentrated and that tend to collapse from "inside-out" (cf. Chapter 18).

(c) Equilibrium corresponds to $R = \text{constant} \equiv R_0$, i.e.,

$$P_{\text{ext}} = P_0(R_0),$$

which equals equation (24.17) of the text if we write R_0 for R. Consider now small-amplitude motions about the equilibrium point

$$R(t) = R_0 + R_1(t),$$

where we may consider $R_1 \ll R_0$. With M, a, Φ and P_{ext} held constant, show that the linearized form of the equation of motion for the cloud radius reads

$$\frac{d^2R_1}{dt^2} + \omega_0^2 R_1 = 0,$$

where we have defined

$$\omega_0^2 \equiv -\frac{4\pi R_0^2}{\gamma M}P_0'(R_0),$$

with $P_0'(R_0)$ denoting the derivative of $P_0(R)$ evaluated at R_0. Argue now that the original equilibrium state is stable only if $P_0'(R_0) < 0$, i.e., only on the side of the $P_{\text{ext}} = P_0(R_0)$ curve which increases with decreasing R_0 (see discussion in Chapter 24). On the other side of the pressure maximum, $\omega_0^2 < 0$, and small departures from equilibrium grow exponentially in time (with the interesting behavior astronomically being gravitational collapse, although explosive growth could also occur in principle).

(d) Except when the cloud exists right at the margin of stability, $P_0'(R_0) = 0$, argue that $|\omega_0| \sim v_{\text{eff}}/R_0$, where v_{eff} is an effective propagation speed in the cloud with magnitude $\sim (P_0/\rho_0)^{1/2}$, where ρ_0 is the average density of the cloud. Interpret this result in terms of the fundamental oscillation frequency of a stable cloud compared to a natural dynamical rate.

3. Consider the case of shock propagation perpendicular to a frozen-in magnetic field. (In the notation of the text, take $u_\parallel = 0$ and $B_\perp = 0$.)

(a) Assume ideal MHD and demonstrate that the jump conditions read

$$\rho_1 u_1 = \rho_2 u_2 = \rho_3 u_3,$$

$$\rho_1 u_1^2 + P_1 + \frac{1}{8\pi}B_1^2 = \rho_2 u_2^2 + P_2 + \frac{1}{8\pi}B_2^2 = \rho_3 u_3^2 + P_3 + \frac{1}{8\pi}B_3^2,$$

$$\frac{\gamma}{\gamma-1}\frac{P_1}{\rho_1} + \frac{1}{2}u_1^2 + \frac{1}{4\pi}B_1^2 = \frac{\gamma}{\gamma-1}\frac{P_2}{\rho_2} + \frac{1}{2}u_2^2 + \frac{1}{4\pi}B_2^2,$$

$$B_1 u_1 = B_2 u_2 = B_3 u_3,$$

where 1 represents the preshock state; 2, the post-viscous-shock state, and 3, the downstream state after the gas has radiatively relaxed to the same temperature as the initial preshock state:

$$P_3/\rho_3 = P_1/\rho_1.$$

(b) Introduce the two characteristic velocities, a_0 and c_0, through the relations,

$$a_0^3 = \frac{u_1 B_1^2}{4\pi\rho_1} = u_1^2 v_{\mathrm{A}1}^2,$$

$$c_0^2 \equiv \frac{2\gamma}{\gamma-1}\frac{P_1}{\rho_1} + \left(\frac{\gamma-1}{\gamma+1}\right)\left(u_1^2 + \frac{B_1^2}{2\pi\rho_1}\right)$$

$$= \frac{2}{\gamma+1}a_{s1}^2 + \frac{\gamma-1}{\gamma+1}u_1^2 + 2\left(\frac{\gamma-1}{\gamma+1}\right)v_{\mathrm{A}1}^2,$$

where $a_s^2 \equiv \gamma P/\rho$ and $v_{\mathrm{A}}^2 \equiv B^2/4\pi\rho$. Although we have defined a_0 and c_0 in terms of the upstream state 1, we could as well have used the state 2 inasmuch as a_0^3 and c_0^2 are conserved across the viscous shock. Algebraically manipulate the jump conditions for 1 to 2 to derive the relation,

$$1 - \frac{c_0^2}{u_1 u_2} - \frac{2-\gamma}{\gamma+1}a_0^3\left(\frac{u_2+u_1}{u_2^2 u_1^2}\right) = 0.$$

Solve this condition as a quadratic equation for $B_2/B_1 = u_1/u_2 = \rho_2/\rho_1$, and show that the solution for $\rho_2/\rho_1 \geq 1$ reads

$$\frac{\rho_2}{\rho_1} = \frac{1}{2} - \left[1 + \left(\frac{\gamma+1}{2-\gamma}\right)\frac{u_1 c_0^2}{a_0^3}\right]$$

$$+ \frac{1}{2}\left\{\left[1 + \left(\frac{\gamma+1}{2-\gamma}\right)\frac{u_1 c_0^2}{a_0^3}\right]^2 + 4\left(\frac{\gamma+1}{2-\gamma}\right)\right\}^{1/2}.$$

Show, in particular, that the shock has zero jump, ρ_2/ρ_1, when the upstream velocity is magnetosonic:

$$u_1^2 = a_{s1}^2 + v_{rmA1}^2.$$

(c) Write the solution for ρ_2/ρ_1 in a somewhat different form by introducing the dimensionless ratio of upstream magnetic-to-gas pressure,

$$\alpha_1 \equiv B_1^2/8\pi P_1,$$

and the upstream Mach number,

$$M_1 \equiv u_1/a_{s1}.$$

With these definitions, verify the identities,

$$u_1^3/a_0^3 = u_1^2/v_{\mathrm{A}1}^2 = \frac{\gamma}{2\alpha_1}M_1^2,$$

$$u_1 c_0^2/a_0^3 = c_0^2/v_{\mathrm{A}1}^2 = \frac{\gamma}{(\gamma+1)\alpha_1} + \frac{\gamma}{2\alpha_1}\left(\frac{\gamma-1}{\gamma+1}\right)M_1^2 + 2\left(\frac{\gamma-1}{\gamma+1}\right).$$

With the notation

$$\mathcal{M} \equiv \gamma[2(\alpha_1+1) + (\gamma-1)M_1^2]$$

show now that our expression for ρ_2/ρ_1 becomes

$$\frac{\rho_2}{\rho_1} = \frac{1}{4(2-\gamma)\alpha_1}\left\{-\mathcal{M} + \left[\mathcal{M}^2 + 8\gamma\alpha_1(2-\gamma)(\gamma+1)M_1^2\right]^{1/2}\right\}.$$

Use the momentum flux constancy to obtain also the pressure jump condition

$$\frac{P_2}{P_1} = 1 + \gamma M_1^2\left(1 - \frac{\rho_1}{\rho_2}\right) - \alpha_1\left[\left(\frac{\rho_2}{\rho_1}\right)^2 - 1\right].$$

In this notation, compressive shocks appear only when

$$M_1^2 \geq 1 + \frac{2}{\gamma}\alpha_1 \equiv M_{\mathrm{cr}}^2.$$

Compute M_{cr} for $\gamma = 5/3$ when $\alpha_1 = 0$, 0.5, 1.0, 5.0, 20, and 40. Discuss qualitatively the potential importance of magnetic fields for the existence of supersonic (but sub-magnetosonic) turbulence in the interstellar medium. Notice for $M_1^2 \gg \alpha_1$, $\rho_2/\rho_1 \to (\gamma+1)M_1^2/[2+(\gamma-1)M_1^2$, the value appropriate for nonmagnetic shocks.

(d) Algebraically eliminate u_3, P_3 and B_3 in the jump conditions that relate the 3 state to the 1 state, and derive

$$\frac{\rho_1^2 u_1^2}{\rho_3} + \frac{P_1}{\rho_1}\rho_3 + \frac{B_1^2}{8\pi}\left(\frac{\rho_3}{\rho_1}\right)^2 = \rho_1 u_1^2 + P_1 + \frac{B_1^2}{8\pi}.$$

Write $y \equiv \rho_3/\rho_1$ and show that the above equation becomes

$$(y-1)\left[\frac{\alpha_1}{\gamma M_1^2}y(y+1) + \frac{1}{\gamma M_1^2}y - 1\right] = 0,$$

where we have introduced α_1 and M_1 as in part c. If we factor out the trivial root $y = 1$, we obtain a quadratic equation for y, which we may solve to obtain $y = \rho_3/\rho_1 = u_1/u_3 = B_3/B_1 = P_3/P_1$ (for an isothermal magnetic shock):

$$y = \frac{1}{2\alpha_1}\left\{-(1+\alpha_1) + \left[(1+\alpha_1)^2 + 4\gamma\alpha_1 M_1^2\right]^{1/2}\right\}.$$

Consider, for example, $\gamma = 5/3$ and $M_1 = 5$. Compute numerically for ρ_2/ρ_1, P_2/P_1, and $\rho_3/\rho_1 = P_3/P_1$ for $\alpha_1 = 0$, 1, 5, 10, and 20, and arrange your results in the form of a table. Comment on the ability of the presence of a strong magnetic field (α_1 comparable to M_1^2) to reduce the asymptotic density compression in a shock wave considerably below its nonmagnetic value, with the effect becoming particularly important for radiating ("isothermal") shocks. We may associate the considerable stiffness given the magnetized medium in the present geometry with an effective exponent $\gamma_{\text{mag}} = 2$ for the one-dimensional compression of a magnetic field (where $P_{\text{mag}} \propto B^2 \propto \rho^2$). In such a situation, a frozen-in magnetic field is able eventually to absorb most of the momentum transfer behind a shock wave.

Bibliography

G.I. Barenblatt, 1979, *Similarity, Self-Similarity, and Intermediate Asymptotics*, Consultants Bureau, Plenum.

C.M. Bender and S.A. Orszag, 1978, *Advanced Mathematical Methods for Scientists and Engineers*, McGraw-Hill.

J. Binney and S. Tremaine, 1987, *Galactic Dynamics*, Princeton University Press.

J.C. Brandt and P.W. Hodge, 1964, *Solar System Astrophysics*, McGraw-Hill.

S. Chapman and T.G. Cowling, 1961, *The Mathematical Theory of Nonuniform Gases*, Cambridge University Press.

H.B. Callen, 1960, *Thermodynamics*, Wiley.

S. Chandrasekhar, 1939, *Stellar Structure*, University of Chicago Press.

S. Chandrasekhar, 1961, *Hydrodynamic and Hydromagnetic Stability*, Oxford University Press.

S. Chandrasekhar, 1960, *Plasma Physics*, University of Chicago Press.

R. Courant and K.O. Friedrichs, 1948, *Supersonic Flow and Shock Waves*, Interscience.

J.P. Cox and R.T. Giuli, 1968, *Principles of Stellar Structure*, Gordon and Breach.

P. Garabedian, 1964, *Partial Differential Equations*, Wiley.

H. Goldstein, 1959, *Classical Mechanics*, Addison-Wesley.

K. Huang,, 1963, *Statistical Mechanics*, Wiley.

J.D. Jackson, 1962, *Classical Electrodynamics*, Wiley.

H. Jeffreys and B.S. Jeffreys, 1962, *Mathematical Physics*, Cambridge University Press.

L.D. Landau and E.M. Lifshitz, 1959, *Fluid Mechanics*, Pergamon Press.

L.D. Landau, E.M. Lifshitz, and L.P. Pitaeviski, 1984, *Electrodynamics of Continuous Media*, Pergamon Press.

E.M. Lifshitz and L.P. Pitaevskii, 1981, *Physical Kinetics*, Pergamon Press.

C.C. Lin, 1966, *Theory of Hydrodynamical Stability*, Cambridge University Press.

L.M. Milne-Thompson, 1960, *Theoretical Hydrodynamics*, Macmillan.

D. Mihalas and B.W. Mihalas, 1984, *Foundations of Radiation Hydrodynamics*, Oxford University Press.

C.W. Misner, K.S. Thorne, and J.A. Wheeler, 1973, *Gravitation*, Freeman.

D.C. Montgomery and D.A. Tidman, 1964, *Plasma Kinetic Theory*, McGraw-Hill.

R.W. Noyes, 1982, *The Sun, Our Star*, Harvard University Press.

E.N. Parker, 1963, *Interplanetary Dynamical Processes*, Wiley.

E.N. Parker, 1979, *Cosmical Magnetic Fields: Their Origin and Their Activity*, Oxford University Press.

W.H. Press, B.P. Flannery, S.A. Teukolsky, and W.T. Vetterling, 1986, *Numerical Recipes: The Art of Scientific Computing*, Cambridge University Press.

M. Schwarzschild, 1958, *Structure and Evolution of the Stars*, Princeton University Press.

S.L. Shapiro and S.A. Teukolsky, 1983, *Black Holes, White Dwarfs, and Neutron Stars*, Wiley.

L. Spitzer, 1962, *Physics of Fully Ionized Gases*, Interscience.

L. Spitzer, 1978, *Physical Processes in the Interstellar Medium*, Interscience.

T.H. Stix, 1962, *Theory of Plasma Waves*, McGraw-Hill.

D.A. Tidman and N.A. Krall, 1971, *Shock Waves in Collisionless Plasmas*, Interscience.

G.E. Uhlenbeck and G.W. Ford, 1963, *Statistical Mechanics*, American Mathematical Society.

G.B. Whitham, 1974, *Linear and Nonlinear Waves*, Wiley.

Ya. B. Zeldovich and Yu. P. Raizer, 1966, *Physics of Shock Waves and High Temperature Hydrodynamic Phenomena*, Academic Press.

INDEX

Note: An "f" following some page numbers refers to a figure.

C

D

E

F

J

K

L

M

N

Q

R

S

T

U

V

W

X